XXI International School on Nuclear Physics and Applications and International Symposium on Exotic Nuclei (ISEN 2015)

Journal of Physics: Conference Series Volume 724

Varna, Bulgaria
6 – 12 September 2015

ISBN: 978-1-5108-2675-5
ISSN: 1742-6588

Printed from e-media with permission by:

Curran Associates, Inc.
57 Morehouse Lane
Red Hook, NY 12571

Some format issues inherent in the e-media version may also appear in this print version.

Printed by Curran Associates, Inc. (2016)

For permission requests, please contact the Institute of Physics
at the address below.

Institute of Physics
Dirac House, Temple Back
Bristol BS1 6BE UK

Phone: 44 1 17 929 7481
Fax: 44 1 17 920 0979

techtracking@iop.org

Additional copies of this publication are available from:

Curran Associates, Inc.
57 Morehouse Lane
Red Hook, NY 12571 USA
Phone: 845-758-0400
Fax: 845-758-2633
Email: curran@proceedings.com
Web: www.proceedings.com

TABLE OF CONTENTS

XXI International School on
Nuclear Physics, Neutron Physics and Applications
&
International symposium on exotic nuclei (2015)

Dedicated to the 60th anniversary of JINR-DUBNA

September 7 - 12, 2015
Varna, Bulgaria

Organised by:

Institute for Nuclear Research and Nuclear Energy,
Sofia, Bulgaria

Joint Institute for Nuclear Research, Dubna, Russia

Supported by:

Bulgarian Nuclear Regulatory Agency

PREFACE

The present volume contains the lectures and short talks given at the XXI International School on Nuclear Physics and Applications and International Symposium on Exotic Nuclei (ISEN-2015). The School and the Symposium were held from 6–12 September 2015 in "Club Hotel Bolero" located in 'Golden Sands' (Zlatni Pyasaci) Resort on the Black Sea coast, near Varna, Bulgaria. The School and the Symposium were organized by the Institute for Nuclear Research and Nuclear Energy of the Bulgarian Academy of Sciences and the Flerov Laboratory of Nuclear Reaction if the Joint Institute for Nuclear Research (Dubna). The co-organizer of the School was the Bulgarian Nuclear Regulatory Agency.

According to a long-standing tradition the School has been held biannually since 1973. The School's program has been restructured according to our enlarged new international links and today it is more similar to an international conference than to a classical nuclear physics school. In 2015, the International Symposium on Exotic Nuclei (ISEN-2015) was scheduled from 8–11 September under the auspices of the School. The Symposium was dedicated to the investigation of nuclei in extreme states and, in particular, at the limits of nuclear stability (from very light neutron- and proton-rich up to superheavy nuclei). This new format attracted many young scientists and students from around the world. We had the pleasure of welcoming more than 60 distinguished scientists as lecturers. Additionally, 20 young colleagues received the opportunity to present a short contribution. In all, 90 participants enjoyed the scientific presentations and discussions as well as the relaxing atmosphere at the beach and during the pleasant evenings.

The program of the School and the Symposium ranged from the latest results in fundamental areas such as nuclear structure and reactions to frontier issues of the application of nuclear methods. The main topics were the following:

- Exotic Nuclei and their Properties
- Methods of production of light exotic nuclei and study of their properties
- Superheavy elements. Synthesis and properties
- Rare processes and decays
- Radioactive beams. Production and research programmes
- Evolution of Shell Structure
- Collective Modes of the Nucleus
- Nuclear Astrophysics
- Applications of Exotic Beams in Materials Research
- Current and Future Facilities

Several colleagues contributed to the organization of the School and the Symposium. We would like to thank them and especially the Scientific Secretary of the School, Dr. Elena Stefanova, and the members of the Organizing Committee, Dr. Dimitar Tarpanov and Peter Zhivkov, for their cordiality and high-level assistance. We are also grateful to colleagues from the Dubna Institute for their collaboration.

Sofia, 20 May 2016

Co-chair persons of the Organizing Committee

Prof. Dr. Sc. Ch. Stoyanov

Prof. Dr. Sc. S. Dimitrova

Prof. Dr. Sc. Yu. Penionzhkevich

List of Participants

Prof. A. A. Cowley	aac@sun.ac.za
Dr. A. Nurmukhanbetova	anurmukhanbetova@nu.edu.kz
Prof. Alexander Volya	avolya@fsu.edu
Prof. Alexandrina Petrovici	spetro@nipne.ro
Prof. Alexey A. Ogloblin	ogloblina@bk.ru
Dr. Alexey P. Severyukhin	sever@theor.jinr.ru
Mr. Alexey Voinov	voinov_2000@mail.ru
Dr. Alla S. Demyanova	a.s.demyanova@bk.ru
Prof. Amiram Leviatan	ami@phys.huji.ac.il
Dr. Andreas Best	andreas.best@lngs.infn.it
Prof. Anton N. Antonov	antonovshumen@gmail.com
Ms. Aysegul Ertoprak	aysegulertoprak@gmail.com
Prof. Baki Akkus	akkus@istanbul.edu.tr
Dr. Boris Chernyshev	chernyshev@mephi.ru
Dr. Bozkurt Kutsal	b_kutsal@yahoo.com
Dr Brigitte Roussière	roussier@ipno.in2p3.fr
Dr. Bruce Marsh	bruce.marsh@cern.ch
Prof. Charlotte Elster	elster@ohio.edu
Prof. Chavdar Stoyanov	stoyanov@inrne.bas.bg
Mr. Christoph Kremer	ckremer@ikp.tu-darmstadt.de
Dr. Clementina Agodi	agodi@lns.infn.it
Dr. David Verney	verney@ipno.in2p3.fr
Dr. Deyan T. Yordanov	yordanov@ipno.in2p3.fr
Ms. Diana Kocheva	diana.kocheva@gmail.com
Dr. Dimitar Tarpanov	dimitert@yahoo.co.uk
Dr. Dimitre Kadrev	kadrev@inrne.bas.bg
Dr. Elena Stefanova	elenas@inrne.bas.bg
Dr. Elena Zemlyanaya	elena@jinr.ru
Dr. Elena Lawrie	elena@tlabs.ac.za
Dr. Evgeny Konobeevski	konobeev@inr.ru
Dr. Fadi Ibrahim	ibrahim@ipno.in2p3.fr
Dr. Feodor F. Karpeshin	fkarpeshin@gmail.com
Mr. Firdous Ahmad Khan	29firdous11@gmail.com
Ms. Gaukhar Kabdrakhimova	gaukharkd@gmail.com
Mr. Georgi Gavrilov	georgi.gavrilov@gmail.com
Prof. Georgi Rainovski	rig@phys.uni-sofia.bg
Dr. Gerfried Kumbartzki	kum@physics.rutgers.edu
Prof. Dr. Giacomo De Angelis	giacomo.deangelis@lnl.infn.it

Mr. Giovanni De Gregorio	degregorio@na.infn.it
Dr. Huben Ganev	huben@theor.jinr.ru
Mrs. Irina Zarubina	zarubina@lhe.jinr.ru
Dr. Isabelle Deloncle	isabelle.deloncle@csnsm.in2p3.fr
Dr. János Timár	timar@atomki.mta.hu
Dr. Jérémy Bonnard	jeremy.bonnard@pd.infn.it
Prof. Jon Billowes	j.billowes@manchester.ac.uk
Jozsef Cseh	cseh@atomki.mta.hu
Dr. Judit Darai	darai@atomki.mta.hu
Prof. Kairat Kuterbekov	kkuterbekov@gmail.com
Dr. Kalin Gladnishki	kag@phys.uni-sofia.bg
Mrs.Katharine Henninger	k.henninger@gsi.de
Dr. Kobus Lawrie	lawrie@tlabs.ac.za
Mr. Konstantin Shegunov	kshegunov@gmail.com
Dr. Korneliya Spasova	kyspasova@yahoo.com
Dr. Krassimira Marinova	marinova@nrmail.jinr.ru
Mrs. Larisa Korotkova	korotkovalara@gmail.com
Prof. Madan M Sharma	madan.sharma@ku.edu.kw
Dr. Manuela Cavallaro	manuela.cavallaro@lns.infn.it
Dr. Marcella Grasso	grasso@ipno.in2p3.fr
Dr. Mariana Petris	mpetris@nipne.ro
Prof. Marilena Avrigeanu	Marilena.Avrigeanu@nipne.ro
Dr. Mario Cromaz	mcromaz@lbl.gov
Dr. Mariya Yavahchova	m.yavahchova@gmail.com
Mrs. Martha Liliana Cortes Sua	m.l.cortes@gsi.de
Dr. Martin Ivanov	martin@inrne.bas.bg
Mr. Michael Thürauf	mthuerauf@ikp.tu-darmstadt.de
Dr. Mitko Gaidarov	gaidarov@inrne.bas.bg
Prof. Naftali Auerbach	auerbach@post.tau.ac.il
Nessipbek Mynbayev	nmynbayev@nu.edu.kz
Dr. Nguyen Van Giai	nguyen@ipno.in2p3.fr
Prof. Nicolae Sandulescu	sandulescu@theory.nipne.ro
Prof.Dr. Norbert Pietralla	pietralla@ikp.tu-darmstadt.de
Dr. Pavel Zarubin	zarubin@lhe.jinr.ru
Prof. Pedro Sarriguren	p.sarriguren@csic.es
Mr. Petar Kirilov Zhivkov	petar.zhivkov@gmail.com
Mr. Philipp Ries	pries@ikp.tu-darmstadt.de
Prof. Pierre Capel	pierre.capel@ulb.ac.be
Prof. R. V. Jolos	jolos@theor.jinr.ru
Dr. Radomira Lozeva	radomira.lozeva@iphc.cnrs.fr
Dr. Ralitsa Stanoeva	rstanoeva@swu.bg
Mr. Robert Stegmann	rstegmann@ikp.tu-darmstadt.de
Dr. S. Lukyanov	lukyan@nrmail.jinr.ru
Dr. Sarkis A. Karamian	karamian@nrmail.jinr.ru
Dr. Sergey Zemlyanoy	zemlya@jinr.ru
Dr. Serkan Akkoyun	serkan.akkoyun@gmail.com

Prof. Sevdalina Dimitrova	sevdim@inrne.bas.bg
Mr. Shailesh Kumar Singh	shailesh@iopb.res.in
Prof. Sherry Yennello	yennello@comp.tamu.edu
Simone Ceruti	simone.ceruti@mi.infn.it
Dr. Stoyanka Ilieva	silieva@ikp.tu-darmstadt.de
Prof. Sydney Gales	sydney.gales@eli-np.ro
Prof. Thorsten Kröll	tkroell@ikp.tu-darmstadt.de
Prof. Viacheslav V. Samarin	samarin@jinr.ru
Prof. Victor Voronov	voronov@theor.jinr.ru
Dr. Vladimir Anatolevich Maslov	maslov_vova@mail.ru
Dr. Vladimir Wagner	wagner@ujf.cas.cz
Prof. W. von Oertzen	oertzen@helmholtz-berlin.de
Mr. Waseem Raja	waseemjnv@gmail.com
Dr. X. Roca-Maza	xavier.roca.maza@mi.infn.it
Prof. Yuri Penionzhkevich	pyuer@mqail.ru
Prof. Zeblon Vilakazi	zeblon.vilakazi@wits.ac.za

Journal of Physics: Conference Series **724** (2016) 011004 doi:10.1088/1742-6596/724/1/011004

Peer review statement

All papers published in this volume of *Journal of Physics: Conference Series* have been peer reviewed through processes administered by the proceedings Editors. Reviews were conducted by expert referees to the professional and scientific standards expected of a proceedings journal published by IOP Publishing.

Journal of Physics: Conference Series **724** (2016) 011005 doi:10.1088/1742-6596/724/1/011005

NUMEN Project @ LNS : Heavy Ions Double Charge Exchange as a tool towards the *0νββ* Nuclear Matrix Element

C. Agodi[1)], F. Cappuzzello[1),2)], D. L.Bonanno[3)], D. G.Bongiovanni[3)], V. Branchina[2)], S.Calabrese[1)], L. Calabretta[1)], A.Calanna[1)], D. Carbone[1)], M. Cavallaro[1)], M. Colonna[1)], A. Foti[2),3)], P. Finocchiaro[1)], V. Greco[1),2)], G.Lanzalone[5)], D. Lo Presti[2),3)], F.Longhitano[3)], A.Muoio[4)], L. Pandola[1)], D. Rifuggiato[1)], S. Tudisco[1)]

1) INFN - Laboratori Nazionali del Sud, Catania, Italy
2) Dipartimento di Fisica e Astronomia, Università di Catania, Catania, Italy
3) INFN - Sezione di Catania, Catania, Italy
4) Università di Messina, Messina, Italy
5) Università di Enna, Enna, Italy
Corresponding author: agodi@lns.infn.it

Abstract. The NUMEN Project, proposed at INFN Laboratori Nazionali del Sud (LNS) in Catania, has the aim to access the nuclear matrix elements, entering the expression of the life time of double beta decay, by relevant cross sections of double charge exchange reactions. The basic point, on which it is based this innovative technique, is the coincidence of the initial and final state wave-functions in the two classes of processes and the similarity of the transition operators. A key aspect of the Project is the use of MAGNEX large acceptance magnetic spectrometer, for the detection of the ejectiles, and of the INFN LNS K800 Superconducting Cyclotron (CS), for the acceleration of the required high resolution and low emittance heavy-ion beams.

1. Introduction

Neutrinoless double beta decay (0νββ) is potentially the best resource to probe the Majorana or Dirac nature of neutrino and to extract its effective mass. Presently, this physics case is one of the most important research "beyond the Standard Model" and might guide the way towards a Grand Unified Theory of fundamental interactions.

Since the ββ decay process involves transitions in atomic nuclei, nuclear structure issues must be accounted for. The 0νββ decay rate $[T_{1/2}]^{-1}$ can be factorized as a phase-space factor $G_{0\nu}$, the nuclear matrix element (NME) $M_{0\nu}$ and a term $f(m_i,U_{ei})$ containing the masses m_i and the mixing coefficients U_{ei} of the neutrino species:

$$[T_{1/2}]^{-1} = G_{0\nu} |M_{0\nu}|^2 |f(m_i, U_{ei})|^2$$

The NME is defined as the transition amplitude from the initial to the final nuclear state of the ββ process through the 0νββ decay operator. Thus, if the NMEs are established with sufficient precision, the neutrino masses and the mixing coefficients can be extracted from 0νββ decay rate measurements.

The evaluation of the NMEs is presently limited to state of the art model calculations based on different methods (QRPA, shell-model, IBM etc.) [1], [2], [3], [4]. However the ambiguities in the models are still too large and the constraints too loose to reach accurate values of the NMEs. Discrepancy factors higher than two are presently reported in literature [5].

In this picture, we suggest an innovative technique in the NUMEN (Nuclear Matrix Elements of Neutrinoless Double Beta Decay) Project [6]: the idea is to use for the first time Heavy-Ions Double Charge Exchange (HI-DCE) cross section measurements towards the *0νββ* NME determination.

2. DCE reactions and neutrinoless double beta decays

There are a number of important similarities among DCE and 0νββ decay processes, despite they are mediated by different interactions. The basic point is the coincidence of the initial and final state wave-functions in the two classes of processes, and the similarity of the transition operators, which in both cases present a superposition of Fermi, Gamow-Teller and rank-two tensor component. Moreover a large linear momentum (~100 MeV/c) is available in the virtual intermediate channel in both processes; the two processes are non-local and are characterized by two vertices localized in a pair of valence nucleons; both processes take place in the same nuclear medium; a relevant off-shell propagation through virtual intermediate channels is present in the two cases.

The description of NMEs extracted from DCE and *0νββ* presents the same degree of complexity, with the advantage for DCE to be "accessible" in laboratory. However a simple relation between DCE cross sections and *ββ*-decay half-lives is not trivial and needs to be explored.

The use of modern high resolution and large acceptance spectrometers, like MAGNEX, togheter with high resolution and low emittance heavy-ion beams, like that produced at LNS K800 Superconducting Cyclotron (CS), are crucial for the experimental challenges, looking forward to extract quantitative information from DCE reactions. Moreover the measurement of DCE high resolution energy spectra and accurate cross sections at very forward angles are key points to identify the transitions of interest [7]. The concurrent measurement of the other relevant reaction channels allows to isolate the direct DCE mechanism from the competing transfer processes. These are at least of 4th-order and can be effectively minimized by the choice of the proper projectile-target system and incident energy [8].

First experimental results [9], obtained at the INFN-LNS laboratory in Catania, for the ^{40}Ca(^{18}O,^{18}Ne)^{40}Ar reaction at 270 MeV, in a wide range of transferred momenta, give encouraging indication on the capability to access quantitative information towards the determination of the Nuclear Matrix Elements for *0νββ* decay.

On the basis of the above mentined ground-breaking achievement, we propose an ambitious project, NUMEN, with the aim to go deep inshight in the HI-DCE studies on nuclei of interest in *0νββ* decay, looking forward at the *0νββ* NME dermination.

3. Toward the"hot" cases

The availability of the MAGNEX spectrometer [10] for high resolution measurements of very suppressed reaction channels was essential for the first pilot experiment [9]. However with the present set-up it is difficult to suitably extend this research to the "hot" cases, where ββ decay studies are and will be concentrated. The present limit is the low beam current we have experienced both for the CS accelerator and for the MAGNEX focal plane, that must be sensibly overcome. For a systematic study of the many "hot" cases of ββ decays an upgraded set-up, able to work with two orders of magnitude more current than the present, is necessary. This goal can be achieved by a substantial change in the technologies used in the beam extraction and in the detection of the ejectiles.

For the spectrometer the main foreseen upgrades are: the substitution of the present Focal Plane Detector (FPD) [11] gas tracker with a GEM tracker system; the substitution of the wall of silicon pad stopping detectors with a wall of telescopes based on SiC-CsI detectors; the enhancement of the maximum magnetic rigidity; the introduction of an array of detectors for measuring the coincident γ-rays.

In this framework we propose four phases in the NUMEN project, looking forward to do, in the same time, both the experimental and the up-grade activity, as indicated in the following Phases of the project.

3.1 Phase1: the experiment feasibility

The pilot experiment :^{40}Ca(^{18}O,^{18}Ne)^{40}Ar reaction at 270 MeV, with the first experimental data on heavy-ion double charge-exchange reactions in a wide range of transferred momenta, was already done. The results demostrate the technique feasibility [9].

3.2 Phase2: toward "hot" cases optimizing experimental conditions and getting first results

The necessary work for the upgrading of both the accelerator and MAGNEX will be carried out still preserving the access to the present facility. Due to the relevant technological challenges connected, in which test with and without beam will be crucial, the Phase2 is foreseen to have a duration of a 3-4 years. In the meanwhile, experiments with integrated charge of tens of mC (about one order of magnitude more than that collected in the pilot experiment) will be performed. These will require several weeks data taking for each reaction, since thin targets (a few 1018 atoms/cm2) are mandatory in order to achieve enough energy and angular resolution in the energy spectra and angular distributions. The attention will be focused on a few favorable cases, with the goal to achieve conclusive results for them.

3.3 Phase3: the facility upgrade

Once all the building block for the upgrade of the accelerator and spectrometer facility will be ready at the LNS a Phase3, connected to the disassembling of the old set-up and re-assembling of the new one, will start.

3.4 Phase4: the experimental campaign

The Phase4 will consist of a series of experimental campaigns at high beam intensities (some pμA) and long experimental runs in order to reach in each experiment integrated charge of hundreds of mC up to C, for the experiments in coincidences, spanning all the variety of candidate isotopes for $0\nu\beta\beta$ decay, like:

^{48}Ca, ^{82}Se, ^{96}Zr, ^{100}Mo, ^{110}Pd, ^{124}Sn, ^{128}Te, ^{130}Te, ^{136}Xe, ^{148}Nd, ^{150}Nd, ^{154}Sm, ^{160}Gd, ^{198}Pt.

4. Perspectives

To perform the experimental campaign that we propose it is necessary the LNS CS upgrade to give high beam intensity and the upgrade of the detection system. Actually, we require a new focal plane detector, suitable to resist to high rates, and a modular gamma detector system that, together, allows us to complete the last phase of measurements, spanning among all the nuclei of interest for our studies. Once selected the optimal experimental condition for the different cases, as planned in the NUMEN project Phase2, with the upgrades mentioned above, the Phase4 will be devoted to collect data. These data will be addressed to give, with an accurate analysis, a rigorous determination of the absolute cross sections values and their uncertainties for all the system of interest. This is the way the NUMEN Project aims to access NME, entering the expression of the life time of double beta decay, that is its most ambitious goal.

Journal of Physics: Conference Series **724** (2016) 012001 doi:10.1088/1742-6596/724/1/012001

References

[1] E. Caurier, et al., PRL 100 (2008) 052503.
[2] J. Suhonen and M.Kortelainen, Int. Journ. of Mod. Phys. E 17 (2008) 1.
[3] N. L. Vaquero, et al., PRL 111 (2013) 142501.
[4] J. Barea, et al., PRC 87 (2013) 014315.
[5] J. Barea, et al., PRL 109 (2012) 042501
[6] C.Agodi et al. Nuclear and Particle Physics Proceedings (2015) pp.28-30
[7] H. Matsubara, et al., Few-Body Systems 54 (2013) 1433.
[8] D. M. Brink, Phys. Lett. B 40 (1972) 37
[9] F.Cappuzzello et al. accepted EPJA sept. 2015
[10] F. Cappuzzello et al., MAGNEX: an innovative large acceptance spectrometer for nuclear reaction studies in: Magnets: Types, Uses and Safety, Nova Publisher Inc., New York, 2011, pp 1-63.
[11] M. Cavallaro, et al., Eur. Phys. J. A (2012) 48: 59

Journal of Physics: Conference Series **724** (2016) 012002 doi:10.1088/1742-6596/724/1/012002

Electric Dipole States and Time Reversal Violation in Nuclei.

N. Auerbach

School of Physics and Astronomy, Tel Aviv University

Tel Aviv 69978, Israel

Abstract: The nuclear Schiff moment is essential in the mechanism that induces a parity and time reversal violation in the atom. In this presentation we explore theoretically the properties and systematics of the isoscalar dipole in nuclei with the emphasis on the low-energy strength and the inverse energy weighted sum which determines the Schiff moment. We also study the influence of the isovector dipole strength distribution on the Schiff moment. The influence of a large neutron excess in nuclei is examined. The centroid energies of the isoscalar giant resonance (ISGDR) and the overtone of the isovector giant dipole resonance (OIVGDR) are given for a range of nuclei.

1. Introduction

There is a close connection between collective nuclear motion and symmetries in the nuclear Hamiltonian. In some instances collective excitations are the result of the existence of certain symmetries in the system. One of the best examples is the isobaric analog resonance that results from the charge symmetry of the nuclear force. In other instances however collective excitations, such as giant resonances, serve as intermediate states in the process of breaking symmetries. In this paper we will discuss the role the isoscalar and isovector dipole strength in determining the amount of time reversal violation in spherical nuclei. In particular the contribution of low-lying dipole strength will be examined.

The study of giant resonances is not new in nuclear physics. However there is a continuous research of this subject as new resonances are found and new properties of the resonances are discovered. Moreover it is now clear that many of the resonances play an important role in determining properties of nuclear structure and reactions and contribute to the understanding of nuclear phenomena. For example, a whole class of isovector giant resonances helps in the study of the symmetry energy, the determination of isospin mixing, etc. In a recent paper [1] it was pointed out that the isoscalar dipole (ISD) resonance [2, 3] could have a sizeable contribution to the nuclear Schiff moment [4] whose operator is the same as the operator commonly used in the study of the ISD strength distribution. In recent years new resonances were studied at relatively low excitation energies. Concentrations of low-lying isovector dipole strength was observed in many nuclei throughout the periodic table. These enhancements were termed as isovector pygmy dipole resonances (PDRs). Also in the case of the isoscalar dipole, low-lying strength was observed [3].

2. The isoscalar dipole and the Schiff moment

The operator for the ISD is:

$$\mathbf{D} = \sum_i (r_i^2 - \frac{5}{3} < r^2 >)\mathbf{r}_i \tag{1}$$

The existence of a static dipole moment in an atom means that the time reversal symmetry (as well as parity) are not conserved. Measurements of a static dipole in a neutral atom will indicate that these symmetries are broken. However, since the mean value of the nuclear dipole moment is screened by the electrons in the atom, the atomic electric dipole moment is generated by the nuclear *Schiff moment* [4]. This moment exists because of the finite size of the nucleus. Apart from a normalization constant the Schiff operator is identical to the isoscalar dipole moment.

$$\mathbf{S}=\frac{1}{10}\sum_i (r_i^2-\frac{5}{3}<r^2>)\mathrm{r}_i \tag{2}$$

In the neighboring odd-even nucleus with one nucleon in orbit j the particle +core states are: $|0^+j,J=j\rangle$, and $|1^-j',J=j\rangle$. The two states form a parity doubldet, they have the same spin but opposite parity. A weak parity and time reversal violiting interaction V_{PT} will mix the two doublet states. Taking for the weak interaction a form:

$$V_{PT}=\mathrm{f(r)}\mathbf{r}\cdot\boldsymbol{\sigma} \tag{3}$$

where f(r) is a function of r. Choosing for f(r) a function proportional to:

$\left(\mathrm{r}^2-\frac{5}{3}\langle\mathrm{r}^2\rangle\right)$ one can write $V_{PT}\sim\boldsymbol{S}\cdot\boldsymbol{\sigma}$.

One then finds [1] that the Shiff moment is proportional to the inverse energy weighted sum of the Schiff operator:

$$|\mathbf{S}|\sim\sum_i\frac{\left|\langle 1_i^-|S|0^+\rangle\right|^2}{\Delta E_i} \tag{4}$$

(A precise expressionfor S can be found in reference [1].) ΔE_i are the excitation energies of the 1_i^- states. Experiments indicate some of the dipole strength is found at low-energies thus enhanscing the inverse energy weighted sum. One therefore expects that the above mechanism will contribute significantly to the nuclear Schiff moment. It is important to study the distribution of ISD in spherical nuclei. Such a sudy is presented in the next section.

3. RPA calculations

A fully self-consistent HF-based RPA calculations of the strength distributions of the electric isoscalar and isovector dipole excitations in a wide range of spherical nuclei was performed, using the effective Skyrme type interaction. In order to insure self-consistency the calculations were carried out using a large p-h space and included all the terms of the p-h residual interaction (time-even and time-odd) which are associated with the energy density functional used in the HF calculations.

For a given probing operator F_L, we have calculated the strength function

$$S(E)=\sum_j|\langle 0|F_L|j\rangle|^2\delta(E_j-E_0). \tag{5}$$

Here, $|0\rangle$ is the RPA ground state and the sum is over all RPA excited states $|j\rangle$ with the corresponding excitation energies E_j. We adopt the single particle scattering operator

$$F_L=\sum_{iM} f(r_i)Y_{LM}(i)\,, \tag{6}$$

for isocalar ($T = 0$) excitations and

$$F_L = \frac{Z}{A}\sum_n f(r_n)Y_{LM}(n) - \frac{N}{A}\sum_p f(r_p)Y_{LM}(p)\,, \quad (7)$$

for isovector excitations ($T = 1$). We then determine the energy moments of the strength function,

$$m_k = \int_0^\infty E^k S(E)\, dE\,. \quad (8)$$

The centroid energy, E_{CEN}, is then obtained from

$$E_{CEN} = \frac{m_1}{m_0}\,. \quad (9)$$

We will present the results for the isoscalar dipole using:

$$f(r) = r^3 - (5/3)\langle r^2\rangle r,$$

as the probing operator. Note, for the ISD the second term in f(r) is added in order to insure zero contribution from the spurious state, associated with the center of mass motion.
We also performed calculations for the isovector dipole (IVD) resonance with an operator

$$\mathbf{D} = \sum_i (r_i^2 - \frac{5}{3} < r^2 >)\mathrm{r}_i\, \boldsymbol{\tau}_{zi} \quad (10)$$

However as shown in [1] the isovector part of the Schiff moment has other contributions that are not proportional to the *strength* of the IVD but involve the amplitudes of the IVD. Therefore when considering the IVD strength contribution to the atomic dipole measurements one should keep this in mind. Apart from the role played by the ISD and the IVD strength distributions in the calculation of the nuclear Schiff moment, there is general interest in the properties of the isocalar giant dipole resonance (ISGDR) and the overtone of the isovector giant dipole resonance (OIVGDR) and in particular in the low energy components the ISD and IVD strength distributions.

4. Results

In this section the results of the HF based RPA calculations of the strength distribution, S(E), the strength function divided by the energy, S(E)/E and the inverse energy weighted sum of the strength distribution, m_{-1}, of the ISD and IVD, for a wide range of nuclei, are presented. We should stress that the reason we present in each case the inverse energy weighted strength distribution, in addition the strength distribution, is the fact the Schiff moment depends on the inverse energy weighted sum [1]. We are concentrating on the low-energy region where the S(E)/E is large. We also include results for the centroid energies of isoscalar giant dipole resonance (ISGDR) and the overtone of the isovector giant dipole resonances (OIVGDR). The calculations were carried out for 18 frequently used Skyrme type interactions. Using the probing operator of Eqs. (1), we display in Figures 1 and 2 the HF-based RPA results for the strength function S(E) of the ISD in ^{40}Ca, ^{48}Ca, ^{56}Ni, ^{68}Ni and ^{78}Ni (Figure 1) and in ^{90}Zr, ^{96}Zr, ^{104}Zr, ^{144}Sm, and ^{208}Pb (Figure 2), calculated using the KDE0v1 Skyrme interaction [5] that is representative of the strength distributions for the rest of the interactions. *Note the strong enhancement in the ISD strength distributions at low energy in the neutron rich nuclei.*

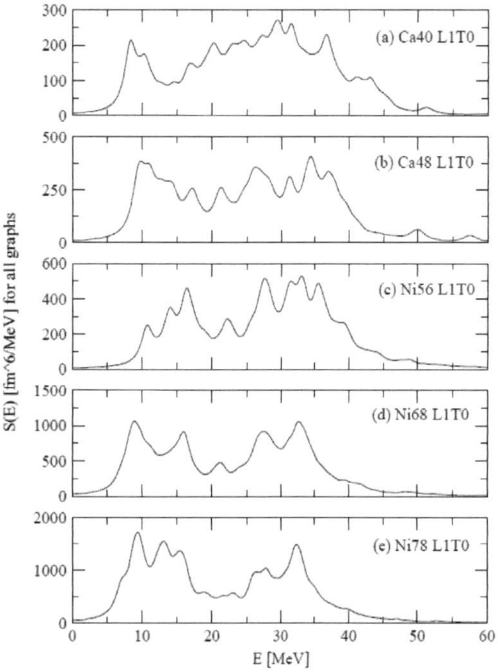

Fig 1. HF-based RPA results for the distribution of the strength function, S(E), of the isoscalar dipole, for the ^{40}Ca, ^{48}Ca, ^{56}Ni, ^{68}Ni and ^{78}Ni nuclei, calculated using the KDE0v1 Skyrme interaction [5]. An excitation energy range of 0-60 MeV and Lorenzian smearing of a 2 MeV width were used in the calculation.

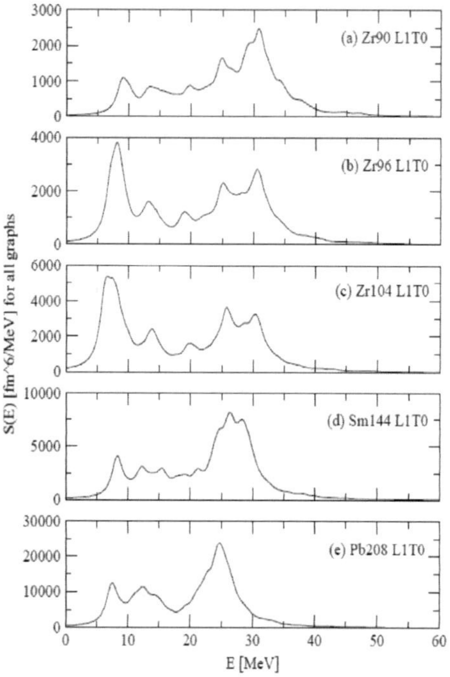

Fig. 2. Same as Fig. 1 but for the ^{90}Zr, ^{96}Zr, ^{96}Zr, ^{104}Zr, ^{144}Sm, and ^{208}Pb nuclei.

In Figures 3 and 4 are displayed the HF-based RPA results for the strength function divided by the energy, S(E)/E, of the ISD, obtained in ^{40}Ca, ^{48}Ca, ^{56}Ni, ^{68}Ni and ^{78}Ni (Figure 3) and in ^{90}Zr, ^{96}Zr, ^{104}Zr, ^{144}Sm, and ^{208}Pb (Figure 4) calculated using the KDE0v1 Skyrme interaction [5]. Note the relatively large contribution to moment m_{-1} from the low energy region, in particular in the case of neutron rich nuclei.

In Table 1 we display the values of the centroid energies, m_1/m_0 (in MeV) for the ISGDR, obtained using the probing operator of Eqs.(1) and the OIVGDR, obtained using the probing operator of Eqs.(10)), for a wide range of nuclei. The KDE0v1 Skyrme interaction [5] was used in the calculations. The ISGDR and OIVGDR are calculated over the excitation energy ranges, 16 – 40 and 20 – 60 MeV, respectively. Note the significant enhancement in the values of the inverse energy moment m_{-1}, in neutron rich isotopes which is associated with the increase in the ISD and IVD strengths at low energy.

Considering, for example, the case of 208Pb, we find that the centroid energies of the ISGDR and the OIVGDR are 24.5 and 35.2 MeV, respectively. The values of the inverse energy moment are m_{-1}= 17737 and 1806 $fm^6\ MeV^{-1}$1 for the ISD and the IVD in 208Pb, respectively. The value of m_{-1} for the IVD is only about 10% of that of the ISD. The findings for this nucleus are consistent with the estimate found for the Schiff moment in ref. [1]. Using the expressions for the weak interaction part from the above reference we find that the contribution to the Schiff moment considered here, is as large as the result obtained using the single particle models (see for example [6] and references therein)

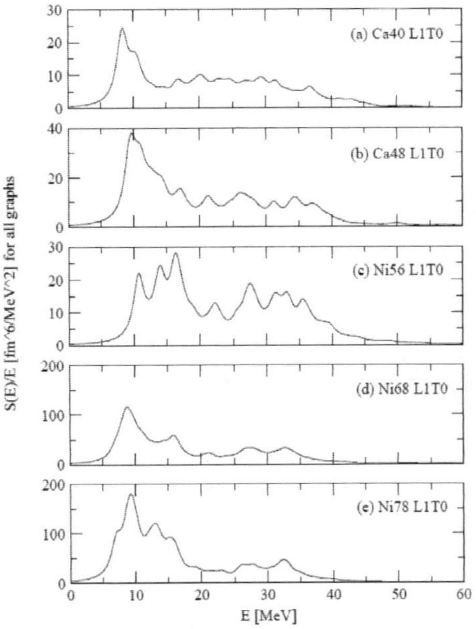

Fig. 3. Self-consistent HF-based RPA results for the distribution of the strength function divided by the energy, S(E)/E, of the isoscalar dipole, for the ^{40}Ca, ^{48}Ca, ^{56}Ni, ^{68}Ni and ^{78}Ni nuclei, calculated using the KDE0v1 Skyrme interaction [5]. An excitation energy range of 0-60 MeV and Lorenzian smearing of a 2 MeV width were used in the calculation.

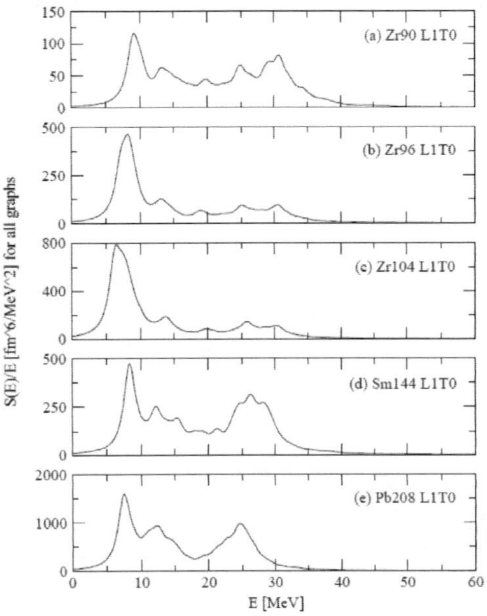

Fig.4. The same as Fig.3 but for nuclei listed in the figure.

5. Summary

We have presented results of HF-RPA calculations with Skyrme type interactions, for the strength functions and corresponding inverse energy moments of the isocalar dipole (ISD) and isovector dipole (IVD) for various nuclei using a probing operator which is the same as the Schiff operator up to a normalization. We see that the contribution of the IVD to the Schiff strength distribution in the even-even nucleus is smaller by an order of magnitude than that of the ISD. We find that in exotic nuclei with a large neutron excess the ISD and the IVD strength distributions is pushed to lower energies and thus increasing considerably the inverse energy moment m_{-1} of the strength distribution. In particular this means that the ISD contribution to the Schiff moment in the odd-even nucleus might be further enhanced. We also find that the contribution of the isovector inverse energy Schiff distribution in the even-even nucleus is smaller than that of the isoscalar by an order of magnitude.

The distributions of low-lying dipole strength (both isoscalar and isovector) are presently the subject of a large number of experimental studies. Our work provides a theoretical description of these distributions for a wide range of nuclei. Moreover, the various studies in the past, (except reference [1]), have not addressed the question, of the relation between the low-lying dipole strength and the Schiff moment, a quantity that is central to atomic studies of time reversal violation

[1, 4]. This relation is emphasized and discussed in this presentation. Future experiments that will measure static electric dipole moments using exotic atoms with large neutron excess spherical nuclei will need to use theoretical input in order to relate the results obtained to the limits of time reversal conservation. The results of the present study will be of help in this respect.

	ISGDR	OIVGDR
^{40}Ca	27.9	38.4
^{48}Ca	29.0	39.6
^{56}Ni	28.5	39.5
^{68}Ni	28.4	38.5
^{78}Ni	28.2	39.6
^{90}Zr	28.0	38.2
^{96}Zr	27.4	37.5
^{104}Zr	26.9	37.5
^{100}Sn	27.7	38.5
^{144}Sm	26.3	36.9
^{208}Pb	24.5	35.2

Table 1. Values of the centroid energies, m_1/m_0 (in MeV) for the ISGDR, using the probing operator of Eqs. (1), and the OIVGDR, using the probing operator of Eqs. (10), for a wide range of nuclei, calculated using the KDE0v1Skyrme interaction [5]. The ISGDR and OIVGDR are calculated over the energy ranges of 16 – 40 MeV and 20 – 60 MeV, respectively.

Acknowledgements: This work was done in collaboration with M. R. Anders , S. Shlomo, Ch. Stoyanov, and V. Zelevinsky. I wish to thank the Bulgarian and Israeli Academies of Science for support.

References

[1] Auerbach N and Zelevinsky VG, 2012 *Phys. Rev.* **C86** 045501

[2] Morsch HP, Rogge M, Turek P, and Mayer-Boericke C, 1980 *Phys. Rev. Lett.* **45** 337; Harakeh MN and Dieperink AEL, 1981 *Phys. Rev.* C**23** 2329; Van Giai N and Sagawa H, 1981 *Nucl. Phys.* A **371** 1

[3] Vretenar D, Niu YF, Paar N, and Meng J , 2012 *Phys. Rev.* C **85** 044317; Piekarewicz J, 2006 *Phys. Rev.* C **73** 044325; Litvinova E, Ring P, Tselyaev V, and. Langanke K, 2009 *Phys. Rev.* C **79** 054312 ;
Roca-Maza X, Pozzi G, Brenna M, Mizuyama K and Col`o G, 2012 *Phys. Rev.* C **85**, 024601 ; Carbone A, Colò G, Bracco A, Li-Gang Cao, Bortignon PF, Camera F, and Wieland O, 2010 *Phys. Rev.* C **81** 041301(R); Garg U, 2004 *Nucl. Phys.* **A731**, 3; Shlomo S and. Sanzur AI, *Phys. Rev.* C **65** 044310

[4] Auerbach N, Flambaum VV and Spevak V, 1996 *Phys. Rev. Lett.* **76**, 4316; Spevak V, Auerbach N, and Flambaum VV, 1997 *Phys .Rev.* C**56** 1357; Auerbach N and Zelevinsky VG, 2008 *J. Phys.G: Nucl. Part. Phys.* **35**, 093101

[5] Argawal BK, S. Shlomo S, and V. Kim Au V, 2008 *Phys. Rev.* C **72**, 014310

[6] Ginges JSM and Flambaum VV, 2004 *Phys. Reps.* **397**, 63

On deuteron interactions within surrogate reactions and nuclear level density studies

M. Avrigeanu and V. Avrigeanu
Horia Hulubei National Institute for Physics and Nuclear Engineering, P.O. Box MG-6, 077125 Bucharest-Magurele, Romania
E-mail: marilena.avrigeanu@nipne.ro

Abstract. The consequences which the suitable consideration of direct interaction, namely the breakup, stripping and pick-up processes, may have for the analysis of the deuteron surrogate reactions as well as for studies of nuclear-level density models on the basis of nucleon spectra analysis in deuteron-induced reactions are discussed. An increased attention is paid to the deuteron breakup which is the dominant mechanism for the interaction processes at low incident energies around Coulomb barrier, particularly on actinides.

1. Introduction

The description of deuteron-nucleus interaction represents an important test for both the appropriateness of reaction mechanism models and evaluation of nuclear data requested especially by the ITER [1], IFMIF [2], and SPIRAL-2 [3] research programmes. The present work continues the series of recent deuteron-induced reaction analyses [4, 5, 6, 7, 8] pointing out the major effects of the direct interaction (DI) contributions responsible for the contradictory results reported so far in testing various approaches of deuteron surrogate reactions and the nuclear level density.

2. Neutron capture analysis by deuteron surrogate reactions

The surrogate reaction method is an indirect measurement technique used to overcome the difficult problems of preparing and handling the highly radioactive targets required for cross-section measurements. Its main point concerns the study the outgoing channel of interest by means of an alternative "surrogate" reaction that involves a projectile-target combination more accessible experimentally (e.g., [9, 10, 11, 12, 13, 14, 15]). A particular feature is that, since last decade, this method has been involved mainly in the investigation of the neutron-induced (n,γ), and (n,f) reaction cross sections, by using an appropriate stable beam and target combination ([9] and Refs. therein).

The "desired" (n,γ) cross section for a target nucleus A is given within the Hauser-Feshbach formalism, in terms of compound nucleus (CN) formation cross section $\sigma_n^{CN}(E_{ex},J,\pi)$ and the branching ratio $G_\gamma^{CN}(E_{ex},J,\pi)$ toward the desired outgoing channel of γ-ray decay, by [10]:

$$\sigma_{n\gamma}(E_n) = \Sigma_{J,\pi}\, \sigma_n^{CN}(E_{ex},J,\pi) G_\gamma^{CN}(E_{ex},J,\pi), \qquad (1)$$

where E_n is the neutron incident energy, and J,π are the spin and parity of the excited state E_{ex} of the decaying compound nucleus (A+1).

CN formation cross section is usually obtained on the basis of a neutron optical model potential, while the G_γ^{CN} branching ratio requires accurate information on the Hauser-Feshbach model ingredients of the all competing decay channels. Such difficulties should be avoided by using alternative surrogate reactions. Among them, the deuteron surrogate reaction $(d,p\gamma)$ produces the same excited nucleus $(A+1)$, decaying through the desired γ channel.

The probability for CN formed in the (d,p) surrogate reaction, with the same specific excitation energy, spin, and parity values as in the desired reaction, to decay through the γ channel is [10]:

$$P_{d,p\gamma}(E_{ex}) = \Sigma_{J,\pi}\, F_{d,p}^{CN}(E_{ex},J,\pi) G_\gamma^{CN}(E_{ex},J,\pi), \tag{2}$$

where $F_{d,p}^{CN}(E_{ex},J,\pi)$ is the probability for the formation of this excited surrogate CN (A+1).

A specific point of the surrogate method is the experimental determination of $P_{d,p\gamma}(E_{ex})$, by measuring the total number of the (d,p) processes, and the number of $p-\gamma$ coincidences:

$$P_{d,p\gamma}^{exp}(E_{ex}) = \frac{N_{p,\gamma}^{coincidences}(E_{ex})}{N_{d,p}^{surrogateevents}(E_{ex})}. \tag{3}$$

The use of the measured $P_{d,p\gamma}(E_{ex})$, together with the calculated $F_{d,p}^{CN}(E_{ex},J,\pi)$, determines the branching ratios $G_\gamma^{CN}(E_{ex},J,\pi)$, leading to the determination of the desired cross section given by Eq. (1). Actually this is the theoretical frame of the surrogate reactions, before to apply approximations which simplify the analysis.

First, an approximation of the surrogate method considers similar $J-\pi$ distributions in both desired and surrogate reactions [9]:

$$F_{d,p}^{CN}(E_{ex},J,\pi) \sim F_n^{CN}(E_{ex},J,\pi) \equiv \frac{\sigma_n^{CN}(E_{ex},J,\pi)}{\Sigma_{J',\pi'}\sigma_n^{CN}(E_{ex},J',\pi')}, \tag{4}$$

where $F_n^{CN}(E_{ex},J,\pi)$ is the probability for the formation of the CN in the desired reaction.

Second, another approximation is related to the Weisskopf-Ewing (WE) limit of the Hauser-Feshbach formalism and considers the decay probabilities to be independent of $J-\pi$:

$$G_\gamma(E_{ex},J,\pi) = G_\gamma(E_{ex}), \tag{5}$$

so that the desired cross section becomes finally:

$$\sigma_{n\gamma}^{WE}(E_n) = \sigma_n^{CN}(E_n) P_{d,p\gamma}^{exp}(E_{ex}). \tag{6}$$

3. Effects of the deuteron surrogate reaction approximations

The validation of the deuteron surrogate method has got a great importance due to the possibility of using $(d,p\gamma)$ as a surrogate reaction for neutron capture while the importance of (n,γ) reaction for basic and applied nuclear physics is well known. On the other hand, large discrepancies have already been pointed out by the validation tests comparing well known (n,γ) cross sections with those provided by deuteron surrogate reaction $(d,p\gamma)$ [11, 12, 13, 14, 15] so that strong question mark concerns the suitability of the associated theoretical frame.

In this respect Allmond *et al.* [13] reported a 23% deviation between the known ratio ^{235}U$(n,\gamma)/^{235}$U(n,f) and the measured surrogate ratio ^{235}U$(d,p\gamma)/^{235}$U(d,pf). This large discrepancy has revealed "breakdown of the Bohr CN and Weisskopf-Ewing approximation" [13], requesting an improved reaction model for the (d,p) surrogate reaction. A similar request for an improved reaction model of the (d,p) surrogate process has followed the validation test performed by Hatarik *et al.* [14] for the ^{171,173}Yb$(d,p\gamma)$ surrogate reactions, by comparison with known neutron capture cross sections. The large discrepancy between evaluated ^{92}Mo(n,γ)

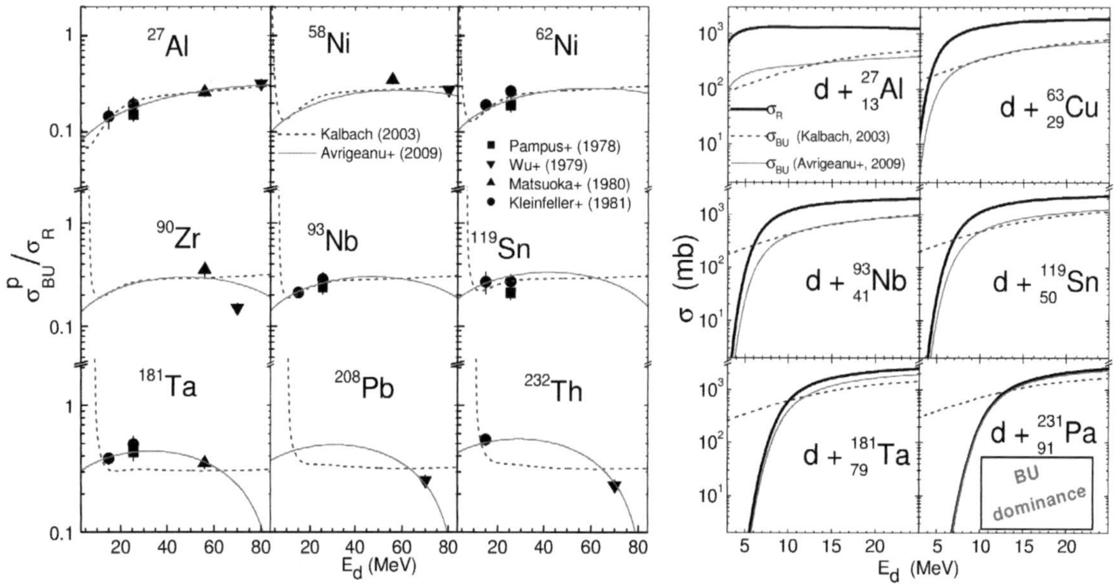

Figure 1. Left-side: Comparison of experimental [16] total proton–emission breakup fraction and the corresponding parametrization of Refs. [7, 8] (solid curves), and [17] (dashed curves) for deuteron interactions with target nuclei from ^{27}Al to ^{232}Th. Right-side: The energy dependence of the total breakup cross sections given by Avrigeanu *et al.* [7, 8] (solid thin curve) and Kalbach [17] (dashed curve) parametrizations for deuteron interactions with ^{27}Al, ^{63}Cu, ^{93}Nb, ^{119}Sn, ^{181}Ta, and ^{231}Pa, while σ_R is shown by solid thick curves.

reaction cross sections and the corresponding ^{92}Mo$(d,p\gamma)$ surrogate cross sections found by Goldblum *et al.* [15] points out the failure of the modeling the deuteron surrogate reactions through the Weisskopf-Ewing approximation. Wilson *et al.* [11] performed a "stringent test of the applicability" of the deuteron surrogate method for the well known ^{232}Th(n,γ) cross sections. A large overestimation of the (n,γ) reaction cross sections by the $(d,p\gamma)$ surrogate reaction results for the low neutron energy range has thus been reported. In addition to the criticism of the Weisskopf-Ewing approximation used in the surrogate formalism, the effect of the breakup process is mentioned by Ducasse *et al.* [12] as another source of the validation failure in the case of ^{238}U$(d,p\gamma)$ reaction.

Actually, the apparent discrepancies evidenced by validation tests [11, 12, 13, 14, 15] are obviously the results of weak points corresponding to the approximations of the (d,p) interaction process analysis in the theoretical frame of the surrogate method. One approximation, being even a contradiction in the terms of the surrogate reaction method, concerns the fact that the direct nucleon-transfer (d,p) reaction forms an excited CN [11, 12, 13, 14, 15]. Therefore, an advanced assessment of the reaction mechanisms involved in deuteron surrogate reactions, populating a highly excited nucleus, should include the direct reactions (DR, e.g., stripping), statistical processes, e.g., pre-equilibrium emission (PE), and CN processes, as well as the deuteron breakup (BU) particularly for deuteron interaction processes [4, 5, 8]. As a matter of fact, BU has the strongest effects in the case of the deuteron surrogate reactions at low incident energies, for heavy targets nuclei (actinides), as it has been pointed out for the low-energy deuteron interaction with ^{231}Pa target nucleus [6].

The deuteron breakup in the Coulomb and nuclear fields of the target nucleus involves two distinct processes, namely the elastic breakup (EB) in which the target nucleus remains in its ground state and none of the deuteron constituents interacts with it, and the inelastic breakup or breakup fusion (BF), where one of these deuteron constituents interacts with the target nucleus while the remaining one is detected. Actually, there are overall two opposite effects of BU on the deuteron activation cross sections that should be considered. First, the deuteron total-reaction cross section σ_R, that is shared among different outgoing channels, is reduced by the value of the total breakup cross section σ_{BU}. At the same time, the BF component, where one of deuteron constituents interacts with the target nucleus leading to a secondary composite nucleus, brings contributions to different reaction channels [4, 5, 6, 7, 8].

A detailed discussion of the work of Wilson *et al.* [11], using the surrogate reaction ^{232}Th$(d,p\gamma)^{233}$Th for an indirect measurement of the well-known ^{232}Th$(n,\gamma)^{233}$Th reaction cross sections for incident-neutron energies between 0 and around 1 MeV, is involved hereafter to underline the breakup effects which are present in the deuteron surrogate experiments. Wilson *et al.* reported a good agreement between indirect and direct (n,γ) cross-section measurements only in the range 500 keV–1 MeV while large discrepancies have been found outside this range.

Firstly, this measurement [11] involved 12 MeV deuterons incident on ^{232}Th target nucleus, while the decay probabilities $P_{d,p\gamma}(E_{ex})$ of the excited nucleus ^{233}Th have been measured at excitation energies between the corresponding neutron binding energy S_n=4.786 MeV and 1 MeV above it. The protons from the $(d,p\gamma)$ reaction which correspond to this excitation have had energies between ~8.7 and 9.7 MeV while their maximum energy has been around 14.5 MeV. At the same time, the BF protons have had a maximum energy of 9.673 MeV in the center-of-mass system, with a twofold outcome for these BF protons: they match the proton emission involved in the surrogate-reaction analysis, but have energies lower than the protons which populate the excited nucleus ^{233}Th below S_n and were considered to prove the lack of any BU effect (Fig. 6 of Ref. [11]).

Secondly, the BF protons with energies between ~8.7 and 9.7 MeV correspond to BF neutrons with energies between around 1 MeV and 0, respectively, i.e., very much alike to the desired neutron capture process. Moreover, these BF neutrons interact with ^{232}Th target nucleus, populating the same analyzed the ^{233}Th CN, at the same excitation energies of interest. The γ-ray decay of the ^{233}Th compound nuclei populated through the BF enhancement contributes thus, together with the companion BF protons, to the measured $p-\gamma$ coincidence events.

Thirdly, in addition to the BF, stripping, and PE contributions to the population of the excited nucleus ^{233}Th, one has to take into account the considerable amount of incident deuterons leakage through these processes [4, 5, 6, 7, 8], which strongly diminishes the probability $F^{CN}_{d,p}(E_{ex},J,\pi)$ for forming the compound nucleus ^{233}Th, given by Eq. (2). The importance of the total (EB+BF) *proton*-emission breakup fractions σ^p_{BU}/σ_R is shown in the left-side of Fig. 1 by means of the comparison of experimental systematics [16], measured for target nuclei from Al to Th, and the calculated predictions of the empirical Kalbach's [17] and Avrigeanu *et al.* [7, 8] parametrizations.

The *total* breakup cross sections predicted by Kalbach [17] and Avrigeanu *et al.* [7, 8] for deuterons interaction with target nuclei from Al to Pa and the total deuteron cross section are compared in the right-side of Fig. 1. Except the differences for incident energies lower than ~10 MeV, where Kalbach's parametrization [17] predicts too high values for the breakup cross sections, both parametrizations predict an increasing role of BU with increased target-nucleus mass/charge, pointing out the dominance of the breakup mechanism at the deuteron incident energies below and around the Coulomb barrier for the target nucleus ^{231}Pa [6]. This conclusion is obviously in line with the experimental total proton-emission BU fraction data for deuterons on ^{232}Th [16] shown on the left-side of Fig. 1. Therefore, the dominance of BU for the actinides nuclei at energy around Coulomb barrier should be particularly considered in the case of (d,x) surrogate reactions analysis [8]. Actually, it is obvious that the neglection of the breakup mechanism (Fig. 1) strongly affects the validation test, being the main reason of its failure.

Moreover, the other assumption concerning the equality of the branching ratios for the deuteron surrogate and the neutron-induced reactions should be considered with increased caution in the analysis due to the population and decay differences between the excited and compound nuclei, respectively, formed in surrogate and desired reactions [12, 18].

Finally, one should be more careful in assuming that the failure of the surrogate-method validation tests follows the use of the too weak Weisskopf-Ewing approximation [11, 12, 13, 14, 15]. It is also even the use of the Hauser-Feshbach formalism alone, within deuteron-induced reactions analysis, which can not lead to expected good results in the absence of the unitary account of the involvement of all BU+DR+PE+CN reaction mechanisms [8].

4. Nuclear level density study with deuteron-induced reactions

The results rather contradictory which are obtained from the investigation of the level density models in deuteron induced reactions at low energies, around the Coulomb barrier, stand for another example of an approach of the deuteron interaction process which is too simplified.

Thus, we would like to comment the assumption of lower noncompound components of the neutron spectra measured at backward angles with 5–9 MeV deuterons on ^{27}Al, ^{63,65}Cu, ^{89}Y [19], and ^{54,56,58}Fe [20] as well as the experimental CN fractions of σ_R obtained by too simplified analysis of the neutron angular distributions resulted from the deuteron interactions with the same target nuclei. Hence, there have been obtained contradictory results as (i) finding that more than 70% of σ_R are determined by the CN mechanism, and (ii) failure to describe the proton, and alpha-emission spectra.

Conversely, we compared the experimental CN fractions from the above-mentioned studies [19, 20] with the results of the unitary analysis of deuteron interactions with the same target nuclei ^{27}Al, ^{63,65}Cu, ^{89}Y, and ^{54,56,58}Fe [4, 5, 7] and found apparent discrepancies in both the numerical values and the slope of their energy dependence [21]. While Refs. [4, 5, 7] include a detailed description of the model assumptions and consistent parameters sets we would mention here only the main lines of these analyses concerning the treatment of nuclear reaction mechanisms involved within the deuteron interaction process. Thus, in order to obtain the CN fractions of σ_R it has been taken into account the decrease of the deuteron total-reaction cross section, that is subsequently shared among different outgoing PE and CN channels, by DI through appropriate treatment of each one. The BU cross sections have been obtained from the empirical parametrization [7, 8] of both the EB, and the total BU including also the BF, while the contributions of (d,p) and (d,n) stripping and (d,t) and (d,α) pick-up direct reactions were calculated using the distorted-wave Born approximation (DWBA) method [4, 5, 7].

The unitary and consistent consideration of the BU, DR, PE, and CN processes has been proved successful by the agreement of the calculated and all available deuteron activation data for ^{27}Al, $^{54,56-58,nat}$Fe, ^{63,65,nat}Cu, and ^{93}Nb target nuclei [4, 5, 7]. The calculated CN fractions have been thus validated.

In what follows, we enlarge our previous analysis [21] by comparison of the ratios of CN cross sections to σ_R with the similar values obtained by analysis of measured neutron angular distributions in deuteron-induced reactions on ^{54,56,58}Fe [20] (upper left-side of Fig. 2). The similar fractions for the DI and PE components are also included in this figure in order to point out the need of their consideration. Moreover, the corresponding excitation functions for each DI process along a wider deuteron energy range are also shown in Fig. 2 for a better understanding of their importance [5]. There are two main features pointed out, on the whole, by the comparison shown in the left-side of Fig. 2. First, there is a significant difference between the larger CN contributions obtained from the measured neutron angular distributions [20], and the calculated values following the unitary analysis of every deuteron-interaction mechanism. The lower values of the calculated CN fractions are mainly due to the significant BU component,

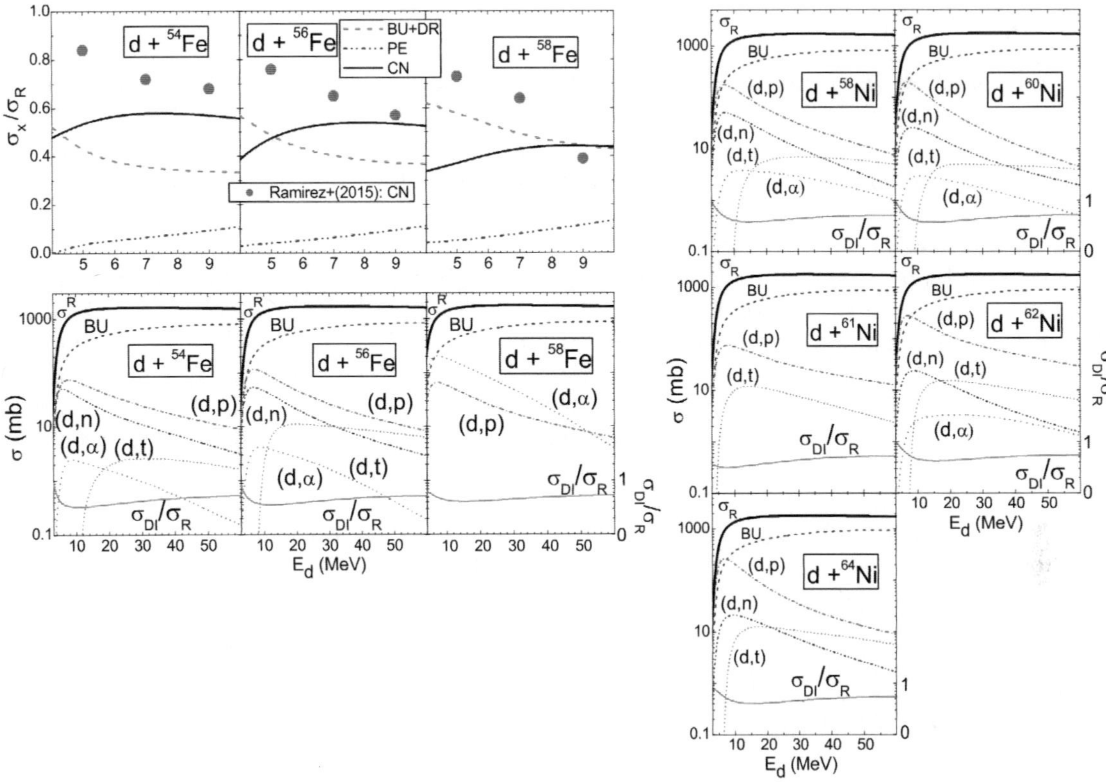

Figure 2. Left-side top: Comparison of the experimental *CN* fractions of the total-reaction cross section σ_R, obtained by analysis of measured neutron angular distributions in deuteron-induced reactions on ^{54,56,58}Fe target nuclei around the Coulomb barrier [20] (circles), and the calculated values [5, 21] of these fractions (solid curves) as well as for the *DI*=*BU*+*DR* (dashed curves) and *PE* (dotted curves) processes. Left-side bottom: The corresponding calculated excitation functions [5, 21] of either *DI* fractions of σ_R (thin solid curves), or the distinct *BU* (dashed curves) and (d,p), (d,n), (d,α), and (d,t) DR (dash-dotted, dash-dot-dotted, dotted, and short dotted curves, respectively). Right-side: Some as left-side bottom but for deuteron interactions with ^{58,60,61,62,64}Ni.

which has been not taken into account by any of the analyses performed within Refs. [19, 20].

Second, a particular note should concern the opposite energy dependence of the experimental and calculated *CN* fractions for low deuteron incident energies. Contrary to the expectation of Ramirez *et al.* [19, 20] of a decreasing *CN* contribution with the deuteron-energy increase, this contribution is slightly increasing with energy since both the major *BU* and especially the *DR* component have a slower increase with energy than σ_R. The energy dependence of the *DI* fractions of σ_R for ^{54,56,58}Fe target nuclei is better understood by analyzing the behavior of the excitation functions for competing *BU* and *DR* processes through an enlarged incident energy domain, as shown in the lower left-side of Fig. 2. Most significant in this respect are the maxima of the (d,p) and (d,n) stripping excitation functions around 6–8 MeV, after which lead to the statistical emission increase. The energy dependence of the *DI* fractions of σ_R found for Fe isotopes [5] is also confirmed by the similar analysis of the nuclear reaction mechanisms involved in the deuteron interaction with Ni isotopes shown in the right-side of Fig. 2.

5. Conclusions

The present work has concerned a deeper analysis of the key role of the direct interaction, particularly of the breakup mechanism, in deuteron-induced reactions, whose neglect led to above-mentioned contradictory results. The opposite effects of the breakup mechanism, namely the enhancement of the counted $protons - \gamma$ coincidences as well as the decrease of the CN cross section due to initial deuteron-flux leakage through breakup but also DR and PE processes may explain [7, 8, 21] the failure of the validation tests of the surrogate method using deuteron-induced reactions.

Similarly, there are significant differences between the CN components established by analysis of measured neutron angular distributions, in deuteron-induced reactions around the Coulomb barrier [19, 20], and the results following the unitary analysis of the deuteron interaction mechanisms [4, 5, 7, 21]. Actually, the latter may account for the failure to describe the neutron spectra from 7.5 MeV deuteron interaction with ^{63,65}Cu, or proton spectra from 6 MeV deuterons on ^{89}Y [19] as well as protons and alpha spectra for 9 MeV deuteron on ^{56,58}Fe [20].

Acknowledgments

This work was partly supported by Fusion for Energy (F4E-GRT-168-02), and by a grant of Unitatea Executiva pentru Finantarea Invatamantului Superior, a Cercetarii, Dezvoltarii si Inovarii (PN-II-ID-PCE-2011-3-0450).

References

[1] http://www.iter.org/proj
[2] http://www.ifmif.org/b/
[3] http://pro.ganil-spiral2.eu/spiral2/instrumentation/nfs
[4] P. Bém *et al.* 2009 *Phys. Rev.* C **79** 044610; E. Šimečková *et al.* 2011 *Phys. Rev.* C **84** 014605; M. Avrigeanu *et al.* 2013 *Phys. Rev.* C **88** 014612
[5] M. Avrigeanu and V. Avrigeanu 2014 *Nucl. Data Sheets* **118** 301; M. Avrigeanu *et al.* 2014 *Phys. Rev.* C **89** 044613
[6] M. Avrigeanu, V. Avrigeanu, and A.J. Koning 2012 *Phys. Rev.* C **85** 034603
[7] M. Avrigeanu *et al.* 2009 *Fusion Eng. Design* **84** 418; M. Avrigeanu and V. Avrigeanu 2010 *EPJ Web of Conferences* **2** 01004; 2012 **21** 07003; 2010 *J. Phys: Conf. Ser.* **205** 012014; 2014 **533** 012004; 2011 *J. Korean Phys.Soc.* **59** 903
[8] M. Avrigeanu and V. Avrigeanu 2015 in *Proc. 14th Int. Conf. on Nuclear Reaction Mechanisms, 15-19 June 2015, Varenna, Italy* (to be published in CERN Proceedings series)
[9] J. Escher *et al.* 2012 *Rev. Mod. Phys.* **84** 353; J.E. Escher and F.S. Dietrich 2010 *Phys. Rev. C* **81** 024612
[10] N.D. Scielzo *et al.* 2010 *Phys. Rev. C* **81** 034608
[11] J.N. Wilson *et al.* 2012 *EPJ Web of Conferences* **21** 01003; J.N. Wilson *et al.* 2012 *Phys. Rev. C* **85** 034607
[12] Q. Ducasse *et al.* 2014 *Nucl. Data Sheets* **119** (2014) 233; Q. Ducasse *et al.* 2015 *in 5th Workshop on Nuclear Level Density and Gamma Strength*, 18–22 May 2015, Oslo, Norway, `http://tid.uio.no/workshop2015/talks/Oslo2015_Ducasse.pdf`
[13] J.M. Allmond *et al.* 2009 *Phys. Rev.* C **79** 054610
[14] R. Hatarik *et al.* 2010 *Phys. Rev.* C **81** 011602
[15] B.L. Goldblum *et al.* 2012 *Phys. Rev.* C **85** 054616
[16] J. Pampus *et al.* 1978 *Nucl. Phys.* A **311** 141; J. R. Wu, C. C. Chang, and H. D. Holmgren 1979 *Phys. Rev.* C **19** 370 ; N. Matsuoka *et al.* 1980 *Nucl. Phys.* A **345** 1; J. Kleinfeller *et al.* 1981 *Nucl. Phys.* A **370** 205; M. G. Mustafa, T. Tamura, and T. Udagawa 1987 *Phys. Rev.* C **35** 2077
[17] C. Kalbach Walker 2003 "Deuteron Breakup and ^{21}Na Production", in *Triangle University Nuclear Laboratory Progress Report* **XLII (2002-2003)**; www.tunl.duke.edu/publications/tunlprogress/2003/
[18] Satoshi Chiba and Osamu Iwamoto 2010 *Phys. Rev.* C **81** 044604
[19] S.I. Al-Quraishi *et al.* 2000 *Phys. Rev.* C **62**, 044616; A.P.D. Ramirez *et al.* 2013 *Phys. Rev.* C **88**, 064324; Y. Byun *et al.* 2014 *Phys. Rev.* C **90**, 044303
[20] A.P.D. Ramirez *et al.* 2015 *Phys. Rev.* C **92**, 014303
[21] M. Avrigeanu and V. Avrigeanu 2015 *Phys. Rev.* C **92** 021601; *in 5th Workshop on Nuclear Level Density and Gamma Strength*, 18–22 May 2015, Oslo, Norway, `http://tid.uio.no/workshop2015/talks/Oslo2015_MAvrigeanu.pdf`

Journal of Physics: Conference Series **724** (2016) 012004 doi:10.1088/1742-6596/724/1/012004

A constrained-path quantum Monte-Carlo approach for the nuclear shell model

Jérémy Bonnard[1,2] **and Olivier Juillet**[2]

[1] INFN, Sezione di Padova, I-35131 Padova, Italy
[2] LPC Caen, ENSICAEN, Université de Caen, CNRS/IN2P3, Caen, France

Abstract. A new QMC approach for the shell model yielding nearly exact spectroscopy of nuclei is presented. The originality of the formalism lies in the use of a variational symmetry-restored wave function to 'steer' the Brownian motion, and to control the sign/phase problem that generally makes the traditional QMC samplings totally ineffective by causing a prohibitive growth of the statistical errors. Tests of convergence and proof-of-principle results are reported.

1. Introduction

The shell model with configuration mixing is one of the most powerful theoretical frameworks for studying the nuclear structure [1]. In this picture, valence protons and neutrons beyond an inert magic core are confined in one or more active shells and interact through an effective two-body residual interaction. The wave functions of the nucleus are then determined by diagonalizing the Hamiltonian in the set of all the accessible configurations and used to compute physical observables. Unfortunately, the shell model is strongly restricted in its applicability by the combinatorial scaling of the dimension of the many-body basis with the size of the single-particle Hilbert space and the number of valence nucleons.

Quantum Monte-Carlo (QMC) methods represent attractive techniques to overcome such limitations by offering a systematic alternative to the diagonalization of the Hamiltonian. Indeed, *via* a stochastic reformulation of the imaginary-time Schrödinger equation, the many-body problem is reduced to a set of numerically tractable one-body problems describing independent particles that randomly walk in fluctuating external fields. In this way, exact correlated wave functions are reconstructed by averaging independent-particle states called *walkers*.

The shell model Monte Carlo method (SMMC) remains to date the principal QMC approach of the shell model [2], and has enabled to investigate systems out of reach of conventional calculations without severe truncations of the configuration space. It relies on the standard sampling by auxiliary fields of the path integrals resulting from the Hubbard-Stratonovich transformation [3] of the imaginary-time propagator. With schematic residual interactions, the SMMC approach accurately reproduces the properties of even-even or $N = Z$ odd-odd nuclei at zero and finite temperature. However, with realistic effective interactions or for other kind of nuclei, the method is plagued by the so-called fermion sign or phase problem whose signature is an exponential collapse of the signal-to-noise ratio as the temperature is decreased. To circumvent this pathology and guarantee convergent simulations, a class of modified Hamiltonians where interactions causing the problem are artificially reduced is usually sampled. The physical observables are then recovered by extrapolating these non-physical SMMC results to recover the original Hamiltonian [4].

Besides the phase problem, another limitation of the SMMC approach lies in its incapacity to achieve detailed spectroscopy of nuclei, even though informations on excited states can be extracted from response functions. In contrast, low-lying states can be reconstructed by means of the so-called Monte-Carlo shell model technique [5] that does not utilize the stochastic process to sample eigenstates, but instead to generate a subspace into which the Hamiltonian matrix is diagonalized. Given that such calculations do not suffer from explicit manifestations of sign or phase problems, they may be successfully performed for very large configuration spaces and provides variational estimates of nuclear properties at low excitation energy [6].

In this work, we present a new QMC scheme to obtain yrast states in the shell model framework [7]. The formalism involves a variational symmetry-restored wave function assuming a twofold role. First, it guides the underlying Brownian motion to improve the efficiency of the sampling. Second, it constrains the stochastic paths to control sign or phase problems, in the spirit of the fixed-node approximation, widely applied in *ab initio* QMC calculations [8] where the random walk takes place within the coordinate space.

2. Phaseless quantum Monte-Carlo method

2.1. Guided dynamics

Implementing a QMC approach based on auxiliary fields requires to write the two-body Hamiltonian $\hat{H}$ in a quadratic form of one-body operators $\hat{T}$ and $\{\hat{O}_s\}$:

$$\hat{H} = \hat{T} - \sum_s \omega_s \hat{O}_s^2, \quad \hat{T} = \sum_{i,j} T_{ij} \hat{c}_i^\dagger \hat{c}_j, \quad \hat{O}_s = \sum_{i,j} (O_s)_{ij} \hat{c}_i^\dagger \hat{c}_j, \tag{1}$$

with $\hat{c}_i^\dagger$ ($\hat{c}_i$) the creation (annihilation) operator of a fermion in the single-particle state $|i\rangle$ of a finite-size discrete orthonormal basis. As discussed in Ref. [9], there is substantial freedom in doing so. For the shell model, we choose the isospin density decomposition given in Ref. [9].

Zero-temperature QMC methods rely on the imaginary-time propagation to project an initial wave function $|\Phi_0\rangle$ onto the lowest-energy eigenstate $|\Psi_G\rangle$ of $\hat{H}$ not orthogonal to $|\Phi_0\rangle$:

$$\lim_{\tau \to +\infty} \mathrm{e}^{-\tau \hat{H}} |\Phi_0\rangle \propto |\Psi_G\rangle. \tag{2}$$

Thanks to the Trotter-Suzuki breakup [10] and the Hubbard-Stratonovich transformation [3], the propagator for a short step $\Delta\tau$ may then be reformulated stochastically,

$$\mathrm{e}^{-\Delta\tau \hat{H}} = \int \mathrm{d}\vec{\eta}\, p(\vec{\eta}) \hat{\mathcal{U}}(\vec{\eta}), \tag{3}$$

where $\vec{\eta}$ is a vector of random variables $\{\eta_s\}$, the so-called auxiliary fields (AF), distributed according to a normal Gaussian distribution p. As exponential of one-body operators, the stochastic propagator

$$\hat{\mathcal{U}}(\vec{\eta}) = \mathrm{e}^{-\frac{\Delta\tau}{2}\hat{T}} \mathrm{e}^{\sum_s \eta_s \sqrt{2\omega_s \Delta\tau} \hat{O}_s} \mathrm{e}^{-\frac{\Delta\tau}{2}\hat{T}}, \tag{4}$$

transforms a Slater determinant $|\Phi_\tau\rangle$ at τ to a new one $|\Phi_{\tau+\Delta\tau}\rangle = \hat{\mathcal{U}}|\Phi_\tau\rangle$, used to sample the exact state at $\tau + \Delta\tau$. Consequently, the wave function resulting from the evolution during $\Delta\tau$ of $|\Phi_\tau\rangle$ is reinterpreted as the coherent statistical average $\mathbb{E}[\cdot]$ of independent-fermion states $|\Phi_{\tau+\Delta\tau}\rangle$, i.e. $\exp(-\Delta\tau\hat{H})|\Phi_\tau\rangle = \mathbb{E}[|\Phi_{\tau+\Delta\tau}\rangle]$, and finally the eigenstate of $\hat{H}$ is recovered as follow,

$$|\Psi_G\rangle \underset{\tau \to +\infty}{\propto} \mathbb{E}[|\Phi_\tau\rangle]\,. \tag{5}$$

In order to improve the efficiency of the standard AFQMC scheme described above, Zhang and Krakauer have proposed to borrow the idea of importance sampling [11], by generating walkers $|\Phi\rangle$ according to their overlap $\langle\Psi_T|\Phi\rangle$ with a trial wave function $|\Psi_T\rangle$ not orthogonal to the ground state. By directly including the complex importance function within the random walk, the exact state at a given τ is therefore reconstructed as

$$\mathrm{e}^{-\tau\hat{H}}|\Phi_0\rangle = \mathbb{E}\left[\Pi_\tau \frac{|\Phi_\tau\rangle}{\langle\Psi_T|\Phi_\tau\rangle}\right], \tag{6}$$

the factor Π being introduced to ensure the exactness of the stochastic scheme. The equations of motion may be derived by the usual Hubbard-Stratonovich procedure provided that the $\hat{O}_s$ operators in Eq. (1) are previously shifted by their local estimates $\langle\hat{O}_s\rangle_{\Psi_T\Phi_\tau} = \langle\Psi_T|\hat{O}_s|\Phi_\tau\rangle/\langle\Psi_T|\Phi_\tau\rangle$:

$$\mathrm{e}^{-\Delta\tau\hat{H}}\Pi_\tau \frac{|\Phi_\tau\rangle}{\langle\Psi_T|\Phi_\tau\rangle} = \int \mathrm{d}\vec{\eta}\, p(\vec{\eta})\, \Pi_\tau \Delta\Pi_\tau \frac{\hat{\mathcal{U}}_T(\vec{\eta})|\Phi_\tau\rangle}{\langle\Psi_T|\Phi_{\tau+\Delta\tau}\rangle}. \tag{7}$$

The stochastic propagator $\hat{\mathcal{U}}_T$ that transforms $|\Phi_\tau\rangle$ into $|\Phi_{\tau+\Delta\tau}\rangle$ reads

$$\hat{\mathcal{U}}_T(\vec{\eta}) = \mathrm{e}^{-\frac{\Delta\tau}{2}\hat{T}}\mathrm{e}^{\sum_s \Delta h_{T,s}\hat{O}_s}\mathrm{e}^{-\frac{\Delta\tau}{2}\hat{T}}, \quad \Delta h_{T,s} = 2\Delta\tau\omega_s\langle\hat{O}_s\rangle_{\Psi_T\Phi_\tau} + \eta_s\sqrt{2\omega_s\Delta\tau}. \tag{8}$$

Compared to the standard AFQMC diffusion (4), the trial state $|\Psi_T\rangle$ now takes part in the dynamics *via* the local estimates of the $\hat{O}_s$, and so guides the Brownian motion towards a region of the Slater determinants manifold where the importance of the walker in the sampling is expected to be large. The variation $\Delta\Pi_\tau$ of the weight Π_τ,

$$\Delta\Pi_\tau = \frac{\langle\Psi_T|\hat{\mathcal{U}}_T|\Phi_\tau\rangle}{\langle\Psi_T|\Phi_\tau\rangle} \exp\Big(-\Delta\tau\sum_s \omega_s\langle\hat{O}_s\rangle^2_{\Psi_T\Phi_\tau} - \sum_s \eta_s\sqrt{2\omega_s\Delta\tau}\langle\hat{O}_s\rangle_{\Psi_T\Phi_\tau}\Big), \tag{9}$$

contains the remaining scalar terms from the rewriting of $\hat{H}$, and may be recast in a more concise form when $\Delta\tau \to 0$. In this limit, the quantities $\eta_s\sqrt{\Delta\tau}$ indeed become infinitesimal increments $\mathrm{d}W_s$ of Wiener processes in the Ito formalism of stochastic calculus [12] that obey $\mathbb{E}[\mathrm{d}W_s] = 0$, $\mathrm{d}W_s\mathrm{d}W_{s'} = \delta_{ss'}\mathrm{d}\tau$. Thus, expanding $\hat{\mathcal{U}}_T$ as well as the exponential scalar term in Eq. (9) yield to the first order

$$\Delta\Pi_\tau = \exp\Big(-\Delta\tau\langle\hat{H}\rangle_{\Psi_T\Phi_\tau}\Big), \tag{10}$$

for $\Delta\tau$ sufficiently small. Similarly, one may also check that in this context the one-body representation of $\hat{\mathcal{U}}_T$ governing the dynamics of the occupied orbitals transforms into the stochastic differential equations (3) of Ref. [7].

2.2. Phase problem

In general, any QMC sampling is plagued by sign or phase problems as soon as the phase of the overlap $\langle\Psi_G|\Phi\rangle$ varies during the random walk. Indeed, extracting $|\Psi_G\rangle$ is strongly compromised when a large proportion of walkers corresponding to different phases mutually cancel their contributions to the sampling, leading to a mean overlap $\mathbb{E}[\langle\Psi_G|\Phi\rangle] \approx 0$. Hence, such trajectories severely degrade the signal-to-noise ratio: They are useless for reconstructing the ground-state wave function and they only increase the statistical error. In the framework of the QMC scheme (6), the overlap $\langle\Psi_G|\Phi\rangle$ is given by

$$\langle\Psi_G|\Phi\rangle \underset{\tau\to+\infty}{\propto} \langle\Psi_T|\mathrm{e}^{-\tau\hat{H}}|\Phi\rangle \underset{\tau\to+\infty}{=} \mathbb{E}[\Pi_\tau], \tag{11}$$

with $\Pi_0 = \langle\Psi_T|\Phi\rangle$. Managing the phase problem requires to resort to approximations that constrain the random walk through a change $\Pi \to \widetilde{\Pi}$ of the weight of the realizations. First, by taking the real part of the local energy in Eq. (10), we guarantee that the overlaps $\langle\Psi_G|\Phi\rangle$ and $\langle\Psi_T|\Phi\rangle$ have the same phase. Then, we need to prevent the Brownian motion from populating symmetrically the complex $\langle\Psi_T|\Phi\rangle$ plane to avoid the mean overlap merging with the origin. To this end, we apply the *phaseless* approximation [11], in which the constrained weight $\widetilde{\Pi}$ evolves according to the dephasing $\Delta\theta_\tau = \arg\langle\Psi_T|\Phi_{\tau+\Delta\tau}\rangle/\langle\Psi_T|\Phi_\tau\rangle$ of the overlap with the trial state:

$$\Delta\widetilde{\Pi}_\tau = \exp\Big(-\Delta\tau\mathrm{Re}\langle\hat{H}\rangle_{\Psi_T\Phi_\tau}\Big)\max\Big\{0\,;\cos(\Delta\theta_\tau)\Big\}. \tag{12}$$

Owing to the factor $\cos(\Delta\theta_\tau)$, the more the phase of the overlap $\langle\Psi_T|\Phi\rangle$ changes during $\Delta\tau$, the more the weight of the associated walker is reduced, and the walker is discarded when $|\Delta\theta_\tau|$ becomes larger than $\pi/2$. As a result, the centroid of the population persists into the half-plane $\mathrm{Re}\langle\Psi_T|\Phi\rangle > 0$ even though some realizations with gradually reduced weights can reach the other half-plane. In this respect, the constraint (12) provides a good compromise between the necessity in controlling the phase problem and the conservation of the initial form for the probability distribution that ensures an exact reproduction of the ground state. Finally, replacing Π by $\widetilde{\Pi}$ in Eq. (6) allows a phase problem-free sampling of an approximate ground state $|\widetilde{\Psi}_G\rangle$. Note that the approximation (12) also removes the risk of infinite variance on the error $|\Psi_G\rangle - |\Phi\rangle$, which would ensue from an accumulation of realizations around $\langle\Psi_T|\Phi\rangle = 0$.

2.3. Trial state: Yrast spectroscopy

The phaseless QMC method has been widely applied in quantum chemistry [11], with a simple mean-field wave function for $|\Psi_T\rangle$. Here, we propose for the shell model a more sophisticated approach allowing to achieve the ground state in each angular-momentum channel J, M, which is now also employed to improve the accuracy of QMC simulations for the ground state of the Hubbard model [13]. Starting from a Slater determinant $|\Phi_0\rangle$, a satisfying trial state $|\Psi_T^{JM}\rangle$ to initiate, guide, and constrain the dynamics may be defined by mixing all the possible values of the spin projection K in the intrinsic frame:

$$|\Psi_T^{JM}\rangle = \sum_{K=-J}^{J} C_K^J \hat{P}_{MK}^J|\Phi_0\rangle, \tag{13}$$

where $\hat{P}_{MK}^J$ are projection operators onto the desired spin [14]. Restricting $|\Phi_0\rangle$ to be factorized into a product of independent-neutron and -proton wave functions, the ansatz (13) also has a good isospin projection. Moreover, when valence spaces containing only one major shell are considered, the parity is a good quantum number and no further restoration is needed. Let us remark that, as a superposition of symmetry-related Slater determinants, $|\Psi_T^{JM}\rangle$ is no longer an independent-particle state so that it can absorb correlations beyond the mean-field level.

The energy E_G^J of the yrast state $|\Psi_G^{JM}\rangle$ can now be determined after a large enough imaginary time as

$$E_G^J \underset{\tau\to+\infty}{=} \frac{\langle\Psi_T^{JM}|\hat{H}\mathrm{e}^{-\tau\hat{H}}|\Psi_T^{JM}\rangle}{\langle\Psi_T^{JM}|\mathrm{e}^{-\tau\hat{H}}|\Psi_T^{JM}\rangle} \underset{\tau\to+\infty}{=} \frac{\langle\Psi_T^J|\hat{H}\mathrm{e}^{-\tau\hat{H}}|\Phi_0\rangle}{\langle\Psi_T^J|\mathrm{e}^{-\tau\hat{H}}|\Phi_0\rangle}, \tag{14}$$

under the proviso that $|\Psi_T^{JM}\rangle$ has a non-zero overlap with the exact state. In Eq. (14), $|\Psi_T^J\rangle = \sum_{K,K'} C_K^{J*} C_{K'}^J \hat{P}_{KK'}^J|\Phi_0\rangle$ denotes the state arising from rotational invariance and from the properties $(\hat{P}_{MK}^J)^\dagger = \hat{P}_{KM}^J$ and $\hat{P}_{MK}^J\hat{P}_{K'M'}^{J'} = \delta_{JJ'}\delta_{KK'}\hat{P}_{MM'}^J$. This wave function thus appears as a natural candidate to apply the QMC formalism (6, 8) with the biased weight (12) to constrain stochastic paths. Finally, an approximate yrast state $|\widetilde{\Psi}_{\rm G}^{JM}\rangle$ is obtained with energy

$$\widetilde{E}_G^J \underset{\tau\to+\infty}{=} \frac{\mathbb{E}\left[\widetilde{\Pi}_\tau \mathrm{Re}\langle\hat{H}\rangle_{\Psi_T^J\Phi_\tau}\right]}{\mathbb{E}\left[\widetilde{\Pi}_\tau\right]}, \tag{15}$$

deduced from Eqs. (6) and (14). Note that a similar expression also holds for any scalar observable that commutes with $\hat{H}$. In other cases, the well-known mixed and extrapolated estimators can be utilized to approximate ground-state expectation values of any operator [8].

The quality of the trial wave function obviously affects the efficiency of the phaseless QMC approach: The better the state $|\Psi_T^{JM}\rangle$, the more reduced the artificial bias due to the constraint. In addition, the imaginary time needed to converge depends on the quality of the state used to initiate the propagation. Both these arguments have naturally led us to adopt as trial wave function the symmetry-restored variational solution obtained by energy minimization in the subspace of Slater determinants after quantum number projection, i.e. by a variation-after-projection (VAP) approach [7]. For the shell model, such a method is similar to the so-called VAMPIR (Variation After Mean-field Projection In Realistic model spaces) approach [15].

2.4. Numerical issues

Let us now focus on the simulation of the probability distribution and the computation of observables, especially the energy. A direct evaluation of Eq. (15) turns out generally very ineffective owing to the exponential spread of the weights $\widetilde{\Pi}_\tau$ with τ. This problem may however be circumvented by sampling the realizations $\{|\Phi_\tau\rangle\}$ according to their weight *via* stochastic reconfiguration schemes developed in the framework of Green's-function Monte-Carlo methods. They are based on a birth/death mechanism, applied regularly during the Brownian motion every step $\Delta\tau_{\rm r}$ before the variance of the probability distribution becomes too large. The walkers being cloned or killed depending on their weight, those causing phase problems are automatically filtered ($\widetilde{\Pi} = 0$ in this case). Therefore, the number N_w of realizations is variable, which introduces a bias in the calculated averages. Nevertheless, a stochastic reconfiguration scheme that eliminates this bias by maintaining a fixed number of walkers has been proposed in Ref. [16], and we implement it in our QMC simulations. When the size N_w of the population is large enough, it does not affect the averages, but artificial statistical correlations between the walkers are induced by the birth/death process and have to be taken into account. Instead, we consider an alternative approach that consists in performing several independent simulations with N_w Slater determinants. In this way, the energy (resp. estimator of an observable) is deduced by averaging the energies (resp. estimators) reconstructed for each population, and the associated error bars are determined on the basis of the central limit theorem.

3. Application: proof-of-principle results

The implementation described above of the phaseless QMC scheme for the shell model thus contains three tunable parameters: The imaginary-time step $\Delta\tau$, the number N_w of walkers that compose the populations, and the period of stochastic reconfiguration $\Delta\tau_{\rm r}$. In order to assess the sensitivity of the results with respect to these variables, the QMC energies of the $J = 0$ ground state of ^{28}Mg as obtained with different sets of values $(\Delta\tau, \Delta\tau_{\rm r}, N_w)$ and the USD effective interaction [17] are reported in Fig. 1. They are compared to the exact energies from direct diagonalization carried out by the code ANTOINE [1]. First, we notice that the phase problem is well managed since no explosion of error bars is observed during the imaginary-time evolution. Furthermore, the upper panel shows that a systematic deviation emerges for a time step $\Delta\tau = 0.1$ MeV^{-1}. It is explained by the Trotter-Suzuki breakup [10] whose validity is limited to second order in $\Delta\tau$. In the central panel, the energies for $N_w = 100$ and $N_w = 200$ walkers are in

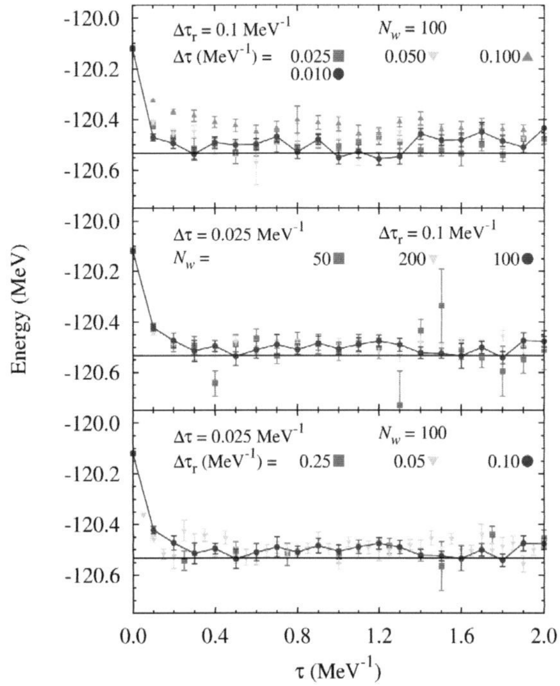

Figure 1. Influence of the simulation parameters: ground-state QMC energies of ^{28}Mg versus imaginary time for different steps $\Delta\tau$ (upper panel), sizes N_w of the populations (central panel), periods of reconfiguration $\Delta\tau_r$ (bottom panel), compared to the exact one (black line).

agreement within statistical errors, while significant fluctuations appear for $N_w = 50$. This ensues from the stochastic reconfiguration scheme that roughly samples the distribution of the weights and entails a bias when the number of realizations is too small. Finally, we see on the bottom panel that the time step of reconfiguration hardly affects the results, even though the error bars are larger for a longer period $\Delta\tau_r = 0.25$ Mev^{-1} as a consequence of a larger dispersion of the weights that exponentially spread with τ. In view of these observations, all the calculations reported here and in Ref. [7] have systematically been performed with the following set of parameters: $\Delta\tau = 0.01$ Mev^{-1}, $\Delta\tau_r = 0.1$ Mev^{-1}, and populations of $N_w = 100$ walkers.

Let us now report proof-of-principle results by addressing a system for which exact results are available. Figure 2 displays the VAP, QMC, and exact yrast spectra of ^{27}Na with the USD interaction. First, we observe that the VAP method offers a good approximation: Yrast energies differ from the exact values by less than about 1 MeV. Then, the phaseless energies agree remarkably well with the values from exact diagonalization: For any angular momentum, the relative errors do not exceed 0.3%, with statistical error bars of about 40-50 keV. The case of ^{27}Na provides a particularly good evidence of the efficiency of the method in controlling the phase problem. Indeed, it represents the most pathological systems for QMC simulations because it combines two sources of phase problems that are an odd number of particles, and a 'bad-sign' interaction [9]. Accordingly, this example clearly shows the ability of the method to provide nearly exact yrast energies, irrespective of the interaction or the kind of nuclei.

4. Conclusions and perspectives

In summary, we have presented a new QMC formalism for the shell model able to almost exactly reproduce the spectroscopy of any type of nucleus with realistic effective interactions. The formalism relies on a trial wave function whose roles are (i) to initiate and guide the underlying Brownian motion in order to improve the efficiency of the sampling (ii) to constrain the stochastic paths in order to control the phase problem thanks to the phaseless approximation.

Recent developments have aimed to extend the approach for computing excited states of a

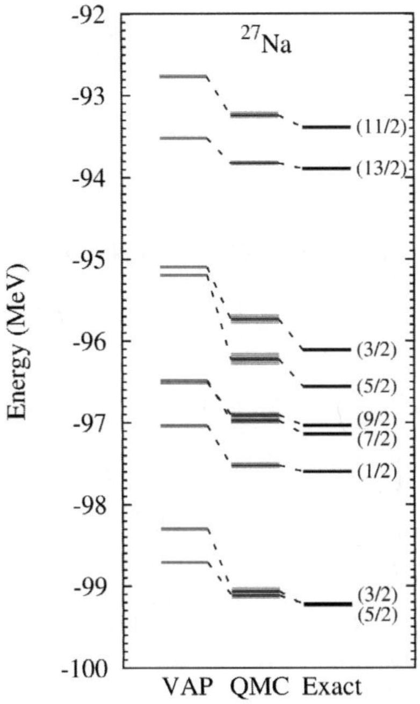

Figure 2. Yrast spectra of ^{27}Na as obtained with the VAP and phaseless QMC methods compared to the exact energy levels. The lighter areas indicate the QMC statistical errors. Preliminary results on non-yrast states are also shown (see section 4).

given angular momentum, achieving thus a complete spectrum of nuclei. To fulfil this objective, the main task is the determination of VAP wave functions orthogonal to the one previously computed for the ground state: For instance, the first excited state of spin J is obtained by minimizing the energy with the wave function $\hat{Q}|\Psi^{JM1}\rangle$, where $|\Psi^{JM1}\rangle$ is a projected Slater determinant of form (13) and $\hat{Q}$ the projector onto the subspace orthogonal to the VAP ground state. This ansatz can easily be generalized for sequentially searching the n-th excited state. In this way, by using such VAP solutions to initiate, guide, and constrain the Brownian motion, the phaseless QMC approach is no longer restricted to yrast spectroscopy. Preliminary results of this extended formalism are shown in figure 2, where the energies of the first excited $5/2^-$ and $3/2^-$ states are reproduced with the same accuracy than the yrast states.

5. References

[1] Caurier E, Martnez-Pinedo G, Nowacki F, Poves A and Zuker A P 2005 *Rev. Mod. Phys.* **77** 427
[2] Koonin S E, Dean D J and Langanke K 1997 *Phys. Rep.* **278** 1
[3] Stratonovich R D 1958 *Sov. Phys. Dokl.* **2** 416
[4] Alhassid Y, Dean D J, Koonin S E, Lang G and Ormand W E 1994 *Phys. Rev. Lett.* **72** 613
[5] Honma M, Mizusaki T and Otsuka T 1996 *Phys. Rev. Lett.* **77** 3315
[6] Otsuka T, Mizusaki T, Shimizu N and Utsuno Y 2001 *Prog. Part. Nucl. Phys.* **47** 319
[7] Bonnard J and Juillet O 2013 *Phys. Rev. Lett.* **111** 012502
[8] Pieper S C and Wiringa R B, 2001 *Ann. Rev. Nucl. Part. Sci.* **51** 53
[9] Lang G H, Johnson C W, Koonin S E and Ormand W E 1993 *Phys. Rev.* C **48** 1518
[10] Trotter H F 1959 *Proc. Am. Math. Soc.* **10** 545
[11] Zhang S and Krakauer H 2003 *Phys. Rev. Lett.* **90** 136401
[12] Gardiner C W 1983 *Handbook of Stochastic Methods* (Berlin: Springer-Verlag)
[13] Shi H, Jiménez-Hoyos C A, Rodríguez-Guzmán R, Scuseria G E and S. Zhang S 2014 *Phys. Rev.* B **89** 125129
[14] Villars F 1966 *Varenna Lectures* vol 36, ed C Bloch (New York: Academic Press) p 1
[15] Schmid K W 2013 *Prog. Part. Nucl. Phys.* **52** 565
[16] Buonaura M C and Sorella S 1998 *Phys. Rev.* B **57** 11446
[17] Wildenthal B H 1984 *Prog. Part. Nucl. Phys.* **11** 5

Recent developments in the eikonal description of the breakup of exotic nuclei

P. Capel[1], F. Colomer[1], H. Esbensen[2], T. Fukui[3], R. C. Johnson[4], F. M. Nunes[5], K. Ogata[3]

[1]Physique Nucléaire et Physique Quantique C.P. 229, Université Libre de Bruxelles (ULB), B-1050, Brussels, Belgium
[2]Physics Division, Argonne National Laboratory, Argonne, Illinois 60439, USA
[3]Research Center for Nuclear Physics, Osaka University, Ibaraki, Osaka 567-0047, Japan
[4]Department of Physics, University of Surrey, Guildford, England GU2 7XH, United Kingdom
[5]National Superconducting Cyclotron Laboratory and Department of Physics and Astronomy, Michigan State University, East Lansing, Michigan 48824, USA

E-mail: pierre.capel@ulb.ac.be

Abstract. The study of exotic nuclear structures, such as halo nuclei, is usually performed through nuclear reactions. An accurate reaction model coupled to a realistic description of the projectile is needed to correctly interpret experimental data. In this contribution, I briefly summarise the assumptions made within the modelling of reactions involving halo nuclei. I describe briefly the Continuum-Discretised Coupled Channel method (CDCC) and the Dynamical Eikonal Approximation (DEA) in particular and present a comparison between them for the breakup of ^{15}C on Pb at 68AMeV. I show the problem faced by the eikonal approximation at low energy and detail a correction that enables its extension down to lower beam energies. A new reaction observable is also presented. It consists of the ratio between angular distributions for two different processes, such as elastic scattering and breakup. This ratio is completely independent of the reaction mechanism and hence is more sensitive to the projectile structure than usual reaction observables, which makes it a very powerful tool to study exotic structures far from stability.

1. Introduction

The development of radioactive-ion beams in the mid 80s has enabled us to study the structure of nuclei away from stability. This has led to the discovery of exotic structures such as halo nuclei [1]. Halo nuclei exhibit an extended matter radius compared to stable nuclei. The unusual size of these neutron-rich nuclei is due to their small binding energy for one or two neutrons. Thanks to this loose binding, the valence neutron(s) tunnel far into the classically forbidden region and form a diffuse halo around a tight and well-bound core. The best known one-neutron halo nuclei are ^{11}Be and ^{15}C, which can be seen as a ^{10}Be or ^{14}C core, respectively, to which one neutron is loosely bound. Archetypical two-neutron halo nuclei are ^{6}He and ^{11}Li, with an α and ^{9}Li core, respectively. Albeit much less probable, proton haloes are also possible; ^{8}B is one of the nuclei that probably exhibit a one-proton halo.

Being located close to the drip lines, halo nuclei exhibit very short lifetime, which impedes the use of regular spectroscopic techniques in their study. To study these fascinating structures, we must then rely on indirect techniques like reaction such as elastic scattering or breakup. In the latter, the halo dissociates from the core during the collision with a target, hence revealing the internal cluster structure of the projectile. In order to infer correct nuclear-structure information from experimental data, an accurate theoretical description of the reaction, coupled to a realistic model of the nucleus under investigation is needed [2]. In this contribution, we review the theoretical framework of the usual reaction models and describe in particular the Continuum Discretise Coupled Channel model (CDCC) [3, 4] and the Dynamical Eikonal Approximation (DEA) (Sec. 2) [6, 7]. In Sec. 3, we compare these two models on the case of the breakup of ^{15}C on Pb at intermediate (68AMeV) and low (20AMeV) energies [8].

Albeit precise, these models rely on inputs, such as optical potentials, which can significantly affect the theoretical predictions. To avoid this dependence a new reaction observable has been recently proposed. It consists of the ratio of angular distributions for two different processes, such as elastic scattering and breakup. Taking this ratio significantly reduces the sensitivity to the reaction process and hence to the model input. Accordingly this ratio is more sensitive to the projectile structure than usual reaction observables and can provide very accurate information about exotic nuclear structures. The ratio method is presented in Sec. 4 before a brief summary in Sec. 5.

2. Reaction modelling

2.1. Theoretical framework

The models of reactions involving one-nucleon halo nuclei usually rely on a three-body framework: a two-body projectile and we a target. The internal structure of the projectile (P) is described by the core (c)-fragment (f) Hamiltonian

$$H_0 = T_r + V_{cf}(\boldsymbol{r}), \tag{1}$$

where $\boldsymbol{r}$ is the c-f relative coordinate, T_r is the kinetic energy term and V_{cf} is a phenomenological potential that simulates the c-f interaction, whose parameters are adjusted to reproduce the energy and quantum numbers of the low-energy levels of the projectile. The eigenstates of H_0 describe the c-f relative motion

$$H_0\ \phi_{lm}(E, \boldsymbol{r}) = E\ \phi_{lm}(E, \boldsymbol{r}), \tag{2}$$

where l and m are the c-f relative orbital angular momentum and its projection, respectively. E is the c-f relative energy. The negative-energy states ($E < 0$) are discrete; they correspond to the bound states of the projectiles. The positive energy state correspond to the continuum of the c-f system, i.e. the states in which the halo has broken up from the core.

The target (T) is usually described as a structureless particle which interacts with the projectile constituents though optical potentials V_{cT} and V_{fT}. Within that framework, studying the reaction reduces to solving the following three-body Schrödinger equation

$$[T_R + H_0 + V_{cT} + V_{fT}]\, \Psi(\boldsymbol{r}, \boldsymbol{R}) = E_T\ \Psi(\boldsymbol{r}, \boldsymbol{R}) \tag{3}$$

under the condition that the projectile is initially in its ground state $\phi_{l_0 m_0}(E_0)$

$$\Psi(\boldsymbol{r}, \boldsymbol{R}) \underset{Z\to-\infty}{\longrightarrow} e^{iKZ+\cdots}\phi_{l_0,m_0}(E_0, \boldsymbol{r}), \tag{4}$$

where the Z axis has been chosen along the incoming beam and the P-T momentum is related to the total energy E_T of the system: $\hbar^2K^2/2\mu_{PT} + E_0 = E_T$, with μ_{PT} the P-T reduced mass.

Various models have been developed to solve this equation numerically. In the following sections, we will describe two of them: the Continuum-Discretised Coupled Channel method (CDCC) and the Dynamical Eikonal Approximation (DEA).

2.2. CDCC
In CDCC, the three-body wave function Ψ is expanded upon the complete set of the eigenstates of H_0:

$$\Psi(\boldsymbol{r},\boldsymbol{R}) = \sum_{lm}\sum_{E_i<0} \chi_{lm}(E_i,\boldsymbol{R})\phi_{lm}(E_i,\boldsymbol{r}) + \int_0^\infty \chi_{lm}(E,\boldsymbol{R})\phi_{lm}(E,\boldsymbol{r})\, dE. \tag{5}$$

To be tractable numerically, the integral over the positive c-f energy must be discretised. This is usually performed dividing the continuum into small bins and defining for each bin i a normalised wave function

$$\tilde{\phi}_i(\boldsymbol{r}) = \frac{1}{N}\int_{E_i-\Delta E_i/2}^{E_i+\Delta E_i/2} w(E)\phi_{lm}(E,\boldsymbol{r})\, dE, \tag{6}$$

where the index i defines not only the energy, but also the partial wave lm of the bin state, w is a weight function and N is a normalisation coefficient.

With this discretisation, introducing the expansion (5) into Eq. (3) becomes tractable and leads to a set of coupled equations in the unknown coefficients χ_i of the expansion (5)

$$[T_R + E_i + V_{ii}]\,\chi_i + \sum_{j\neq i} V_{ij}\chi_j = E_T\chi_i, \tag{7}$$

where the coupling are due to the tidal force resulting from the difference between the c-T and the f-T interactions

$$V_{ij} = \langle\tilde{\phi}_i|V_{cT} + V_{fT}|\tilde{\phi}_j\rangle. \tag{8}$$

2.3. Dynamical eikonal approximation
The DEA is based on the eikonal approximation [5], which states that at sufficiently high energy, the three-body wave function Ψ does not deviate much from the incoming plane wave of Eq. (4). Accordingly, this suggests to perform the following factorisation

$$\Psi(\boldsymbol{r},\boldsymbol{R}) = e^{iKZ}\,\widehat{\Psi}(\boldsymbol{r},\boldsymbol{R}), \tag{9}$$

where $\widehat{\Psi}$ varies smoothly with $\boldsymbol{R}$. Introducing the factorisation (9) into Eq. (3), and neglecting the second-order derivative of $\widehat{\Psi}$ in front of its first-order derivative leads to the DEA equation [6, 7]

$$i\hbar v\frac{\partial}{\partial Z}\widehat{\Psi}(\boldsymbol{r},\boldsymbol{b},Z) = [H_0 - E_0 + V_{cT} + V_{fT}]\widehat{\Psi}(\boldsymbol{r},\boldsymbol{b},Z), \tag{10}$$

where $v = \hbar K/\mu_{PT}$ is the initial P-T relative velocity.

Eq. (10) must be solved for each value of $\boldsymbol{b}$, the transverse component of the P-T relative coordinate $\boldsymbol{R}$, with the condition $\widehat{\Psi} \underset{Z\to-\infty}{\longrightarrow} \phi_{l_0,m_0}(E_0)$. Being simpler to solve than the set of coupled equations (7), it leads to a much faster numerical scheme than CDCC [8]. However, it is valid only at sufficiently high energy, whereas CDCC is not restricted in energy.

3. Coulomb breakup of ^{15}C

3.1. ^{15}C+Pb at 68AMeV
To compare the two aforementioned models, we have performed numerical calculations for the breakup of ^{15}C on Pb at 68AMeV [8], which correspond to the experimental conditions of

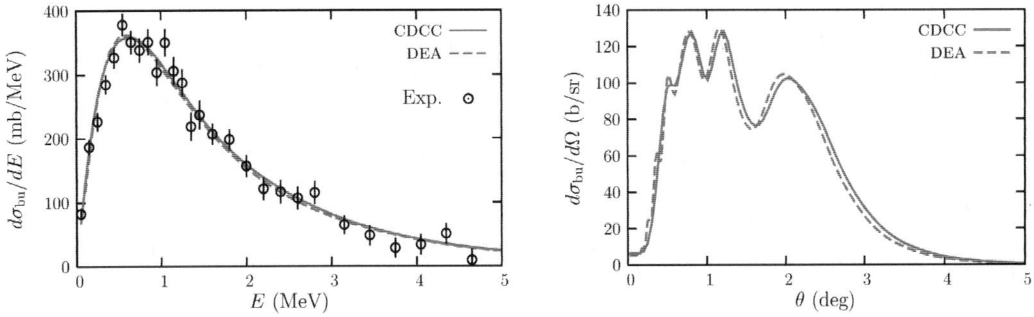

Figure 1. Comparison of the CDCC (solid line) and DEA (dashed line) reaction models on the breakup of ^{15}C on Pb at 68AMeV: energy distribution (left) and angular distribution (right). Calculations are from Ref. [8], data are from Ref. [9].

Ref. [9]. As can be seen in Fig. 1, both models provide nearly identical breakup cross sections. They lead to the same energy distribution (Fig. 1, left) and a mere tiny shift is observed in the angular distribution (Fig. 1, right). This shows that although based on different assumptions, both models provide the same description of the reaction process, and hence lead to the same conclusions about the projectile structure. In particular, the excellent agreement between the calculations and the experimental data of Ref. [9] confirms the halo structure of ^{15}C.

3.2. ^{15}*C+Pb at 20AMeV*

As mentioned above, the eikonal approximation is only valid at high energy [5]. To see what happens when the beam energy is lowered, we have repeated the calculations at 20AMeV [8]. As illustrated in Fig. 2, the DEA predicts a breakup cross section too large with an angular distribution focused too much forward compared to CDCC, i.e. the slight shift observed in Fig. 1 (right) increases at low energy. As analysed in Ref. [8], this is due to the lack of

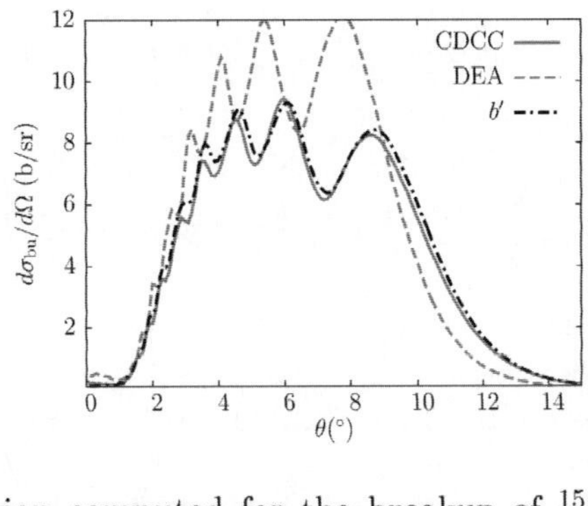

Figure 2. Angular distribution computed for the breakup of ^{15}C on Pb at 20AMeV using CDCC (solid line) and DEA (dashed line). A semiclassical correction (dash-dotted line) corrects perfectly the DEA [10].

Coulomb deflection of the eikonal approximation. Indeed, the factorisation (9) somehow forces the projectile to move forward along a straight line, leading it into the high-field zone of the target. The predicted breakup cross section is hence too large and focused to too forward an angle than if the Coulomb deflection of the projectile off the target were properly taken into account. This does not happen within CDCC, which naturally includes this deflection.

Analysing this effect in more detail in Ref. [10], we have observed that a simple semi-classical approximation [11, 12] corrects efficiently the eikonal framework for its lack of Coulomb deflection. In that correction, the norm of the transverse component $\boldsymbol{b}$ of $\boldsymbol{R}$ is replaced by the distance of closest approached b' between the projectile and the target on a classical Coulomb trajectory. The resulting DEA calculation is now in perfect agreement with the CDCC calculation (see dash-dotted line in Fig. 2). This shows that the range of validity of the eikonal approximation can be reliably extended down to low energy

4. Ratio method

Albeit accurate, the aforementioned models rely on several inputs, including the optical potentials used to simulate the interaction between the projectile components and the target. As shown in Ref. [13], the uncertainty related to this interaction can climb up to a factor 2 in the breakup cross section. This is especially true for V_{cT}, which is usually poorly known since the core of the nucleus is itself radioactive.

To circumvent this problem, a new reaction observable has been suggested [14]. It consists of the ratio of angular distributions for different processes, e.g. elastic scattering and breakup. Within the Recoil Excitation and Breakup model (REB), these distributions elegantly factorise into an elastic scattering cross section for a pointlike projectile and a form factor that accounts for the extension of the halo [15]:

$$\frac{d\sigma_{\mathrm{el}}}{d\Omega} = |F_{00}|^2 \left(\frac{d\sigma}{d\Omega}\right)_{\mathrm{pt}} \tag{11}$$

for the elastic scattering cross section, with the formfactor

$$|F_{00}|^2 = \frac{1}{2l_0+1} \sum_{m_0} \left| \int |\phi_{l_0 m_0}(E_0, \boldsymbol{r})|^2 e^{i\boldsymbol{Q}\cdot\boldsymbol{r}} d\boldsymbol{r} \right|^2, \tag{12}$$

where $\boldsymbol{Q} \propto (\boldsymbol{K} - \boldsymbol{K'})$ is proportional to the exchanged momentum. For breakup at energy E in the c-f continuum, one obtains [16]

$$\frac{d\sigma_{\mathrm{bu}}}{dEd\Omega} = |F_{E0}|^2 \left(\frac{d\sigma}{d\Omega}\right)_{\mathrm{pt}}, \tag{13}$$

with the form factor

$$|F_{E0}|^2 = \frac{1}{2l_0+1} \sum_{m_0} \sum_{lm} \left| \int \phi_{lm}(E, \boldsymbol{r})\, \phi_{l_0 m_0}(E_0, \boldsymbol{r})\, e^{i\boldsymbol{Q}\cdot\boldsymbol{r}} d\boldsymbol{r} \right|^2. \tag{14}$$

Since the pointlike cross section is the same in both factorisations (11) and (13), taking the ratio of these angular distributions should remove most of the angular dependence and lead to an observable that depends only on the projectile wave functions. However, the REB is based on two simplifying assumptions: it assumes the adiabatic approximation and it neglects V_{fT}. To check these predictions we perform DEA calculations, which accounts for the dynamics of the projectile, includes the f-T interaction, and is in excellent agreement with experiment [7].

Various tests have shown that it was best to consider the following ratio [16]

$$d\sigma_{\mathrm{bu}}/d\sigma_{\mathrm{sum}} \overset{(REB)}{=} |F_{E0}|^2 \tag{15}$$

with the summed cross section corresponding to all elastic, inelastic and breakup processes

$$\frac{d\sigma_{\mathrm{sum}}}{d\Omega} = \frac{d\sigma_{\mathrm{el}}}{d\Omega} + \frac{d\sigma_{\mathrm{inel}}}{d\Omega} + \int \frac{d\sigma_{\mathrm{bu}}}{dEd\Omega} dE. \tag{16}$$

The results of these tests are displayed in Fig. 3 for ^{11}Be impinging on Pb at 69AMeV (left) and on C at 67AMeV (right) [14, 16]. This figure shows the breakup cross section (13) (expressed in b/MeV sr, dashed lines), the summed cross section (16) (as a ratio to Rutherford, dotted lines), and their ratio (15) (solid lines). We observe that although both collisions are dominated by very different processes and that, accordingly, their cross sections are very different, the ratio of the DEA cross sections is very smooth, nearly free from any oscillatory pattern, confirming that the ratio removes most of the dependence on the reaction process. Moreover, in both cases it

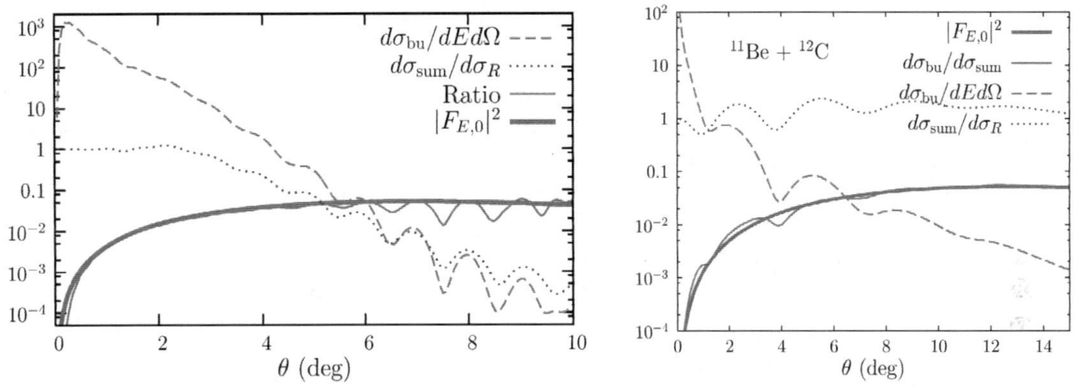

Figure 3. Ratio method illustrated for the collision of ^{11}Be on Pb at 69AMeV (left) and on C at 67AMeV (right). Although the reaction process is very different in each reaction, the ratio (15) are nearly identical and in excellent agreement with the REB prediction.

is in excellent agreement with the REB form factor (14) (thick grey line), showing that the ratio provides information about the projectile structure free from any reaction artefact, contrarily to usual reaction observables. The ratio is especially sensitive to the c-f binding energy and the partial wave of the projectile ground state [14, 16].

More recently, we have tested the validity of the ratio down to low energy. Very surprisingly our calculations have shown that the method remains valid at low energy, especially on light targets [17]. This analysis is illustrated in Fig. 4 for ^{11}Be impinging on C at 20AMeV. Here again

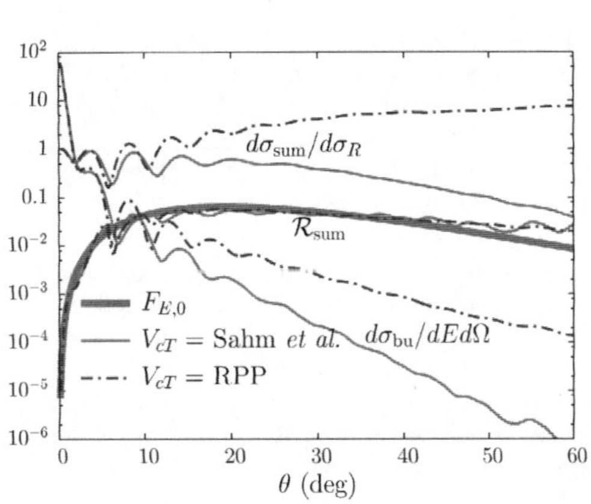

Figure 4. Extension of the ratio method down to low energy (20AMeV) for ^{11}Be on C. Results obtained with different potentials (Sahm *et al* [solid lines] and RPP [dash-dotted lines]) lead to nearly identical ratios.

considering the ratio (15) removes most of the angular dependence leading to an observable in excellent agreement with the REB prediction, even though this beam energy is outside the range of validity of that model. The figure shows also results obtained with different V_{cT}. As observed in Ref. [13], they produce significantly different cross sections. However, the corresponding ratios are nearly identical. These preliminary results confirm the power of the ratio method in the analysis of the nuclear structure away from stability.

5. Summary

Nuclear reactions like elastic scattering and breakup are useful tools to study exotic nuclear structure, such as halo nuclei [1]. Most current models are based on a three-body framework: a two-body projectile impinging on a structureless target. Various techniques have been developed to solve the corresponding three-body Schrödinger equation. In this contribution, we have described two of them: CDCC [3] and DEA [6]. Their comparison shows that at intermediate energy, they provide similar results [8]. At lower energy, the eikonal approximation is no longer valid. However a simple semiclassical correction enables us to extend its domain of validity down to low energies [10].

To avoid the dependence of the reaction models to uncertain inputs such as optical potentials, a new reaction observable is suggested. It consists of the ratio of angular distributions for two different processes such as breakup and elastic scattering, which is predicted by the REB model [15] to depend only on the projectile wave functions. Precise reaction calculations confirm this hypothesis and show the high sensitivity of the ratio to the projectile structure compared to regular reaction observables [16]. This suggests the ratio method to be a very powerful tool to study the structure of nuclei far from stability. Hopefully, this theoretical prediction will be confirmed by actual experimental data.

Acknowledgement

This research was supported in part by the Research Credit No. 19526092 of the Belgian Funds for Scientific Research F.R.S.-FNRS. This paper presents research results of the Belgian Research Initiative on eXotic nuclei (BRIX), program P7/12, on interuniversity attraction poles of the Belgian Federal Science Policy Office.

References

[1] I. Tanihata, J. Phys. G **22**, 157 (1996).
[2] D. Baye and P. Capel *Breakup Reaction Models for Two- and Three-Cluster Projectiles* in *Cluster in Nuclei, Vol. 2*, Lecture Notes in Physics **848**, edited by C. Beck (Springer, Berlin, 2012), p. 121.
[3] M. Kamimura, M. Yahiro, Y. Iseri, Y. Sakuragi, H. Kameyama, and M. Kawai, Prog. Theor. Phys. Suppl. No. 89, 1 (1986).
[4] M. Yahiro, K. Ogata, T. Matsumoto, and K. Minomo, Prog. Theor. Exp. Phys. 2012, 01A206 (2012),
[5] R. J. Glauber, in *High Energy Collision Theory*, Lectures in Theoretical Physics, Vol. 1, ed. W. E. Brittin and L. G. Dunham (Interscience, New York, 1959), p. 315.
[6] D. Baye, P. Capel, and G. Goldstein, Phys. Rev. Lett. **95**, 082502 (2005).
[7] G. Goldstein, D. Baye, and P. Capel, Phys. Rev. C **73**, 024602 (2006).
[8] P. Capel, H. Esbensen, and F. M. Nunes, Phys. Rev. C **85**, 044604 (2012).
[9] T. Nakamura, *et al.*, Phys. Rev. C **79**, 035805 (2009).
[10] T. Fukui, K. Ogata, and P. Capel, Phys. Rev. C **90**, 034617 (2014).
[11] R. A. Broglia and A. Winther, *Heavy Ion Reactions, Lectures Notes, Vol. 1: Elastic and Inelastic Reactions* (Benjamin/Cummings, Reading, England, 1981).
[12] C. A. Bertulani and P. Danielewicz, *Introduction to Nuclear Reactions* (Institute of Physics Publishing, Bristol, 2004).
[13] P. Capel, G. Goldstein, D. Baye, Phys. Rev. C **70**, 064605 (2004).
[14] P. Capel, R. Johnson, and F. Nunes, Phys. Lett. **B705**, 112 (2011).
[15] R. C. Johnson, J. S. Al-Khalili, and J. A. Tostevin, Phys. Rev. Lett. **79**, 2771 (1997).
[16] P. Capel, R. Johnson, and F. Nunes, Phys. Rev. C **88**, 044602 (2013).
[17] F. Colomer, P. Capel, and F. Nunes, *in preparation* (2015).

Neutron decay of the Giant Pairing Vibration in ^{15}C

M. Cavallaro[1a], C. Agodi[a], M. Assié[b], F. Azaiez[b], F. Cappuzzello[a,c], D. Carbone[a], N. de Séréville[b], A. Foti[c,d], L. Pandola[a], J.A. Scarpaci[e], O. Sgouros[f], V. Soukeras[f]
[a] Istituto Nazionale di Fisica Nucleare, Laboratori Nazionali del Sud, I-95125 Catania, Italy.
[b] Institut de Physique Nucléaire, Université Paris-Sud-11-CNRS/IN2P3, 91406 Orsay, France
[c] Dipartimento di Fisica e Astronomia, Università di Catania, I-95125 Catania, Italy.
[d] Istituto Nazionale di Fisica Nucleare, Sezione di Catania, I-95125 Catania, Italy.
[e] Centre de Sciences Nucléaires et de Sciences de la Matières - CSNSM, Université Paris-Sud-11-CNRS/IN2P3, 91405 Orsay, France
[f] Department of Physics and HINP, The University of Ioannina, 45110 Ioannina, Greece

E-mail: manuela.cavallaro@lns.infn.it

Abstract. The neutron decay of the resonant states of light neutron-rich nuclei is an important and poorly explored property, useful to extract valuable nuclear structure information. The neutron decay of the ^{15}C resonances populated via the two-neutron transfer reaction $^{13}C(^{18}O,^{16}O\ n)$ at 84 MeV incident energy is studied using an innovative technique which couples the MAGNEX magnetic spectrometer and the EDEN neutron detector array. The data show that the recently observed ^{15}C Giant Pairing Vibration at 13.7 MeV mainly decays via two-neutron emission.

1. Introduction

The study of the decay modes of nuclear states populated in direct reactions is a powerful tool for understanding their microscopic structure. When the reaction populates unbound or weakly bound neutron-rich nuclei, the decay via neutron emission can be the dominant decay mode of the ground and excited states. In these cases the detection in coincidence of neutrons and charged ejectiles emitted in the reaction and the high-resolution measurement of the neutron energy are crucial tasks for spectroscopic investigations of the residual nuclei populated in the reaction.

The advent of large acceptance magnetic spectrometers has allowed a deeper exploration of stable and unstable nuclei up to high excitation energy with considerable resolution and good statistical significance, leading to the observation of new resonances in the unknown continuum. Examples are the Giant Pairing Vibrations (GPV), which are collective motion in the particle-particle space excited via two-neutron transfer reactions [1]. Signatures of such resonances in light nuclei have been recently shown and discussed in Refs. [2], [3], [4].

The neutron decay of the ^{15}C resonances up to 16 MeV excitation energy, populated via the $^{13}C(^{18}O,^{16}O\ n)$ reaction at 84 MeV incident energy has been studied in ref. [5]. In that paper, a new method to

[1] To whom any correspondence should be addressed.

determine the neutron kinetic energy by time-of-flight (TOF) in exclusive experiments was presented. It involves the use of the MAGNEX magnetic spectrometer and the EDEN neutron detector array. A schematic view of the used facility is shown in Figure 1.

2. The experimental apparatus and the data reduction

The MAGNEX spectrometer consists of two magnetic elements: a large aperture quadrupole and a large bending magnet. It is characterized by large acceptance both in angle (solid angle $\Delta\Omega$ = 50 msr) and momentum (relative momentum with respect to the central trajectory Δp = -14%, +10%). The features and performances of MAGNEX are described in Refs. [6], [7], [9], [10], [11]. The high-order aberrations originated by the large acceptance are calculated and corrected by means of a software ray-reconstruction described in Ref. [12]. The reaction ejectiles, momentum-analyzed by MAGNEX, are detected by the Focal Plane Detector (FPD). A detailed description of the MAGNEX FPD is reported in Ref. [13].

EDEN is an array of 36 cylindrical organic scintillators (NE213) by IPN-Orsay located around the MAGNEX scattering chamber. In the experiment, they were positioned at a distance ranging from 1.8 to 2.4 m from the target, covering a total solid angle of 270 msr. A detailed description of the detectors is given in Ref. [14]. The neutron-gamma discrimination is provided by pulse-shape analysis of the fast and slow components of the scintillation signal [15]. A timing signal is also an output of the module for each EDEN channel.

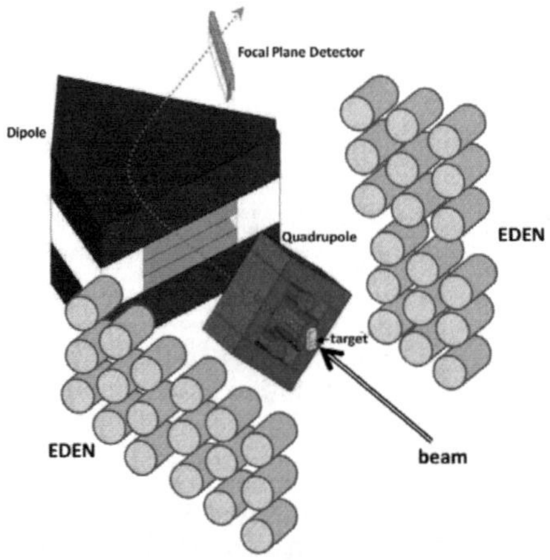

Figure 1. Schematic layout of the MAGNEX-EDEN coupling.

The experiment was performed at INFN-Laboratori Nazionali del Sud using a $^{18}O^{6+}$ beam delivered by the Tandem Van der Graff accelerator at 84 MeV incident energy impinging on a 50 μg/cm2 99% enriched ^{13}C target. The ^{16}O reaction ejectiles were momentum analyzed by MAGNEX. The angular range explored by MAGNEX was $3° < \theta_{lab} < 14°$ in the laboratory. The ejectiles were identified at the FPD as described in Ref. [16] and their momentum reconstructed by the ray-reconstruction technique [12]. The excitation energy spectra of the reaction products were obtained by the missing mass method $E_x = Q_0 - Q$ (where Q_0 is the ground to ground state reaction Q-value).

For each ion reaching the FPD, also the path length along the spectrometer was reconstructed event by event and consequently its time of flight TOF_{ion} extracted. In the TOF technique used, the start signal is given by the detection of the ejectiles at the focal plane of MAGNEX. The stop signal was provided by the EDEN time signal. In more details, the EDEN timing signal is sent to a high-stability delay line which introduces a delay of ΔT_{delay} = 400 ns and then to the stop input of a Time to Digital Converter (TDC). The common start to the TDC is given by the logic OR of the timing signals of the MAGNEX silicon detectors. Thus, once the time-of-flight of the charged ejectiles along the spectrometer is known

(TOF_{ion}) by ray-reconstruction, the time-of-flight of the neutrons from the target point to the EDEN detector (TOF_{EDEN}) can be deduced by the relation:

$$TOF_{EDEN} = TOF_{ion} + T_{TDC} - \Delta T_{delay}$$

Then, the knowledge of the distances and angles of the EDEN array with respect to the target allows to determine the kinetic energy of the neutrons by relativistic relations and energy E_n in the reference frame of the decaying nucleus.

3. Study of the ^{15}C resonances

An example of the excitation energy spectrum (E_x) of ^{15}C measured by MAGNEX is shown in Fig. 2. Several narrow peaks corresponding to known low-lying bound and resonant states of ^{15}C are observed. Among them, the resonance at E_x = 13.7 MeV was associated to the Giant Pairing Vibration in Ref. [15].
Above the one-neutron separation energy of ^{15}C, S_n = 1.218 MeV, the neutron decay of the observed resonances was studied gating on the different peaks of the ^{15}C excitation energy spectrum (E_x) and plotting the correspondent neutron energy spectra (E_n). The background neutron spectra are subtracted as described in Ref. [5].

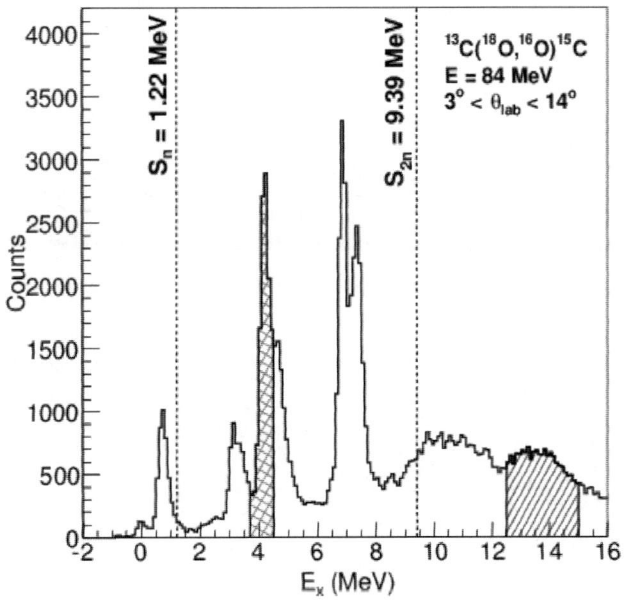

Fig. 2 ^{15}C Excitation energy spectrum for the ^{13}C(^{18}O,^{16}O) reaction at 84 MeV incident energy and 3° < θ_{lab} < 14°. The color filled areas are the two regions selected for the study of the neutron decay spectra in this paper.

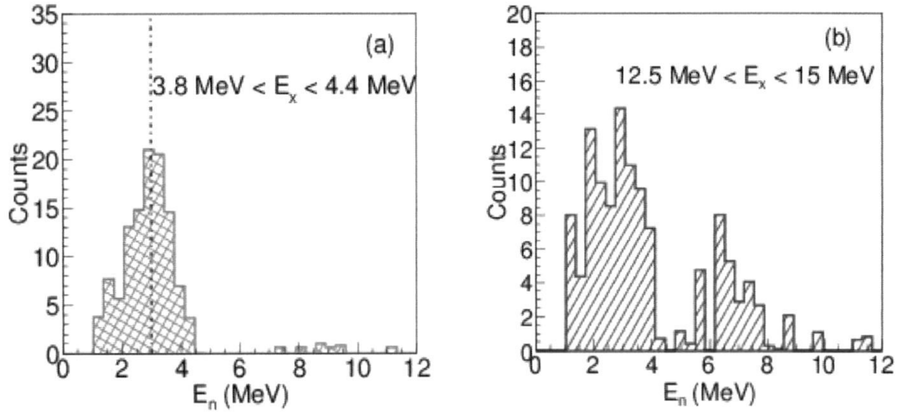

Fig. 3 Neutron energy spectrum gated on the ^{15}C excitation energy regions (a) 3.8 MeV < E_x < 4.4 MeV and (b) 12.5 MeV < E_x < 15 MeV.

Let us consider for example the ^{15}C excited state at E_x = 4.220 MeV. The neutron energies obtained by gating the excitation energy spectrum in the region 3.8 MeV < E_x < 4.4 MeV is shown in Fig. 3(a). A structure in the neutron spectrum at around 3 MeV is well visible. Such an energy corresponds, within the uncertainty, to the energy of the neutrons decaying from the ^{15}C excited state at E_x = 4.220 MeV to the ^{14}C ground state, which is $E_n = E_x - S_n$ = 3 MeV. Similar spectra are obtained gating on the other states below S_n [5]. These results demonstrate that the investigated states in the continuum of ^{15}C mainly decay to the ^{14}C ground state.

The resonance in the ^{15}C spectrum at E_x = 13.7 MeV was associated to the Giant Pairing Vibration in Ref. [15]. Its decay to the ^{14}C ground state and the consequent emission of neutrons of energy distributed around $E_n = E_x - S_n$ = 13.7 - 1.22 = 12.5 MeV is ruled out by the measured neutron spectrum (see figure 3(b)). The most intense neutron distribution in coincidence with the GPV peak of ^{15}C can be explained by the decay to the ^{13}C ground state via a two-neutron emission. In this case, neutron energies ranging from zero to $E_n = E_x - S_{2n}$ = 4.3 MeV are expected and, in fact, observed. The simultaneous measurement of two-neutron coincidences is prevented by the low yields in the present experiment. However a dedicated experiment is foreseen as the next step of our research program.

Acknowledgments

O. Sgouros and V. Soukeras warmly acknowledge financial support from the LLP/ERASMUS Programme - UNIVERSITY OF IOANNINA, year 2012-2013.

References

[1] M. Cavallaro, et al., Phys. Rev. C 88, 054601 (2013).
[2] F. Cappuzzello, et al., Nat. Commun. 6:6743 doi: 10.1038/ncomms7743 (2015).
[3] J. Piekarewicz, NATURE PHYSICS 11, 303 (2015).
[4] D. Carbone, Eur. Phys. J. Plus 130:143 (2015).
[5] M. Cavallaro, et al., Phys. Rev. C (2015) submitted.
[6] A. Cunsolo et al., Nucl. Instrum. Methods A 484, 56 (2002).
[7] A. Cunsolo et al., Nucl. Instrum. Methods A 481, 48 (2002).
[8] A. Cunsolo, et al., Eur. Phys. Journ. S.T. 150, 343 (2007).
[9] M. Cavallaro, et al., Nucl. Inst. and Meth. A 648, 46 (2011).
[10] M. Cavallaro, et al., Nucl. Instr. and Meth. A 637, 77 (2011).

[11] F. Cappuzzello, et al., Nucl. Instr. and Meth. A 638, 74 (2011).
[12] M. Cavallaro, et al., Eur. Phys. J. A, 48: 59 (2012).
[13] H. Laurent, et al., Nucl. Instr. and Meth. A 326, 517 (1993).
[14] M. Cavallaro, et al., Nucl. Instr. and Meth. A 700, 65 (2013).
[15] F. Cappuzzello, et al., Nucl. Instr. and Meth. A 621, 419 (2010).

Study of hydrogen and helium isotopes with A=5,6,7

B.A. Chernyshev[1], Yu.B. Gurov[1], V.S. Karpukhin, L.Yu. Korotkova[1], S.V. Lapushkin[1], M.V. Tel'kushev[1], V.G. Sandukovsky[2]

chernyshev@mephi.ru

[1]*National Research Nuclear University MEPhI, 115409, Moscow, Russia*
[2]*Joint Institute for Nuclear Research, 141980, Dubna, Russia*

The level structure of superheavy hydrogen and heavy helium isotopes with A = 5, 6, 7 was studied in stopped pion absorption by light nuclei. The measurements were performed at low energy pion channel of LANL with two-arm multilayer semiconductor spectrometer. Excited states of the $^{5-7}$H and $^{5-7}$He were observed in several reaction channels on the ^{9}Be and ^{10,11}B nuclei. Several excited levels were found only in our measurements. Candidate on isobar-analog state of the ^{6}H ground states with E_r ~ 5.5 MeV was observed in the reaction ^{10}B(π^-,pt)X.

1. Introduction

Experimental investigation of neutron-rich nuclei near the drip line is mainstream in the development of nuclear physics [1]. Superheavy hydrogen isotopes ^{n}H (n ≥ 4) and heavy helium isotopes ^{n}He (n ≥ 5) are the most interesting ones due to extremely large neutron–to–proton ratio. A relatively small number of nucleons makes possible to find a correct microscopic description of their properties and, therefore, to verify existing nuclear models as well as nucleon–nucleon potentials.

The search for these exotic nuclei and investigation of their level structure have been continued for a long time, but there are still many open questions, especially for superheavy hydrogen isotopes [1-4]. Most of the results have been obtained in the measurements with ion beams, including radioactive beams [1-4].

Another way of the studying is associated with particle (γ, e$^-$, π) beams. Stopped pion absorption is most suitable to the search for light neutron-rich nuclei. Among other things the first observations of the superheavy hydrogen isotope ^{5}H [5] and ground state of lithium isotope ^{10}Li [6] were obtained in this type of reactions.

An opportunity to effectively exploit stopped π^--meson absorption by nuclei in order to form neutron-rich nuclear states is based on the peculiarities of this reaction: a charge reduction of the produced nuclear system and a large energy release (~$m_\pi c^2$), which leads to the formation of fast nuclear particles (n, p, d, t, ^{3,4}He, ...). For channels in which one or two charged particles are detected, a residual nucleus will be a neutron-rich one. Production of studied nuclei in two- and three-body channels appears as a peak in the missing mass spectra for one or two registered particles, respectively. The important advantage of using the stopped pion absorption is the possibility to obtain the information on the wide range of exotic nuclei in one experimental run. Data on the known nuclear states permit to resolve such problems as the energy calibration, the determination of the energy resolution, and the stability of the parameters of the experimental setup during data collecting [7].

In this work experimental results on the level structure of helium isotope with A = 5, 6, 7 obtained in the stopped pion absorption are analyzed with the aim of funding isobar-analog states (IAS) of superheavy hydrogen isotopes $^{5-7}$H. Search for the helium states was performed in the following channels of the stopped pion absorption ^{10}B(π^-,dt)X and ^{11}B(π^-,tt)X – for ^{5}He, ^{11}B(π^-,dt)X, ^{10}B(π^-,pt)X, and ^{10}B(π^-,dd)X – for ^{6}He, ^{10}B(π^-,pd)X, ^{11}B(π^-,pt)X and ^{11}B(π^-,dd)X – for ^{7}He.

2. Experiment details

The experiment was performed on the meson factory LAMPF in Los Alamos, USA using the two arm semiconductor spectrometer [6]. Layout of the setup is presented in Fig. 1. The beam of negatively charged pions with energy of 30 MeV passed through the beryllium moderator and stopped in a thin target (≈ 24 mg·cm^{-2}). The measurements were carried out on the targets ^{9}Be, ^{10,11}B and ^{12,14}C.The pion stopping rate in the target was ~ 6 × 10^4 s^{-1}. The collimators C_1 and C_2 were used to suppress the background from secondary particles emitted by the moderator and chamber walls.

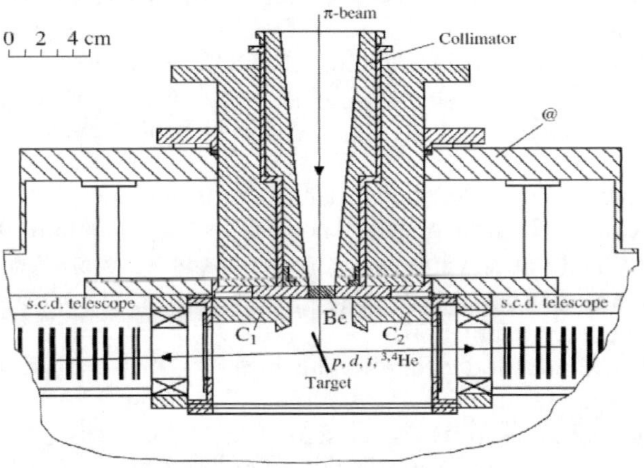

Fig. 1. Layout of two-arm spectrometer.

Charged particles emitted after pion absorption were detected by two semiconductor telescopes located at the angle 180° with respect to each other. Each telescope included two Si(Au) semiconductor detectors (s.c.d.) with thicknesses of 100 and 450 μm and 14 Si(Li) s.c.d. with thickness of ~ 3mm. The diameter of the sensitive area of all detectors was 32 mm (the working area – 8 cm^2). The total sensitive thickness of each telescope was 43 mm. This thickness permitted to measure the total absorption of charged particles up to kinematical limits of the reaction. As a result, a high energy resolution was achieved throughout a whole range of measured energies of detected particles. The energy resolution (FWHM) for single-charged particles (p, d, t) was better than 0.5 MeV [7]. The error of absolute energy calibration did not exceed 100 keV [7].

A search for the neutron-rich states was carried out in the peaks of the missing mass spectrum (*MM*). In the measurements of any pairs of singly charged particles the *MM* resolution was 1 MeV [7]. The error of the *MM* absolute calibration (δ*MM*) did not exceed 100 keV [7].

The spectrometer and experimental technique are described in more detail in [7, 8].

3. Results and discussion

Search for ^{5}He states was carried out in the reactions: ^{10}B(π^-,dt)X and ^{11}B(π^-,tt)X. The *MM* spectra measured in these reactions are shown in Figs. 2 and 3. The sum of α-particle and neutron masses is taken to be a reference point. To separate ^{5}H states and determine its parameters we used the least square approximation while fitting of the experimental spectra by the sum of the Breit–Wigner shaped resonances and *n*–particle distributions over phase space ($n \geq 4$). We used the Breit-Wigner distributions with energy-dependent widths parameterized for states near the threshold [9].

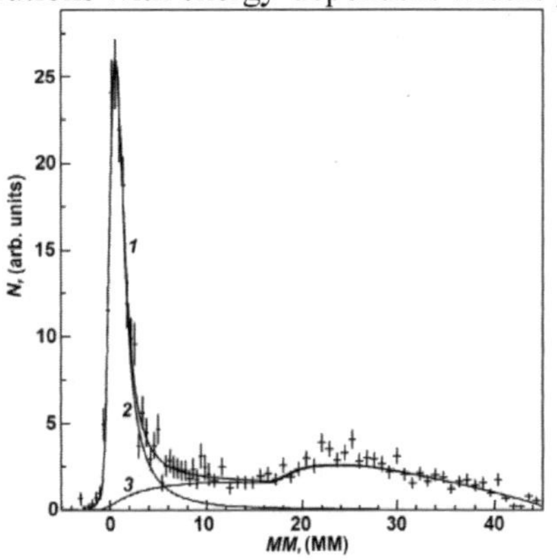

Fig. 2. *MM* spectrum for the reaction ^{10}B(π^-,dt)X. Dots with error bars are the experimental data. The solid lines are: 1 – the fit, 2 – Breit–Wigner distribution for ground state, 3 – sum of phase-space distributions.

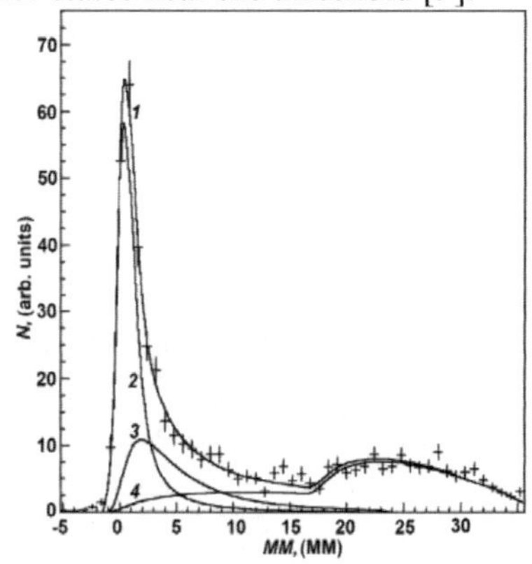

Fig. 3. *MM* spectrum for the reaction ^{11}B(π^-,tt)X. Dots with error bars are the experimental data. The solid lines are: 1 – the fit, 2 (3) – Breit–Wigner distribution for ground (first excited) state, 4 – sum of phase-space distributions.

In both spectra peaks due to three-body channels with ^{5}He in the final states are observed. The peak in Fig.2 is due to the ^{5}He ground state. The peak in Fig.3 is due to the superposition of the ground and the first excited states. Values of resonance parameters of these states (E_r – resonance energy, Γ – level width): (0.80(3) MeV, 0.7(1) MeV) and (2.1(1) MeV, 5.6(3) MeV) are in a coincident with data of the compilation [3].

We did not obtained statistically significant evidence on high excited states of ^{5}He. In the Fig. 2 some excess of the experimental events exists in the range from 22 MeV to 28 MeV. This range corresponds to resonance energies of ^{5}H ground state: 1 MeV $\leq E_r(^5\text{H}) \leq$ 5 MeV. The results of direct measurements of this value in ion reactions lay on the same interval [10, 11] and stopped pion absorption [7]. However it is necessary to point out that several states with isospin T = 1/2 lie in the region 21 MeV $\leq E_r(^5\text{H}) \leq$ 24 MeV [3]. Thus, we did not obtain reliable indication on the existence of IAS of ^{5}H with T = 3/2.

Search for ^{6}He excited states was carried out in the following reactions: ^{11}B(π^-,dt)X, ^{10}B(π^-,pt)X, ^{10}B(π^-,dd)X. The *MM* spectra for these reactions are shown in Fig. 4. The mass of ^{6}He ground state is taken to be a reference point. Description of the spectra was carried out similarly to the case of ^{5}He. Peaks formed due to the appearance of the ground state and the excited states in the three-body reaction channels are well seen. Average values of parameters of ^{6}He excited states (E_x – excitation energy, Γ -width), obtained in this paper are presented in the table 1.

Two high excited states with E_x > 20 MeV are observed in ^{10}B(π^-,pt)X channel only. The parameters of level with E_x = 22.1(1.0) MeV coincide within the limits of experimental error with the parameters of the resonance in the t + t system (E_x = 20.9(3) MeV, Γ = 3.2(1.5) MeV) formed in the reaction ^{9}Be(π^-, *tt*)*t* [12]. The last one has an isospin T = 1, and therefore it is not IAS of ^{6}H. The highest excited ^{6}He state with E_x = 27.0(8) MeV may have T = 1 or 2. In the latter case this state is isobar-analog states of ^{6}H with resonance energy $E_r \approx$ 5.5 MeV. This value is close to resonance energy of ^{6}H ground state (6.6(7) MeV) obtained in measurements of stopped pion absorption [7]. These magnitudes are considerably higher than the results obtained in reaction with heavy ions

[13,14]. In these measurements the resonance energy was ≈ 2.6 MeV. Also, it should be underlined that statistics of the data obtained in this work and in [7] is much more than statistics in works [13, 14]. In our measurements, there were no indications on the existence states with E_x > 30 MeV observed in [15]. These states may be IAS of the excited states of ^{6}H.

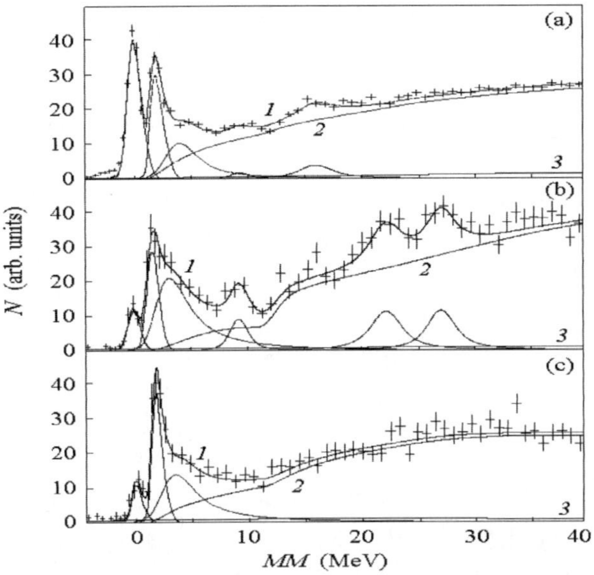

Fig. 4. *MM* spectrum for the reaction a – ^{11}B(π^-,dt)X; b – ^{10}B(π^-,pt)X; c – ^{10}B(π^-,dd)X. Dots with error bars are the experimental data. The solid lines are the Breit–Wigner distributions, 1 – fit, 2 – sum of phase-space distributions and 3 – background coming from random coincidences.

Table 1. Parameters of excited states of ^{6}He and ^{7}He measured in this work

^{6}He			^{7}He		
E_x, MeV	Γ MeV	reaction	E_x, MeV	Γ MeV	reaction
1.80(3)	0.11(2)	1,2,3	3.1(1)	≤0.5	4,5,6
3.5(2)	3.1(4)	1,2,3	4.9(2)	≤0.5	4,5,6
9.3(2)	1.0(4)	1,2	6.7(2)	≤0.5	4,5
15.(9)	3.2(7)	1	16.9(5)	1.0(3)	5,6
22.1(1.0)	2.7(1.4)	2	19.8(3)	1.5(3)	4,5,6
27.0(8)	2.5(1.1)	2	24.8(4)	4.6(7)	4,5,6

1 – ^{11}B(π^-,dt)X, 2 – ^{10}B(π^-,pt)X, 3– ^{10}B(π^-,dd)X, 4 – ^{11}B(π^-,pt)X, 5 – ^{11}B(π^-,dd)X, 6 – ^{10}B(π^-,dd)X,

Search for ^{7}He excited states was carried out in the following reactions: ^{11}B(π^-,pt)X, ^{11}B(π^-,dd)X, ^{10}B(π^-,pd)X. The *MM* spectra for these reactions are shown in Fig. 5. The sum of ^{6}He and neutron masses is taken to be a reference point. Peaks formed due to the appearance of the ground state and the excited states in the three-body reaction channels are well seen. Average values of parameters of ^{6}He excited states, obtained in this paper, are presented in the table 1. High excited state with E_x = 24.8 MeV is observed in the all studied channels.

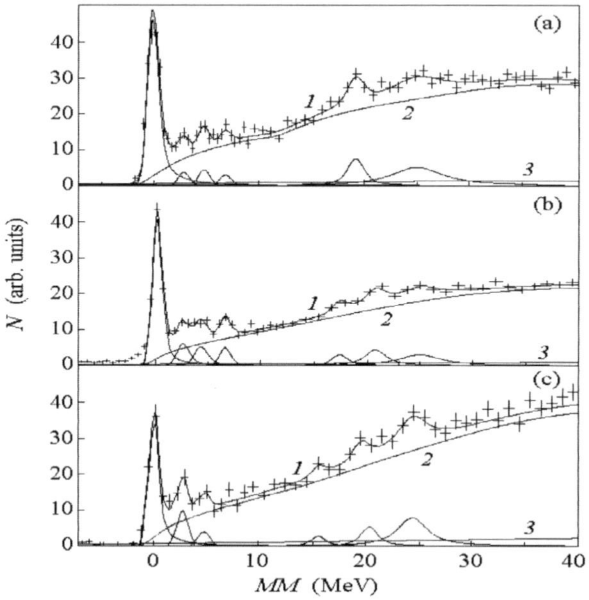

Fig. 5. *MM* spectrum for the reaction a – $^{11}B(\pi^-,pt)X$; b – $^{11}B(\pi^-,dd)X$; c – $^{10}B(\pi^-,pd)X$. Dots with error bars are the experimental data. The solid lines are the Breit–Wigner distributions, 1 – fit, 2 – sum of phase-space distributions and 3 – background coming from random coincidences.

In the region of high excitation energies above the threshold of the decay $^7He \rightarrow t + t + n$ (E_x = 11.9 MeV), relatively narrow states of ^7He were detected in our measurements for the first time. The nature of the observed states is unclear. It only can be noted that the level with E_x = 19.8(3) MeV lies near the threshold of the decay $^7He \rightarrow t + p + 3n$ (18.56 MeV). However, this state is not an IAS of the superheavy hydrogen isotope ^7H, because the isospin of the final state of the $^{11}B(\pi^-, dd)^7He$ reaction can be only $T = 3/2$.

4. Conclusions

Excited states of ^{5-7}He were observed in three-body channels of stopped pion absorption by boron isotopes $^{10,11}B$. ^6He excited state with E_x = 27.0(8) MeV observed in $^{10}B(\pi^-,pt)X$ channel is an IAS candidate for ^6H with $E_r \sim 5.5$ MeV.

This work was supported in part by the grant of the Ministry of Education and Science of Russian Federation № 1729.

References

1. Tanihata I, Savajols H, and Kanungo R 2013 *Prog. Part. Nucl. Phys.* **68**, 215
2. Jonson B 2004 *Phys. Rep.* **389**, 1
3. Tilley D, Cheves C M, Godwin J L, Hale H M, Hofmann H M, Kelley J H, Sheu H R, and Weller H R 2002 *Nucl. Phys. A*, **708**, 3
4. Tilley D, Kelley J H, Godwin J L, Milliner D J, Pursell J E, Sheu H R, and Weller H R 2002 *Nucl. Phys. A*, **745**, 155
5. Gornov M G *et al.* 1991 *Nucl. Phys. A*, **531**, 613
6. Amelin A I *et al.* 1990 Yad. Fiz. **52**, 1231
7. Gurov Yu B, Chernyshev B A, Lapushkin S V, and Sandukovsky V G 2009 *Phys. Part. Nucl.* **40**, 558
8. Gornov M G et al. 2000 Nucl. Instrum. Methods Phys. Res. A, **446**, 461

Journal of Physics: Conference Series **724** (2016) 012007 doi:10.1088/1742-6596/724/1/012007

9. Lane A M and Thomas R G 1958 *Rev. Mod. Phys.* **30**, 257
10. Korsheninnikov A A *et al.* 2001 *Phys. Rev. Lett.* **87**, 092501
11. Meister M *et al.* 2003 *Nucl. Phys. A* **723**, 13
12. Gurov Yu B, Karpukhin V S, Lapushkin S V, Laukhin I V, Pechkurov V A, Poroshin N O, Sandukovsky V G, Tel'kushev M V, and Chernyshev B A 2006 *Pis'ma Zh. Eksp. Teor. Fiz.* **84**, 3
13. Aleksandrov D V *et al.* 1984 *Sov. J. Nucl. Phys.* **62**, 18
14. Belozerov A V, Borcea C, Dlouhy Z, Kalinin A M, Kalpakchieva R, Nguen Hoai Chau, Oganessian Yu Ts, Penionzhkevich Yu E 1986 *Nucl. Phys. A* **460**, 352
15. Franke R, Kochskamper K, Steinheuer B, Wingender K, Von Wilsch W, Mahner H *Nucl. Phys. A* **433**, 351

Inelastic scattering of ^{72,74}Ni off a proton target

M L Cortés[1,2], P Doornenbal[3], A Obertelli[4], N Pietralla[1], V Werner[1], G Authelet[4], H Baba[3], D Calvet[4], F Château[4], A Corsi[4], A Delbart[4], J-M Gheller[4], A Gillibert[4], T Isobe[3], V Lapoux[4], C Louchart[1], M Matsushita[5], S Momiyama[3,6], T Motobayashi[3], M Niikura[6], H Otsu[3], C Péron[4], A Peyaud[4], E C Pollacco[4], J-Y Roussé[4], H Sakurai[3,6], C Santamaria[4], M Sasano[3], Y Shiga[3,7], S Takeuchi[3], R Taniuchi[3,6], T Uesaka[3], H Wang[3], K Yoneda[3], F Browne[8], L X Chung[9], Zs Dombradi[10], S Franchoo[11], F Giacoppo[12], A Gottardo[11], K Hadynska-Klek[12,13], Z Korkulu[10], S Koyama[3,6], Y Kubota[3,5] , J Lee[14], M Lettmann[1], R Lozeva[15], K Matsui[3,6], T Miyazaki[3,6], S Nishimura[3], L Olivier[11], S Ota[5], Z Patel[16], E Sahin[12], C M Shand[16], P-A Söderström[3], I Stefan[11], D Steppenbeck[5], T Sumikama[17], D Suzuki[11], Zs Vajta[10], J Wu[3,18] and Z Xu[14]

[1] Institut für Kernphysik, Technische Universität Darmstadt, 64289 Darmstadt, Germany
[2] GSI Helmoltzzentrum für Schwerionenforschung GmbH, 64291 Darmstadt, Germany
[3] RIKEN Nishina Center, 2-1 Hirosawa, Wako, Saitama 351-0198, Japan
[4] CEA, Centre de Saclay, IRFU/Service de Physique Nucléaire, F-91191 Gif-sur-Yvette, France
[5] Center for Nuclear Study, University of Tokyo, RIKEN campus, Wako, Saitama 351-0198, Japan
[6] Department of Physics, University of Tokyo, 7-3-1 Hongo, Bunkyo, Tokyo 113-0033, Japan
[7] Department of Physics, Rikkyo University, 3-34-1 Nishi-Ikebukuro, Toshima, Tokyo 172-8501, Japan
[8] School of Computing Engineering and Mathematics, University of Brighton, Brighton BN2 4GJ, United Kingdom
[9] Institute for Nuclear Science & Technology, VAEC, P.O. Box 5T-160, Nghia Do, Hanoi, Vietnam
[10] MTA Atomki, P.O. Box 51, Debrecen H-4001, Hungary
[11] Institut de Physique Nucléaire Orsay, IN2P3-CNRS, 91406 Orsay Cedex, France
[12] Department of Physics, University of Oslo, N-0316 Oslo, Norway
[13] Istituto Nazionale di Fisica Nucleare, Laboratori Nazionali di Legnaro, Legnaro, Padova, Italy
[14] Department of Physics, The University of Hong Kong, Pokfulam, Hong Kong
[15] IPHC, CNRS/IN2P3 and University of Strasbourg, F-67037 Strasbourg, France
[16] Department of Physics, University of Surrey, Guildford GU2 7XH, United Kingdom
[17] Department of Physics, Tohoku University, Sendai 980-8578, Japan
[18] State Key Laboratory of Nuclear Physics and Technology, Peking University, Beijing 100871, P.R. China

E-mail: `m.l.cortes@gsi.de`

Abstract. Inelastic scattering of ^{72,74}Ni off a proton target was performed at RIBF, RIKEN, Japan. The isotopes were produced by the fission of ^{238}U on a thick Beryllium target and were then selected and identified on an event-by-event basis using the BigRIPS separator. Selected isotopes were focused onto the liquid hydrogen target of the MINOS device and gamma rays from the reactions were measured with the DALI2 array. The energy of the ions in the middle of the target was 213 MeV/u. Outgoing particles were identified using the ZeroDegree spectrometer. Here, we report on the current status of the data analysis and preliminary results for the proton inelastic scattering cross sections for both isotopes.

1. Introduction

The study of very neutron-rich exotic nuclei has recently attracted a great deal of scientific attention and offers a unique opportunity to test the shell evolution as a function of isospin. This evolution reflects properties of the nucleon-nucleon interaction, as well as the importance of the proton-neutron tensor force [1], and thus, allows a better insight into the fundamental properties of the nuclear force. The reduced $E2$ transition strength, $B(E2)$, for the $0^+ \rightarrow 2^+$ transition in even-even nuclei is a fundamental quantity in measuring the degree of collectivity of a nucleus. Moreover, in order to distinguish the relative importance of the contribution of protons and neutrons to the collectivity, neutron and proton multipole matrix elements should be obtained. These matrix elements are particularly important for neutron- or proton-rich nuclei, where the motion of protons and neutrons can be significantly different. The proton multipole matrix element, M_p, reflects only the contribution of the protons to the collectivity, and can be directly related to the $B(E2)$ value. The neutron matrix element, M_n, can be determined by means of a hadronic probe, for example proton inelastic scattering, which is sensitive to both protons and neutrons. Following the approach by Bernstein *et al.* [2, 3], the ratio of neutron-to-proton matrix elements can be expressed as

$$\frac{M_n}{M_p} = \frac{N}{Z}\left[\frac{\delta_{p,p'}}{\delta_C} + \frac{Zb_p}{Nb_n}\left(\frac{\delta_{p,p'}}{\delta_C} - 1\right)\right], \tag{1}$$

where δ_C is the deformation length measured by the electromagnetic probe, $\delta_{p,p'}$ is the deformation length measured with the hadronic probe, and b_p and b_n are the external-field proton and neutron interaction strengths of the hadronic probe. The ratio b_n/b_p is approximately 3 for proton scattering with a relative proton kinetic energy below 50 MeV and approximately 1 at 1 GeV [3, 4]. Thus, by combining measurements of Coulomb excitation, which yields the $B(E2)$ value, and proton inelastic scattering, the ratio of proton-to-neutron matrix elements can be obtained.

Of particular interest is the behavior of proton and neutron collectivity in isotopes with a single closed shell, for example in the Nickel isotopes with $Z = 28$. Because of the difficulty in producing very exotic neutron-rich isotopes, until recently the measurement of the $B(E2)$ value by means of Coulomb excitation was limited to ^{70}Ni. In 2014 a measurement of the $B(E2)$ value for ^{74}Ni yielded a deformation length of $\delta_C = 0.78 \pm 0.13$ fm [5]. In earlier measurements of ^{74}Ni by inelastic scattering on a proton target, a deformation length of $\delta_{p,p'} = 1.07(16)$ fm was deduced [6]. The combination of these two measurements resulted in $M_n/M_p = 2.4 \pm 0.8$, as reported in Ref. [5]. This value is larger than the $M_n/M_p = N/Z$, which is expected for a pure isoscalar collective excitation, and may indicate an enhancement of the neutron contribution to the structure of neutron-rich Nickel isotopes towards the expected shell closure at ^{78}Ni.

In this proceeding, we report on a second measurement of the proton inelastic scattering cross-section for ^{74}Ni performed at intermediate energies, as well as the first measurement of proton inelastic scattering of ^{72}Ni. We describe the performed experiment, the status of the data analysis and preliminary results. The deformation length previously reported for ^{74}Ni is preliminarily confirmed within uncertainty and, using the $B(E2)$ value reported in Ref. [5], the ratio of proton-to-neutron matrix elements has been calculated.

2. Experimental setup

The experiment was performed at the Radioactive Isotope Beam Factory (RIBF), operated by the RIKEN Nishina Center and the Center for Nuclear Study of the University of Tokyo as a part of the first SEASTAR campaign [7]. A detailed description of the experimental conditions can be found in a previous article [8]. A ^{238}U primary beam was accelerated to an energy of 345 MeV/nucleon and impinged on a 3-mm-thick Beryllium target at the entrance of the BigRIPS separator [9]. The $B\rho - \Delta E - B\rho$ technique was used to select the secondary isotopes of interest around ^{71,73}Co, in this case ^{72,74}Ni. Secondary fragments were identified on an event-by-event basis: The atomic number of each ion (Z) was deduced from an energy loss measurement in an ionization chamber placed at the focal plane F7; the mass-to-charge ratio (A/Q) was determined combining the position and angle measurements performed at the focal planes F3 and F5 using Parallel Plate Avalanche Counters (PPACs) and the Time-of-Flight (ToF) measured between two plastic scintillators placed at the focal planes F3 and F7. The MINOS device [10], composed of a 10 cm long liquid hydrogen target surrounded by a Time Projection Chamber (TPC), was placed at the focal plane F8 in order to perform proton inelastic scattering. In the present analysis the information given by the TPC was not used. A case where it was can be seen in Ref [8]. The energy of the Nickel isotopes in the middle of the target was 213 MeV/u. To detect the γ-rays emitted after the reaction, the DALI2 array [11] was placed around the target. It consisted of 186 NaI(Tl) detectors covering angles from 7 to 115 degrees (integrated along the target length) relative to the beam axis. The full-energy-peak efficiency of the array was determined using a detailed `Geant4` [12] simulation and was found to be 14% at 1.33 MeV with an energy resolution of 6.2% (FWHM) for a stationary source. After the target, reaction products were identified using the ZeroDegree spectrometer [9]. In this case, the ionization chamber used to determine the Z of the outgoing ions was placed at the focal plane F11. The position and angle measurements were performed with PPACs placed at the focal planes F9 and F11 and the ToF was measured with plastic scintillators placed at the same focal planes. Figure 1 shows the particle identification performed for the incoming as well as for the outgoing ions. The gates used to select incoming and outgoing ^{72,74}Ni isotopes are shown

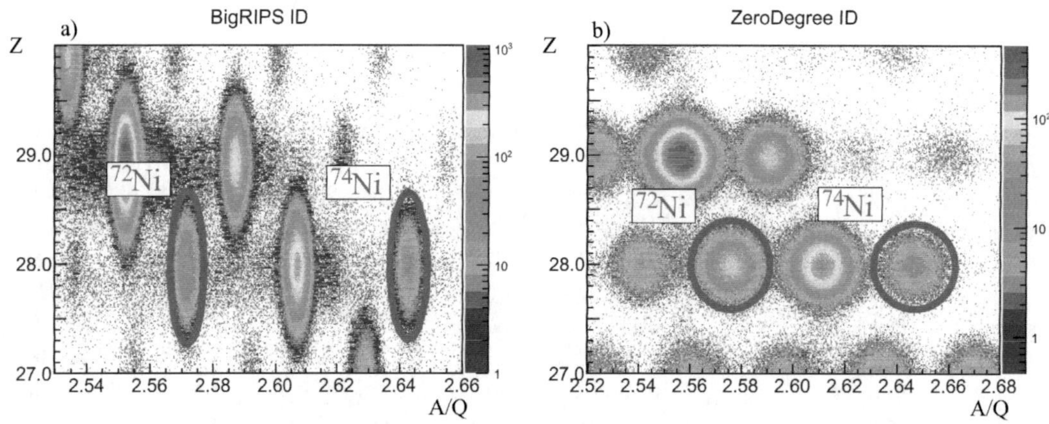

Figure 1. Particle identification obtained for a) the BigRIPS separator and b) the ZeroDegree spectrometer. Gates used to select ^{72,74}Ni isotopes incoming and outgoing are shown.

in both cases. A total of 4.6×10^6 ^{72}Ni and 1.5×10^6 ^{74}Ni isotopes impinging on the liquid hydrogen target were finally registered in the ZeroDegree spectrometer.

3. Results

The γ-rays measured in coincidences with the particle gates shown in Figure 1 were corrected to take into account the Doppler shift. For this purpose, the decay was assumed to occur at the center of the target, and the velocity was assumed to be the average of the measured incoming and outgoing velocities with a constant offset, which considers the energy loss in different detectors along the beam line in the F8 area. Figure 2 shows the Doppler-corrected γ-ray spectra obtained for ^{72,74}Ni. For the case of ^{72}Ni two transitions are observed at energies 1092(4) keV and 830(11) keV, in good agreement with the reported $2^+ \to 0^+$ and $4^+ \to 2^+$ transitions, respectively. For ^{74}Ni a transition with energy 1023(6) keV is clearly seen as well as a noticeable structure at 734(16) keV. The two energies are also in agreement with the reported $2^+ \to 0^+$ and $4^+ \to 2^+$ $\gamma-$ray transitions. To obtain the proton inelastic scattering cross

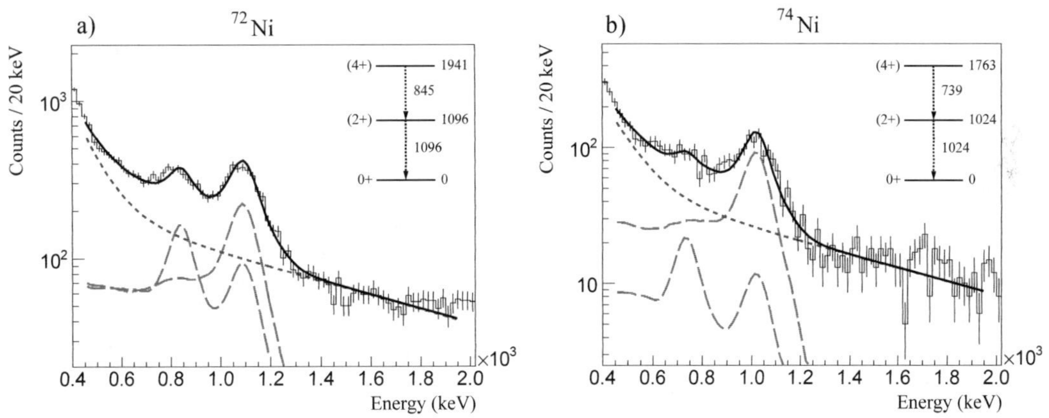

Figure 2. Doppler-corrected γ-ray spectra for a) ^{72}Ni and b) ^{74}Ni, measured with the DALI2 array. Simulated responses for the 2^+ transition and for the $4^+ \to 2^+ \to 0^+$ cascade are shown, as well as the double exponential used as background and the total fit.

sections to excited states, each spectrum was fitted using the simulated response functions of the DALI2 array and a double exponential function to take into account the background. The simulation included the experimentally determined individual resolution of each crystal. The fits are shown in Figure 2 together with the experimental data. The fitting procedure was performed for different add-back conditions and the results obtained agreed within error bars. In each case, the feeding contribution from the 4^+ decay was subtracted. Possible feeding from higher lying states has not been observed but should be considered in more detail. The error in the calculated cross section includes the statistical uncertainty of the fit, 6% uncertainty in the efficiency of the DALI2 array, the statistical uncertainty on the number of incident ions and a 2% uncertainty in the target thickness. Following this procedure we obtained a preliminary cross section for the occurrence of the $2^+ \to 0^+$ transition of $\sigma_{p,p'} = 3.8 \pm 0.5$ mb for ^{72}Ni and $\sigma_{p,p'} = 4.3 \pm 0.6$ mb for ^{74}Ni. In order to obtain deformation lengths from the measured cross sections, the ECIS-97 code [13, 14] was used. The calculation included a first order, harmonic vibrational model together with the KD02 potential [15]. In the model, volume, surface and spin-orbit potentials were chosen to have the same deformation. However, we stress that this is a preliminary approach because the KD02 optical potential has been tested up to 200 MeV only, which is below the energy used in this experiment. Nevertheless, among the available potentials this is the one that better suits the experimental conditions. The preliminary values for the deformation lengths obtained with ECIS-97 are $\delta_{p,p'} = 0.93 \pm 0.07$ fm for ^{72}Ni and $\delta_{p,p'} = 0.98 \pm 0.05$ fm for ^{74}Ni. It is noted that the value obtained for ^{74}Ni agrees within error bars with the previous measurement of $\delta_{p,p'} = 1.07(16)$ fm performed by Aoi *et al.* [6]. Using Eq. (1) and the $B(E2)$ value reported by Marchi *et al.* [5], we obtain a neutron-to-proton matrix elements ratio of $M_n/M_p = 2.2 \pm 0.6$ for ^{74}Ni. In this calculation the values of b_n and b_p were chosen as 0.9 and 0.3 respectively following Ref. [5]. As the energy used in this experiment is higher than the one reported in Ref [5], further calculations of b_n and b_p are still needed. For ^{72}Ni no Coulomb excitation data is available, therefore the determination of the proton-to-neutron matrix elements ratio is not feasible.

4. Summary

We performed proton inelastic scattering for ^{72,74}Ni. Preliminary analysis results in a cross section for the occurrence of the $2^+ \rightarrow 0^+$ transition of $\sigma_{p,p'} = 3.8 \pm 0.5$ mb for ^{72}Ni and $\sigma_{p,p'} = 4.3 \pm 0.6$ mb for ^{74}Ni. Using coupled-channel calculations, the deformation lengths of the first 2+ states of both isotopes were obtained. For the case of ^{74}Ni, combination of our result with existing Coulomb excitation data, yields $M_n/M_p = 2.2 \pm 0.6$, which is consistent with the result obtained using previous (p, p') measurements.

Acknowledgments

The authors are thankful to the accelerator and BigRIPS teams for stable operation, high intensity of the Uranium primary beam, and production of secondary beams during the experiment. Some of us thank N. Van Giai and A. Severyukhin for discussions. MINOS has been supported by the European Research Council through the ERC Grant No. MINOS-258567. A. O. has been supported by the JSPS long-term fellowship L-13520 at the RIKEN Nishina Center. C. S. has been supported by the IPA program at the RIKEN Nishina Center. M. L. C. and V. W. further acknowledge support through the German BMBF Grant No. 05P12RDFN8. The experiment and the data analysis have been supported by the Helmholtz International Center for FAIR funded by the State of Hesse within its LOEWE initiative.

References

[1] Sorlin O and Porquet M G 2008 *Prog. Part. Nucl. Phys.* **61** 602 – 673
[2] Bernstein A M *et al.* 1981 *Phys. Lett. B* **103** 255 – 258
[3] Bernstein A M *et al.* 1983 *Comments Nucl. Part. Phys.* **11** 203 – 215
[4] Riley L A *et al.* 2005 *Phys. Rev. C* **72**(2) 024311
[5] Marchi T *et al.* 2014 *Phys. Rev. Lett.* **113**(18) 182501
[6] Aoi N *et al.* 2010 *Phys. Lett. B* **692** 302 – 306
[7] Doornenbal P and Obertelli A 2013 Riken proposal for scientific program: Shell Evolution And Search for Two-plus states At the RIBF (SEASTAR) Unpublished
[8] Santamaria C *et al.* 2015 *Phys. Rev. Lett.* **115**(19) 192501
[9] Kubo T *et al.* 2012 *Prog. Theor. Exp. Phys.* **2012**
[10] Obertelli A *et al.* 2014 *Eur. Phys. J. A* **50** 8
[11] Takeuchi S *et al.* 2014 *Nucl. Instr. Meth. A* **763** 596 – 603
[12] Agostinelli S *et al.* 2003 *Nucl. Instr. Meth. A* **506** 250 – 303
[13] Raynal J Coupled channel code ECIS97. Unpublished
[14] Raynal J 1981 *Phys. Rev. C* **23**(6) 2571–2585
[15] Koning A and Delaroche J 2003 *Nucl. Phys. A* **713** 231 – 310

Distorted-wave Born approximation study of the ^{11}Li(p,t)^{9}Li reaction

A A Cowley[1,2]

[1] Department of Physics, Stellenbosch University, Private Bag X1, Matieland 7602, South Africa
[2] iThemba Laboratory for Accelerator Based Sciences, P O Box 722, Somerset West 7129, South Africa

E-mail: aac@sun.ac.za

Abstract. The reaction ^{11}Li(p,t)^{9}Li(gs) at an incident energy of 4 MeV is treated in terms of a simplistic distorted-wave Born approximation transfer. The halo neutrons involved in the reaction are treated as a di-neutron cluster transferred in a simultaneous process. This appears to be a good approximation of the mechanism. The dominant contribution to the reaction comes from the known $(1s_{1/2})^2$ structure component of the ground state of ^{11}Li, and the cross section angular distribution seems to be relatively insensitive to the fact that ^{11}Li has an anomalously large radius due to its Borromean halo properties. Significantly this simple treatment of the reaction is in much better agreement with the experimental angular distribution than a more sophisticated calculation.

1. Introduction

Proton-induced two-nucleon transfer represents a convenient method to study an exotic nucleus such as ^{11}Li [1, 2]. The Borromean character of the target system when viewed as a two-neutron halo bound to a ^{9}Li core promises to be of special interest. The short lifetime of about 8 ms of ^{11}Li causes a slight experimental complication to the measurement of cross section angular distributions for a two-neutron pickup reaction, which is then properly written as ^{1}H(^{11}Li, ^{9}Li)^{3}H to reflect the inverse kinematics required for a radioactive beam. In order to emphasize the transfer of the halo component of the nucleus of interest, we write this reaction consistently as ^{11}Li(p,t)^{9}Li. However, this latter alternative convention is a personal preference and it is of no consequence or further significance.

In order to extract information regarding the prominence of two-nucleon correlations in ^{11}Li from experimental measurements of the (p,t) reaction, the distorted-wave Born approximation (DWBA) promises to be a useful tool. Although DWBA theory has been thoroughly studied over many years, several issues remain somewhat obscure. Of these, the prominence of a sequential [3, 4] pickup process is the most controversial. The motivation for the importance of a sequential, two-step mechanism in nucleon-induced two-nucleon transfer is inspired by concerns that the absolute magnitude of the cross sections predicted by the DWBA seems to be much too low when compared with some experimental values. Also, in the ^{208}Pb(p,t)^{206}Pb(3^+) reaction to the ground state, which involves an unnatural parity transition, a zero-range first order DWBA is forbidden. Nevertheless, Nagarajan *et al.* [5] show that a careful finite range, simultaneous transfer DWBA estimation for this reaction provides the correct absolute magnitude. On the other hand, Igarashi *et al.* [6] argue that addition of a two-step sequential transfer mechanism to the direct pickup in DWBA is essential to describe the cross section magnitude of the ^{208}Pb(p,t)^{206}Pb(3^+) reaction. The controversy of sequential versus a simultaneous mechanism has not been satisfactorily clarified up to date, and this may partly be because the absolute cross section magnitude of a reaction such as (p,t) suffers from severe momentum mismatch between the incident and exit channels. This causes the cross section to be proportional to the square of the momentum wave function of the bound di-nucleon system which participates in the pickup process. At the large momenta involved in the transfer, the magnitude of the wave functions of the neutrons constituting the picked-up pair are not known well, hence a large uncertainty in cross section magnitude follows.

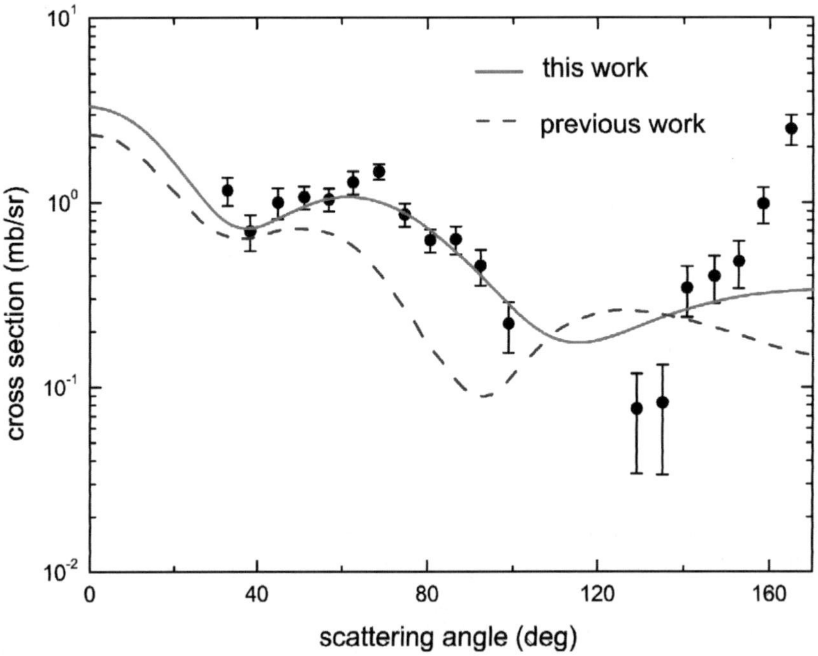

Figure 1. Angular distribution for the reaction ^{11}Li(p,t)^{9}Li(gs) at an incident energy of 4.4A MeV. The solid curve represents a DWBA calculation as described in the text and the dashed curve (*reproduced with permission*) is a preliminary results of Thompson [7]. Experimental data were measured at TRIUMF and are available in Ref. [7].

2. Sophisticated and alternative simplistic DWBA approaches

Because earlier sophisticated DWBA approaches, in which all the known properties of ^{11}Li were incorporated, were not particularly successful [1, 7] , the present work approaches the problem differently. Instead of *ab initio* attempting to reproduce the absolute magnitude of the ^{11}Li(p,t)^{9}Li cross section, we at first explore a simplistic zero-range, simultaneous di-neutron cluster transfer process. The di-neutron cluster is assumed to be bound in a standard best-value geometry as suggested by Meijer *et al.* [8, 9, 10] for stable target nuclei. Thus the di-neutron bound state is calculated with a geometry of radius parameter r_0=1.15 and diffuseness

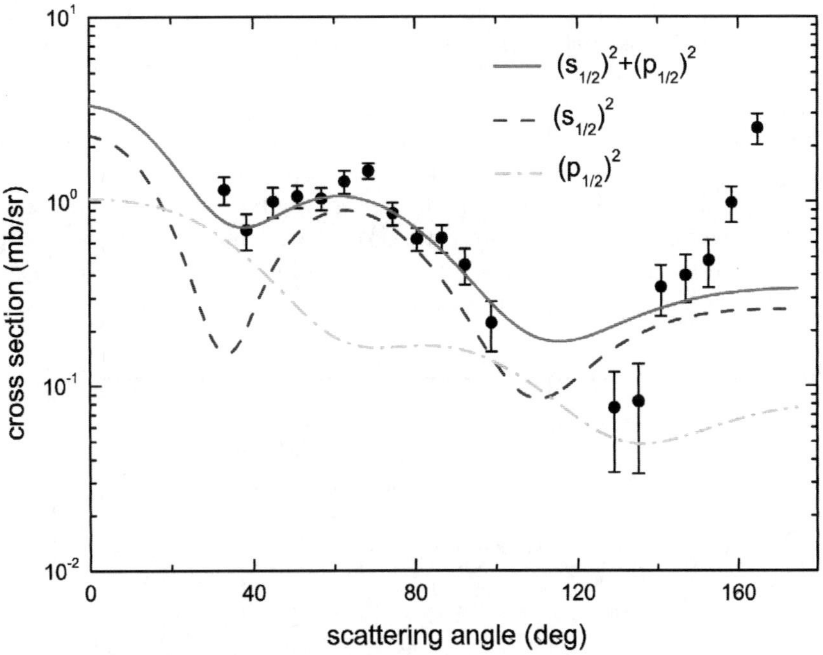

Figure 2. Contributions from angular momentum transfers of L=0 and L=2 to the ^{11}Li(p,t)^{9}Li(gs) cross section at an incident energy of 4.4A MeV. These components originate from $(s_{1/2})^2$ and $(p_{1/2})^2$ configurations. Normalizations have been chosen independently and arbitrarily in order that the incoherent sum of the two contributions should best reproduce the experimental data. (From figure in Ref. [11]).

Table 1. Optical potential parameters used in the DWBA calculations.

Channel	Ref.	V	r_v	a_v	W	W_D	r_W	a_W	V_{SO}	r_{SO}	a_{SO}
p+^{11}Li	[1]	54.06	1.17	0.75	2.37	16.87	1.32	0.82	6.2	1.01	0.75
t+^{9}Li	[1]	1.42	1.16	0.78	28.2	0	1.88	0.61	0		
bound state	[8]	adjusted	1.15	0.78							

Definition of optical potentials in Table 1 (with strengths in MeV and geometrical sizes in fm):

$$V(r) = -Vf(r,r_v,a_v) - Wf(r,r_v,a_v) + W_D 4a_W \tfrac{d}{dr} f(r,r_W,a_W) + W_{SO}\sigma.\ell \bar{\lambda}^2 \tfrac{1}{r}\tfrac{d}{dr} f(r,r_{SO},a_{SO})$$

$$\text{with } f(r,r_i,a_i) = [1+\exp(\tfrac{r-r_i A^{\frac{1}{3}}}{a_i})]^{-1}$$

a=0.76 in a Woods-Saxon well, adjusted to reproduce the cluster binding energy with quantum numbers based on oscillator-well energy conservation. Consequently this means that in such an approach the extended spatial extent of ^{11}Li is not taken into account. Distorted waves, derived from optical potentials for the incident and exit channels, are identical to those of the earlier sophisticated calculations [1, 7], and parameter values are provided in Table 1. As was already mentioned, possible sequential pickup of the two neutrons is excluded and core excitations are also neglected. Of course, in our approach, all information regarding absolute magnitude of the cross section scale is lost, but the shape reproduction of the angular distribution now becomes crucial as a criterion for successful reproduction of the experimental data.

As shown in Fig. 1 for the ^{11}Li(p,t)^{9}Li reaction at an incident energy of 4.4MeV, the simplistic DWBA calculation from the present work gives a result which is in vastly superior agreement in angular-distribution shape to the earlier preliminary work of Thompson [7]. This is also true when compared with published results [1] at a lower incident energy of 3 MeV. Although we have arbitrarily normalized the DWBA prediction to the experimental distribution, as the loss of information on cross-section magnitude inherent to the simplistic calculation requires, it should be noted that the sophisticated calculation cannot be further improved by a similar adjustment. The surprising superiority of a standard, simplistic DWBA calculation over a more sophisticated treatment needs to be understood, and this is investigated further in the following sections.

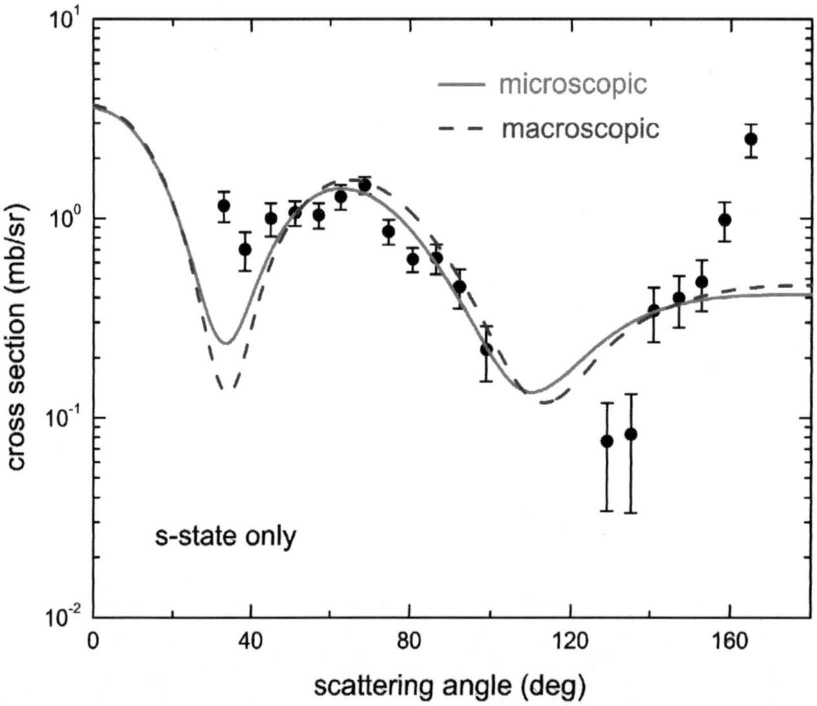

Figure 3. Comparison between macroscopic and microscopic calculations for ^{11}Li(p,t)^{9}Li(gs) at an incident energy of 4.4A MeV. In the microscopic case a pure $(s_{1/2})^2$ configuration is assumed, whereas for the macroscopic calculation a di-neutron cluster and an angular momentum transfer of L=0 are introduced. The normalizations to the experimental data are arbitrary. (From figure in Ref. [11]).

3. Influence of ground state structure of ^{11}Li and di-neutron approach

As is known, the wave function φ of the two halo neutrons in the ground state of ^{11}Li may be expressed [12, 13] as

$$\varphi = 0.45\left|s^2_{1/2}(0)\right\rangle + 0.55\left|p^2_{1/2}(0)\right\rangle + 0.04\left|d^2_{5/2}(0)\right\rangle, \tag{1}$$

where the base wave functions are indicated in a standard notation.

The component where both neutrons are in the 2s-state dominate, and the contributions of the cross sections of the s- and p-states to the summed value (with a neligible contribution from the d-state occupation) is shown in Fig. 2. Angular momentum transfers are $L = 0$ and $L = 2$ for s- and p-states, respectively. Relative magnitudes are adjusted to reproduce the overall shape of the angular distribution best, and added incoherently. If only the dominant s-state is considered, the comparison of the DWBA prediction and the experimental angular distributions is still reasonable, as shown in Fig. 3. Furthermore, as shown in Fig. 3, a macroscopic treatment of the two halo neutrons as a di-neutron, instead of a realistic microscopic formulation, is of no consequence.

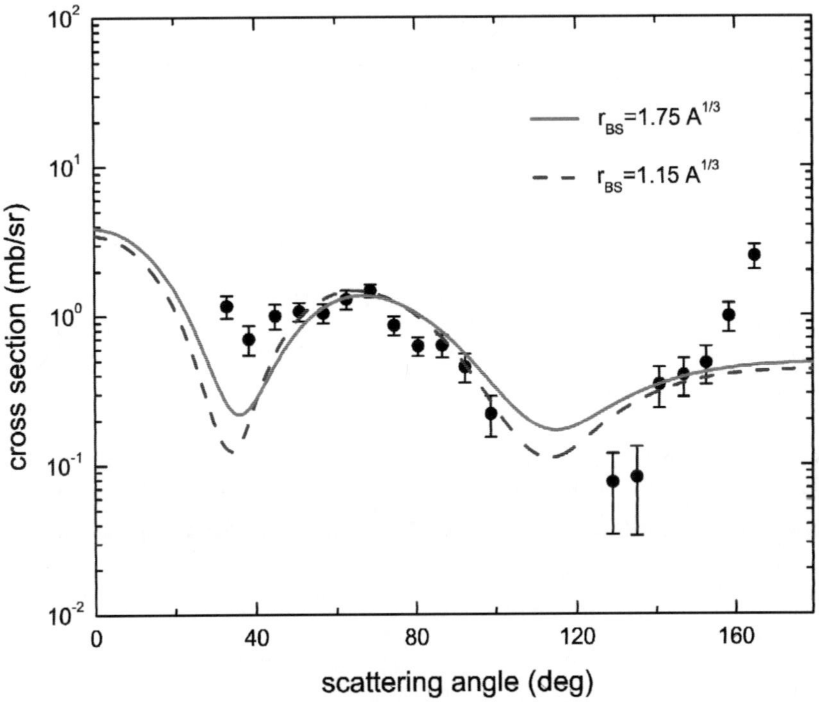

Figure 4. Macroscopic calculations for $^{11}Li(p,t)^{9}Li$(gs) at an incident energy of 4.4A MeV with different values of the bound state radius of the di-neutron cluster. The solid line corresponds to a radius which corresponds to the rms size of ^{11}Li, and the dashed line to a number appropriate for stable nuclei in general.

4. Influence of bound state geometry

Variation of the geometry of the bound state serves as a crude approximation to the influence of the extended range of the halo mass distribution of the ^{11}Li nucleus. This is shown in Fig. 4 where the radius of the bound state is increased from $R = 1.15A^{1/3}$ to $R = 1.75A^{1/3}$, where A is the mass number of the core system ^{9}Li. The latter radius value corresponds to the actual rms radius of ^{11}Li. Clearly no significant difference is observable, and similar insensitivity is shown to the diffuseness parameter.

5. Sensitivity to distorting potentials

As is evident from the discussion thus far, the crucial issue which still needs to be investigated is the sensitivity of the DWBA results to the distorting potentials which are employed in the incident and exit channels. As the global potentials used in the present work were extracted based on target nuclei at or near the line of stability, those parameters are unlikely to be optimized for a nuclear species such as ^{11}Li. In fact, it is known that the present optical parameters do not reproduce elastic scattering of protons from ^{11}Li well in the incident energy range of interest [14].

In Fig. 5 a comparison is shown of a preliminary attempt to use a modified parameter set for the incident channel. The dashed curve corresponds to the standard global proton potentials of Table 1, whereas the solid curve is based on a set [14] which describes elastic scattering of protons from ^{11}Li well, except that the real radius parameter had to be arbitrarily adjusted (from r_0=1.09 to r_0=1.69) for better overall agreement with the experimental data. Hence the only useful purpose that this exercise serves is to suggest that further improvement to our understanding of the ^{11}Li(p,t)^{9}Li reaction should follow from a further investigation of the influence of the distorting potentials.

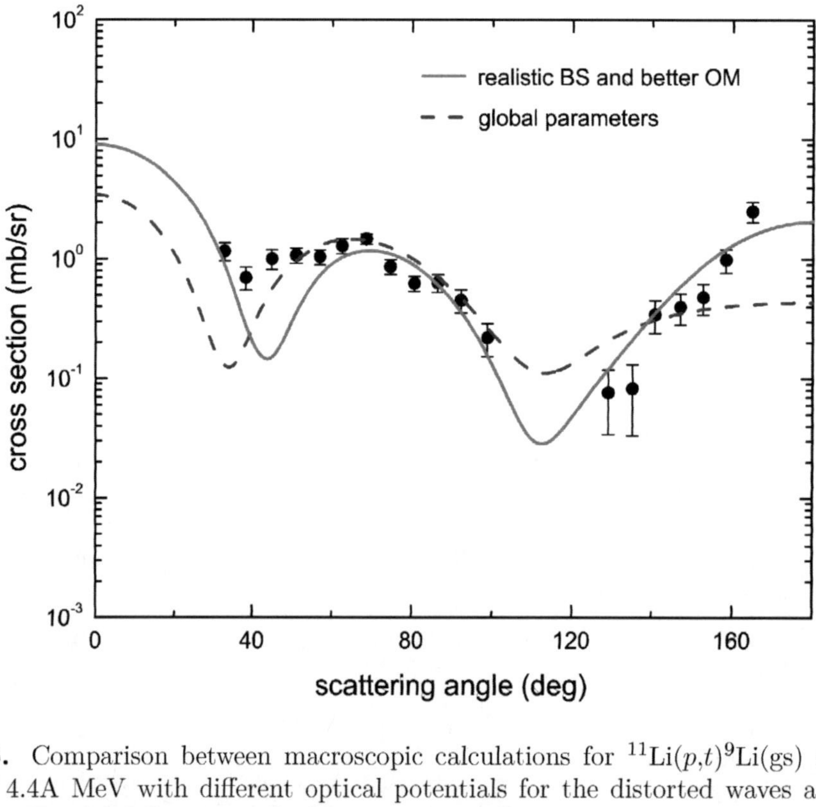

Figure 5. Comparison between macroscopic calculations for ^{11}Li(p,t)^{9}Li(gs) at an incident energy of 4.4A MeV with different optical potentials for the distorted waves an bound state geometry. The solid line is applicable to more realistic optical potentials and bound state geometry, as discussed in the text.

6. Summary and Conclusions

The DWBA prediction of a simplistic calculation for the reaction ^{11}Li(p,t)^{9}Li(gs) at an incident energy of 4.4A MeV was investigated. Such a DWBA was shown to give a reasonably good reproduction of the cross section angular distribution. Results from the present study were found to be superior to those from more sophisticated analyses at the same or lower incident energy. The present results are independent of whether di-neutron cluster transfer is assumed, or a more realistic microscopic two-particle mechanism is considered. The response to the transfer reaction is similar to that of nuclear species of normal rms size, and the extended halo structure of ^{11}Li does not seem to be influence the shape of the (p,t) angular distribution. Alternative optical potentials which specifically attempts to reproduce the elastic scattering of protons from ^{11}Li, as opposed to global optical parameters, appear to hold the promise of even better agreement with existing experimental data.

The simplicity of the theoretical treatment, combined with its ability to reproduce the main characteristics of experimentally observed angular distributions, is encouraging. Clearly this needs to be investigated further.

Acknowledgement

This work was performed with funding from the National Research Foundation (NRF) of South Africa. The financial support is gratefully acknowledged.

References

[1] Tanihata I *et al.*2008 *Phys. Rev. Lett.* **100** 192502
[2] Tanihata I Savajols H and Kanungo R 2013 *Progress in Particle and Nuclear Physics* **68** 215
[3] Satchler G R in Heavy Ion Collisions *Lecture Notes in Physics* **168** (Springer, Berlin, 1982) pp. 44–53
[4] Pinkston W T and Satchler G R 1982 *Nucl. Phys.* **A383** 61
[5] Nagarajan M A, Strayer M R and Werby M F 1977 *Phys. Lett.* **B68** 421
[6] Igarashi M , Kubo K-I and Yagi K 1991 *Physics Reports* **199** 1
[7] Thompson Ian 2011 *Theory and Calculation of Two-Nucleon Transfer Reactions* Lawrence Livermore National Laboratory, Report LLNL-PRES-492069 (unpublished)
[8] De Meijer R J, Put L W and Vermeulen J C 1981 *Phys. Lett.* **107B** 14
[9] De Meijer R J, Put L W, Akkermans J J, Vermeulen J C and Bingham C R 1982 *Nucl. Phys.* **A386** 200
[10] Sens J S and De Meijer R J 1983 *Nucl. Phys.* **A407** 45
[11] Cowley A A 2014 *Proceedings of the 33rd International Workshop on Nuclear Theory (Rila, Bulgaria) 22 - 28 June 2014. (Editors: A Georgieva and N Minkov, Publisher: Heron Press, Sofia, Bulgaria 2014)* pp. 87 - 94
[12] Potel G, Barranco F, Vigezzi E and Broglia R A 2010 *Phys. Rev. Lett.* **105** 172502
[13] Barranco F, Bortignon P F, Broglia R A, Colò G and Vigezzi E 2010 *Eur. Phys. J. A* **11** 385
[14] Roger T 2010 *Study of two neutron transfer reaction and resonant elastic scattering induced by ^{11}Li, with the MAYA active target ISOLDE Seminar 6 October 2010* (unpublished)

Quartet excitations in atomic nuclei

J. Cseh

Institute for Nuclear Research, Hungarian Academy of Sciences,
Debrecen, Pf. 51, Hungary-4001

E-mail: cseh@atomki.mta.hu

Abstract. The recently invented phenomenologic and semimicroscopic algebraic quartet models, as well as their relations to other approaches are discussed. The semimicroscopic model is applied to the ^{20}Ne and ^{28}Si nuclei.

1. Introduction

Quarteting is an important phenomenon in various fields of human culture, among others in the study of nuclear structure. Its remarkable role in nuclei is a consequence of the fact that the effective nucleon-nucleon interaction within an atomic nucleus is attractive and short-ranged. Therefore, the nucleons tend to occupy the same single-particle states. The Pauli-exclusion principle allows two protons and two neutrons to be in the same orbital. They interact with each other very strongly, and form a quartet. These facts are known from the beginning of nuclear research, therefore, several models have been invented for the description of nucleon quartets.

In what follows first we give a very brief review of the history of quartet theories. Then we present the basic concepts of the newly introduced [1] phenomenologic (PAQM) and semimicroscopic (SAQM) algebraic quartet models. In this description two nice characteristic features of the previous frameworks are combined: the exactly defined microscopic (shell model) content of the quartets, and the efficient and easily applicable group-theoretical technology of the spectrum generation. Then applications to the ^{20}Ne and ^{28}Si nuclei are considered, as illustrative examples. Finally some conclusions are drawn.

2. Historical background

An experimental signature of quarteting is that the separation energy of a nucleon in an even-even $N = Z$ nucleus is larger than that of an α-particle. It was already the motivation for Wigner's supermultiplet theory [2], which might be considered as the first quartet model. (Much work has been done on the binding energies and quarteting later on, too, see e.g. [3].)

A quartet model of the nucleons was formulated in [4], (based on the stretched scheme), and then it was generalized in several steps. In [5] quartet excitations were considered from one major shell to the other, and the corresponding energies were determined from mass relationships. In this generalized interpretation a quartet is not related to a specific angular momentum coupling scheme: it is made of 2 protons and 2 neutrons, occupying a fourfold degenerate single-particle state (l, m orbit in $L - S$ coupling, or j, m and $j, -m$ orbits in $j - j$ coupling). The internal binding of a quartet is strong, while the quartet-quartet interaction is relatively weak. Arima and Gillet took into account [6] also pairs of nucleons, as further building blocks, extending the description to even-even nuclei of different Z and N.

In [7] intrashell quartet excitations have been introduced in addition to the intershell excitations of [5]. This concept leads to a quartet shell model, i.e. one assumes the existence of a self-consistent quartet potential well, and its states are used to describe the quartet states in $4n$ nuclei. The $0s$, $0p$, $1s-0d$, ... oscillator shells of the nucleon shell model are replaced by $0s$, $0p$, $1s-0d$, ... quartet shells, having 1, 3, 6, ... single quartet states, respectively. The corresponding energies were determined empirically, too.

A further extension was presented in [8] by incorporating any number of particle-hole excitations (in the language of the nucleon shell model), contrary to the quartet shell model of [5, 7] which had only 0, 4, 8 ... excitation quanta (in terms of nucleon shell model). This considerable extension of the quartet model space appeared due to the conceptual generalization of a quartet. Harvey defined [8] it as 2 protons and 2 neutrons having a quartet-symmetry: permutational symmetry of {4}, and spin-isospin symmetry of {1,1,1,1}.

Interacting boson type quartet models were invented [9, 10] for the description of quarteting in heavy nuclei. In [9] the basic building block quartets are treated as $l=0$ (s) and $l=2$ (d) bosons, and the model has a U(6) group structure, like the interacting boson model of the quadrupole collectivity [11]. This model describes a spectrum of positive parity states. In [10] the alpha-like correlation is treated in terms of bosons of nucleon pairs, but in addition to the s and d bosons another set of basic building blocks of $l=0$ (s^*) boson and $l=1$ (p) boson is included, therefore, negative parity states are also involved. These phenomenological models have the efficiency and elegance of the algebraic methods in generating the spectrum. E.g. they have dynamical symmetries as limiting cases, which provide us with exact solutions for the eigenvalue problem.

In [12] a BCS-like study was carried out for bosons of the proton-neutron interacting boson model [11] and it was concluded that the superfluid condensate is more of a quartet type, rather than separate superfluid phases of proton and neutron pairs.

The new microscopic quartet model of Sandulescu et al. [13, 14], has also concentrated on the study of quartet condensates, and it turned out to be a useful description of the nucleon pairing of isovector as well as isoscalar type (and quarteting). More recently it has been applied for the calculation of excitation spectrum [15], too. This aspect will be further discussed in the concluding section, in comparison with the algebraic quartet models.

3. The algebraic quartet models

The algebraic quartet models are based on the concepts of shell-model-like quarteting of [5, 7, 8]. Their main purpose is the description of the excitation spectrum. The simpler model, called phenomenologic algebraic quartet model (PAQM), has the building blocks very similar to that of the quartet shell model of [7], i.e. the composite nature of the quartet does not appear explicitly. The more involved approach is the semimicroscopic algebraic quartet model (SAQM), which is based on the quartet concept of [8], i.e. each of the four nucleons of the quartet is treated. The novel feature in comparison with the works [5, 7, 8] is that an algebraic framework is formulated for the description of the detailed spectrum, like in the group theoretical approach of the works [9, 10, 11]. On the other hand, the new models are different from the interacting boson type models of [9, 10], because of the nature of their building blocks, and shell-like structure of the model spaces.

Elliott's SU(3) scheme is applied [16, 17] for generating the spectrum both in the phenomenological and in the semimicroscopical descriptions. In the former case structureless quartets are supposed to occupy the single-particle levels of the harmonic oscillator shells, while in the latter model nucleons do so. Therefore, the phenomenological model space has only a spatial part, while the semimicroscopical one contains a space and a spin-isospin components.

In fact, this latter model space is a truncation of that of the $L-S$ coupled no-core shell model [18], based on the spin-isospin formalism.

The physical operators are expressed in terms of the group generators, thus algebraic techniques can be applied in calculating the matrix elements. The building blocks of the description are the nine operators, $\hat{A}_{\alpha\beta}=\frac{1}{2}(\hat{a}^\dagger_\alpha\hat{a}_\beta+\hat{a}_\beta\hat{a}^\dagger_\alpha)$, $\alpha,\beta=x,y,z$, $\hat{a}_\alpha=\sum_j\hat{a}_\alpha(j)$, $\hat{a}^\dagger_\alpha=\sum_j\hat{a}^\dagger_\alpha(j)$, $j=1,...,N$; (here N is the total number of particles), which are number-conserving bilinear products of the creation and annihilation operators of oscillator quanta. They can be rewritten into three spherical tensors: a scalar operator $\hat{n}$, which is the number of oscillator quanta, five components of the quadrupole momentum $\hat{Q}_m$ (acting in a single major shell), and three components of the angular momentum $\hat{L}_m$. The nine operators $\hat{n},\hat{Q}_m,\hat{L}_m$ generate the U(3) group, the eight operators $\hat{Q}_m,\hat{L}_m$ generate the SU(3) group, and the three $\hat{L}_m$ are generators of the SO(3) group.

The basis states are characterized by the representation labels of the group chain:

$$\begin{array}{l} U(3)\supset SU(3)\supset SO(3)\supset SO(2) \\ |[n_1,n_2,n_3],(\lambda,\mu)\ ,K,L\quad ,\ M\ \rangle. \end{array} \tag{1}$$

Here $n=n_1+n_2+n_3$ is the eigenvalue of the $\hat{n}$ operator. The angular momentum content of a (λ,μ) representation is as follows [16, 17]: $L=K,K+1,...,K+max(\lambda,\mu)$, $K=min(\lambda,\mu),min(\lambda,\mu)-2,...,1$ *or* 0, with the exception of $K_L=0$, for which $L=max(\lambda,\mu),max(\lambda,\mu)-2,...,1$ *or* 0. The SU(3) content is given by the U(k) $\supset$ SU(3) decomposition [19], where $k=3,6,10,...$ for the major shell with $1,2,3,...$ quartet excitations. The irreducible representation (irrep) of U(k) is the same as that of the permutational group in the major shell in question. The U(3) symmetry of the whole nucleus is obtained as a direct product of the major shell U(3) irreps. The irreps of the spurious center of mass excitations can be determined easily, due to the fact that the c.m. excitation operator is fully symmetric in particle indices, and has an $\{1,0,0\}$ U(3) irreducible tensor character [20, 21, 22].

The operators of physical quantities are obtained in this description in terms of the generators of the U(3) group. In particular the Hamiltonian can be expanded in terms of the generators of the U(3) group, coupled to spherical scalars. The general solution of the eigenvalue problem then involves two steps: *i)* calculation of matrix elements of the Hamiltonian between the basis states, and *ii)* numerical diagonalization of the energy matrix. In the special case of the dynamical symmetry, i.e. when the Hamiltonian is expressed in terms of the invariant operators of the group chain (1), an analytical solution is available.

The electromagnetic transition operators are obtained as Hermitian combinations of group generators with appropriate tensorial character. The lowest order transition operators are:

$$\hat{T}^{(E0)}=e^{(0)}\hat{n},\ \hat{T}^{(E2)}_m=e^{(2)}\hat{Q}^{(2)}_m,\ \hat{T}^{(M1)}_m=m^{(1)}\hat{L}^{(1)}_m. \tag{2}$$

The phenomenologic and the semimicroscopic algebraic quartet models share the same formalism of the physical operators, their difference shows up between the two model spaces.

3.1. The phenomenologic algebraic quartet model

In this approach an excitation quantum $(\hbar\omega)_q$ between the major shells is expected to be approximately 4 times that of the nucleon shell model: $(\hbar\omega)_q\approx 4(\hbar\omega)$. All the shells have positive parity, due to their quartet nature. If a single quartet state is occupied, then no other particle can be put there, therefore, the permutational symmetry of the quartets has to be that of a single-columned Young diagram: {1,1,...}.

3.2. The semimicroscopic algebraic quartet model

On the semimicroscopic level one takes into account the composition of the quartets explicitly. They are considered [8] as 2 protons and 2 neutrons having permutational symmetry of {4}, and spin-isospin symmetry of {1,1,1,1}. Therefore, in this case the nucleon shell model space is applied, and it is truncated according to these symmetries. Subsequent major shells have opposite parities.

The spectrum is determined by the U(3) spatial and the $U^{ST}(4)$ spin-isospin irreps. In this case, however, the particles are nucleons, not structureless quartets, as in the phenomenologic model. Therefore, the groups describe the symmetries of the many-nucleon systems. Their relevant irreps are obtained in the following way. In each major shell those U(k) (or permutational) symmetries have to be taken into account which result in the required quartet symmetry $\{4\}^{N_q}$ when calculating their outer product with those of the other major shells. (Equivalently, those $U^{ST}(4)$ irreps are relevant, which result in the $\{1,1,1,1\}^{N_q}$ quartet symmetry in the direct products.) These U(k) (or $U^{ST}(4)$) symmetries determine the relevant U(3) representations, and their direct products define the model space.

The model space is much richer than that of the phenomenological model. Especially remarkable is the angular momentum content of the model space; e.g. already in the lowest-lying major shell of the ^{20}Ne L=8 appears, contrary to the L≤2 of the PAQM.

4. Applications

The PAQM states correspond to very highly excited states (due to the large excitation quantum of the quartets: $(\hbar\omega)_q \approx 4(\hbar\omega)$). Therefore, it is hard to find a well-established correspondence between the experimental and model states. In case of the semimicroscopic description, on the other hand, it is much more straightforward.

Here we discuss the application of the SAQM for the description of the ^{20}Ne and ^{28}Si nuclei. In both cases the band structure is known experimentally. The U(3) dynamical symmetry approach is used, i.e. the interactions are written in terms of the invariant operators of group-chain (1):

$$\hat{H} = (\hbar\omega)\hat{n} + a\hat{C}^{(2)}_{SU3} + b\hat{C}^{(3)}_{SU3} + d\frac{1}{2\theta}\hat{L}^2. \qquad (3)$$

The first term is the harmonic oscillator Hamiltonian (linear invariant of the U(3)), with a strength obtained from the systematics [23] $\hbar\omega = 45A^{-\frac{1}{3}} - 25A^{-\frac{2}{3}}$ MeV. The second-order invariant of the SU(3) ($\hat{C}^{(2)}_{SU3}$) represents the quadrupole-quadrupole interaction, while the third-order Casimir-operator ($\hat{C}^{(3)}_{SU3}$) distinguishes between the prolate and oblate shapes. θ is the moment of inertia calculated classically for the rigid shape determined by the U(3) quantum numbers (for a rotor with axial symmetry) [1], and the a, b and d parameters were fitted to the experimental data. The $B(E2)$ value is given as [1, 24]:

$$B(E2, I_i \to I_f) = \frac{2I_f+1}{2I_i+1}\alpha^2 |\langle(\lambda,\mu)KI_i,(11)2||(\lambda,\mu)KI_f\rangle|^2 C(\lambda,\mu), \qquad (4)$$

where $\langle(\lambda,\mu)KI_i,(11)2||(\lambda,\mu)KI_f\rangle$ is the SU(3) $\supset$ SO(3) Wigner coefficient [25], and α^2 is a parameter fitted to the experimental value of the $2^+_1 \to 0^+_1$ transition.

The experimental data of ^{20}Ne are taken from [26], but for the band assignment of the highly-excited alpha-cluster states also the conclusions of [27] are taken into account. The lower part of Figure 1. shows the states with definite band-assignment. All the bands with K^π values of [26] are included, except the one of the very uncertain (and somewhat contradictory) 0^+_7 band. In case of the 7^- state of the 0^- band, and the 6^+ and 8^+ states of the 0^+_6 band, which have more than one experimental candidates, the average energies are indicated. (In [26] there are only three states, which are not included here, for not having corresponding states in the model

Table 1. The parameters of the SAQM for the description of the experimental spectra. (Note that $\hbar\omega$ is calculated from the systematics [23].) The coefficients of the Hamiltonian are in MeV, while α^2 is measured in W.u.

Nucleus	$\hbar\omega$	a	b	d	α^2
^{20}Ne	13.18	-0.421	0.000909	0.635	2.31
^{28}Si	12.11	-0.133	0.000444	1.003	0.366

spectrum: a 6^+ state in the 0_2^+ band, a 9^- state in the 1^- band, and a 8^+ state in the 0_6^+ band. Each of them has uncertain band assignment.)

The experimentally identified bands are described by the lowest-lying model bands with the appropriate spin-parity content, in both nuclei.

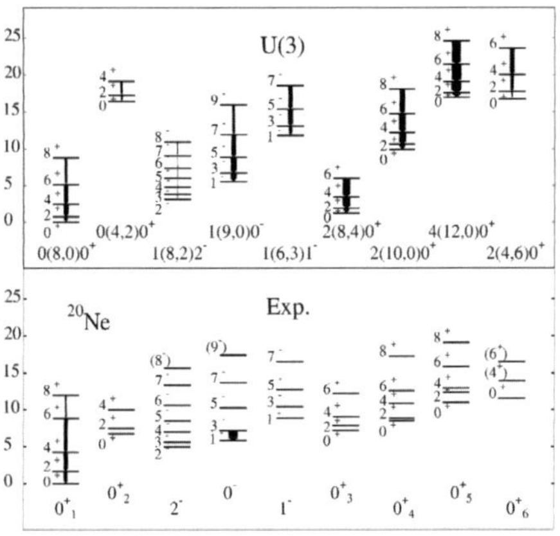

Figure 1. The spectrum of the semimicroscopic algebraic quartet model in comparison with the experimental data of the ^{20}Ne nucleus. The experimental bands are labeled by the K^π, and the model states by the $n(\lambda,\mu)K^\pi$ quantum numbers. The spin-parity in parenthesis indicates uncertain band assignment. The width of the arrow between the states is proportional to the strength of the $E2$ transition.

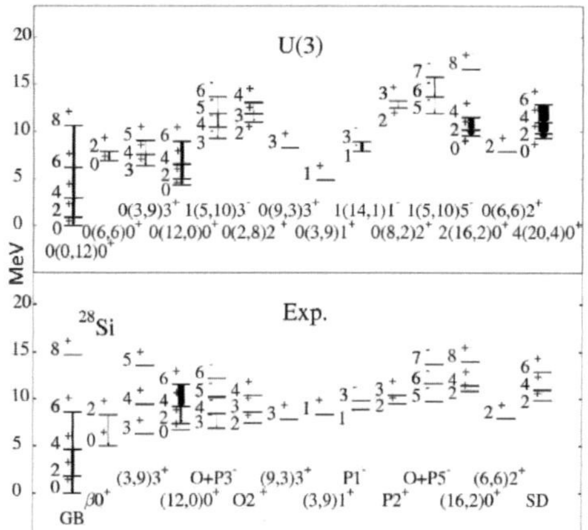

Figure 2. The spectrum of the semimicroscopic algebraic quartet model in comparison with the experimental data of the ^{28}Si nucleus. The experimental bands are labeled by the available quantum numbers, and the model states by the $n(\lambda,\mu)K^\pi$ labels. β means β-instabil, while O, P and SD stand for oblate, prolate and superdeformed respectively.

The reason for choosing the ^{28}Si nucleus as another illustrative example is manyfold. i) It has a well-established band structure in the low-energy region, and to several bands SU(3) quantum numbers could be associated as a joint conclusion of experimental and theoretical investigations [28]. ii) More recently a new candidate was proposed for the superdeformed (SD) band [29]. Theoretical studies predicted the SD band [30, 31] in line with the experimental observation. iii) There are two cluster configurations: ^{24}Mg+^{4}He, and ^{16}O+^{12}C, belonging to reaction channels in which fine-resolution measurements revealed a rich spectrum of resonances.

The lower part of Figure 2. shows the experimental bands of the ^{28}Si nucleus, as established in [28] together with the recently found superdeformed (SD) band [29]. Please, note that the

SU(3) quantum numbers are associated to the experimental bands without any reference to the quartet or cluster studies.

5. Summary and conclusions

In this contribution we have presented algebraic quartet models based on the shell-like quarteting concept both on the phenomenologic and on the semimicroscopic levels. The spectrum is generated in both cases by Elliott's U(3) formalism. The model space of the PAQM is identical with that of [7], while the SAQM model space is similar to that of [8]. The main difference between the approach of [5, 7, 8] and the present one is that in [5, 7, 8] the interaction matrix elements are obtained empirically from the binding energies, while here we construct all the physical operators algebraically. This enables us to calculate the complete spectrum in an easy way.

In [15] a recently developed microscopic quartet model is applied for the description of the excitation spectrum. In that approach the wavefunction of a quartet is determined from a shell model calculation, which serves as an input for the quartet analysis. The present algebraic quartet model does not use such an input, i.e. it is self-contained, and easily applicable, though not fully microscopic: the parameters of its interactions are fitted to the experimental data. The combination of the two methods seems to be promising for finding the exact territory of the relevance of quarteting.

As for the connection to other models, the transparent symmetry properties of the present approach is very helpful. Via its obvious shell-model relation, the connection of the quartet model to the cluster and collective models is also well-defined (see e.g. [32, 33], and references therein for a recent discussion). In this respect the models with algebraic structure are relevant, in particular, the microscopic cluster model applying U(3) basis [34, 35], and the semimicroscopic algebraic cluster model [36], as well as the symplectic shell model [37], and the contracted symplectic model [38] of the quadrupole collectivity.

The semimicroscopic algebraic quartet model can be considered as an effective model [39], which is obtained as a symmetry-governed truncation of the no-core shell model with the $U^{ST}(4)\otimes U(3)$ formalism [18].

Especially promising can be the application of the present semimicroscopic quartet model in combination with the concept of the multichannel dynamical symmetry [40], when the spectra of different cluster configurations are obtained from the quartet spectrum by simple projections. E.g. preliminary studies [41] show that the complete high-lying spectrum of the $^{16}O+^{12}C$ resonances can be obtained as a prediction from the presently determined quartet spectrum of the ^{28}Si nucleus. In particular, the resonance spectrum corresponds to the $^{16}O+^{12}C$ cluster states in the second (superdeformed) minimum of the energy surface.

Acknowledgments

This work was supported by the OTKA (Grant No. K106035) and by the international collaborations: MTA-BAS (NKM-6/2014), as well as Al-Farabi Kazakh National University (Grant No: GF3). The technical help of Mr. G. Riczu is kindly acknowledged.

References

[1] Cseh J 2015 *Phys. Lett.* B **743** 213
[2] Wigner E P1937 *Phys. Rev.* **51** 106
[3] Van Isacker P, Warner D D and Brenner D S 1995 *Phys. Rev. Lett.* **74** 4607
[4] Danos M and Gillet V 1967 *Phys. Rev.* **161** 1034
[5] Arima A, Gillet V and Ginocchio J 1970 *Phys. Rev. Lett.* **25** 1043
[6] Arima A and Gillet V 1971 *Ann Phys.* **66** 117
[7] Satpathy L, Schmid K W and Faessler A 1972 *Phys. Rev. Lett.* **28** 832
[8] Harvey M 1973 *Nucl. Phys.* A **202** 191
[9] Dukelsky J, Federman P, Perazzo R P J and Sofia H M 1982 *Phys. Lett.* B **115** 359
[10] Iachello F and Jackson A D 1982 *Phys. Lett.* B **108** 151
[11] Iachello F and Arima A 1987 The Interacting Boson Model, Cambridge University Press, Cambridge
[12] Gambhir Y K, Ring P and Schuck P 1983 *Phys. Rev. Lett.* **51** 1235
[13] Sandulescu N, Negrea D, Dukelsky J and Johson C W 2012 *Phys. Rev.* C **85** 061303(R)
[14] Sambataro M and Sandulescu N 2013 *J. Phys.* G **40** 055107
[15] Sambataro M and Sandulescu N 2015 *Phys. Rev. Lett.* **115** 112501
[16] Elliott J P 1958 *Proc. Roy. Soc.* **245** 128; 562
[17] Harvey M 1968 *Adv. Nucl. Phys.* **1** 67
[18] Dytrych T, Sviratcheva K D, Draayer J P, Bahri C and Vary J P 2008 *J. Phys.* G **35** 123101
[19] Draayer J P, Lebschber Y, Park S C and Lopez R 1989 *Comput. Phys. Commun.* **56** 279
[20] Elliott J P and Skyrme T H K 1955 *Proc. Roy. Soc.* A **232** 561
[21] Brink D M and Nash G F 1963 *Nucl. Phys.* **40** 608
[22] Hecht K T 1971 *Nucl Phys.* A **170** 34
[23] Blomqvist J and Molinari A 1968 *Nucl. Phys.* A **106** 545
[24] Sun Y et al. 2002 *Nucl. Phys.* A **703** 130
[25] Akiyama Y and Draayer J P 1973 *Comp. Phys. Comm.* **5** 405
[26] Tilley D R et al. 1998 *Nucl Phys.* A **636** 249
[27] Richards H T 1984 *Phys. Rev.* C **29** 276
[28] Sheline R K et al. 1982 *Phys. Lett.* B **119** 263
[29] Jenkins D G et al. 2012 *Phys. Rev.* C **86** 064308
[30] Taniguchi Y, Kanada-En'yo Y and Kimura M 2009 *Phys. Rev.* C **80** 044316
[31] Darai J, Cseh J and Jenkins D G 2012 *Phys. Rev.* C **86** 064309
[32] Cseh J 2015 *J. Phys. Conf. Ser.* **580** 012046
[33] Darai J and Cseh J 2015 *J. Phys. Conf. Ser.* **580** 012057
[34] Horiuchi H 1977 *Prog. Theor. Phys. Suppl.* **62** 90
[35] Hecht K T, Reske E J, Seligman T H and Zahn W 1981 *Nucl. Phys.* A **356** 146
[36] Cseh J 1992 *Phys. Lett.* B **281** 173 ; Cseh J and Lévai G 1994 *Ann. Phys.* (NY) **230** 165
[37] Rosensteel G and Rowe D J 1977 *Phys. Rev. Lett.* **38** 10; 1980 *Ann. Phys.* (N.Y.) **126** 343;
Rowe D J 1985 *Rep. Prog. Phys.* **48** 1419
[38] Rowe D J and Rosensteel G 1982 *Phys. Rev.* C **25** 3236;
Castanos O and Draayer J P 1989 *Nucl. Phys.* A **491** 349
[39] Rowe D J, Thiamowa G and Wood J L 2006 *Phys. Rev. Lett.* **97**, 202501
[40] Cseh J 1994 *Phys. Rev.* C **50** 2240; Cseh J and Katō K 2013 *Phys. Rev.* C **87** 067301
[41] Cseh J and Riczu G 2015 arXiv 1504.04480 [nucl-th]

Nuclear Spectroscopic Properties within a Selfconsistent Multiphonon Approach

G De Gregorio[1,2], F Knapp[3], N Lo Iudice[1,2], P Veselý[4], F.Andreozzi[1,2], , A Porrino[1,2]

[1] Dipartimento di Fisica, Università di Napoli Federico II, Napoli, Italy
[2] Istituto Nazionale di Fisica Nucleare (INFN), Sezione di Napoli, Napoli, Italy
[3] Faculty of Mathematics and Physics, Charles University in Prague, Czech Republic
[4] Institute of Nuclear Physics, Academy of Sciences of the Czech Republic, 250 68 Řež, Czech Republic

E-mail: degregorio@na.infn.it

Abstract. The equation of motion phonon method is adopted to investigate the dipole response in neutron rich nuclei. The calculation is carried out in a space spanned by one- and two-phonon basis states. It makes direct use of a nucleon-nucleon optimized chiral potential complemented with a density dependent term simulating a three-body force and is fully selfconsistent. The inclusion of the two-phonon states induces a pronounced fragmentation of both giant and pygmy resonances consistently with recent experiments.

1. Introduction

In recent years, an intense experimental and theoretical effort has been concentrated on the electric dipole response in neutron rich nuclei and, more specifically, on the low-energy levels detected around the neutron threshold and interpreted as a manifestation of a soft mode, dubbed pygmy dipole resonance (PDR), promoted by a translational oscillation of the neutron excess against a N=Z core [1].

A comprehensive and updated list of experimental investigations can be found in a recent review [2]. We mention here a radioactive beam experiment on the unstable tin isotopes around ^{132}Sn [3], the (γ, γ') [4–6] combined with the $(\alpha, \alpha'\gamma)$ experiments [7–9], which detected rich low lying spectra of weakly excited discrete levels below the neutron threshold in chains of stable nuclei over different mass regions.

^{208}Pb deserves a special mention. A recent proton scattering experiment on ^{208}Pb has extracted the electric dipole spectrum below and above the neutron threshold [10, 11] with great accuracy. The data so obtained, combined with the ones made available by previous (γ, γ') and (n, γ) [12–14] and photo-absorption experiments [15, 16], allow a detailed investigation of the properties of the dipole response and, in particular, of the low-lying soft mode. The nucleus, therefore, represents a testing ground for the theoretical models.

The mode was investigated in a considerable number of theoretical approaches. Many calculations were carried out in Hartree-Fock-Bogoliubov (HFB) plus quasi-particle RPA (QRPA) using Skyrme or Gogny forces, others were performed in relativistic RPA (RRPA) using density functionals derived from meson-nucleon Lagrangians treated in mean field approximation. An exhaustive list of references can be found in the reviews [2, 17, 18] .

Several extensions of (Q)RPA were adopted to study the fragmentation of the mode. We mention the QRPA plus phonon coupling [19], second RPA [20], the quasi-particle-phonon model (QPM) [6, 12, 21], and the relativistic quasi-particle time-blocking approximation (RTBA) [22–24].

The fragmentation was studied also within an equation of motion phonon method (EMPM) [25–27]. In such an approach, one constructs and solves iteratively a set of equations of motion to generate a multiphonon basis built of phonons obtained in Tamm-Dancoff approximation (TDA). Represented in such a basis the Hamiltonian matrix has a simple structure and can be diagonalized in large configuration and phonon spaces. The method holds for a Hamiltonian of general form, treats one-phonon and multiphonon states on the same footing and fulfills the Pauli principle.

It was first implemented numerically to study the dipole response using a modified oscillator single particle basis and a V_{lowk} potential derived from the CD-Bonn NN interaction [28]. We report here on calculations carried out for the doubly magic nuclei ^{132}Sn and ^{208}Pb in a space spanned by one-plus two-phonon basis states [29, 30].

The calculations are fully self-consistent and make use of a nucleon-nucleon (NN) chiral potential $V_\chi = NNLO_{opt}$ optimized so as to minimize the effects of the three-body forces [31]. A review of the derivation and properties of chiral potentials can be found in Ref. [32]. As we shall see, it was necessary to add to V_χ a phenomenological, density dependent, potential V_ρ, derived from a contact three-body interaction [33], in order to obtain a more realistic Hartree-Fock spectrum. This corrective term was already introduced for other realistic potentials [34, 35].

2. Outline of the EMPM method

The main goal of the EMPM method consists in generating a basis of n-phonon states $|n;\beta>$, of energies E_β, having the form

$$|n;\beta>=\sum_{\lambda\alpha} C^{\beta}_{\lambda\alpha}\left\{O^{\dagger}_{\lambda}\times \mid n-1,\alpha>\right\}^{\beta}, \tag{1}$$

where

$$O^{\dagger}_{\lambda}=\sum_{ph} c^{\lambda}_{ph}(a^{\dagger}_{p}\times b_{h})^{\lambda} \tag{2}$$

is the TDA particle-hole (p-h) phonon operator of energy E_λ acting on the $(n-1)$-phonon states $\mid n-1,\alpha>$ of energies E_α.

The procedure leading to such a basis starts with constructing the equations of motion

$$<n,\beta \parallel [H,O^{\dagger}_{\lambda}] \parallel n-1,\alpha>=\Big(E_\beta-E_\alpha\Big)<n,\beta\parallel O^{\dagger}_{\lambda}\parallel n-1,\alpha>. \tag{3}$$

After expanding the commutator and expressing the p-h operators, present in the expanded commutator, in terms of the phonon operators $O^{\dagger}_{\lambda}$ upon inversion of Eq. (2), one obtains [27] the generalized eigenvalue equation

$$\sum_{\lambda'\alpha'\lambda_1\alpha_1}\Big[\mathcal{A}^{\beta}(\lambda\alpha,\lambda_1\alpha_1)-E_\beta\delta_{\lambda_1\lambda}\delta_{\alpha_1\alpha}\Big]\mathcal{D}^{\beta}(\lambda_1\alpha_1,\lambda'\alpha')C^{\beta}_{\lambda'\alpha'}=0. \tag{4}$$

Here

$$\mathcal{D}^{\beta}(\lambda_1\alpha_1;\lambda'\alpha')=\Big[<n-1,\alpha_1\mid\times O_{\lambda_1}\Big]_{\beta}\Big[O^{\dagger}_{\lambda'}\times\mid n-1,\alpha'>\Big]_{\beta} \tag{5}$$

is the metric matrix and $\mathcal{A}^\beta$ is a matrix of the simple structure

$$\mathcal{A}^\beta(\lambda\alpha,\lambda'\gamma) = (E_\lambda + E_\alpha)\delta_{\lambda\lambda'}\delta_{\alpha\gamma} + \sum_\sigma W(\beta\lambda'\alpha\sigma;\gamma\lambda)\mathcal{V}^\sigma_{\lambda\alpha,\lambda'\gamma}, \tag{6}$$

where W is a Racah coefficient and $\mathcal{V}^\sigma$ a phonon-phonon potential given by

$$\mathcal{V}^\sigma_{\lambda\alpha,\lambda'\gamma} = \sum_{rstq} \rho_{\lambda\lambda'}([q\times t]^\sigma)F^\sigma_{qtrs}\rho^{(n)}_{\alpha\gamma}([r\times s]^\sigma). \tag{7}$$

We have denoted by $\rho_{\lambda\lambda'}$ and $\rho^{(n)}_{\alpha\alpha'}$, respectively, the $n = 1$ (TDA) and the n-phonon ($n > 1$) density matrices

$$\rho_{\lambda\lambda'}([r\times s]^\sigma) = \langle\lambda' \parallel \left[a^\dagger_r \times b_s\right]^\sigma \parallel \lambda\rangle, \tag{8}$$

$$\rho^{(n)}_{\alpha\alpha'}([r\times s]^\sigma) = \langle n;\alpha' \parallel \left[a^\dagger_r \times b_s\right]^\sigma \parallel n;\alpha\rangle. \tag{9}$$

The formal analogy between the structure of the phonon matrix $\mathcal{A}^\beta(\lambda\alpha,\lambda'\alpha')$ and the form of the TDA matrix $A^\lambda(ph;p'h')$ was pointed out in Ref. [27]. The first is deduced from the second by replacing the TDA p-h energies with the sum of phonon energies $(E_\lambda + E_\alpha)$ and the TDA p-h interaction with the phonon-phonon interaction $\mathcal{V}^\sigma$.

Eq. (4) represents the eigenvalue equation in the overcomplete basis $\left\{O^\dagger_\lambda \times \mid n-1,\alpha >\right\}^\beta$. The redundant states are eliminated by the procedure outlined in [25, 26], based on the Cholesky decomposition method.

Since recursive formulas hold for all quantities entering $\mathcal{A}$ and $\mathcal{D}$, the eigenvalue equations are solved iteratively starting from the TDA phonons and, thereby, yield a set of orthonormal multiphonon states $\{|0>, |1,\lambda>, \ldots |n,\alpha> \ldots\}$.

Represented in such a basis, the Hamiltonian matrix is composed of a sequence of diagonal blocks, one for each n, mutually coupled by off-diagonal terms $\langle n' \mid H \mid n\rangle$ which are non vanishing only for $n' = n\pm1, n\pm2$ and are computed by means of recursive formulas. A matrix of such a simple structure can be easily diagonalized yielding eigenfunctions of the form

$$\mid \Psi_\nu\rangle = \sum_{n\alpha} C^{(\nu)}_\alpha \mid n;\alpha\rangle. \tag{10}$$

3. Calculations and results

The Hamiltonian we used is composed of an intrinsic kinetic operator T_{int} plus the two-body potential

$$V = V_\chi + V_\rho. \tag{11}$$

As pointed out in the introduction, $V_\chi = NNLO_{opt}$ is the NN optimized chiral potential [31] and V_ρ is a corrective repulsive, density dependent, two-body potential simulating a three-body contact force [33] which improves the description of bulk properties in closed shell nuclei [36].

We generate a HF basis in a configuration space including up to the $N_{max} = 12$ harmonic oscillator major shells, sufficient for reaching a good convergence. The benefits coming from adding the density dependent potential can be noticed already within the HF context. In fact, the inclusion of V_ρ enhances the diffuseness of the charge density, which thereby approaches the empirical charge distribution, and yields a neutron skin radius compatible with the one deduced from the experimental data.

Moreover, as illustrated in Figure 1, it induces a compression of the single particle spectrum [29] thereby filling partially the gap with the empirical levels [37]. Such a compression improves also the (Q)TDA and (Q)RPA descriptions of the dipole response [29].

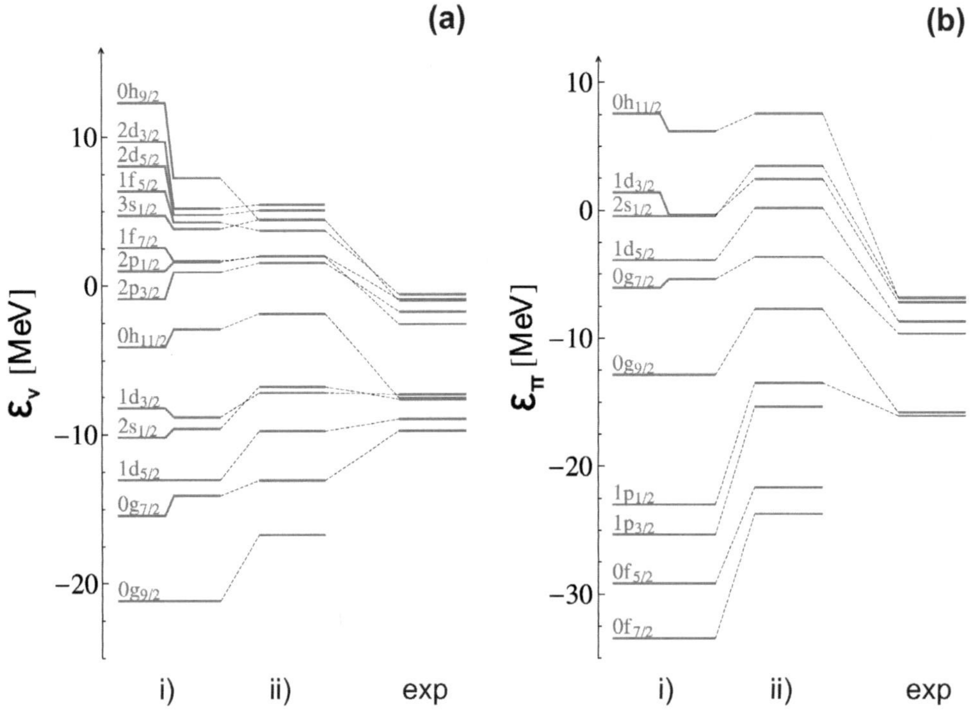

Figure 1. Neutron and proton single particle spectra in ^{132}Sn with i) $V = V_\chi$ and ii) $V = V_\chi + V_\rho$ (from [29]). The empirical (exp) single particle levels are taken from [37]

Using the HF basis, we first create the TDA one phonon states $\mid \lambda\rangle$ and, then, adopt the EMPM procedure outlined in Sect. 2 to produce an orthonormal correlated two-phonon basis $\mid \alpha\rangle$. The one- + two-phonon basis is used to diagonalized the residual Hamiltonian and yield the final eigenvalues and eigenvectors.

The inclusion of the two-phonon states has a strong damping effect on the response. As shown in Figure 2, the cross section is severely quenched and reshaped by the one- to two-phonon coupling in both ^{132}Sn and ^{208}Pb. It has a smoother behavior and follows more closely the experimental points.

In ^{132}Sn, the calculation yields a small peak in the cross section around $\sim$ 10 MeV, fairly close in position and height to the one at $\sim$ 9.8 MeV observed experimentally [3]. The EMPM strength collected by the low-lying states up to $\sim$ 11 MeV exhausts $\sim$ 5.8% of the TRK sum rule. This value is within the error of the measured fraction 4(3)% [3]. A similar peak is produce around $\sim$ 7 MeV also in the cross section of ^{208}Pb and overlaps with the measured peak.

The phonon coupling enhances greatly the fragmentation of the strength [29]. As compared with TDA, the EMPM spectrum is much more dense and is composed of peaks of considerably shorter height in both GDR and PDR regions.

Figure 3 shows that also in the low energy sector the level density is greatly enhanced by the phonon coupling. It is even higher than the density of the observed levels. The calculation [30], in fact, yields about 20 1^- peaks between $\sim$ 4.5 MeV and $\sim$ 7 MeV, twice as much as the ones detected experimentally [10, 11].

On the other hand, the experimental strength distribution is reproduced only qualitatively. Figure 4 shows that significant discrepancies exist. Some EMPM transitions result to be too strong compared to the measured ones, while the large majority of them are considerably weaker.

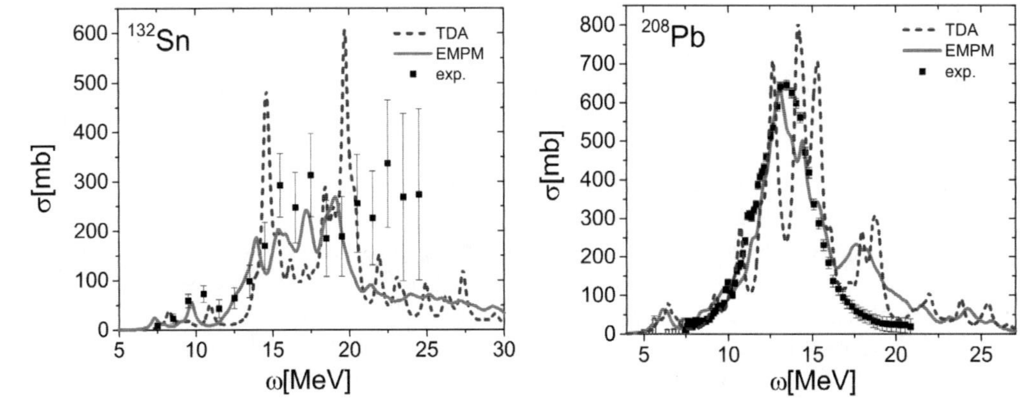

Figure 2. EMPM and TDA versus experimental cross sections in ^{132}Sn [29] and ^{208}Pb [30]. A Lorentzian width $\Delta = 0.5$ MeV was used to compute the cross sections. The experimental data are taken from [3] for ^{132}Sn and from [10, 11] for ^{208}Pb.

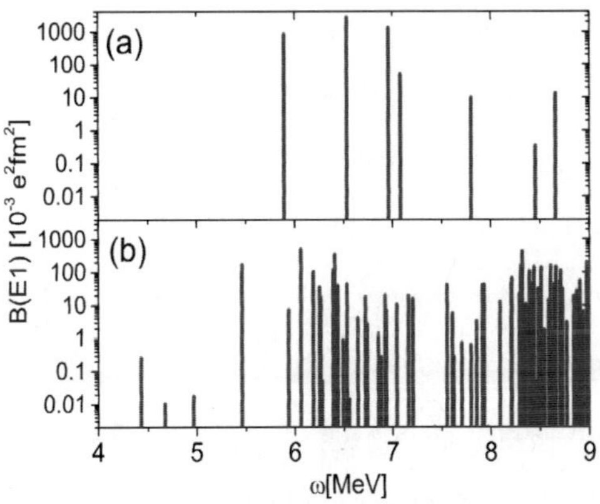

Figure 3. TDA (a) versus EMPM (b) $E1$ low-lying spectra in ^{208}Pb [30].

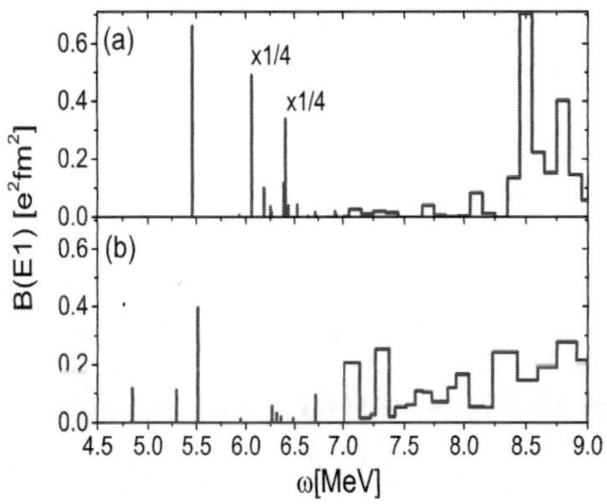

Figure 4. EMPM versus experimental $E1$ low-lying spectra in ^{208}Pb [30]. The experimental data are taken from [11].

In any case, one can infer from a comparison with Figure 6 of Ref. [11] that our computed spectrum is comparable with the ones computed in QPM and RTBA and, especially, with the spectrum obtained in shell model by using empirical single particle energies and a phenomenological potential in a restricted configuration space [14].

The analysis of the structure of the wavefunctions shows that the low-lying excitations are mainly promoted by the valence neutrons hinting at their pygmy nature, a suggestion supported by the transition densities and the excitation mechanism [29].

In fact, the theoretical low-lying levels are excited by both isoscalar and isovector probes [29, 30], in analogy with the experimental spectrum produced by (γ, γ') and $(\alpha, \alpha'\gamma)$ experiments in the open shell ^{124}Sn [8, 9, 38]. On the other hand, our ^{132}Sn and ^{208}Pb do not exhibit any splitting between isoscalar and isovector dipole modes at low-energy, at variance with the conclusions drawn from the analysis of the observed spectra in ^{124}Sn [7].

4. Conclusions

According to our EMPM study, the two-phonon states exert a strong damping and fragmentation of the TDA peaks in the GDR region thereby yielding a smooth cross section consistently with experiments. The same two-phonon states greatly enhance the density of low-lying levels more in qualitative agreement with recent experiments.

Serious discrepancies between theoretical and experimental spectra remain. They may be cured by trying to improve the HF description of the single particle spectra, which amounts to improving the nucleon-nucleon potential. The optimized chiral potential $V_\chi = NNLO_{opt}$[31] we used, though representing a promising starting point, is not sufficient. In fact, we had to add a phenomenological density dependent potential which simulates a repulsive three-body contact force. Such a potential however, is phenomenological and contains an unconstrained coupling constant. Moreover, it fills only partially the gap between HF and empirical single particle energies.

It is also desirable to try to enlarge the space by including the three-phonon states which are known to couple strongly to the one-phonon subspace [27]. Such a coupling is supposed to induce a further redistribution of the dipole strength. In this perspective, we are trying to improve the efficiency of the codes and to achieve reliable truncations of the phonon space.

Acknowledgments

This work was partly supported by the Czech Science Foundation (project no. P203-13-07117S).

References

[1] Mohan R, Danos M and Biedenharn L C 1971 *Phys. Rev. C* **3**(5) 1740–1749

[2] Savran D, Aumann T and Zilges A 2013 *Progress in Particle and Nuclear Physics* **70** 210 – 245

[3] Adrich P, Klimkiewicz A, Fallot M, Boretzky K, Aumann T, Cortina-Gil D, Datta Pramanik U, Elze T W, Emling H, Geissel H, Hellström M, Jones K L, Kratz J V, Kulessa R, Leifels Y, Nociforo C, Palit R, Simon H, Surówka G, Sümmerer K and Waluś W (LAND-FRS Collaboration) 2005 *Phys. Rev. Lett.* **95**(13) 132501

[4] Hartmann T, Babilon M, Kamerdzhiev S, Litvinova E, Savran D, Volz S and Zilges A 2004 *Phys. Rev. Lett.* **93**(19) 192501

[5] Isaak J, Savran D, Fritzsche M, Galaviz D, Hartmann T, Kamerdzhiev S, Kelley J H, Kwan E, Pietralla N, Romig C, Rusev G, Sonnabend K, Tonchev A P, Tornow W and Zilges A 2011 *Phys. Rev. C* **83**(3) 034304

[6] Savran D, Elvers M, Endres J, Fritzsche M, Löher B, Pietralla N, Yu Ponomarev V, Romig C, Schnorrenberger L, Sonnabend K and Zilges A 2011 *Phys. Rev. C* **84**(2) 024326

[7] Savran D, Babilon M, van den Berg A M, Harakeh M N, Hasper J, Matic A, Wörtche H J and Zilges A 2006 *Phys. Rev. Lett.* **97**(17) 172502

[8] Endres J, Litvinova E, Savran D, Butler P A, Harakeh M N, Harissopulos S, Herzberg R D, Krücken R, Lagoyannis A, Pietralla N, Ponomarev V Y, Popescu L, Ring P, Scheck M, Sonnabend K, Stoica V I, Wörtche H J and Zilges A 2010 *Phys. Rev. Lett.* **105**(21) 212503

[9] Endres J, Savran D, Butler P A, Harakeh M N, Harissopulos S, Herzberg R D, Krücken R, Lagoyannis A, Litvinova E, Pietralla N, Ponomarev V, Popescu L, Ring P, Scheck M, Schlüter F, Sonnabend K, Stoica V I, Wörtche H J and Zilges A 2012 *Phys. Rev. C* **85**(6) 064331

[10] Tamii A, Poltoratska I, von Neumann-Cosel P, Fujita Y, Adachi T, Bertulani C A, Carter J, Dozono M, Fujita H, Fujita K, Hatanaka K, Ishikawa D, Itoh M, Kawabata T, Kalmykov Y, Krumbholz A M, Litvinova E, Matsubara H, Nakanishi K, Neveling R, Okamura H, Ong H J, Özel-Tashenov B, Ponomarev V Y, Richter A, Rubio B, Sakaguchi H, Sakemi Y, Sasamoto Y, Shimbara Y, Shimizu Y, Smit F D, Suzuki T, Tameshige Y, Wambach J, Yamada R, Yosoi M and Zenihiro J 2011 *Phys. Rev. Lett.* **107**(6) 062502

[11] Poltoratska I, von Neumann-Cosel P, Tamii A, Adachi T, Bertulani C A, Carter J, Dozono M, Fujita H, Fujita K, Fujita Y, Hatanaka K, Itoh M, Kawabata T, Kalmykov Y, Krumbholz A M, Litvinova E, Matsubara H, Nakanishi K, Neveling R, Okamura H, Ong H J, Özel-Tashenov B, Ponomarev V Y, Richter A, Rubio B, Sakaguchi H, Sakemi Y, Sasamoto Y, Shimbara Y, Shimizu Y, Smit F D, Suzuki T, Tameshige Y, Wambach J, Yosoi M and Zenihiro J 2012 *Phys. Rev. C* **85**(4) 041304

[12] Ryezayeva N, Hartmann T, Kalmykov Y, Lenske H, von Neumann-Cosel P, Yu Ponomarev V, Richter A, Shevchenko A, Volz S and Wambach J 2002 *Phys. Rev. Lett.* **89**(27) 272502

[13] Enders J, von Brentano P, Eberth J, Fitzler A, Fransen C, Herzberg R D, Kaiser H, Käubler L, vonNeumann Cosel P, Pietralla N, Ponomarev V, Richter A, Schwengner R and Wiedenhöver I 2003 *Nucl. Phys. A* **724** 243 – 273
[14] Schwengner R, Massarczyk R, Brown B A, Beyer R, Dönau F, Erhard M, Grosse E, Junghans A R, Kosev K, Nair C, Rusev G, Schilling K D and Wagner A 2010 *Phys. Rev. C* **81**(5) 054315
[15] Veyssiere A, Beil H, Bergere R, Carlos P and Lepretre A 1970 *Nucl. Phys. A* **159** 561 – 576
[16] Schelhaas K, Henneberg J, Sanzone-Arenhövel M, Wieloch-Laufenberg N, Zurmühl U, Ziegler B, Schumacher M and Wolf F 1988 *Nucl. Phys. A* **489** 189 – 224
[17] Paar N, Vretenar D, Khan E and Colò G 2007 *Report on Progress in Physics* **70** 691
[18] Paar N 2010 *J. Phys. G: Nucl. Part. Phys.* **37** 064014
[19] Sarchi D, Bortignon P and Colò G 2004 *Phys. Lett. B* **601** 27 – 33
[20] Gambacurta D, Grasso M and Catara F 2011 *Phys. Rev. C* **84**(3) 034301
[21] Tsoneva N and Lenske H 2008 *Phys. Rev. C* **77**(2) 024321
[22] Litvinova E, Ring P and Vretenar D 2007 *Phys. Lett. B* **647** 111 – 117
[23] Litvinova E, Ring P and Tselyaev V 2008 *Phys. Rev. C* **78**(1) 014312
[24] Litvinova E, Ring P and Tselyaev V 2010 *Phys. Rev. Lett.* **105**(2) 022502
[25] Andreozzi F, Knapp F Lo Iudice N, Porrino A and Kvasil J 2007 *Phys. Rev. C* **75**(4) 044312
[26] Andreozzi F, Knapp F, Lo Iudice N, Porrino A and Kvasil J 2008 *Phys. Rev. C* **78**(5) 054308
[27] Bianco D, Knapp F, Lo Iudice N, Andreozzi F and Porrino A 2012 *Phys. Rev. C* **85**(1) 014313
[28] Bianco D, Knapp F, Lo Iudice N, Andreozzi F, Porrino A and Veselý P 2012 *Phys. Rev. C* **86**(4) 044327
[29] Knapp F, Lo Iudice N, Veselý P, Andreozzi F, De Gregorio G and Porrino A 2014 *Phys. Rev. C* **90** 014310
[30] Knapp F, Lo Iudice N, Veselý P, Andreozzi F, De Gregorio G and Porrino A 2015 *Phys. Rev. C* **in press**
[31] Ekström A, Baardsen G, Forssén C, Hagen G, Hjorth-Jensen M, Jansen G R, Machleidt R, Nazarewicz W, Papenbrock T, Sarich J and Wild S M 2013 *Phys. Rev. Lett.* **110**(19) 192502
[32] Machleidt R and Entem D 2011 *Physics Reports* **503** 1 – 75
[33] Waroquier M, Heyde K and Vincx H 1976 *Phys. Rev. C* **13**(4) 1664–1673
[34] Hergert H, Papakonstantinou P and Roth R 2011 *Phys. Rev. C* **83**(6) 064317
[35] Bianco D, Knapp F, Lo Iudice N, Veselý P, Andreozzi F, De Gregorio G and Porrino A 2014 *J. Phys. G: Nucl. Part. Phys.* **41** 025109
[36] Günther A, Roth R, Hergert H and Reinhardt S 2010 *Phys. Rev. C* **82**(2) 024319
[37] Isakov V, Erokhina K, Mach H, Sanchez-Vega M and Fogelberg B 2002 *The European Physical Journal A - Hadrons and Nuclei* **14** 29–36
[38] Govaert K, Bauwens F, Bryssinck J, De Frenne D, Jacobs E, Mondelaers W, Govor L and Yu Ponomarev V 1998 *Phys. Rev. C* **57**(5) 2229–2249

Valence space techniques and QRPA vibrational mass parameters

I. Deloncle[1,2] **and F. Lechaftois** [2] **and S. Péru** [2]
[1]CSNSM, IN2P3/CNRS, 91405 Orsay Campus, FRANCE
[2]CEA, DAM, DIF 91680 Arpajon, FRANCE
E-mail: isabelle.deloncle@csnsm.in2p3.fr

Abstract. The vibrational mass parameters entering the quadrupolar 5DCH Hamiltonian are commonly obtained neglecting beyond mean-field correlations and the dynamical mean-field rearrangement. The Quasiparticle Random Phase Approximation (QRPA) framework would allow to avoid these disadvantages [1], if the computation time, when using density dependent force, was not prohibitive. Here, a significant time reduction is obtained by applying valence space (VS) techniques (energy cut-off and inert core) in QRPA calculations. The VS techniques allow to probe the physical content of the mass parameter. The QRPA mass parameter exhibit robustness toward VS limitations contrarily to the intrinsic QRPA outputs, that show deceptive appearance when an inert core is used. Excited states energy, and associated transition probabilities, should not be considered for optimizing the valence space limits.

1. Introduction

In the 5-Dimensional Collective Hamiltonian (5DCH) theory the Bohr Hamiltonian is built microscopically, without any free parameter. Starting from only one ingredient – a density-dependent force – the low-energy quadrupolar dynamics, involving five collective degrees of freedom, is obtained from self-consistent mean-field solutions. The reduction of the computation time, relatively to the GCM process, obtained thanks to the GOA approximation makes 5DCH a possible cornerstone for the construction of an "universal" approach describing, on the same foot and from only one nucleon-nucleon interaction, nuclei along the whole chart. Indeed, 5DCH calculations can be undertaken in light nuclei –as soon as the concept of mean-field is meaningful (for N, Z $\gtrsim$ 10)– as well as in the heaviest ones for which the computation time is manageable. It is worth recalling that this approach is not limited to axial symmetry, and implies a β,γ mapping. Yet, several drawbacks of 5DCH are known [1, 2], giving rise to a dependence of the result reliability on the nuclear deformation. Possible cure of 5DCH could be obtained by adding beyond mean-field correlations to the 5DCH vibrational mass parameters [1, 2, 3, 4]. In principle that could be achieved in Quasiparticle Random Phase Approximation (QRPA), but not in practice due to the prohibitive QRPA time consumption, in particular when using a density-dependent force. On another hand, the strong variations with the deformation of the vibrational mass parameters [5, 6] impose to compute them for a large amount of β,γ values. The time related cost is thus a key issue in the development of 5DCH cures, and our recent work [16], in that respect, represents an important breakthrough. Indeed, for the first time a vibrational mass parameter is obtained, in a reasonable computational time, within a HFB+QRPA approach built on a Gogny interaction (D1M). In our work, the computing time reduction is reached by setting limitations on the valence space available for the quasi-particle (qp) excitations entering the QRPA calculations. The technique of the cut-off in the 2qp energy, that limits the upper single level involved in the excitation, has already been validated for phonon energy, with Skyrme [7] as well as with D1S [8]. Another way of limiting the valence space is used in standard shell model: the introduction of an inert core that put aside the lowest single-particle levels. We apply the cut-off technique and, for the first time, the inert core one, to the determination of mass parameters. Both the optimal 2qp energy cut-off value and the inert core size are determined according to the validity of the resulting mass parameters. With the same limitations the inaccuracy of calculated built-in QRPA outputs, namely phonon energies and reduced transition probabilities establish the robustness of mass parameters.

2. Formalism for vibrational mass parameters

2.1. From TDHFB to Inglis-Belyaev

In 5DCH calculations, the potential energy V, the three moments of inertia $\mathfrak{J}_i$ (with i∈ (x,y,z)) and the three vibrational mass parameters $B_{\mu\nu}$ (with$(\mu,\nu) \in$ (0,2)) of the Bohr Hamiltonian

$$\mathcal{H} = V + \tfrac{1}{2}\sum_i \mathfrak{J}_i\omega_i^2 + \tfrac{1}{2}(B_{00}\dot{q_0}^2 + B_{22}\dot{q_2}^2 + B_{02}\dot{q_0}\dot{q_2}) \tag{1}$$

are determined microscopically at few tens of points of the sextant (β, $0° \leq \gamma \leq 60$ °) from statical constrained Hartree-Fock-Bogolyubov (CHFB) calculations. Usually the vibrational mass parameters are computed at the cranking order using the Inglis-Belyaev formula. This formula is established from time dependent Hartree-Fock (TDHFB) framework when assuming adiabaticity of the collective motion (ATDHFB) –that allows a perturbative development of the generalised density, stopped at the first (cranking) order– and neglecting the modification, time-odd, of the mean-field brought by the perturbated density. The dynamics is "restored" at the 5DCH diagonalisation stage. Indeed, in the Gaussian Overlap Approximation (GOA) of the Generalized Coordinate Method, the eigenstates of the 5DCH Hamiltonian are sought as superposition of CHFB states from different deformation. Different excited single-particle level configurations (or quasiparticle occupation (u,v)) are thus explored with deformation. At the deformation where a vibrational mass parameter B presents a peak, the $\hbar^2/2B$ term of the requantized 5DCH Hamiltonian is very small. In this way, configurations with high potential energy from that deformation can mix with configurations from near-CHFB minimum with similar requantized "mechanical" energy. Deformed low-energy 5DCH excited states can be obtained despite a spherical HFB minimum as in Tin isotopes [9, 4] or in ^{32}Mg [10] for example. The mass parameters being determinative for the dynamics, their computation should include as much physics as possible, such as dynamical mean-field rearrangement or beyond the mean-field correlations. The mean-field being a statical and particle independent (mostly uncorrelated) view of the nucleus, it can only partly include (in averaging) the correlations brought by the interaction. The Quasiparticle Random Phase Approximation (QRPA) formalism takes into account 2qp correlations excluded from HFB formalism (residual interaction). It is a possible path toward enriched vibrational mass parameter.

2.2. From TDHFB to QRPA

QRPA is another limit (small amplitude one) of TDHFB allowing to study the harmonic answer of a nucleus subjected, at its minimum of potential energy, to a small impulsion. The answer is sought as coherent (or correlated) superpositions of 2qp excitations, the phonons defined by:

$$\theta_n^+ = \sum_{kk'}(X_{kk'}^n \alpha_k^+ \alpha_{k'}^+ - Y_{kk'}^n \alpha_k \alpha_{k'}), \tag{2}$$

$$\theta_n^+ \,|\mathrm{QRPA}\rangle = |\mathrm{n}\rangle \tag{3}$$

In the creation of a phonon, 2qp are deleted and 2qp are created together, with the X and Y amplitudes respectively, which are solutions of the equation:

$$\begin{pmatrix} A & B \\ B^* & A^* \end{pmatrix} \begin{pmatrix} X_n \\ Y_n \end{pmatrix} = \omega_n \begin{pmatrix} X_n \\ -Y_n \end{pmatrix} \tag{4}$$

In Equation 4 the matrix elements of A and B are calculated from the residual interaction. Nearly forty years ago, Vautherin [11] established the equivalence of the ATHFB mass parameter (related to the nuclear polarizability) and the cubic inverse energy weighted RPA sum rule :

$$B_{\mu\nu} = \frac{\hbar^2}{2} \frac{\mathcal{M}_{-3}(Q_{\mu\nu})}{\mathcal{M}_{-1}(Q_{\mu\nu})}, \quad \mu, \nu \in (0,2) \tag{5}$$

The kth order moment of the $Q_{\mu\nu}$ strength distribution is given by:

$$\mathcal{M}_{\mathrm{k}}(Q_{\mu\nu}) = \sum_n \omega^{\mathrm{k}} |\langle \phi \,|\theta_n^+ Q_{2\mu}\,|\,\phi\rangle\langle\phi\,|\theta_n^+ Q_{2\nu}\,|\,\phi\rangle| \tag{6}$$

Unfortunately, as already mentioned, the calculation time prevents the use of QRPA as it is for the 5DCH mass parameter. In the following, we restrict the number of 2qp excitations entering in Eq. (2) in order to evaluate Eq. (5).

3. Results

The study is performed on the $^{110-144}$Sn isotopes. A large variety of underlying single-particle level spectrum is then explored with N, as it would as a function of deformation in one nucleus. The $^{110-144}$Sn isotopes have their minimum of HFB energy at zero deformation, where the QRPA calculations will be performed. It allows to decorrelate the notion of valence space from deformation. Moreover, at zero deformation our study can be realized on B_{00} only, since the $B_{\mu\nu}$ fulfil the relation:

$$B_{00} = \frac{\hbar^2}{2} \frac{\mathcal{M}_{-3,00}}{[\mathcal{M}_{-1,00}]^2} = 2B_{20} = 4B_{22}. \tag{7}$$

The quasiparticles, inputs of the QRPA process, have been determined with the axial HFB code of ref.[12]. In HFB and in QRPA calculations, the effective interaction is D1M [13] and an harmonic oscillator basis with 11 major shells provides stabilized results in the $^{100-144}$Sn isotopes. The axial QRPA calculation have been achieved with the multi threads version [8] of the code detailed in [14]. The restriction on the quasiparticles is done by defining in the whole matrix QRPA sub-matrices, for which the diagonalization, fastest part of the QRPA process, is performed.

3.1. Mass parameter with 2qp energy cut-off

Here, the valence space is reduced by an energy cut-off set on the 2qp excitations (i.e the sum of the energies of the two qp involved in the excitation) keeping only 2qp excitations with energy below or equal that of the cut-off. One then assumes that high energy 2qp excitation play a minor role in the nuclear polarizability. A high enough cut-off value should provide a reasonable $B_{00_{\mathrm{cut}}}^{\mathrm{no\ core}}$, approximation of the value obtained without any valence space restriction, $B_{00_\emptyset}^{\mathrm{no\ core}}$, our absolute reference (full QRPA). (The symbol $\emptyset$ indicates calculations without any cut-off.)
In Figure 1 is drawn $R_{abs}(B_{00})=B_{00_{\mathrm{cut}}}^{\mathrm{no\ core}}/B_{00_\emptyset}^{\mathrm{no\ core}}$ as a function of A. The choice of a 2qp energy cut-off above 50 MeV is immediate. With cut-off energy below 50 MeV the $B_{00_{\mathrm{cut}}}^{\mathrm{no\ core}}$ values lie between 1.5 up to 13.5 times the $B_{00_\emptyset}^{\mathrm{no\ core}}$ one. It can be seen in the insert of the Figure 1,

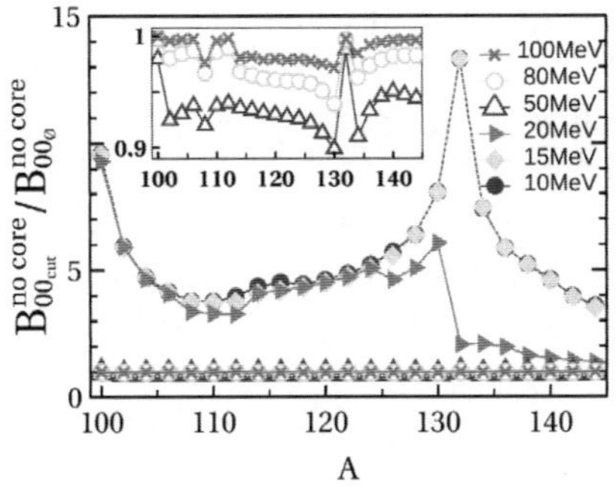

Figure 1. For different 2qp energy cut-off values and as a function of mass, ratio $R_{abs}(B_{00})$ (see text). A zoom on the ordinate around the value of 1.0 is shown in the insert.

that better than 90% in mean of the reference value (with the exceptio of ^{130}Sn) is reached with energy cut-offs from 50-MeV. It would be interesting to get more information in the intermediate 20 - 50 MeV energy range, by setting for example a 40 MeV cut-off. This is in progress.
Figure 1 presents effects due to the underlying single-particle level density. The curves obtained with small energy cut-offs present typical shell effects, large peaks at shell closures. Between A=100 and A=132, 2qp benefit by the shells lying between the gaps, to generate excitations with little energy. This leads to the parabolic shape, with a minimum around N=62 ($\nu g_{7/2}$ occupancy) of the 10, 15 and 20 MeV curves. They are superimposed, thus 2qp excitations with at most 10 MeV give the main low energy contribution for A=100-132. However, from A=126, 2qp excitations of 15-20 MeV increase. Better values are obtained in $^{126-130}$Sn with a 20 MeV cut-off. In $^{132-144}$Sn (from N $\geq$82), after a stiff drop (A=130-132, not observed at A=100-102), the 2qp excitations of 15-20 MeV energy become essential, leading to a mass parameter twice the reference only. The previous features indicate that 2qp excitations with "low" energy carry fine structure information, they can not be neglected. The use of a high- and low-cut filter for computing vibrational mass parameter is thus excluded.
It is worth noting that with Tamm Dancoff approximation, we obtained $B_{00_{\mathrm{TDA}}}$ greater than $B_{00_\emptyset}^{\mathrm{no\ core}}$ in all Sn isotopes[16]. Their ratio (1.3 in mean) presents also variations as a function of A, but a minimum at N=82. Neglecting 2qp excitations with energy higher than 20 MeV, or ground-state correlations results in overestimating B_{00}. Inversely here, from 50 MeV cut-off, the obtained mass parameters are satisfactory underestimate (better than 90%) of the expected value (see insert of Fig 1). A fine agreement is obtained at shell closures, structure effects are still visible. In our previous work $B_{00}^{Inglis\text{-}Belyaev}$ commonly used in 5DCH, was found to underestimate 1.3 times, in mean, $B_{00_\emptyset}^{\mathrm{no\ core}}$, with a larger discrepancy at shell closure.
The valuable results obtained with a 50-MeV energy cut-off validates the hypothesis of the high energy 2qp excitations playing only a minor role in the mass parameter value. Moreover, a 50-MeV energy cut-off provides a reduction in computation time of a factor 16.

3.2. Mass parameter with inert core and 2qp energy cut-off

Never applied in QRPA, but largely and commonly used in standard shell model calculation, is the technique consisting in freezing deep qp lying at the bottom of the mean-field, in an inert (non excitable) core. We directly imported this technique in our QRPA calculations and build submatrices by selecting qp that are rows and columns of the whole QRPA matrices.
We performed the calculations with five different cores, ^{40}Ca, ^{48}Ca, ^{56}Ni, ^{70}Ca, and ^{78}Ni. The results obtained with the two first cores are sufficient to draw conclusions. On Figure 2 are reported $R_{abs}(B_{00})$, ratios of $B_{00_{\mathrm{cut}}}^{^{40}\mathrm{Ca}}$ (a)) or $B_{00_{\mathrm{cut}}}^{^{48}\mathrm{Ca}}$ (b)) over $B_{00_\emptyset}^{\mathrm{no\ core}}$, for $^{100-144}$Sn and as a function of 2qp energy cut-off. A common feature appears at a first glance in the results drawn in Figure 2a) and b): the dramatic change in the slope and values of the curves when the energy cut-off reaches 50 MeV. Below, the mass parameter are overestimated, the slopes

exhibit important variations according to the neutron number. Above, only slight variations as a function of N are observed (see inserts). However, in the left panel insert, the curves merge toward each other when converging to relevant no cut-off values – lying within 90 percent of the $B_{00_\emptyset}^{\text{no core}}$–. In the right panel the "inter convergence" of the curves toward each other is less pronounced, and their asymptotic values are less satisfactory, ranging between 1.05 and 1.25. For the heavier inert cores (not shown here), the curves diverge between themselves. Such effect may be of great interest to check the pertinence of the convergence process. In Table 1, are

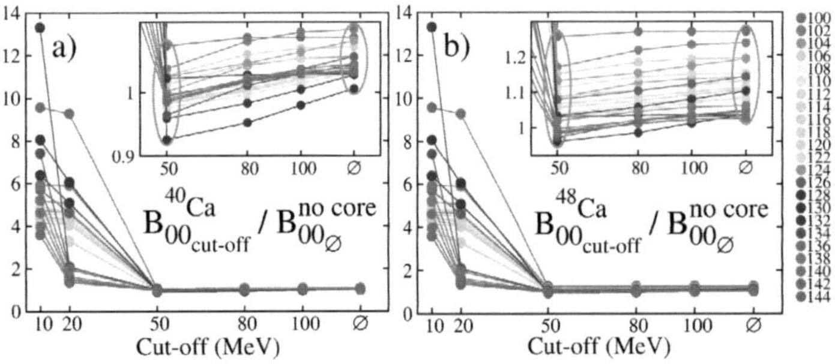

Figure 2. As a function of the cut-off value, ratios $R_{abs}(B_{00})$ (see text) in $^{100-140}$Sn isotopes with a) ^{40}Ca (left) or b) ^{48}Ca as inert core. In the insert, ellipses allow to highlight the convergence of the curves toward themselves.

reported the mean value and the standard deviation of the distribution of the asymptotic values of the ratio $R_{abs}(B_{00})=B_{00_\emptyset}^{\text{core}}/\,B_{00_\emptyset}^{\text{no core}}$ obtained for different cores without energy cut-off. All the

Table 1. Mean value and standard deviation of the distribution over the $^{100-140}$Sn isotopes of the asymptotic values of $R_{abs}(B_{00})$ according to different inert cores.

	^{40}Ca	^{48}Ca	^{56}Ni	^{70}Ca	^{78}Ni
mean val.	1.04	1.11	1.24	1.53	1.70
stdev	0.02	0.07	0.09	0.29	0.34

asymptotic values are, from ^{56}Ni, large overestimations (by a factor at least 1.15) of the mass parameter. In this paper, overestimation was previously mentioned when neglecting ground-state correlations or 2qp excitations of less than 50 MeV. Here, of course, it is due to the inert core insertion, which reduces *de facto* ground-state correlations, since many qp do participate anymore to Eqs (2) and (4). Frozen in "large" inert core (N, Z $\geq$ 28), there are qp, not so deep, that would have participated to in-shells excitations with little energy. Indeed, their excitation toward shells above the Fermi level would have required very high energy.
With a ^{48}Ca inert core, few Sn isotopes have a mass parameter slightly underestimated when using a 50-MeV energy cut-off, see Figure 2. They also have asymptotic values lying very near the reference. For these isotopes it should be possible to take benefit of the ^{48}Ca inert core to reduce more drastically the computation time. Of course, for systematic calculations, the results summed up in Table 1 and Figure 2 rule out the use of any core heavier than ^{40}Ca in combination with a 50-MeV cut-off. Fortunately, the computation time is significantly reduced, by a factor 2, when using this inert core. Using both a 50-MeV energy cut-off and the ^{40}Ca inert core the total gain in computation time reaches up to a factor 30. If results using these two limitations are robust, it should give room enough for facing calculations of QRPA mass parameters, with a valence space, in most of the nuclei at any deformation. In order to check the robustness of our mass parameter, and to assess our hypothesis, calculations of intrinsic QRPA outputs, i.e. phonon energy and their reduced transition probability with valence space have been undertaken.

3.3. QRPA outputs with 2qp energy cut-off

In Figure 3 are drawn as a function of the 2qp energy cut-off, the variation of the 2_1^+ and 3_1^- energies, and of their associated reduced transition probability, relatively to the reference values obtained without cut-off (and no core). Opposite trends are observed for the excited state energies and the transition probabilities. At 50 MeV, neither the energies nor the

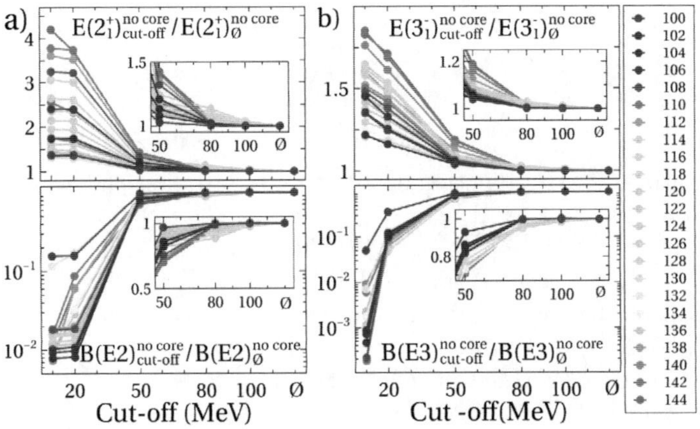

Figure 3. As a function of the cut-off value, ratio $R_{abs}(E(2_1^+))$ (a)) and $R_{abs}(E(3_1^-))$ (b)) (see text). The color to mass correspondence is given in the palette at the right. A zoom on the ordinate is shown in the insert.

transition probabilities are converged. The convergence is only obtained from 100 MeV (80 MeV respectively) cut-off for the 2_1^+ (3_1^- resp.) energy and transition probability. The 2qp excitations with energy lower or equal to 50 MeV, that make 90% of the vibrational mass parameter B_{00}, represent of the order of 70% only of what is needed to build the first excited states. The lack of high-energy 2qp excitations has heavy consequences. Inversely, it confirms their secondary role in the polarizability. In Figure 3, one can also notice, on the one hand, the greater sensitivity of the transition probability, as compared to the excited state energy, to missing 2qp excitations with energy comprised between 20 and 50 MeV, and on the other hand, the substantial improvement brought by 2qp excitation of energy between 10 and 20 MeV in $E(3_1^-)$) and B(E3) only.

3.4. QRPA outputs with inert core

In Table 2, are reported the mean value and the standard deviation of the asymptotic values (i.e without energy cut-off) of the ratios $R_{abs}(X)=X^{core}_{\emptyset} / X^{no\ core}_{\emptyset}$, where X stands for $E(2_1^+)$ or $E(3_1^-)$, obtained with different cores. Once more the results for the 2_1^+ differ from those for the 3_1^-. The latter leads to better convergence properties, though not sufficients to give valuable results with any of the cores used here. It is worth noting that we obtained satisfactory values for both state energies when using a much smaller inert core, ^{16}O.

Table 2. Mean value and standard deviation of the distribution over the $^{100-144}Sn$ isotopes of the asymptotic values of (a)) $R_{abs}(E(2_1^+))$ and (b)) $R_{abs}(E(3_1^-))$ for different inert cores.

		^{40}Ca	^{48}Ca	^{56}Ni	^{70}Ca	^{78}Ni
$E(2_1^+)$	mean val.	1.39	1.47	1.65	1.70	1.79
	stdev	0.31	0.36	0.49	0.54	0.61
$E(3_1^-)$	mean val.	1.16	1.19	1.27	1.30	1.34
	stdev	0.06	0.07	0.10	0.11	0.13

As a last result is shown in Figure 4 the remarkable deceptive convergence scheme of the relative ratio $R_{rel}(E(2_1^+))=E(2_1^+)^{core}_{cut-off}/\ E(2_1^+)^{core}_{\emptyset}$ obtained with ^{78}Ni as inert core. The convergence toward the asymptotic no cut-off value obtained with a given core has the appearance of a fast and efficient process, despite large differences obtained between the asymptotic no cut-off value with a core and the absolute value (no cut-off, no core). These results are astonishing since

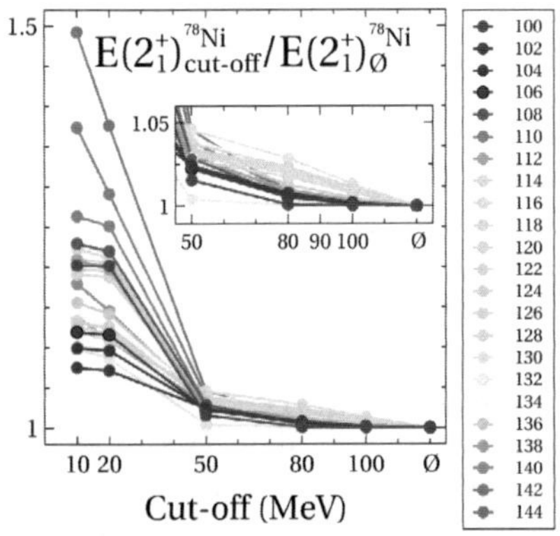

Figure 4. As a function of the cut-off value, ratios $R_{rel}(E(2_1^+))$ (see text) in $^{100-140}$Sn isotopes for the ^{78}Ni inert core. The color to mass correspondence is given by the palette at the right. A zoom on ordinate is provided in the insert.

in standard shell models calculations larger cores are commonly used. It would be possible to perform the required normalizations (on each multipolarity) using some absolute values – get full QRPA calculations for few points being feasible –. That would be, however, counter productive since time-consuming. Moreover it would lead to the loss of the "universal" and coherent character of the 5DCH approach, which consists in using one density-dependent force only to describe nuclei over the whole chart.

4. Conclusions

The introduction of the valence space notion in QRPA calculations, via 2qp energy cut-off and inert core, is an important breakthrough for the 5DCH framework. It offers a glimpse of a possible path toward valuable vibrational mass parameter calculations on the required few tens of (β,γ) points. The vibrational mass parameters appear very robust against valence space limitations, allowing a factor of 30 to be reached in the computation time reduction. On the contrary, unreliable values are obtained for intrinsic QRPA outputs in the same approximations, that exhibit deceptive convergence.

[1] J.-P. Delaroche, M. Girod, J. Libert, H. Goutte, S. Hilaire, S. Péru, N. Pillet and G.F. Bertsch, Phys. Rev. C **81**, 014303 (2010).
[2] G.F. Bertsch, M. Girod, S. Hilaire, J.-P. Delaroche, H. Goutte and S. Péru, Phys. Rev. Lett. **99**, 032502 (2007).
[3] M. Bender and P.-H. Heenen, Phys. Rev. C **78**, 024309 (2008).
[4] S. Péru and M. Martini, Eur. Phys. J. A **50**, 88 (2014).
[5] B. Mohammed-Azizi, Electronic J. Theor. Phys. **9**, No. **27** (2012) 143-158.
[6] E.Kh. Yuldashbaeva, J. Libert, P. Quentin and M. Girod, Phys. Lett. B **461**, 1 (1999).
[7] K. Yoshida and N. Van Giai, Phys. Rev. C **78**, 064316 (2008).
[8] S. Péru, G. Gosselin, M. Martini, M. Dupuis, S. Hilaire and J.-C. Devaux, Phys. Rev. C **83**, 014314 (2011).
[9] J. Libert, B. Roussière and J. Sauvage, Nucl. Phys. A 786, 47 (2007)
[10] S. Péru, M. Girod and J.-F. Berger, E. P. J. A **9**, 35 (2000)
[11] D. Vautherin, Phys. Lett. Vol. 69B Numb. 4, 393 (1977)
[12] S. Hilaire and M. Girod, Eur. Phys. J. A **33**, 237 (2007).
[13] S. Goriely, S. Hilaire, M. Girod and S. Péru, Phys. Rev. Lett. **102**, 242501 (2009).
[14] S. Péru and H. Goutte, Phys. Rev. C **77**, 044313 (2008).
[15] J. Libert, M. Girod and J.-P. Delaroche, Phys. Rev. C **60**, 054301 (1999).
[16] F. Lechaftois, I. Deloncle and S. Péru, Phys. Rev. C **92**, 034315 (2015).

Rotational bands in ^{11}B and identification of diluted states

A S Demyanova[1], A N Danilov[1], A A Ogloblin[1], S A Goncharov[2], T L Belyaeva[3] and W Trzaska[4]
[1]National Research Centre Kurchatov Institute, Moscow 123182, Russia,
[2]Lomonosov Moscow state University, Moscow, 119991, Russia,
[3]Universidad Autonoma del Estado de Mexico, Toluca, 50000, Mexico,
[4]JYFL, Jyvaskyla, FIN-40500, Finland

E-mail: a.s.demyanova@bk.ru

Abstract. Differential cross-sections of the ^{11}B + α inelastic scattering at $E(\alpha)$ = 65 leading to the most of the known ^{11}B states at the excitation energies up to 14 MeV were measured. The data analysis was done by DWBA and in some cases by the modified diffraction model allowing determining the radii of the excited states. The radii of the states with excitation energies less than ~ 7 MeV with the accuracy not less than 0.1-0.15 fm coincide with the radius of the ground state. This result is consistent with the traditional view of the shell structure of the low-lying states in ^{11}B. Most of the observed high-energy excited states are distributed among four rotational bands. The moments of inertia of band states are close to the moment of inertia of the Hoyle state of ^{12}C. The calculated radii, related to these bands, are 0.7 - 1.0 fm larger than the radius of the ground state, and are close to the radius of the Hoyle state. These results are in agreement with existing predictions about various cluster structure of ^{11}B at high excitation energies. The state with the excitation energy 12.56 MeV, $J^\pi = 1/2^+$, $T = 1/2$ and the root mean square radius R ~ 6 fm predicted in the frame of the alpha condensate hypothesis was not found.

1. Introduction

During long time ^{11}B nucleus was considered as good example of shell effects in light nuclei. Up to excitation energies ~ 7 MeV states of entire spectrum were described by different variants of shell models. Recently, however, a number of theoretical and experimental works appeared what resulted in increased interest to the ^{11}B structure. This interest was primarily connected with the predictions [1-3] about cluster configurations of various types co-existed in ^{11}B and the appearance of some experimental data indicating a greater probability of such a possibility [4].

Particular attention was drawn to the idea that there may be states in ^{11}B, analogues of the famous 0^+_2 state with excitation energy 7.65 MeV in ^{12}C nucleus (the so-called Hoyle state). The Hoyle state properties, consisting of three weakly interacting alpha - clusters were crucial for verification and justification of the theory of alpha - particle condensation in nuclei [5], the main of which was the suggestion about its abnormally large radius. Accordingly, the Hoyle state analogues in ^{11}B (in which one proton is removed from alpha cluster, and having, therefore, the structure of α + α + t), must also have increased size.

Initially it was assumed [1] that the Hoyle state analogue in ^{11}B is the state $3/2^-$ with excitation energy 8.56 MeV, which is not described by any variant of shell model. However, other possibilities were started to be discussed afterwards. For example, in [3] it was suggested that the true analogue of the Hoyle state is the state with excitation energy 12.56 MeV, despite contradictory data about its quantum numbers. Moreover, in [3] it was made even more ambitious assumption that this state has a "giant" radius $R_{rms} \approx 6$ fm, comparable to the radius of the uranium nucleus (!). The radius of 8.56 MeV state was still considered to be abnormally large, and it was predicted that this state is base for rotational band.

In [2, 4] other unusual states in ^{11}B were predicted, such as quasimolecular states, with the structure of "two alpha – particle core plus three nucleons moving on their orbits". It is expected that the sequence of such states forms the positive parity rotational band, based on the $1/2^+$, $E^* = 6.79$ MeV state. It has been suggested [3] that this state also has an increased radius, 0.6 - 0.8 fm larger than the radius of the ground state.

Thus, recently coexistence of several types of structures was predicted in ^{11}B, and this nucleus suddenly become in the focus of the entire problem of nucleon clustering in nuclei.

There are a lot of experimental studies of ^{11}B (see. e.g., [6] and references therein), but they did not affect the excitation energy region of interest for the problem. Due to the fact that many questions about ^{11}B states remained open, we have undertaken a new study of inelastic ^{11}B + α scattering at $E(\alpha)$ = 65 MeV. We measured the differential cross sections in a wide range of ^{11}B excitation energies and analyzed them together with the data available at other energies. Some preliminary results have been reported in [7].

2. Experiment and data analysis

The measurements were carried out at the University of Jyvaskyla K130 cyclotron (Finland) using the LSC (Large scattering chamber). The standard ΔE-E method was used. 4 telescopes were used simultaneously, each telescope made measurements for 2 degrees. The target was a self-supporting film of ^{11}B with 0.275 mg/cm^2 thickness. ^{12}C and ^{10}B were the only impurities in the target. The beam intensity was about 30 nA (for the particles). In the experiment, we used a system of beam monochromatisation, which allowed us to obtain a complete energy resolution about 150 keV.

The measured differential cross sections were analyzed by the Distorted Wave Born Approximation method (DWBA). To determine the radius of the ^{11}B nucleus it was supposed to use Modified Diffraction Model (MDM), described in details in [8], which allowed us to obtain self-consistent data on the mean-square radii of a large number of excited states. MDM allows determining radii of the nucleus excited states through the difference between diffraction radii of excited and ground states using the expression (1):

$$R^* = R_0 + [R^*_{dif} - R_{dif}(0)] \qquad (1)$$

where R_0 - is the mean square radius of the ground state of the nucleus, R^*_{dif} and $R_{dif}(0)$ - diffraction radii calculated from positions of minima and maxima of experimental inelastic and elastic scattering angular distributions, respectively.

3. Results and discussion

3.1. Low-lying states

The differential cross sections for low-lying 2.12, 4.44, 5.02 и 6.74 + 6.79 MeV states in ^{11}B were measured. It was shown that in the above unresolved doublet of excited states 6.74 MeV state dominates. The DWBA calculations were fulfilled.

Mean square radii of low-lying excited states of ^{11}B were received using MDM based on experimental data of current work and data [6] at 40 and 50 MeV. As it was expected, ^{11}B has equal non-enhanced radii in all low-lying states. The radii of low-lying (E^* < 7 MeV) states practically coincide with the radius of the ground state.

The received results are consistent with traditional notions about shell structure of the low-lying states in the ^{11}B nucleus.

Journal of Physics: Conference Series **724** (2016) 012013 doi:10.1088/1742-6596/724/1/012013

3.2. $3/2_3^-$, 8.56 MeV state and rotational band based on it

As it was mentioned in introduction, one of the main aims of this work was to get information about "abnormal" properties of 8.56 MeV state, and especially to determine its radius. Differential cross section with excitation of this state is shown of figure 1.

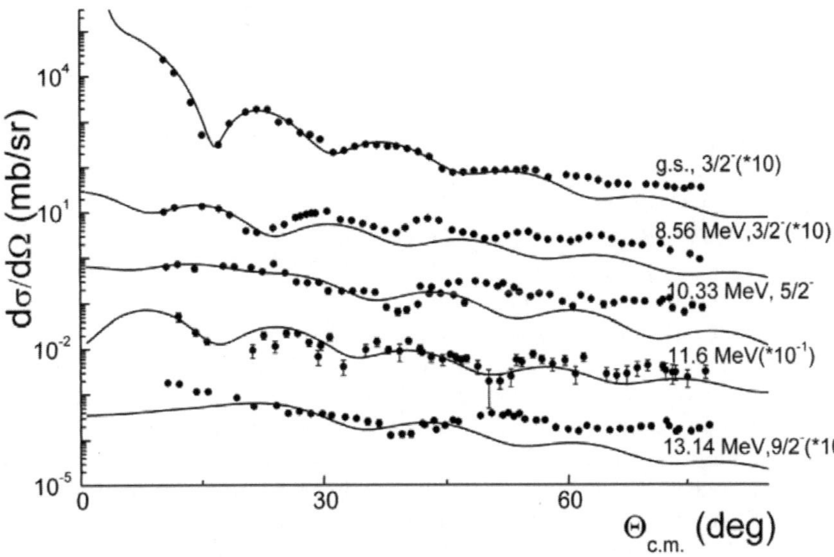

Figure 1. Differential cross sections of ^{11}B + α scattering at E_α=65 MeV. Points correspond to experimental data and solid curves - to calculations: optical model for elastic scattering; DWBA for inelastic scattering with excitation of the $3/2^-$, 8.56 MeV state (angular momentum transfer L=0, 2); $5/2^-$, 10.33 MeV (L=4); 11.6 MeV; $9/2^-$, 13.14 MeV (L=4).

Diffractive structure of angular distributions is shown clearly, however, the difference between DWBA calculations and experimental data takes place in the region of the second minimum. Compared with the elastic scattering, observed minima and maxima are shifted toward smaller angles, which is an indication of the increased radius of the 8.56 MeV state.

AMD - calculations [2] predict the existence of a rotational band, based on the 8.56 MeV state (see figure 2). There and in [4], it was suggested that the band is formed by sequence of states: 10.33 ($5/2^-$) - 11.60 - 13.14 ($9/2^-$) MeV. However, before these predictions, it was considered [9] that the spin - parity of the 11.60 MeV state is $5/2^+$ and therefore it cannot be a member of this band.

As it was shown in [10], 10.33 and 13.14 MeV states are excited by angular momentum transfer L = 4. MDM calculations give $R_{\rm dif}$= 4.8 fm for these states. This diffraction radius is close to diffraction radius of recently observed state in ^{12}C with E^* = 13.75 MeV (this state is excited by angular momentum transfer L = 4 in ^{12}C (α, α') scattering) and identified [11], as member of rotational band, based on the Hoyle state. It was shown [11] that the radius of the 13.75 MeV state is 0.8 fm larger than the radius of the 14.08 MeV state. Thus an estimation can be done, that the radii of above mentioned excited states of ^{11}B have radii 0.8 fm larger than radius of the ground state.

So 10.33 ($5/2^-$) и 13.14 ($9/2^-$) MeV states have properties, similar to 8.56 MeV state and can be members of its rotational band.

Separate discussion is required for question about spin-parity of the 3rd predicted state of rotational band - the 11.60 MeV state, according to [2, 4], $J^\pi = 7/2^-$. But, in review [9] $J^\pi = 5/2^+$.

In [10] we received using DWBA that this state is excited by angular momentum transfer $L = 1$ and $L = 3$, so $J^\pi = 5/2^+$. If so, the question about the fact of the existence of this band arises. However, this result is not definitive. A significant contribution to the cross section the 11.44 MeV state can give, which has not been completely separated from the 11.60 MeV state, and for which there is no data on its spin and parity. In case $J^\pi = 7/2^-$, this state should be excited by $L = 2$ and the use of MDM provides for the latter value $R_{dif} = 6.11 \pm 0.35$ fm.

3.3. Rotational bands for higher excited states

Following rotational bands were predicted [2, 4] in high excitation energy region in ^{11}B. Main part of these states was observed in our experiment:

K = 3/2⁻: 8.56 ($3/2^-$) – 10.33 ($5/2^-$) – 11.60 – 13.14 ($9/2^-$) MeV,
K = 1/2⁺: 6.79 ($1/2^+$) – 9.88 ($3/2^+$) – 11.60 ($5/2^+$) – 13.16 ($7/2^+$) MeV,
K = 3/2⁺: 7.98 ($3/2^+$) – 9.27 ($5/2^+$) – 10.60 ($7/2^+$) – 12.63 ($9/2^+$) MeV,
K = 5/2⁺: 7.29 ($5/2^+$) – 9.19 ($7/2^+$) – 11.27 ($9/2^+$) MeV.

These rotational bands are shown on figure 2 together with band in ^{12}C, based on the Hoyle state. Received from our experiment data about angular momentums transfer with excitation states, belonging to specified bands, are in agreement with known spin-parities of ^{11}B states. However they could not be determined unambiguously for the states 6.79, 9.88, 10.33, 13.14 and 13.16 MeV because of an insufficient energy resolution. Several special features in the $J(J + 1)$ dependence of the excitation energies in figure 2 can be seen. Firstly, moments of inertia of the band states are very high and comparable. The largest of them ($2I/\hbar^2 \approx 4.0$, by the energy difference between the excitation energies 11.60 and 10.33 MeV) are observed for the higher members of the rotational band K = $3/2^-$, for which cluster structure 2α + t is predicted. It is interesting that it is much larger than the moment of inertia of its analogue - the Hoyle state, for which $2I/\hbar^2 = 2.7$.

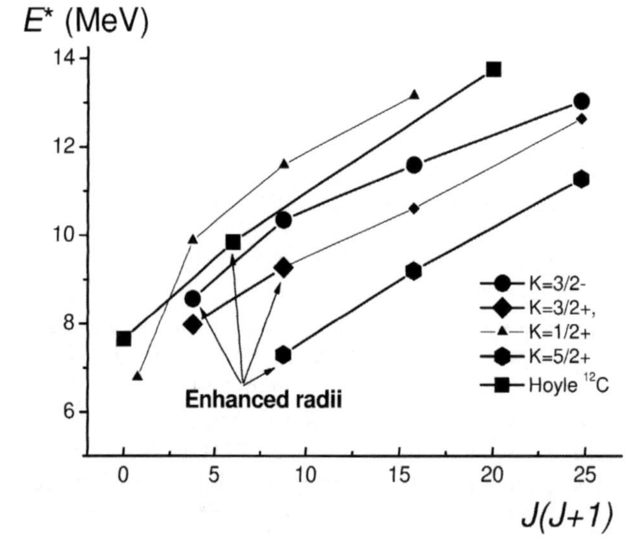

Figure 2. Predicted [2, 4] rotational bands for higher excited states. For comparison, rotational band [11], based on the Hoyle state (0^+_2. 7.65 MeV) of ^{12}C, is shown. Large points correspond to enhanced radii.

Secondly, there is a clear correlation between the moments of inertia and values of radii obtained using MDM from scattering data. Low-lying states of ^{11}B have "normal" radii and "reduced" moments of inertia about $2I/\hbar^2 \approx 1.1$. These values are close to values for the first excited state of ^{12}C, 4.44 MeV. "Big" moments of inertia correspond to increased radii. The radii of the states, located at excitation energies above 7 MeV in ^{11}B, were obtained using MDM. The increased radii were found, at least, for one of the members of each band, and in most cases they are about 0.7 - 1.0 fm larger than the radius of the ground state of ^{11}B. This leads to the conclusion that all states belonging to the bands under consideration have abnormal size. Theoretical works [1 - 4] suggest a significant deformation of the rotational states of ^{11}B with $E^* > 7$ MeV and it allows the increase of their radii. Radii and moments of inertia of these states are close to the corresponding values of the Hoyle state in ^{12}C nucleus.

Concerning the K=1/2$^+$ band: this band produces the largest number of questions. In the measured spectra there were not observed any sign of contribution from the first member of the band, the 6.79 MeV state, to the unresolved group with the 6.74 MeV state. The second member, the 3/2$^+$, 9.88 MeV state was not observed at all. Spin-parity of the third member of the band, the 11.60 MeV state has not been established unambiguously. The estimated fourth band member, the 13.16 MeV state was not separated from the 13.14 MeV state. In the angular distribution the 13.16 state probably manifests itself, but probably with angular momentum transfer $L = 1$, which would mean that it cannot have spin 7/2$^+$, and it therefore doesn't belong to this band.

3.4. Radii of the states in ^{11}B

The levels of ^{11}B could be divided in 2 groups from the point of view of their radii values (see figure 3). The resulting radii of states with excitation energies up to ~ 7 MeV within the errors do not differ from the radius of the ground state, which is consistent with the traditional view of the ^{11}B shell structure.

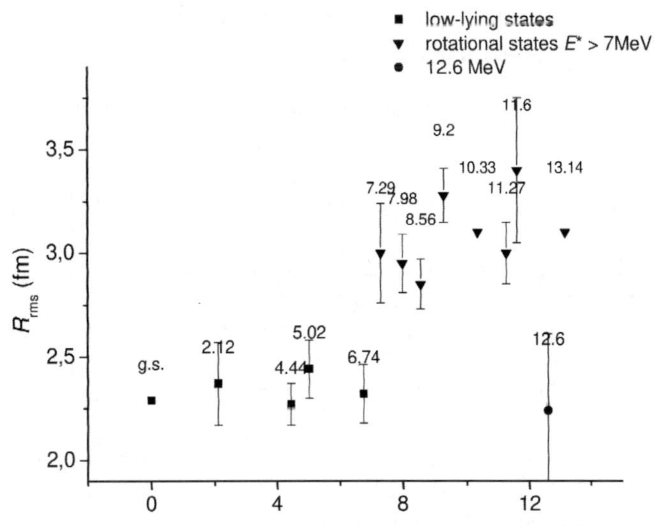

Figure 3. The results of the MDM analysis. The values of RMS radii for ^{11}B states.

At higher excitation energies formation of states belonging to rotational bands based on the $3/2^-$ (8.56 MeV), $3/2^+$ (7.98 MeV), $5/2^+$ (7.29 MeV) and maybe $1/2^+$ (6.79 MeV) states was observed. All these bands are characterized by large moments of inertia ($2I/\hbar^2$ = 3-4). The radii of rotational states, defined using MDM, were determined 0.7-1.0 fm larger than the radius of the ^{11}B ground state. The moments of inertia and the obtained values of the radii are close to the corresponding parameters of the Hoyle state in the ^{12}C and the rotational band, built on it, indicating cluster structure of these ^{11}B states. Particularly, the state with excitation energy 8.56 MeV can be considered as an analogue of the Hoyle state. The predictions of the theory of the coexistence of different structures in the ^{11}B, including cluster, are confirmed. Regarding the proposed band, based on the 6.79 MeV state, a number of uncertainties stay, including the fact of such band existence.

3.5. 12.56/12.63 MeV state

Special attention should be provided to the problem of the 12.56 / 12.63 MeV state. Before the work [4], the only state at excitation energies 12.0 - 12.9 MeV was the state at 12.56 MeV with spin - parity $J^\pi = 1/2^+$, and isospin $T = 3/2$ [9], the isobar - analogue of the ground state of the ^{11}Be. In [3] it has been suggested that this state really has a spin - parity $J^\pi = 1/2^+$, and isospin $T = 1/2$ and should be a "true" analogue of the Hoyle state in ^{12}C nucleus. The value of the isospin $T = 1/2$ is supported by the observation of the state decay with the emission of alpha - particles [12 - 14].

In reaction ^{7}Li (α,α) [4] a strong resonance was indeed found, but with an excitation energy 12.63 ± 0.04 MeV. The state with excitation energy 12.56 MeV was not observed. In addition to the obvious value of the isospin $T = 1/2$, the 12.63 MeV state was received having the spin - parity $J^\pi = 3/2^+$ or $9/2^+$. Possible value of the spin $9/2^+$ made it possible [4] to interpret the 12.63 MeV state, as the last member of the rotational band with $K = 3/2^+$.

In our experiment state with an excitation energy 12.6 MeV also was observed, so its isospin is $T = 1/2$. It is possible the state we observed is the state from [4] with excitation energy 12.63 MeV. Most probable value of the spin-parity according to our data is $J^\pi = 3/2^+$. Analysis within MDM gave a "normal" value of the mean square radius (see figure 3), so prediction [3] has not been confirmed.

4. Conclusions

The ^{11}B nuclear has a number of excited states with the radii exceeding that of the ground state (size isomers). They are grouped in rotational bands whose moments of inertia correlate with enhanced radii values.

Radii and moments of inertia of the low-lying ($E^* < 7$ MeV) states of ^{11}B are close to those of the ground state.

Radii and moments of inertia of rotational states with $E^*>7$MeV are similar to those of the Hoyle state rotational band indicating their predicted cluster structure.

The ^{11}B state with $E^* = 12.6$ MeV was observed, and thus its isospin is $T = 1/2$. However, the radius of this state, in contradiction with some predictions, is not significantly larger than the radius of the ground state, and thus the hypothesis about the presence of "giant" states in some light nuclei has not been confirmed.

Established numerous correlations between the values of moments of inertia of rotational band states and their radii are further confirmation of MDM ability to determine radii of excited states using experimental differential scattering cross sections.

5. References

[1] Kanada-En'yo Y 2007 *Phys. Rev.* C **75** 024302
[2] Suhara T and Kanada-En'yo Y 2012 *Phys. Rev.* C **85** 054320
[3] Yamada T and Funaki Y 2010 *Phys. Rev.* C **82** 064315
[4] Yamaguchi H, Hashimoto T, Hayakawa S, Binh D N, Kahl D, Kubono S, Wakabayashi Y, Kawabata T and Teranishi T 2011 *Phys. Rev.* C **83** 034306

[5] Tohsaki A Horiuchi H Schuck P and Röpke G 2001 *Phys. Rev. Lett.* **87** 192501
[6] Burtebaev N, Baktybaev M K, Duisebaev B A, Peterson R J, Sakuta S B 2005 *Physics of Atomic Nuclei* **68** 1303
[7] Danilov A N, Demyanova A S, Ogloblin A A, Dmitriev S V, Belyaeva T L, Goncharov S A, Gurov Yu B, Maslov V A, Sobolev Yu G, Trzaska W G et al. 2014 *EPJ Web of Conferences* **66** 03007
[8] Danilov A N, Belyaeva T L, Demyanova A S, Goncharov S A and Ogloblin A A 2009 *Phys.Rev.* C **80** 054603
[9] Ajzenberg-Selove F 1990 *Nucl. Phys* A **506** 1
[10] Danilov A N, Demyanova A S, Dmitriev S V, Ogloblin A A, Belyaeva T L, Goncharov S A, Gurov Yu B, Maslov V A, Sobolev Yu G, Trzaska W G et al. 2015 *Physics of Atomic Nuclei* **78** No. 6 777
[11] Ogloblin A A, Demyanova A S, Danilov A N, Dmitriev S V, Belyaeva T L, Goncharov S A, Maslov V A, Sobolev Yu G, Trzaska W G and Khlebnikov S V 2014 *EPJ Web of Conferences* **66** 02074
[12] Charity R J, Komarov S A, Sobotka L G, Clifford J, Bazin D, Gade A, Lee Jenny, Lukyanov S M, Lynch W G, Mocko M et al. 2008 *Phys. Rev* C **78** 054307
[13] Soić N, Freer M, Donadille L, Clarke N M, Leask P J, Catford W N, Jones K L, Mahboub D, Fulton B R, Greenhalgh B J et al. 2004 *Nucl. Phys* A **742** 271
[14] Curtis N, Ashwood N I, Catford W N, Clarke N M, Freer M, Mahboub D, Metelko C J, Pain S D, Soić N and Weisser D C 2005 *Phys Rev* C **72** 044320

Acknowledgments
The work was supported in part by grants of Russian Foundation for Basic Researches 14-02-00560 and the mobility grant from the Academy of Finland.

Separable Potentials for (d,p) Reaction Calculations

Ch. Elster [(a)], **L. Hlophe**[(a)], **V. Eremenko**[(a,e)] **F.M. Nunes**[(b)], **I.J. Thompson**[(c)], **G. Arbanas**[(d)], **J.E. Escher**[(c)]

[(a)] INPP, Ohio University, Athens, OH, USA
[(b)] NSCL, Michigan State University, East Lansing, MI, USA
[(c)] Lawrence Livermore National Laboratory, Livermore, CA, USA
[(d)] Oak Ridge National Laboratory, Oak Ridge, TN, USA
[(e)] SINP, Lomonosov Moscow State University, Moscow, Russia

E-mail: elster@ohio.edu

Abstract. An important ingredient for applications of nuclear physics to e.g. astrophysics or nuclear energy are the cross sections for reactions of neutrons with rare isotopes. Since direct measurements are often not possible, indirect methods like (d, p) reactions must be used instead. Those (d, p) reactions may be viewed as effective three-body reactions and described with Faddeev techniques. An additional challenge posed by (d, p) reactions involving heavier nuclei is the treatment of the Coulomb force. To avoid numerical complications in dealing with the screening of the Coulomb force, recently a new approach using the Coulomb distorted basis in momentum space was suggested. In order to implement this suggestion, one needs to derive a separable representation of neutron- and proton-nucleus optical potentials and compute their matrix elements in this basis.

1. Introduction

Nuclear reactions are an important probe to learn about the structure of unstable nuclei. Due to the short lifetimes involved, direct measurements are usually not possible. Therefore indirect measurements using (d, p) reactions have been proposed (see e.g. Refs. [1, 2, 3]). Deuteron induced reactions are particularly attractive from an experimental perspective, since deuterated targets are readily available. From a theoretical perspective they are equally attractive because the scattering problem can be reduced to an effective three-body problem [4]. Traditionally deuteron-induced single-neutron transfer (d, p) reactions have been used to study the shell structure in stable nuclei, nowadays experimental techniques are available to apply the same approaches to exotic beams (see e.g. [5]). Deuteron induced (d, p) or (d, n) reactions in inverse kinematics are also useful to extract neutron or proton capture rates on unstable nuclei of astrophysical relevance. Given the many ongoing experimental programs worldwide using these reactions, a reliable reaction theory for (d, p) reactions is critical.

One of the most challenging aspects of solving the three-body problem for nuclear reactions is the repulsive Coulomb interaction. While the Coulomb interaction for light nuclei is often a small correction to the problem, this is certainly not the case for intermediate mass and heavy systems. Over the last decade, many theoretical efforts have focused on advancing the theory for (d, p) reactions (e.g. [6, 7]) and testing existing methods (e.g. [8, 4, 9]). Currently, the most complete implementation of the theory is provided by the Lisbon group [10], which solves the Faddeev equations in the Alt, Grassberger and Sandhas [11] formulation. The method introduced in

[10] treats the Coulomb interaction with a screening and renormalization procedure as detailed in [12, 13]. While the current implementation of the Faddeev-AGS equations with screening is computationally effective for light systems, as the charge of the nucleus increases technical difficulties arise in the screening procedure [14]. Indeed, for most of the new exotic nuclei to be produced at the Facility of Rare Isotope Beams, the current method is not adequate. Thus one has to explore solutions to the nuclear reaction three-body problem where the Coulomb problem is treated without screening.

In Ref. [6], a three-body theory for (d,p) reactions is derived with explicit inclusion of target excitations, where no screening of the Coulomb force is introduced. Therein, the Faddeev-AGS equations are cast in a Coulomb-distorted partial-wave representation, instead of a plane-wave basis. This approach assumes the interactions in the two-body subsystems to be separable. While in Ref. [6] the lowest angular momentum in this basis ($l = 0$) is derived for a Yamaguchi-type nuclear interaction is derived as analytic expression, it is desirable to implement more general form factors, which are modeled after the nuclei under consideration.

In order to bring the three-body theory laid out in Ref. [6] to fruition, well defined preparatory work needs to be successfully carried out. Any momentum space Faddeev-AGS type calculation needs as input transition matrix elements in the different two-body subsystems. In the case of (d,p) reactions with nuclei these are the t-matrix elements obtained from the neutron-proton, the neutron-nucleus and proton-nucleus interactions. Since it is essential to use separable interactions when solving the Faddeev equations in the Coulomb basis, those need to be developed not only in the traditionally employed plane wave basis, but also the basis of Coulomb scattering states.

In this contribution major developments needed to provide reliable input to a Faddeev-AGS formulation of (d,p) reactions in the Coulomb basis are summarized. Those are the derivation of separable representations of neutron-nucleus and proton-nucleus optical potentials. Here it is important that those representations not only describe the cross sections for elastic scattering accurately, but also allow to represent a wide variety of nuclei. Furthermore, it should be straightforward to generalize the representation to account for excitations of the nuclei.

2. Separable Representation of Nucleon-Nucleus Optical Potentials

Separable representations of the forces between constituents forming the subsystems in a Faddeev approach have a long tradition in few-body physics. There is a large body of work on separable representations of nucleon-nucleon (NN) interactions (see e.g. Refs. [15, 16, 17, 18, 19]) or meson-nucleon interactions [20, 21]. In the context of describing light nuclei like ^{6}He [22] and ^{6}Li [23] in a three-body approach, separable interactions have been successfully used. A separable nucleon-^{12}C optical potential was proposed in Ref. [24], consisting of a rank-1 Yamaguchi-type form factor fitted to the positive energies and a similar term describing the bound states in the nucleon-^{12}C configuration. However, systematic work along this line for heavy nuclei, for which excellent phenomenological descriptions exist in terms of Woods-Saxon functions [25, 26, 27, 28] has not been carried out until recently [29]. The separable representation of two-body interactions suggested by Ernst-Shakin-Thaler [30] (EST) is well suited for achieving this goal. We note that this EST approach has been successfully employed to represent NN potentials [15, 16]. However, the EST scheme derived in Ref. [30], though allowing energy dependence of the potentials [31, 32], assumes that they are Hermitian. Therefore, we generalized the EST approach in Ref. [29] in order to be applicable for optical potentials which are complex. Here it is important that the reaction matrix elements constructed from these separable potentials satisfy reciprocity relations.

In analogy to the procedure followed in Ref. [30] we define a complex separable potential of arbitrary rank in a given partial wave as

$$\mathbf{U} = \sum_{i,j} u|f_{l,k_{E_i}}\rangle\langle f_{l,k_{E_i}}|M|f^*{}_l, k_{E_j}\rangle\langle f^*_{l,k_{E_j}}|u. \tag{1}$$

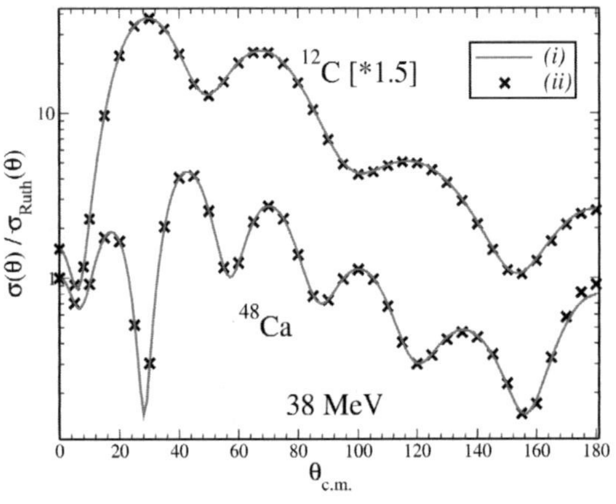

Figure 1. The unpolarized differential cross section for elastic scattering of protons from ^{12}C (upper) and ^{48}Ca (lower) divided by the Rutherford cross section as function of the c.m. angle calculated for E_{lab} = 38 MeV. The ^{12}C cross section is scaled by a factor 1.5. The solid lines (i) depict the cross section calculated in momentum space based on the rank-4 separable representation of the CH89 [25] phenomenological optical potential, while the cross symbols (ii) represent the corresponding coordinate space calculations.

Here $f_{l,k_E}(r)$ is the unique regular radial wave function corresponding to u and $f^*_{l,k_E}(r)$ is the unique regular radial wavefunction corresponding to u^*, where u is the potential for which the separable representation is constructed. The EST scheme guarantees that at the fixed set of energies E_i (support points) the wave functions obtained with the original potential u and those obtained with the separable representation U are identical. The matrix M is defined and constrained by

$$\delta_{ik} = \sum_j \langle f_{l,k_{E_i}}|M|f^*_{l,k_{E_j}}\rangle\langle f^*_{l,k_{E_j}}|u|f_{l,k_{E_k}}\rangle = \sum_j \langle f^*_{l,k_{E_i}}|u|f_{l,k_{E_j}}\rangle\langle f_{l,k_{E_j}}|M|f^*_{l,k_{E_k}}\rangle. \tag{2}$$

The corresponding separable partial wave t-matrix must be of the form

$$t(E) = \sum_{i,j} u|f_{l,k_{E_i}}\rangle \tau_{ij}(E)\langle f^*_{l,k_{E_j}}|u \ , \tag{3}$$

with the following restrictions

$$\delta_{nj} = \sum_i \langle f^*_{l,k_{En}}|u - ug_0(E)u|f_{l,k_{E_i}}\rangle \ \tau_{ij}(E), \tag{4}$$

$$\delta_{ik} = \sum_j \tau_{ij}(E) \ \langle f^*_{l,k_{E_j}}|u - ug_0(E)u|f_{l,k_{E_k}}\rangle, \tag{5}$$

which are used to obtain the matrix $\tau_{ij}(E)$. In general, optical potentials are energy dependent. Though the wave functions $f_{l,k_{E_i}}$ carry part of this energy dependence, the EST scheme needs to be extended to exactly take into account energy dependent potentials [32, 33].

Extending the EST separable representation to the Coulomb basis involves replacing the neutron-nucleus half-shell t-matrix in Eq. (3) by Coulomb distorted scattering states $|f^c_{l,p}\rangle$, which defines the Coulomb-distorted separable nuclear t-matrix

$$\tau_l^{CN}(E) = \sum_{i,j} u^s_l|f^c_{l,k_{E_i}}\rangle \ \tau^c_{ij}(E) \ \langle f^{c\,,\,*}_{l,\,k_{E_j}}|u^s_l. \tag{6}$$

Here $|f^c_{l,k_{E_i}}\rangle$ and $|f^{c\,,\,*}_{l,k_{E_i}}\rangle$ are the regular radial Coulomb scattering wave functions, corresponding to u^s_l and $(u^s_l)^*$ at energy E_i. The constraints are similar to those of Eqs. (4) and (5). The

Coulomb Green's function is given as $g_c(E+i\varepsilon) = (E+i\varepsilon - H_0 - v^C)^{-1}$ with H_0 being the free Hamiltonian and v^C the point Coulomb potential. It remains to calculate the half-shell t-matrix at the support points in the Coulomb basis. Here we follow the method suggested in [34] and successfully applied in [35], and note that in this case the Coulomb Green's function behaves like a free Green's function.

For studying the quality of the representation of proton-nucleus optical potentials we consider p+^{12}C and p+^{48}Ca elastic scattering and show the unpolarized differential cross sections divided by the Rutherford cross section as function of the c.m. angle $\theta_{c.m.}$ in Fig. 1. First, we observe very good agreement in both cases of the momentum space calculations using the separable representation with the corresponding coordinate space calculations. Second, we want to point out that we used for the separable representation of the proton-nucleus partial-wave t-matrices the same support points as in the neutron-nucleus case. This makes the determination of suitable support points E_i for a given optical potential and nucleus quite efficient.

3. Separable Representation of Multi-Channel Optical Potentials

For representing the NN interaction with a separable interaction, the EST scheme had to be extended to include channel coupling [15, 16] e.g. in the deuteron channel. For describing (d,p) reactions within a Faddeev-AGS approach, this is a natural channel coupling to include. However, for (d,p) on nuclei, specifically exotic nuclei, which are often deformed, one also has to consider possible excitations of those nuclei. In the formulation of Ref. [6] the three-body theory to include excitations is layed out. Again, we need to construct separable representations for optical potentials that include excitations of the nucleus, e.g. rotational degrees of freedom.

To extend the EST scheme to non-Hermitian coupled channel potentials, we define for a fixed total angular momentum J

$$\begin{aligned} U^J &= \sum_{\rho\sigma}\sum_{mn}\left(\sum_{\nu} u\left|\Psi^{(+)J}_{\nu\rho,k^\rho_m}\mathcal{Y}^{JM}_{\nu}\right\rangle\right)\lambda^{\rho\sigma}_{mn}\left(\sum_{\nu}\left\langle\Psi^{(-)J}_{\nu\sigma,k^\sigma_n}\mathcal{Y}^{JM}_{\nu}\right|u\right) \\ &\equiv \sum_{\rho\sigma}\sum_{mn} T(E_m)\left|\mathcal{Y}^{JM}_{\rho}\phi_{l_\rho k_m}\right\rangle\lambda^{\rho\sigma}_{mn}\left\langle\phi_{l_\sigma k^\sigma_n}\mathcal{Y}^{JM}_{\sigma}\right|T(E_n), \end{aligned} \tag{7}$$

where $\langle\Psi^{(+)J}_{\nu\sigma,k^\sigma_n}|$ are the solutions of the coupled-channel Lippmann-Schwinger equation corresponding to u and $\langle\Psi^{(-)J}_{\nu\sigma,k^\sigma_n})$ those corresponding to $u^\dagger$ with incoming boundary conditions. The indices ρ and σ indicate the coupling between channels, while m and n refer to the rank of the separable representation. The constraints necessary to satisfy the EST conditions become

$$\begin{aligned} \delta_{mi}\delta_{\rho\gamma} &= \sum_{\sigma n}\sum_{\nu}\lambda^{\rho\sigma}_{mn}\left\langle\phi_{l_\sigma k^\sigma_n}\mathcal{Y}^{JM}_{\sigma}\right|T(E_n)\left|\mathcal{Y}^{JM}_{\gamma}\Psi^{(+)J}_{\nu\gamma,k^\gamma_i}\right\rangle, \\ &= \sum_{\sigma n}\sum_{\nu}\left\langle\Psi^{(-)J}_{\nu\gamma,k^\gamma_i}\mathcal{Y}^{JM}_{\nu}\right|T(E_n)\left|\mathcal{Y}^{JM}_{\upsilon}\phi_{l_\sigma k^\sigma_n}\right\rangle\lambda^{\sigma\rho}_{nm}. \end{aligned} \tag{8}$$

The multi-channel separable t-matrix then takes the form

$$\begin{aligned} t^J_{\alpha\beta}(k',k;E) &= \sum_{\rho\sigma}\sum_{mn} T^J_{\alpha\rho}(k',k^\rho_m;E_m)\tau^{\rho\sigma}_{mn}(E)T^J_{\sigma\beta}(k,k^\sigma_n;E_n), \\ &= \sum_{ab} T^J_{\alpha a}(k',k_a;E_a)\tau_{ab}(E)T^J_{b\beta}(k,k_b;E_b), \end{aligned} \tag{9}$$

where the quantities $T^J_{\alpha\rho}(k',k_a;E_a)$ represent the half-shell transition matrix elements in channels α and ρ. The coupling matrix $\tau^{\rho\sigma}_{mn}(E)$ depends now on the rank, indicated by the indices m and n as well as on the channels indicated by the superscripts. Simplyfing the quadrupel sum of the first row in Eq. (9) over rank as well as channel indices to a double sum over leads to the second row. Explicitly, the coupling matrix $\tau(E)$ is calculated as

$$
\begin{aligned}
[\tau(E)]^{-1}_{ab} &= \Big\langle \phi_{l_a k_a} \mathcal{Y}^{JM}_{\alpha_a} \Big| T(E_a) \Big| \mathcal{Y}^{JM}_{\alpha_b} \phi_{l_b k_b} \Big\rangle \qquad (10) \\
&+ \sum_\beta \int_0^\infty dp\, p^2 \Big\langle \phi_{l_a k_a} \mathcal{Y}^{JM}_{\alpha_a} \Big| T(E_a) \Big| \mathcal{Y}^{JM}_{\beta} \phi_{l_\beta p} \Big\rangle G_\beta(E_b) \Big\langle \phi_{l_\beta p} \mathcal{Y}^{JM}_{\alpha_a} \Big| T(E_b) \Big| \mathcal{Y}^{JM}_{\alpha_b} \phi_{l_b k_b} \Big\rangle \\
&- \sum_\beta \int_0^\infty p^2 dp\, \Big\langle \phi_{l_a k_a} \mathcal{Y}^{JM}_{\alpha_a} \Big| T(E_a) \Big| \mathcal{Y}^{JM}_{\beta} \phi_{l_\beta p} \Big\rangle G_\beta(E) \Big\langle \phi_{l_\beta p} \mathcal{Y}^{JM}_{\beta} \Big| T(E_b) \Big| \mathcal{Y}^{JM}_{\alpha_b} \phi_{l_b k_b} \Big\rangle .
\end{aligned}
$$

Here $G_\beta(E) = (E - \varepsilon_\beta - E^\beta_p + i\epsilon)^{-1}$ is the channel Green's function with $E^\beta_p = p^2/2\mu_\beta$, with μ_β being the reduced mass in the channel β.

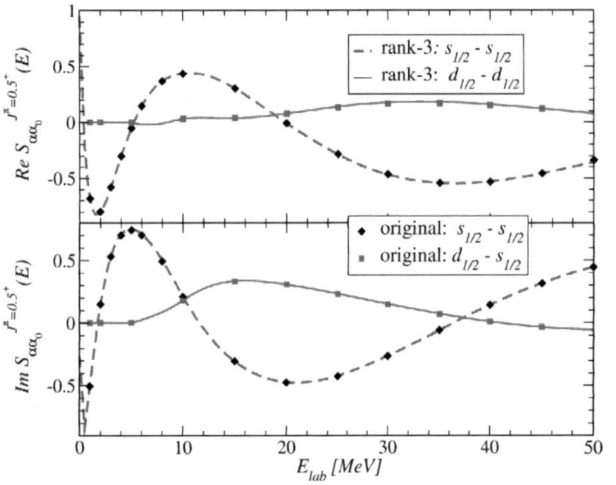

Figure 2. The separable representation of rank 3 for the s-d multichannel S-matrix $S^J_{\alpha\alpha_0}(E)$ for $n+^{12}$C scattering as function of the laboratory kinetic energy for $J^\pi = \frac{1}{2}^+$. The $0^+ \otimes s_{1/2} - 0^+ \otimes s_{1/2}$ and $2^+ \otimes d_{1/2} - 0^+ \otimes s_{1/2}$ S-matrix elements are indicated by the (green) dashed and (red) solid lines. The EST support points are given at 6, 20, and 40 MeV. The S-matrix elements obtained using the original potential are shown for comparison.

As example for constructing a separable optical potential including rotational excitations we use the potential from Ref. [36] which describes elastic and inelastic scattering of neutrons from ^{12}C in the energy range 16.5 to 22 MeV. This potential includes the excitation $I^\pi = 2^+$ and $I^\pi = 4^+$ at 4.44 MeV and 14.08 MeV in ^{12}C, and takes into account quadrupole and octupole deformations. In Fig. 2 we focus on the on-shell properties of the potential up to 50 MeV laboratory kinetic energy. The EST support points are chosen to be at 6, 20, and 45 MeV. In Fig. 2 the multi-channel S-matrix elements are depicted for calculations based on the original potential and on its separable representation. The calculations indicate that the extension of the EST formulation which includes excitations of the nucleus represents the original optical potential very well, and thus those potentials can serve as input for (d,p) reaction calculations.

4. Summary and Outlook

In a series of steps we developed the input that will serve as a basis for Faddeev-AGS three-body calculations of (d,p) reactions, which will not rely on the screening of the Coulomb force. To achieve this, Ref. [6] formulated the Faddeev-AGS equations in the Coulomb basis using separable interactions in the two-body subsystems. For this ambitious program to have a chance of being successful, the interactions in the two-body subsystems, namely the NN and the neutron- and proton-nucleus systems, need to developed so that they separately describe the observables of the subsystems. While for the NN interaction separable representations are available, this is was not the case for the optical potentials describing the nucleon-nucleus interactions. Furthermore, those interactions in the subsystems need to be available in the Coulomb basis.

We developed separable representations of phenomenological optical potentials of Woods-Saxon type for neutrons and protons. First we concentrated on neutron-nucleus optical potentials and generalized the Ernst-Shakin-Thaler (EST) scheme [30] so that it can be applied to complex)¿potentials [29]. In order to consider proton-nucleus optical potentials, we further extended the EST scheme so that it can be applied to the scattering of charged particles with a repulsive Coulomb force [37]. While the extension of the EST scheme to charged particles led to a separable proton-nucleus t-matrix in the Coulomb basis, we had to develop methods to reliably compute Coulomb distorted neutron-nucleus t-matrix elements [38]. Here we also show explicitly that those calculations can be carried out numerically very accurately by calculating them within two independent schemes. We also showed that the scheme can be further extended to take channel coupling into account.

Our results demonstrate, that our separable representations reproduce standard coordinate space calculations of neutron and proton scattering cross sections very well, and that we are able to accurately compute the integrals leading to the Coulomb distorted form factors. Now that these challenging form factors have been obtained, they can be introduced into the Faddeev-AGS equations to solve the three-body problem without resorting to screening. Our expectation is that solutions to the Faddeev-AGS equations written in the Coulomb-distorted basis can be obtained for a large variety of $n+p+A$ systems, without a limitation on the charge of the target. From those solutions, observables for (d,p) transfer reactions should be readily calculated. Work along these lines is in progress.

Acknowledgments

This material is based on work in part supported by the U. S. Department of Energy, Office of Science of Nuclear Physics under program No. DE-SC0004084 and DE-SC0004087 (TORUS Collaboration), under contracts DE-FG52-08NA28552 with Michigan State University, DE-FG02-93ER40756 with Ohio University; by Lawrence Livermore National Laboratory under Contract DE-AC52-07NA27344 and the U.T. Battelle LLC Contract DE-AC0500OR22725. F.M. Nunes acknowledges support from the National Science Foundation under grant PHY-0800026. This research used resources of the National Energy Research Scientific Computing Center, which is supported by the Office of Science of the U.S. Department of Energy under Contract No. DE-AC02-05CH11231.

References

[1] Escher J E, Burke J T, Dietrich F S, Scielzo N D, Thompson I J and Younes W 2012 *Rev. Mod. Phys.* **84**(1) 353–397 URL `http://link.aps.org/doi/10.1103/RevModPhys.84.353`
[2] Cizewski J *et al.* 2013 *J. Phys. Conf* **420** 012058
[3] Kozub R, Arbanas G, Adekola A, Bardayan D, Blackmon J *et al.* 2012 *Phys.Rev.Lett.* **109** 172501
[4] Nunes F and Deltuva A 2011 *Phys.Rev.* **C84** 034607
[5] Schmitt K, Jones K, Bey A, Ahn S, Bardayan D *et al.* 2012 *Phys.Rev.Lett.* **108** 192701
[6] Mukhamedzhanov A, Eremenko V and Sattarov A 2012 *Phys.Rev.* **C86** 034001
[7] Deltuva A 2013 *Phys.Rev.* **C88** 011601 (*Preprint* `1307.0997`)
[8] Deltuva A, Moro A, Cravo E, Nunes F and Fonseca A 2007 *Phys.Rev.* **C76** 064602 (*Preprint* `0710.5933`)
[9] Upadhyay N, Deltuva A and Nunes F 2012 *Phys.Rev.* **C85** 054621 (*Preprint* `1112.5338`)
[10] Deltuva A and Fonseca A 2009 *Phys.Rev.* **C79** 014606
[11] Alt E, Grassberger P and Sandhas W 1967 *Nucl. Phys. B* **2** 167
[12] Deltuva A, Fonseca A and Sauer P 2005 *Phys.Rev.* **C71** 054005
[13] Deltuva A, Fonseca A and Sauer P 2005 *Phys.Rev.* **C72** 054004
[14] Nunes F and Upadhyay N 2012 *J. Phys. G: Conf. Ser.* **403** 012029
[15] Haidenbauer J and Plessas W 1983 *Phys.Rev.* **C27** 63–70
[16] Haidenbauer J, Koike Y and Plessas W 1986 *Phys.Rev.* **C33** 439–446
[17] Berthold G, Stadler A and Zankel H 1990 *Phys.Rev.* **C41** 1365–1383
[18] Schnizer W and Plessas W 1990 *Phys.Rev.* **C41** 1095–1099
[19] Entem D, Fernandez F and Valcarce A 2001 *J.Phys.* **G27** 1537–1546

[20] Ueda T and Ikegami Y 1994 *Prog.Theor.Phys.* **91** 85–104
[21] Gal A and Garcilazo H 2011 *Nucl.Phys.* **A864** 153–166
[22] Ghovanlou A and Lehman D R 1974 *Phys. Rev.* **C9** 1730–1741
[23] Eskandarian A and Afnan I R 1992 *Phys. Rev.* **C46** 2344–2353
[24] Miyagawa K and Koike Y 1989 *Prog. Theor. Phys.* **82** 329
[25] Varner R, Thompson W, McAbee T, Ludwig E and Clegg T 1991 *Phys.Rept.* **201** 57–119
[26] Weppner S, Penney R, Diffendale G and Vittorini G 2009 *Phys.Rev.* **C80** 034608
[27] Koning A and Delaroche J 2003 *Nucl.Phys.* **A713** 231–310
[28] Becchetti FD J and Greenlees G 1969 *Phys.Rev.* **182** 1190–1209
[29] Hlophe L *et al.* (The TORUS Collaboration) 2013 *Phys.Rev.* **C88** 064608 (*Preprint* 1310.8334)
[30] Ernst D J, Shakin C M and Thaler R M 1973 *Phys.Rev.* **C8** 46–52
[31] Ernst D, Londergan J, Moniz E and Thaler R 1974 *Phys.Rev.* **C10** 1708–1721
[32] Pearce B 1987 *Phys.Rev.* **C36** 471–474
[33] Hlophe L and Elster C *in preparation.*
[34] Elster C, Liu L C and Thaler R M 1993 *J.Phys.* **G19** 2123–2134
[35] Chinn C R, Elster C and Thaler R M 1991 *Phys.Rev.* **C44** 1569–1580
[36] Olsson B, Trostell B and Ramstrom E 1989 *Nucl.Phys.A* **469** 505
[37] Hlophe L, Eremenko V, Elster C, Nunes F M, Arbanas G, Escher J E and Thompson I J 2014 *Phys. Rev.* **C90** 061602 (*Preprint* 1409.4012)
[38] Upadhyay N *et al.* (TORUS Collaboration) 2014 *Phys. Rev.* **C90** 014615

Structure of Exotic Mg Isotopes and Temperature Dependence of the Symmetry Energy of Finite Nuclei

M K Gaidarov[1], P Sarriguren[2], A N Antonov[1], D N Kadrev[1], E Moya de Guerra[3]

[1]Institute for Nuclear Research and Nuclear Energy, Bulgarian Academy of Sciences, Sofia 1784, Bulgaria
[2]Instituto de Estructura de la Materia, IEM-CSIC, Serrano 123, E-28006 Madrid, Spain
[3]Departamento de Fisica Atomica, Molecular y Nuclear, Facultad de Ciencias Fisicas, Universidad Complutense de Madrid, E-28040 Madrid, Spain

E-mail: `gaidarov@inrne.bas.bg`

Abstract. We study various ground-state properties of neutron-rich and neutron-deficient Mg isotopes with A=20–36 in the framework of the self-consistent deformed Skyrme-Hartree-Fock plus BCS method. The nuclear symmetry energy is investigated for the same isotopic chain following the theoretical approach based on the coherent density fluctuation model. The results of the calculations show that the behavior of the nuclear charge radii and the nuclear matter properties in the Mg isotopic chain is closely related to the nuclear deformation. The temperature dependence of the symmetry energy for isotopic chains of even-even Ni (A=58–82), Sn (A=124–152), and Pb (A=202–214) nuclei is investigated in the framework of the local density approximation. The results for the thermal evolution of the symmetry energy coefficient show that for all isotopic chains considered and for both Skyrme forces used in the calculations the symmetry energy coefficient decreases with the increase of the mass number in the temperature interval T=0–5 MeV.

Introduction

The study of exotic nuclei is one of the main topics of modern nuclear structure physics. The recent developments of radioactive beams enable us to study such unstable nuclei toward the neutron and proton drip lines. lose to them, a large variety of formerly un nown nuclear configurations has been observed. The magic numbers in such exotic systems can be a quite intriguing issue. ew magic numbers appear and some others disappear in moving from stable to exotic nuclei in a rather novel manner due to specific components of the nucleon-nucleon interaction (see, for example, ef. [1]).

ow-lying states of neutron-rich nuclei around the neutron number N 20 attract a great interest, as the spherical configurations associated with the magic number disappear in the ground states. For ^{32}Mg, from the observed population of the excited 0_2^+ state (found at 1.0 Me) in the (t,p) reaction on ^{30}Mg, it is suggested [2] that the 0_2^+ state is a spherical one coexisting with the deformed ground state and that their relative energies are inverted at N 20. ery recently, a new signature of an existence of "island of inversion" [3] has been experimentally tested by measuring the charge radii of all magnesium isotopes in the *sd* shell at -
[ ] showing that the borderline of this island lies between ^{30}Mg and ^{31}Mg.

n the light of the new precise spectroscopic measurements of the neutron-rich ^{32}Mg nucleus that lies in the much explored "island of inversion" at N 20, in the present wor (see also ef. [ ]) we aim to perform a systematic study of the nuclear ground-state properties of neutron-rich and neutron-deficient Mg isotopes with A 20–3 , such as charge and matter rms radii, two-neutron separation energies, neutron, proton, and charge density distributions, neutron (proton) rms radii and related with them thic ness of the neutron (proton) s ins. The new data for the charge rms radii [ ] is a challenging issue to test the applicability of the mean-field description to light nuclei, thus expecting to understand in more details the nuclear structure revealed by them. Following our recent wor s [,], we analy e the correlation between the s in thic ness and the characteristics related to the density dependence of the nuclear symmetry energy for the same Mg isotopic chain. uch an analysis may probe the accurate account for the effects of interactions in our method within the considered Mg chain, where the brea down of the shell model could be revealed also by the nuclear symmetry energy changes. special attention is paid to the neutron-rich ^{32}Mg nucleus by performing additional calculations modifying the spin-orbit strength of the effective interaction, to chec theoretically the possible appearance of the "island of inversion" at N 20.

s nown, the nuclear symmetry energy is a measure of the energy gain in converting isospin asymmetric nuclear matter (M) to a symmetric system. ts value depends on the density and temperature. The need of information for the symmetry energy in finite nuclei, even theoretically obtained, is a ma or issue because it allows one to constrain the bul and surface properties of the nuclear energy-density functionals (Fs) quite effectively. Therefore, in the present paper the thermal behavior of the symmetry energy is investigated for i, n, and b isotopic chains using different model local density distributions for these nuclei and the symmetry energy coe cient for a specific nucleus is calculated in the local density approximation ( ) [ ].

Theoretical framework

The results for the nuclear properties of Mg isotopes have been obtained from self-consistent deformed artree-Foc calculations with density dependent yrme interactions [ ] and pairing correlations. airing between li e nucleons has been included by solving the B equations at each iteration with a fixed pairing strength that reproduces the odd-even experimental mass differences [10]. e consider the yrme forces y , 3, and because they are among the most extensively used and are considered as standard references.

The spin-independent proton and neutron densities are given by [11, 12]

$$\rho(\vec{R}) \quad \rho(r,z) \quad \sum_i 2v_i^2 \rho_i(r,z)\,, \tag{1}$$

where r and z are the cylindrical coordinates of $\vec{R}$, v_i^2 are the occupation probabilities resulting from the B equations and ρ_i are the single-particle densities. The mean square radii for protons and neutrons are defined as

$$\langle r_{p,n}^2\rangle \quad \frac{\int R^2 \rho_{p,n}(\vec{R})d\vec{R}}{\int \rho_{p,n}(\vec{R})d\vec{R}}\,, \tag{2}$$

and the rms radii for protons and neutrons are given by

$$r_{p,n} \quad \langle r_{p,n}^2\rangle^{1/2}\,. \tag{3}$$

aving the neutron and proton rms radii [q. (3)], the neutron s in thic ness is usually estimated as their difference

$$R \quad \langle r_n^2\rangle^{1/2} - \langle r_p^2\rangle^{1/2}. \qquad (\ )$$

The mean square radius of the charge distribution in a nucleus can be expressed as

$$\langle r_{ch}^2\rangle \quad \langle r_p^2\rangle \quad \langle r_{ch}^2\rangle_p \quad (N/Z)\langle r_{ch}^2\rangle_n \quad r_{CM}^2 \quad r_{SO}^2, \qquad (\;)$$

where $\langle r_p^2\rangle$ is the mean square radius of the point proton distribution in the nucleus (2), $\langle r_{ch}^2\rangle_p$ and $\langle r_{ch}^2\rangle_n$ are the mean square charge radii of the charge distributions in a proton and a neutron, respectively, r_{CM}^2 is a small correction due to the center of mass motion, and the last term r_{SO}^2 is a tiny spin-orbit contribution to the charge density. orrespondingly, we define the charge rms radius $r_c \quad \langle r_{ch}^2\rangle^{1/2}$.

e calculate the symmetry energy s of Mg isotopes using the coherent density fluctuation model (FM) [13, 1]. The FM allows us to ma e a transition from the properties of nuclear matter to those of finite nuclei. The analysis of the infinite nuclear matter has been carried out on the basis of the Bruec ner F [1 , 1]. t can be shown in the FM that under some approximation the properties of finite nuclei can be calculated using the corresponding ones for M, folding them by the weight function $|f(x)|^2$. The latter can be obtained by using a nown density distribution $\rho(r)$ for a given nucleus (see, e.g. [ ])

$$|f(x)|^2 \quad -\frac{1}{\rho_0(x)}\left.\frac{d\rho(r)}{dr}\right|_{r=x}, \qquad (\;)$$

where $\rho_0(x) \quad 3A/\ \pi x^3$ and with the normali ation $\int_0^\infty dx|f(x)|^2 \quad 1$. n the FM the symmetry energy for finite nuclei is expressed by that of M $s^{ANM}(x)$

$$s \quad \int_0^\infty dx|f(x)|^2 s^{ANM}(x). \qquad (\;)$$

For finite systems, different definitions of the symmetry energy coe cient are adopted in the literature yielding different values. n our case we calculate the symmetry energy coe cient for a specific nucleus within the and it is given as

$$e_{sym}(A,T) \quad \frac{1}{I^2A}\int \rho(r)e_{sym}[\rho(r),T]\delta_l^2(r)d^3r. \qquad (\;)$$

n q. ( ) $e_{sym}[\rho(r),T]$ is the symmetry energy coe cient at temperature T of infinite matter at the value of the local density $\rho(r)$, $\delta_l(r) \quad [\rho_n(r)-\rho_p(r)]/\rho(r)$ is the isospin asymmetry of the local density, where $\rho_n(r)$ and $\rho_p(r)$ are the neutron and proton densities, $\rho(r) \quad \rho_n(r) \quad \rho_p(r)$ and $I \quad (N-Z)/A$. The symmetry energy coe cient can be approximated by

$$e_{sym}(\rho,T) \quad \frac{e(\rho,\delta,T)-e(\rho,\delta \quad 0,T)}{\delta_l^2}, \qquad (\;)$$

where $e_{sym}(\rho,\delta,T)$ is the energy per nucleon in an asymmetric infinite matter, while $e(\rho,\delta \quad 0,T)$ is that one of symmetric nuclear matter. For an infinite system the energy per nucleon is calculated as $e \quad \varepsilon(r)/\rho$. e use for the total energy density of the system $\varepsilon(r)$ the yrme energy density functional with two yrme-class effective interactions, M* and y . The symmetry coe cient $e_{sym}(\rho,T)$ can then be computed from q. ( ). n contrast to the methodology employed in ef. [ ], where the inetic energy density $\tau_q(r)$ (with $q \quad (n,p)$ referring to neutrons or protons) entering the expression for $\varepsilon(r)$ is considered in the Thomas-Fermi approximation at finite temperature, we use in our theoretical scheme $\tau_q(r)$ [1]

$$\tau_q(r) \quad \frac{2m}{\hbar^2}\varepsilon_{K_q} \quad \frac{3}{}(3\pi^2)^{2/3}\left[\rho_q^{5/3} \quad \frac{\pi^2 m_q^2}{3\hbar^4}\frac{1}{(3\pi^2)^{4/3}}\rho_q^{1/3}T^2\right], \qquad (10)$$

which is valid at low T. The temperature-dependent densities are calculated through the FBT code that solves the nuclear yrme- artree-Foc -Bogoliubov (FB) problem by using the cylindrical transformed deformed harmonic-oscillator basis [1]. n addition, two other density distributions of ^{208} b [1], namely the Fermi-type density determined within the extended Thomas-Fermi (TF) method [20] and the symmetri ed-Fermi local density obtained within the rigorous density functional approach (F) [21], are used.

Results and discussion

The evolution of the quadrupole parameter β of the ground states as a function of the mass number A of neutron-rich and neutron-deficient Mg isotopes is presented in Fig. 1. s expected, the semi-magic ^{20}Mg isotope (N) is spherical. s the number of neutrons increases we start populating first the $d_{5/2}$ orbital between ^{22}Mg and ^{26}Mg, leading to prolate and oblate defomations in the energy profiles. rolate shapes with $\beta \sim 0.$ are ground states in ^{22}Mg and ^{24}Mg, whereas an oblate shape at around $\beta \sim -0.2$ is developed in ^{26}Mg in competition with a prolate shape at around $\beta \sim 0.3$. ue to the con unction of the N Z 12 deformed shell effects, the nucleus ^{24}Mg is the most deformed of the isotopic chain. e obtain rather flat profiles of the corresponding potential-energy curves around sphericity in ^{28,30}Mg, whose correct description would need beyond mean-field techniques involving configuration mixing of the shape fluctuations. ^{32}Mg becomes also spherical, thus showing that the magic number N 20 exists for the ground-state of ^{32}Mg in the mean-field theories [22]. For heavier isotopes the prolate deformation grows with the increase of the neutron number for $^{32-36}$Mg. n general, we find almost identical values of the quadrupole parameter β with the three yrme parametri ations.

The results for the symmetry energy s [q. ( )] as a function of the mass number A for the whole Mg isotopic chain (A 20–3) are presented in Fig. 2. t is seen that the and 3 forces yield values of s comparable with each other that lie above the corresponding symmetry energy values when using y set. lthough the values of s slightly vary within the Mg isotopic chain (23–2 Me) when using different yrme forces, the curves presented in Fig. 2 exhibit the same trend. t is useful to search for possible indications of an "island of inversion" around N 20 revealed also by the symmetry energy. e would li e to note that the modification of the spin-orbit strength of the y effective interaction by increasing it by 20 corresponding to prolate deformed ground-state of ^{32}Mg leads to a smaller value of s 23. Me compared with the one for the spherical case s 2 . Me . ndeed, the results shown in Fig. 2 are related to the evolution of the quadrupole parameter β as a function of the mass number A that is presented in Fig. 1. lthough the considered Mg chain does not contain a double-magic isotope, it is worth mentioning that we find maximum values within a plateau around the semi-magic isotope A 32 that resembles the sharp pea observed in previous wor s including double-magic nuclei [,].

The charge radius is related to the deformation and the isotope shifts of charge radii can be used to investigate the deformations in the isotopic chains. ur results for the squared charge radii differences $\delta < r_c^2 >^{26,A}$ $< r_c^2 >^{A} - < r_c^2 >^{26}$ ta ing the radius of ^{26}Mg as the reference are compared in Fig. 3 with the experimental data [ ]. n general, different yrme forces do not differ much in their predictions of charge rms radii of magnesium spanning the complete *sd* shell. The trend of the behavior of the experimental points and theoretical values strongly corresponds to the neutron shell structure. For $^{21-26}$Mg isotopes the charge distribution is compressed due to the filling of the $d_{5/2}$ orbital and the charge radii do not fluctuate too much. The addition of more neutrons on either $s_{1/2}$ or $d_{3/2}$ in the range $^{28-30}$Mg results in a fast increase of the radius. Finally, for isotopes beyond ^{30}Mg, where the "island of inversion" does exist in terms of the rms charge radius [ ], the theoretical results underestimate clearly the experimental points. bviously, an additional treatment is needed to understand in more details this specific region. e note the intermediate position of ^{27}Mg, where a minimum is observed in Fig. 3, since one of

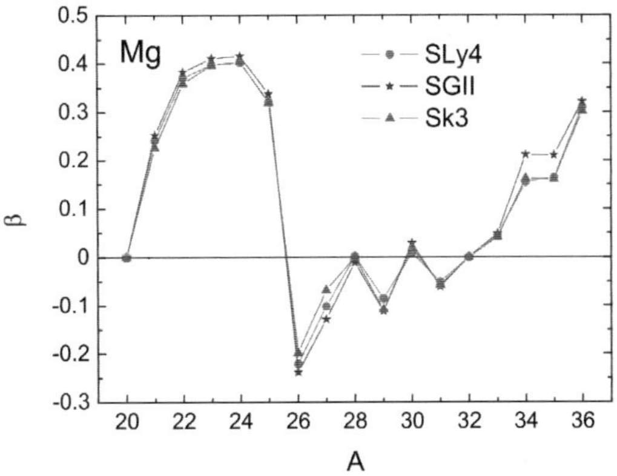

Figure The quadrupole parameter β of the ground state as a function of the mass number A for Mg isotopes (A 20–3) in the cases of y , , and 3 forces.

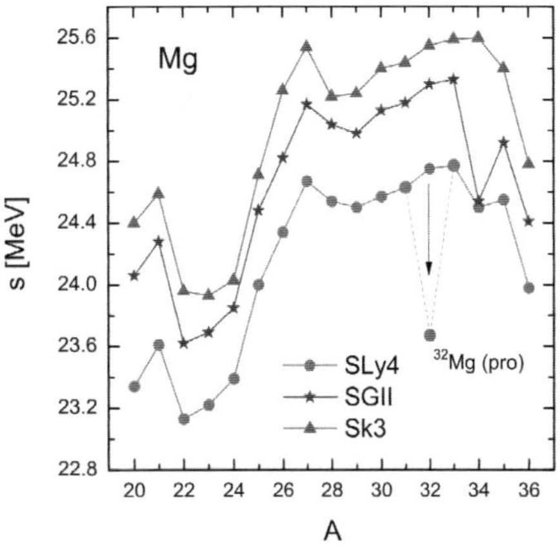

Figure The symmetry energies s for Mg isotopes (A 20–3) calculated with y , , and 3 forces.

the neutrons added to ^{25}Mg fills the last $d_{5/2}$ hole and the other one populates the $s_{1/2}$ subshell.

e plot in Fig. the differences between the rms radii of neutrons and protons R $r_n - r_p$. The latter is a simple measure of a neutron (proton) s in emergence in Mg isotopes from the considered isotopic chain. e can see from Fig. that R increases monotonically with neutron excess in the chain of Mg isotopes. From the yrme F analysis of the rms radii of proton and neutron density distributions ens e *et al.* [23] have found a proton s in in the neutron-deficient Mg isotopes, while at the neutron dripline the Mg isotopes develop extended neutron s ins. t can be also seen from Fig. that, in particular when using y force, a rather pronounced proton s in in ^{20}Mg develops which exceeds the neutron density by 0.33 fm. The latter value almost coincides with the value of 0.3 fm that has been obtained in ef. [23] using a standard yrme interaction. n the neutron-rich Mg isotopes our calculations predict quite massive neutron s ins whose thic ness for ^{36}Mg nucleus reaches 0.3 fm when the same y parametri ation is used.

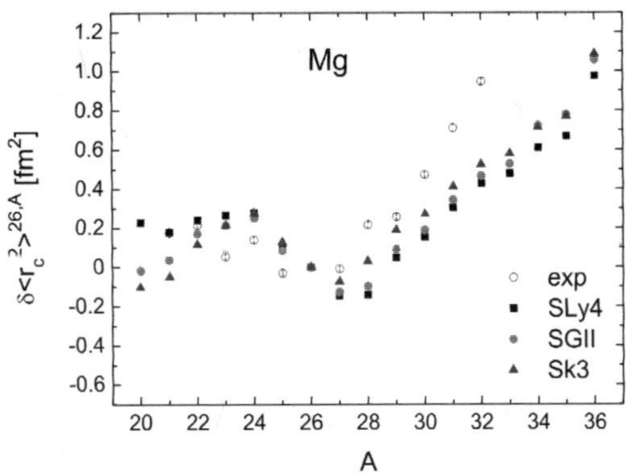

Figure Theoretical (with different yrme forces) and experimental [ ] isotope shifts $\delta < r_c^2 >$ of magnesium isotopes relative to ^{26}Mg.

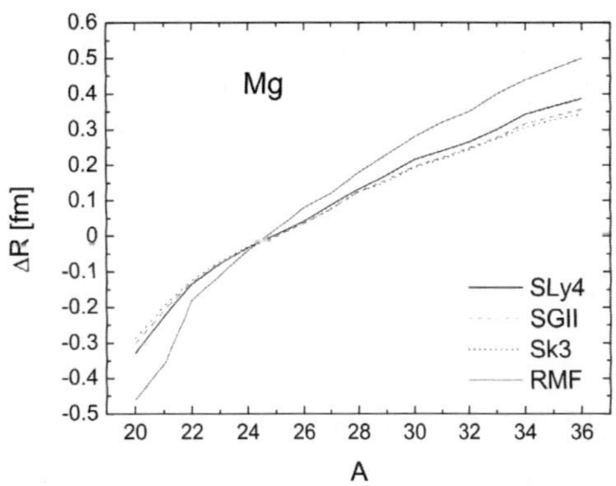

Figure ifference between neutron and proton rms radii r_{np} of Mg isotopes calculated by using y , , and 3 forces. The MF calculation results are from ef. [22].

n Fig. a comparison between the results for the symmetry energy coe cient for ^{208} b obtained by using three different densities, namely within the TF, F and FB methods (with M* and y forces), is made. The results for the thermal evolution of the symmetry energy coe cient in the interval T 0– Me show that its values decrease with temperature being larger in the case of symmetri ed-Fermi density of ^{208} b obtained within the F . Figure illustrates the isotopic evolution of the symmetry energy coe cient on the example of the b chain (A 202–21) in the case of y yrme interaction. smooth decrease of e_{sym} is observed with the increase of the mass number. e also would li e to note the lac of in for this b isotopic chain when loo ing at the values of e_{sym} at ero temperature. This result confirms our previous observations when studying the density dependence of the symmetry energy for b isotopes [,].

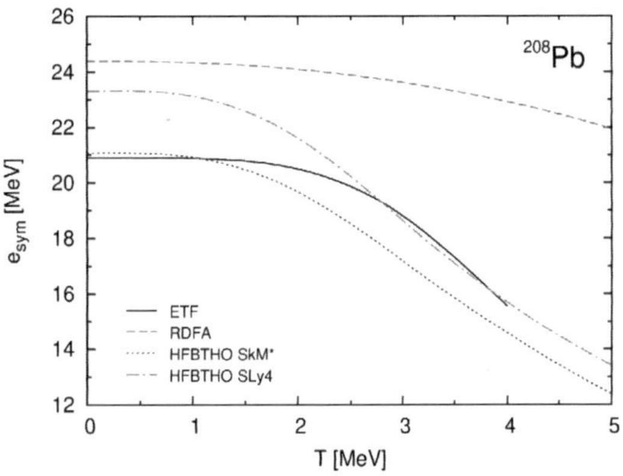

Figure 5 omparison of the results for the symmetry energy coe cient e_{sym} for ^{208} b calculated with TF, F and FB (with M* and y forces) densities.

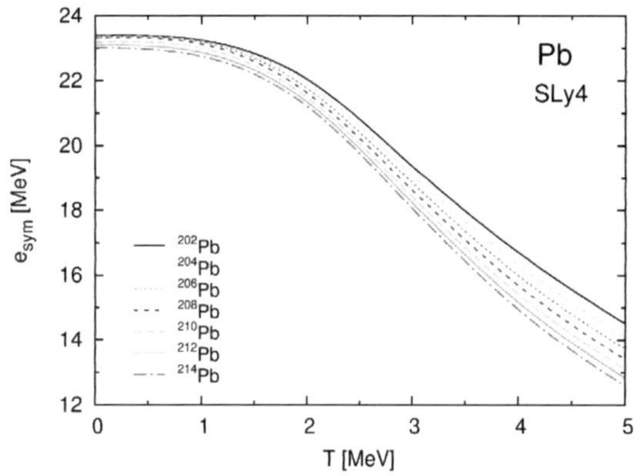

Figure 6 The temperature dependence of the symmetry energy coe cient e_{sym} obtained for several nuclei from b isotopic chain (A 202–21) in FB method with y force.

Conclusions

The results of the present wor can be summari ed as follows

i) The charge and mass radii follow the trends observed in the experiment, with fluctuating values up to A 2 , and smoothly increasing values with A, beyond A 2 . These global properties are found to be rather similar with the three yrme forces.

ii) The values of the symmetry energy s vary roughly between 23 and 2 Me . More dramatic is the considerable change in the trend of symmetry energy evolution with the mass number, when we include the results for the prolate (β 0.3) ground-state of ^{32}Mg obtained with the spin-orbit modified y effective interaction.

iii) The TF and F results for the density distributions demonstrate a smooth function of r at any temperature T, while the yrme FB densities have a stronger T-dependence. n general, the density distributions decrease with the temperature.

iv) t is observed that for all isotopic chains considered and for both yrme forces used in the calculations the symmetry energy coe cient decreases with the increase of the mass number in the same temperature interval.

To conclude, we would li e to note that further study is necessary to prove theoretically the existence of an "island of inversion" probed by the X- experiment. n particular, it is worth to perform calculations by including effects of tensor and three-body forces and exploring novel energy density functionals.

5 Acknowledgments

Three of the authors (M. . ., . . ., and . .) are grateful for support of the Bulgarian cience Fund under ontract o. F -T02 1 . . . . than s for the partial support from ontract o. F - 02 of the Bulgarian cience Fund. .M. . and . . ac nowledge support from M (pain) under ontracts F 2011–23 and F 201 – 1 1– .

References

[1] Otsuka T, Fujimoto R, Utsuno Y, Brown B A, Honma M and Mizusaki T 2001 *Phys. Rev. Lett.* **87** 082502
[2] Wimmer K *et al.* 2010 *Phys. Rev. Lett.* **105** 252501
[3] Warburton E K, Becker J A and Brown B A 1990 *Phys. Rev.* C **41** 1147
[4] Yordanov D T *et al.* 2012 *Phys. Rev. Lett.* **108** 042504
[5] Gaidarov M K, Sarriguren P, Antonov A N and Moya de Guerra E 2014 *Phys. Rev.* C **89** 064301
[6] Gaidarov M K, Antonov A N, Sarriguren P and Moya de Guerra E 2011 *Phys. Rev.* C **84** 034316
[7] Gaidarov M K, Antonov A N, Sarriguren P and Moya de Guerra E 2012 *Phys. Rev.* C **85** 064319
[8] Agrawal B K, De J N, Samaddar S K, Centelles M and Viñas X 2014 *Eur. Phys. J.* A **50** 19
[9] Vautherin D 1973 *Phys. Rev.* C **7** 296
[10] Audi G, Wapstra A H and Thibault C 2003 *Nucl. Phys.* A **729** 337
[11] Sarriguren P, Gaidarov M K, Moya de Guerra E and Antonov A N 2007 *Phys. Rev.* C **76** 044322
[12] Moya de Guerra E, Sarriguren P, Caballero J A, Casas M and Sprung D W L 1991 *Nucl. Phys.* A **529** 68
[13] Antonov A N, Nikolaev V A and Petkov I h 1979 *Bulg. J. Phys.* **6** 151; 1980 *Z. Phys.* A **297** 257; *ibid* 1982 **304** 239; 1985 *Nuovo Cimento* A **86** 23; Antonov A N *et al.*, *ibid* 1989 **102** 1701; Antonov A N, Kadrev D N and Hodgson P E 1994 *Phys. Rev.* C **50** 164
[14] Antonov A N, Hodgson P E and Petkov I h 1988 *Nucleon Momentum and Density Distributions in Nuclei* (Clarendon Press, Oxford); 1993 *Nucleon Correlations in Nuclei* (Springer-Verlag, Berlin-Heidelberg-New York)
[15] Brueckner K A, Buchler J R, Jorna S and Lombard R J 1968 *Phys. Rev.* **171** 1188
[16] Brueckner K A, Buchler J R, Clark R C and Lombard R J 1969 *Phys. Rev.* **181** 1543
[17] Lee S J and Mekjian A 2010 *Phys. Rev.* C **82** 064319
[18] Stoitsov M, Schunck N, Kortelainen M, Michel N, Nam H, Olsen E, Sarich J and Wild S 2013 *Computer Physics Communications* **184** 1592
[19] Antonov A N, Kanev J, Petkov I h and Stoitsov M V 1989 *Nuovo Cimento* **101A** 525
[20] Brack M, Guet C and Håkansson H-B 1985 *Phys. Rep.* **123** 275
[21] Stoitsov M V, Petkov I and Kryachko E S 1987 *C. R. Bulg. Acad. Sci.* **40** 45; Stoitsov M V 1987 *Nuovo Cimento* **98A** 725
[22] Ren , hu Y, Cai Y H and Xu G 1996 *Phys. Lett.* B **380** 241
[23] Lenske H and Schrieder G 1998 *Eur. Phys. J.* A **2** 41

The proton-neutron symplectic model of nuclear collective motions

H G Ganev

Joint Institute for Nuclear Research, Dubna, Russia

Abstract. The proton-neutron symplectic model of nuclear collective motion is presented. It is shown that it appears as a natural multi-major-shell extension of the generalized proton-neutron $SU(3)$ scheme which includes rotations with intrinsic vortex as well as monopole, quadrupole and dipole giant resonance vibrational degrees of freedom.

Introduction

ymmetry plays an important role in our understanding of collective structure of atomic nuclei. n nuclear physics there are two approaches which exploit the symmetry. The first, standard, approach appears in the form of exact or dynamical symmetries in which the basis states and physical observables li e the amiltonian and the transition operators are expressed as tensor operators with respect to the symmetry under consideration. This yields exact solutions for the eigenvalues and eigenfunctions and allows the calculation of required matrix elements by using a generali ed igner- c art theorem. The most popular examples of this approach are the $SU(3)$ lliott model [1] and the nteracting Boson Model [2]. There is, however, another non-standard approach for exploiting the symmetry in which one first identifies the generators of possible collective flows and then the algebra they close under commutation [3, , ,]. This reveals directly the dynamical content of a certain algebraic model. The amiltonian of nuclear system is then assumed to be a function of these operators. long these lines, some algebraic models of collective motions in nuclei have been proposed based on the algebras of $SL(3,R)$ [ ], $Rot(3)$ [ ], $CM(3)$ [ ] and $Sp(,R)$ [10] groups, respectively. The $Rot(3)$ algebra is ust the rigid rotor algebra proposed by i [ ]. The $SL(3,R)$ model of eaver and Biedenharn ta es into consideration, beyond the rotational, also vibrational degrees of freedom. The $CM(3)$ model has been shown to be a microscopic version of the Bohr-Mottelson model [11] augmented by the vortex spin degrees of freedom. Finally, in order to ma e the $CM(3)$ model compatible with the microscopic shell model structure, owe and osensteel extended the latter to the one-component symplactic model with $Sp(,R)$ dynamical symmetry [10]. The $Sp(,R)$ group is shown to be the full dynamical group of collective excitations in the one-component many-particle nuclear system.

ecently, the non-standard symmetry approach to nuclear collective motion was applied to the two-component many-particle nuclear systems. s a result, the proton-neutron symplectic model was formulated by considering the symplectic geometry and possible collective flows in the many-particle nuclear system [12]. This allowed to reveal some new features and forms of collective excitations which are missing in the microscopic theory of the one-component nuclear systems [12, 13].

The Proton-Neutron Symplectic Model

The collective observables of the proton-neutron symplectic model (M) are spanned by the following one-body operators [12]

$$Q_{ij}(\alpha,\beta) \quad \sum_{s=1}^{m} x_{is}(\alpha)x_{js}(\beta), \tag{1}$$

$$S_{ij}(\alpha,\beta) \quad \sum_{s=1}^{m} \Big(x_{is}(\alpha)p_{js}(\beta) \quad p_{is}(\alpha)x_{js}(\beta) \Big), \tag{2}$$

$$L_{ij}(\alpha,\beta) \quad \sum_{s=1}^{m} \Big(x_{is}(\alpha)p_{js}(\beta) - x_{js}(\beta)p_{is}(\alpha) \Big), \tag{3}$$

$$T_{ij}(\alpha,\beta) \quad \sum_{s=1}^{m} p_{is}(\alpha)p_{js}(\beta), \qquad (\,)$$

which are expressed in terms of position $x_{is}(\alpha)$ and momentum $p_{jt}(\beta)$ $(i,j \quad 1,2,3 \;\; \alpha,\beta \quad p,n \;\; s,t \quad 1,\ldots,m)$ coordinates of the $m \quad A-1$ relative acobi vectors. The mass quadrupole tensor operators (1) determine the shape and orientation of the nucleus. The shear momentum operators (2) are the infinitesimal generators of deformation and irrotational-flow rotations, which are related to the high-lying excitations associated with the giant resonance vibrational degrees of freedom. The operators (3) are the infinitesimal operators of rigid rotations in the -dimensional space and correspond to the low-lying nuclear excitations. Finally, the generators ( ) are the momentum tensor operators among which are the inetic energies of the proton and neutron systems. ll these operators close under commutation the non-compact symplectic algebra $sp(12,R)$, which is the dynamical algebra of collective excitations in the two-component many-particle nuclear system.

By ta ing different subsets of symplectic generators ones obtains different subalgebras of $sp(12,R)$ which reveal directly the dynamical content of the model. For more details see ef.[12]. n this way, e.g., one obtains the reduction chain

$$sp(12,R) \supset gcm(\;) \supset gcm(3) \supset rot(3) \supset so(3). \qquad (\,)$$

nother interesting example is provided by the following reduction chain

$$sp(12,R) \supset gcm(\;) \supset [R^{21}]so(\;) \supset rot_p(3) \oplus rot_n(3) \supset rot(3) \supset so(3), \qquad (\,)$$

from which one can see the algebraic structure of two coupled rigid rotors (two rotor model). ere, in contrast to the original version of the Two otor Model of . o udice and F. alumbo [1], we do not have the restriction to the case of two axial rotors.

ynamical group $Sp(12m,R)$ of the whole system allows to separate the many-particle nuclear coordinates, represented by the set of m translationally-invariant acobi vectors, into collective and intrinsic ones. This can be done by the corresponding ic endraht- ubli coordinate transformation which leads to the separation of 21 collective variables $\xi \equiv \{\rho^{(a_0)}, g_6^+\}$ and $\;m-21$ intrinsic coordinates g_m [12]. The six radial variables $\rho^{(a_0)}$ ta e the values $0 \leq \rho^{(a_0)} \leq \infty$, g_6^+ denotes the set of 1 variables parametri ing the group $SO(\;)$, and the set of intrinsic variables g_m parametri es the coset space $O(m)/O(m-\;)$. The change of variables induces a corresponding decomposition of the nuclear wave functions into collective and intrinsic parts, respectively. t can be shown that the simplest inematically correct wave functions can be

Journal of Physics: Conference Series **724** (2016) 012016 doi:10.1088/1742-6596/724/1/012016

classified by the quantum numbers provided by the unitary scheme chain [12, 1]

$$
\begin{array}{ccccc}
U(\ m) & \supset & U(\ ) & \otimes & U(m) \\
[E_0 0 \ldots 0] & & [E_1 \ldots E_6] & & [E_1 \ldots E_6 \dot{0}] \\
 & & \cup & & \beta\cup \\
 & & G & & O(m) \\
 & & & & (\omega_1 \ldots \omega_6) \\
 & & & & \delta\cup \\
 & & & & S_{m+1} \\
 & & & & [f]h
\end{array}
\qquad (\ )
$$

which is embedded in the full many-particle dynamical group $Sp(12m, R)$. Then the wave function can be written in the form [12]

$$\Psi\binom{E_0 E\eta}{\beta\omega\delta[f]h}|x_1^a, \ldots, x_m^a) \equiv \Psi\binom{E_0 E\eta}{\beta\omega\delta[f]h}|\rho^{(a_0)}, g_6^+, g_m) \quad \sum_{\nu^0} \Theta\binom{E_0 E\eta}{\beta\omega\nu^0}|\rho^{(a_0)}, g_6^+) D^{\omega}_{\nu^0,\delta[f]h}(g_m). \qquad (\ )$$

The proton-neutron symplectic model is then given a simple expression as a hydrodynamical model with wave functions comprising collective (irrotational-flow) and intrinsic (vortex) components. ere we want to point out that if the $O(m)$ irreducible representation ω is the scalar representation, then one simply obtains the pure two-fluid irrotational-flow collective model which states are associated with the giant resonance degrees of freedom. For real nuclei ω / 0, then although we are spea ing of collective degrees of freedom, the latter are always coupled to the intrinsic vortex degrees of freedom, as can be seen from ( ). The need of consideration of the intrinsic degrees of freedom and their coupling to the irrotational-flow collective dynamics has been reali ed long time ago. The latter, as we will see further, turns out to be of vital importance for the appearing of the low-lying collective bands.

Representations of the $Sp(12, R)$ Lie algebra

aving a given classical collective model with a certain dynamical algebra we can quanti e it by constructing its irreducible representations in the many-particle ilbert space. To construct the irreducible representations of the $Sp(12, R)$ ie algebra in shell-model terms, it is convenient to introduce the harmonic oscillator raising $b^{\dagger}_{i\alpha,s} \quad \sqrt{\frac{m_\alpha \omega}{2\hbar}}\left(x_{is}(\alpha) - \frac{i}{m_\alpha \omega} p_{is}(\alpha)\right)$ and lowering $b_{i\alpha,s} \quad \sqrt{\frac{m_\alpha \omega}{2\hbar}}\left(x_{is}(\alpha) \quad \frac{i}{m_\alpha \omega} p_{is}(\alpha)\right)$ operators which satisfy the standard boson commutation relations $[b_{i\alpha,s}, b^{\dagger}_{j\beta,t}] \quad \delta_{ij}\delta_{\alpha\beta}\delta_{st}$. Then, a basis for the many-particle reali ation of the $Sp(12, R)$ ie algebra is given by [13]

$$F_{ij}(\alpha,\beta) \quad \sum_{s=1}^{m} b^{\dagger}_{i\alpha,s} b^{\dagger}_{j\beta,s}, \quad G_{ij}(\alpha,\beta) \quad \sum_{s=1}^{m} b_{i\alpha,s} b_{j\beta,s}, \qquad (\ )$$

$$A_{ij}(\alpha,\beta) \quad \frac{1}{2}\sum_{s=1}^{m} (b^{\dagger}_{i\alpha,s} b_{j\beta,s} \quad b_{j\beta,s} b^{\dagger}_{i\alpha,s}). \qquad (10)$$

The F's operators, as we will see, represent the core collective excitations of monopole, quadrupole and dipole type, associated with the giant resonance vibrational degrees of freedom.

Journal of Physics: Conference Series **724** (2016) 012016 doi:10.1088/1742-6596/724/1/012016

The 's are the corresponding con ugate deexcitation operators. The number-conserving operators (10) generate the maximal compact subalgebra $u( ) \subset sp(12, R)$, related to the internal vortex degrees of freedom, responsible for the appearance of the low-lying collective bands.

n $Sp(12, R)$ unitary irreducible representation is characteri ed by the $U( )$ quantum numbers $\sigma \quad [\sigma_1, \ldots, \sigma_6]$ of its lowest-weight state $|\sigma\rangle$, i.e. $|\sigma\rangle$ satisfies

$$G_{ab}|\sigma\rangle \quad 0 \quad A_{ab}|\sigma\rangle \quad 0, \quad a < b \quad A_{aa}|\sigma\rangle \quad \sigma_a|\sigma\rangle \tag{11}$$

for the indices $a \equiv i\alpha$ and $b \equiv j\beta$ ta ing the values $1, \ldots,$. f we introduce the $U( )$ tensor product operators $P^{(n)}(F) \quad [F \times \ldots \times F]^{(n)}$, where $n \quad [n_1, \ldots, n_6]$ is a partition with even integer parts, then by an $U( )$ coupling of these tensor products to the lowest-weight $U( )$ state $|\sigma\rangle$, one constructs the whole basis of states for an $Sp(12, R)$ irrep

$$|\Psi(\sigma n \rho E \eta)\rangle \quad [P^{(n)}(F) \times |\sigma\rangle]^{\rho E}_{\eta}, \tag{12}$$

where $E \quad [E_1, \ldots, E_6]$ indicates the $U( )$ quantum numbers of the coupled state, η labels a basis of states for the coupled $U( )$ irrep E and ρ is a multiplicity index. This is shown schematically in Fig.1. n this way we obtain a basis of $Sp(12, R)$ states that reduces the subgroup chain $Sp(12, R) \supset U( )$.

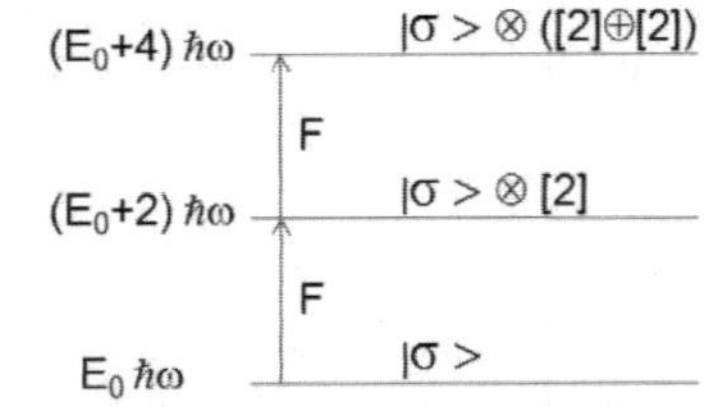

Figure The construction of the symplectic basis by acting with the $U( )$-coupled tensor products $P^{(n)}(F) \quad [F \times \ldots \times F]^{(n)}$ on the lowest-weight state $|\sigma\rangle$.

The shell-model classification of nuclear collective states

ow we can determine the relevant $Sp(12, R)$ irreducible representations appropriate for the description of the low-lying collective states in heavy mass deformed nuclei? The answer is suggested by the shell model, and in particular, from the underlying shell-model structure of the ground state. To reveal this structure we specify the basis of an $Sp(12, R)$ irrep by considering the following reduction of the subgroup $U( ) \subset Sp(12, R)$ [13]

$$\begin{array}{cccccccc} Sp(12,R) \supset & U( ) \supset & SU_p(3) \otimes & SU_n(3) \supset & SU(3) \supset & SO(3) \supset & SO(2). \\ \sigma & n\rho \;\; E \;\; \gamma & (\lambda_p, \mu_p) & (\lambda_n, \mu_n) \;\; \varrho & (\lambda, \mu) \;\; K & L & M \end{array} \tag{13}$$

The chain (13) naturally generali es the lliott's $SU(3)$ model [1] by extending the model space to the direct product space $SU_p(3) \otimes SU_n(3)$ of proton and neutron subsystems. The $SU(3)$ irreps of the two subsystems are subsequently strongly coupled to the $SU(3)$ irrep of the combined proton-neutron system to form composite configurations with different possible deformations. The maximum deformation is obtained for the leading $SU(3)$ irrep. The combined $SU(3)$ algebra is generated by the lliott's quadrupole $\widetilde{Q}_M \quad \widetilde{Q}^p_M \quad \widetilde{Q}^n_M$ and angular momentum $L_M \quad L^p_M \quad L^n_M$ operators, respectively. The chain (13) corresponds to the following choice of the index $\eta \quad \gamma(\lambda_p, \mu_p)(\lambda_n, \mu_n)\varrho(\lambda, \mu)KLM$ labeling the basis states (12) of the $Sp(12, R)$ irrep.

The generators of $Sp(12, R)$ ( )–(10) can be classified as irreducible tensor operators with respect to the different subgroups of the whole chain (13) and hence can be characteri ed by the quantum numbers determining their irreducible representations. For the raising operators one readily obtains the following tensors

$$F^{[2]_6}_{(2,0)(0,0)}\quad {}^{LM}_{(2,0)}(p,p),\qquad F^{[2]_6}_{(0,0)(2,0)}\quad {}^{LM}_{(2,0)}(n,n),\qquad F^{[2]_6}_{(1,0)(1,0)}\quad {}^{LM}_{(2,0)}(p,n),\qquad (1\ )$$

where $L\quad 0, 2\ M\quad -L, \ldots, M$, and

$$F^{[2]_6}_{(1,0)(1,0)}\quad {}^{1M}_{(0,1)}(p,n).\qquad (1\ )$$

e see a multiplication of the standard one-component $Sp(\ , R)$ raising generators [10] which for the two-component system correspond to the creation of monopole and quadrupole pp, nn, and pn pairs. n addition to the $(2,0)$ $SU(3)$ raising generators $F^{[2]_6}_{(p,0)(q,0)}\quad {}^{LM}_{(2,0)}(\alpha,\beta)$ (1) we have also the $(0,1)$ $SU(3)$ tensor operators $F^{[2]_6}_{(1,0)(1,0)}\quad {}^{1M}_{(0,1)}(p,n)$ (1), which are new ones compared to the generators of the $Sp(\ , R)$ model of osensteel and owe [10]. The shell model properties of raising operators (1)–(1) can be determined by considering their $U(3) \supset U(1) \otimes SU(3)$ tensor character. Thus, it becomes obvious that the raising operators of $Sp(12, R)$ with $U(1) \otimes SU(3)$ quantum numbers $N(\lambda, \mu)\quad 2(2,0)$ and their con ugate lowering ones represent $\pm 2\hbar\omega$ inter-shell collective excitations of monopole ($L\quad 0$) and quadrupole ($L\quad 2$) type. dditionally, in contrast to the $Sp(\ , R)$ model, the $Sp(12, R)$ raising operators with $N(\lambda, \mu)\quad 2(0,1)$ together with their con ugate lowering operators correspond to the $\pm 2\hbar\omega$ inter-shell excitations of dipole ($L\quad 1$) type. Thus, the $Sp(12, R)$ collective dynamics covers the nuclear coherent excitations of monopole, dipole and quadrupole type.

The construction of the shell model representations of the proton-neutron symplectic model is given in detail in ef.[13]. For example, the shell model considerations give for ^{154}Sm the symplectic bandhead structure $\sigma\quad [132,\ \ , 0, 0, 0, 0]$. The latter will reduce to a huge amount of $SU_p(3) \otimes SU_n(3)$ irreducible representations among which is the $(12,\ ) \otimes (3\ ,\ )$ one that is composed by the direct product of the leading $SU(3)$ irreps for the proton and neutron subsystems, which are strongly coupled to the final $SU(3)$ symmetry. The maximum deformation, corresponding to the ground state, is obtained by restricting the direct product irrep $(12,\ ) \otimes (3\ ,\ )$ of $SU_p(3) \otimes SU_n(3)$ to the leading irreducible representation $(\ ,1\ )$ of $SU(3)$. The other $SU(3)$ irreps of the combined proton-neutron system (with decreasing deformation), contained in the symplectic $Sp(12, R)$ bandhead structure, are obtained from the direct product of the parent $SU(3)$ irreducible representations

$$\begin{array}{l} (12,\ ) \otimes (3\ ,\ ) \rightarrow (\ ,1\ ), (\ ,1\ ), (\ ,12), \ldots \\ \qquad\qquad\qquad\quad (\ ,1\ ), (\ ,1\ ), (\ ,13), \ldots \\ \qquad\qquad\qquad\quad \vdots \qquad \vdots \qquad \vdots \end{array}\qquad (1\ )$$

The fact that one obtains a nonscalar $Sp(12, R)$ irreducible representation $\langle\sigma\rangle\ /\ 0$ for the real nuclei is a very important feature of the present approach, which is a consequence of the two-component composite character of the nuclear systems. The $Sp(12, R)$ irreducible representation is determined by the symplectic bandhead (or intrinsic) structure, defined by the lowest $U(\ )$ irrep $\sigma\quad [\sigma_1, \ldots, \sigma_6]\ /\ 0$. The latter, in contrast to the $Sp(\ , R)$ case, will contain a plethora of $SU(3)$ multiplets. This is of significant importance in the microscopic nuclear structure theory because the intrinsic base space spanned by the $U(\ )$ irrep $[\sigma_1, \ldots, \sigma_6]$, as can be seen from (1), will contain many $SU(3)$ irreducible representations appropriate for the description of different low-lying collective bands in the spectra of heavy even-even deformed nuclei. n this way, in contrast to the $Sp(\ ,R)$ model, the intrinsic $Sp(12,R)$ bandhead structure provides us with a framewor for the simultaneous shell-model interpretation of the ground state band and the other excited low-lying collective bands without the need of involving the mixing of different symplectic irreps (cf. ef.[1]).

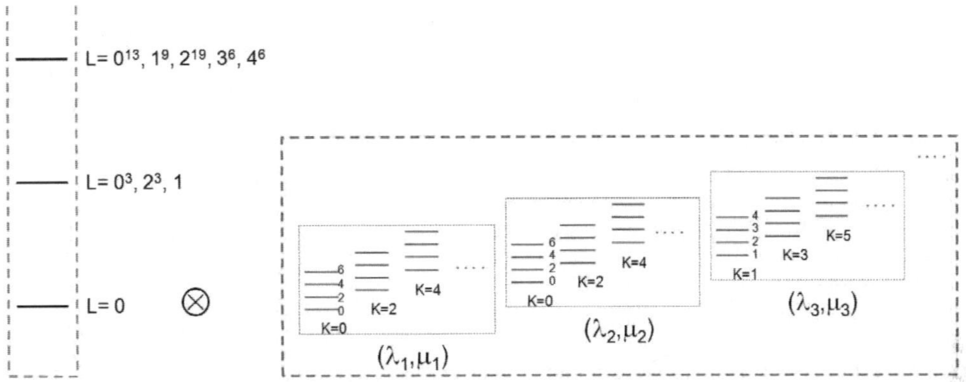

Figure The basis states of the $Sp(12,R)$ irreps are in 1-1 correspondence with the coupled product of the two-fluid irrotational-flow collective model and an $U(\ )$ model, related to the valence shell proton-neutron degrees of freedom, which contains many $SU(3)$ multiplets appropriate for the description of different low-lying collective bands.

n this way, the $Sp(12,R)$ irreducible representations can be represented as a coupled product of two structures a symplectic bandhead related to the proton-neutron valence shell degrees of freedom and a collective structure, associated with the giant resonance degrees of freedom. f the symplectic bandhead is represented by the scalar $Sp(12,R)$ representation, then one obtains simply the two-fluid irrotational-flow collective model. The states of the latter are in 1-1 correspondence with the states of the 21-dimensional harmonic oscillator. For non-trivial bandhead symplectic structure, the states of the $Sp(12,R)$ irreps are in 1-1 correspondence with the states of the coupled product of the giant resonance and the proton-neutron valence shell degrees of freedom, as shown in Fig.2. Moreover, in the limit of large $Sp(12,R)$ representation quantum numbers, the proton-neutron symplectic model contracts to the $U(\ )$-phonon model [1] with the semi-direct product structure $[HW(21)]U(\ )$ which appears as an unification of the two-fluid irrotational-flow collective model and an $U(\ )$ model, containing all the necessary $SU(3)$ multiplets required for the description of the low-lying collective bands. ll this becomes more clear, if we express the basis functions in the form

$$\psi_{\sigma} \quad \sum_{\eta} \psi_{\eta}(F)|\sigma\eta\rangle, \qquad (1 \ )$$

where ψ_{η} is a polynomial in the $Sp(12,R)$ raising operators. q.(1) can be interpreted as a factoring of an arbitrary wave function into collective and intrinsic parts. The bandhead states $|\sigma\eta\rangle$ can be thought of as intrinsic states and the rasing operators ψ_{η} as collective wave functions.

5 Conclusions

The fully-microscopic proton-neutron symplectic model of nuclear collective motion, based on the non-compact $Sp(12,R)$ group, is presented. From a hydrodynamic perspective, it is shown that it appears as a two-fluid irrotational-flow collective model augmented by the intrinsic vortex degrees of freedom. From a shell-model perspective, from other side, the M appears as a natural multi-ma or-shell extension of the generali ed proton-neutron $SU(3)$ scheme which ta es into account the core collective excitations of monopole and quadrupole, as well as dipole type associated with the giant resonance vibrational degrees of freedom.

[1] J. P. Elliott, Proc. R. Soc. London, Ser. A **245**, 128 (1958); **245**, 562 (1958).
[2] F. Iachello and A. Arima, *The Interacting Boson Model* (Cambridge University Press, Cambridge, 1987).
[3] G. Rosensteel and D. J. Rowe, Ann. Phys. **96**, 1 (1976).
[4] G. Rosensteel and D. J. Rowe, Ann. Phys. **123**, 36 (1978).
[5] D. J. Rowe and G. Rosensteel, Ann. Phys. **126**, 198 (1980).
[6] O. L. Weaver, R. Y. Cussion and L. C. Biedenharn, Ann. Phys. (N.Y.) **102**, 493 (1976).
[7] L. Weaver and L. C. Biedenharn, Nucl. Phys. **A185**, 1 (1972).
[8] H. Ui, Prog. Theor. Phys. **44**, 153 (1970).
[9] L. Weaver, R. Y. Cusson and L. C. Biedenharn, Ann. Phys. (N.Y.) **77**, 250 (1973).
[10] D. J. Rowe and G. Rosensteel, Phys. Rev. Lett. **38**, 10 (1977).
[11] A. Bohr and B. R. Mottelson, *Nuclear Structure* (W.A. Benjamin Inc., New York,1975), Vol. II.
[12] H. G. Ganev, Eur. Phys. J. **A50**, 183 (2014).
[13] H. G. Ganev, Eur. Phys. J. **A51**, 84 (2015).
[14] N. Lo Iudice and F. Palumbo, Phys. Rev. Lett. **41**, 1532 (1978); Nucl. Phys. **A326**, 193 (1979).
[15] V. V. Vanagas, *Algebraic foundations of microscopic nuclear theory* (Nauka, Moscow, 1988) (in Russian).
[16] J. Carvalho *et al.*, Nucl. Phys. **A452**, 240 (1986).
[17] H. G. Ganev, Int. J. Mod. Phys. **E 24**, 1550039 (2015).

Lifetimes and electromagnetic transition strength in ^{157}Dy

K A Gladnishki[1], P Petkov[2,3], A Dewald[3], J Jolie[3],O Möller[3], B Saha[3], A Fitzler[3], K Jessen[3], D Tonev[2], T Klug[3], S Heinze[3], P von Brentano[3], G Rainovski[1], M Trichkova[1], D Bazzacco[4], C A Ur[4,7], E Farnea[4,5], M Axiotis[5,6], S Lunardi[4], G de Angelis[5], D R Napoli[5], N Marginean[5,7], T Martinez[5], M A Caprio[8,9]

[1]Faculty of Physics, St. Kliment Ohridski University of Sofia, 1164 Sofia, Bulgaria
[2]Bulgarian Academy of Sciences, Institute for Nuclear Research and Nuclear Energy, 1784 Sofia, Bulgaria
[3]Institut für Kernphysik, Universität zu Köln, D-50937 Köln, Germany
[4]Dipartimento di Fisica and INFN, Sezione di Padova, Padova, Italy
[5]INFN, Laboratori Nazionali di Legnaro, Legnaro, Italy
[6]TANDEM Accelerator Laboratory, Institute of Nuclear Physics,NCSR "Demokritos", Athens, Greece
[7]Horia Hulubei National Institute of Physics and Nuclear Engineering, RO-077125 Bucharest-Magurele, Romania
[8]Wright Nuclear Structure Laboratory, Yale University, New Haven, Connecticut, USA
[9]Department of Physics, University of Notre Dame, Notre Dame, Indiana 46556-5670, USA

E-mail: kag@phys.uni-sofia.bg

Abstract. Excited states in ^{157}Dy have been studied by γ-γ coincidence measurements via the reaction ^{124}Sn(^{36}S,3n) at a beam energy of 155MeV. Lifetimes of the relatively lower-spin states in ^{157}Dy were measured by means of the Recoil Distance Dopplershift technique in the coincidence mode. The experiment was performed at the Laboratori Nazionali di Legnaro with the GASP array and the Cologne plunger device. With the same setup a Doppler-shift attenuation (DSA) lifetime measurement was performed for the higher spin states. The Differential decay-curve method was applied for the lifetime determination.

1. Introduction

The $^{156-158}$Dy isotopes have neutron numbers between 90 and 92 and thus lie close to the onset of prolate deformations taking place around N = 88 in this mass region. From the known properties of the low-spin states of these nuclei it can be deduced that they are already rather well deformed. The β-deformations in ^{156}Dy is 0.29 which is slightly smaller than the deformation in the heaviest stable isotope ^{164}Dy β = 0.35 close to midshell. The structure of the energy levels in these nuclei has been study in detail in the past [1, 2, 3, 4, 5, 6, 7]. An IBA fit [9] helps to position ^{156}Dy in the transitional region between spherical and deformed nuclei. The results of the fit and empirical considerations reveal that in ^{156}Dy the gamma-degree of freedom (or the triaxiality) plays a more important role than in the established X(5) nuclei ^{152}Sm,^{150}Nd and ^{154}Gd. A fit of the data using the General Collective Model points to a deeper collective potential V (β,γ) in ^{156}Dy which may be also a reason for the differences in the spectroscopic properties of ^{156}Dy and the neighboring N=90 isotones exhibiting an X(5) character. A recent work on ^{155}Dy [8] reveals that the different low-lying quasineutron bands are characterized by different quadrupole deformations. This may be explained by configuration dependent particle-core interaction in the transitional ^{155}Dy. At high spin, the new data for ^{155}Dy seem to confirm earlier findings of reduced E2 collectivity related to the phenomenon of band termination for some of the bands. Our study of ^{157}Dy allows to investigate if the shape-transition effects observed in neighboring lighter Dy isotopes continue gradually when going away from the N=82 shell closure towards midshel. In addition, by analyzing the data from our experiment we have the possibility to study the influence of the rotation alignment on the collective properties in ^{157}Dy. The entire set of E2-transition strengths deduced for ^{157}Dy will result in a systematic view of the collective behaviour of this nucleus with increasing rotational frequency and provide strong evidence for deformation changes.

2. Experimental details

The experiments were performed at the XTU Tandem Facility in Laboratori Nazionali di Legnaro, Italy. The excited states in ^{157}Dy were populated using the ^{124}Sn(^{36}S,3n)^{157}Dy reaction. For the RDDS part of the experiment the beam of ^{36}S was accelerated to an energy of 155 MeV and then delivered to the target consisting of 0.9 mg/cm^2 Tin, enriched to 97.7% in ^{124}Sn. The tin material was evaporated onto a backing of 1.8 mg/cm^2 Ta foil. To stop the recoils a 12.0 mg/cm^2 Au foil was used. The levels of interest for DSAM part of the experiment were populated using a beam energy of 145 MeV. A 0.9 mg/cm^2 Sn foil enriched to 95.3% in ^{124}Sn was used as a target. A 13.4 mg/cm^2 Ta foil was used as a backing to stop the recoils. The emitted γ-rays were detected by the GASP detector array [10]. The HpGe detectors from the GASP array are grouped in 7 rings with respect to the beam line. For our analysis four rings where appreciable Doppler-shifts can be observed were used, namely ring 0 (mean angle with respect to the beam axis of 34.6°), ring 1 (59.4°), ring 5 (120.6°) and ring 6 (145.4°). With the same setup we measured DSAM and RDDS lifetimes in ^{156,155}Dy and the results were published in Refs.[8, 9, 11, 12] where additional details on the experimental setup can be found.

3. Data analysis

The procedure described in [13] was used for analysis of RDDS data. This procedure represents a further extension of the Differential decay-curve method (DDCM) [14, 15]. This extennsion mainly concerns taking into account the velocity distribution of the recoils and the finite slowing-down time in the stopper in γ-γ coincidence. An example of such line-shape analysis and lifetime determination is shown on the l.h.s. of Fig. 1 where the Doppler-shift attenuated (DSA) fraction due to emissions during the slowing-down is also displayed. Separate contributions to the RDDS spectra are obtained when setting a gate on part of the "shifted" component of the feeding transition which result from the different possible combinations of the emission times of the feeding transition and depopulating transition. These contributions are disentangled by a fitting procedure whose details are given in Ref. [13].

The approach used for the analysis of the singles DSA line shapes and deriving lifetimes is presented in Refs. [15, 16]. For the description of the stopping process we used a modified version of the computer code DESASTOP [17, 18] by G. Winter. This newer version allows for a numerical treatment of the electron stopping powers at relatively higher ion energies.The gates were set on lower-lying transitions, that's why it turned out that some unknown or side-feeding is present at every investigated level. For this reason we adopted an often applied hypothesis on the time-behavior of the side-feeding, according to which it is the same as that of the known feeding. Additionally, we investigated the influence of the side-feeding on the results obtained for the lifetimes (cf. Eq.8 in Ref.[16]).

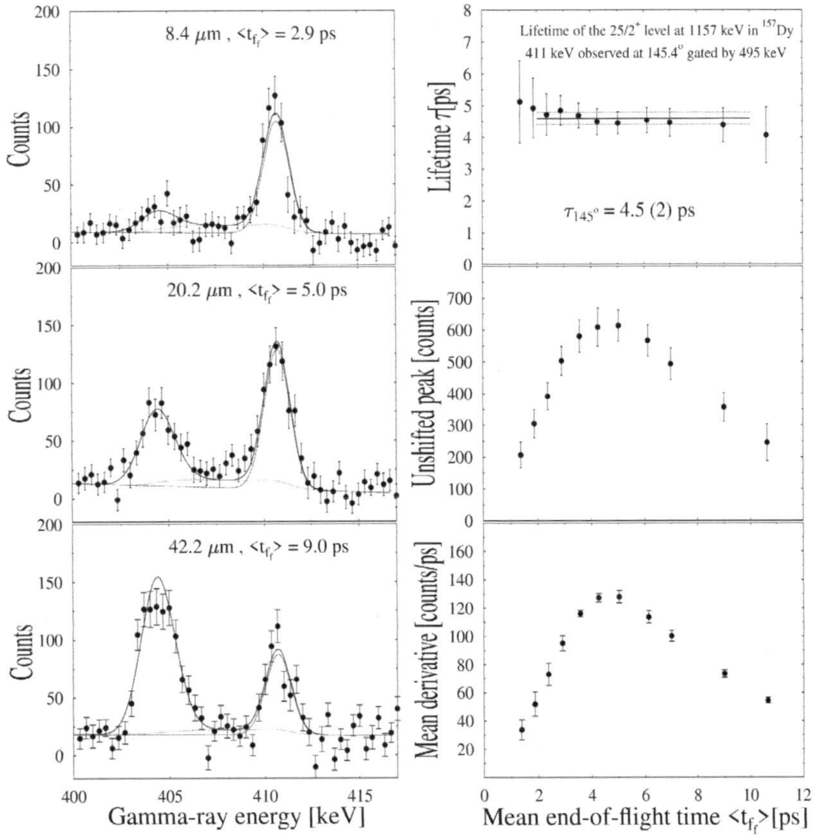

Figure 1. (Color online) Example of lifetime determination for the $25/2^+$ level in the yrast band of ^{157}Dy using the gated spectra measured with the detectors at 145°. The gate is set on the shifted component of the directly feeding 495 keV transition. The left part of the figure presents fits of the line shape of the 411 keV γ-ray transtion at different distances. The solid black line is the full fit. In addition, the fractions of the line shape related to the unshifted peak (dashed green line) and DSA effects (dotted red line) are shown. The right part of the figure illustrates the lifetime determiantion according to an equation given in [13]. On top, the τ-curve derived is displayed together with a fit with a horizontal line within the region of sensitivity.

4. Results and discussion

In the present work, 9 lifetimes were determined in the yrast band of ^{157}Dy, 3 of them for the first time. To describe the yrast band B(E2)'s in the present work, we used the rigid rotor model [19]. The comparison between our preliminary results (from RDDS and DSAM experiments) and the prediction of the rigid rotor model are shown in Fig. 2. The dependence of the transition strengths in the yrast band deduced from our analysis on the spin of the initial level is displayed. Up to spin $I^{\pi}=53/2^+$, the data are consistent with the predictions of the rigid-rotor model as shown by the comparison with two calculations. The latter were performed using Eq. 1 assuming Q_0=6.51 eb and two different values of the projection K=Ω of the angular momentum on the

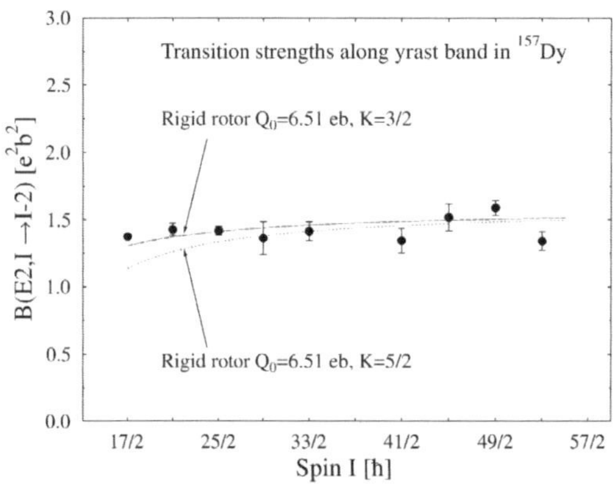

Figure 2. Reduced B(E2,I → I-2) transition strengths in the yrast band. The results of the RDDS measurements as well as results from DSAM data analysis are presented together with calculations within the rigid-rotor model with the indicated parameters

symmetry axis.

$$B(E2, I \rightarrow I-2) = \frac{5}{16\pi} e^2 Q_0^2 < IK20|I-2K >^2 \tag{1}$$

Looking at the evolution of the β deformation in Dy isotopes we can see that it is smoothly rising from $\beta \approx 0.28$ in ^{155}Dy to $\beta \approx 0.30$ in ^{157}Dy as getting closer to midshell. In the framework of the rigid-rotor model the transition quadrupole moments within a band are equal to the intrinsic quadrupole moment Q_0 which is related [20] to the deformation β via Eq. 2. Due to the bigger value for β deformation the value for Q_0 is also rising from ≈ 5.8 eb in ^{155}Dy to ≈ 6.5 eb in ^{157}Dy.

$$Q_0 = \frac{3}{\sqrt{5\pi}} Z R_0^2 \beta (1 + 0.16\beta). \tag{2}$$

The structure of the yrast band is dominated by the $3/2^+[651]$ quasineutron orbital, but with increasing spin, other low-Ω orbitals from $i_{13/2}$ parentage have an appreciable contribution in the wave functions. Both calculations (with K=3/2 and K=5/2) describe reasonably the data.

5. Summary and conclusions

In the present work, RDDS and DSAM lifetime measurements were carried out. For the data analysis, dedicated versions of the Differential decay curve method were employed. The data on the deduced transition strengths were compared to the results of rigid rotor model calculations. Pointing on this result we can say that there is no indication found for a reduction of the collectivity of the E2 transitions in the yrast band with increasing spin in the investigated spin range.

6. Acknowledgments

K.A.G. would like to thank his colleagues from the University of Cologne for their kind hospitality. This work was supported by the partnership agreement between the University of Cologne and University of Sofia and by the SU NSF within contract 152/15.

References

[1] R. M. Lieder *et al.*, Phys.Lett. B **49**, 161 (1974)
[2] H. R. Andrews *et al.*, Nucl. Phys. **A219**, 141 (1974)
[3] D. Ward *et al.*, Nucl. Phys. **A332**, 433 (1979)
[4] Y. El Masri *et al.*, Nucl. Phys. **A271**, 135 (1976)
[5] H. Beuscher *et al.*, Nucl. Phys. **A249**, 379 (1975)
[6] A. W. Sunyar, Proc. Symp. on high-spin phenomena in nuclei (Argonne, 1979) ANL/PHY-79-4, p. 77;
[7] P. Thieberger *et al.*, Phys. Rev. Lett. **28**, 1454 (1972)
[8] P.Petkov *et al.*, Phys. Rev. **C88**, 034323 (2013)
[9] O.Möller *et al.*, Phys. Rev. **C74**, 024313 (2006).
[10] D. Bazzacco, in *Proc. Int. Conf. on Nuclear Structure at High Angular Momentum*, Ottawa, 1992, Chalk River Report (AECL 10613), p.386
[11] P.Petkov *et al.*, Phys. Rev. **C68**, 034328 (2003)
[12] M.Yavahchova *et al.*, J.Phys.: Conf. Ser. **366**, 012050 (2012)
[13] P.Petkov, D.Tonev, J.Gableske, A.Dewald, T.Klemme and P. von Brentano, Nucl. Instr. Meth. Phys. Res. A **431**, 208 (1999)
[14] A.Dewald, S.Harissopulos and P. von Brentano, Z. Phys. **A334**, 163 (1989)
[15] G. Böhm, A. Dewald, P. Petkov and P. von Brentano, Nucl. Instr. Meth. Phys. Res. A **329**, 248 (1993)
[16] P.Petkov *et al.*, Nucl. Phys. A **640**, 293 (1998).
[17] G. Winter, ZfK Rossendorf Report ZfK-497, 1983
[18] G. Winter, Nucl. Instr. Meth. **214**, 537 (1983)
[19] A.Bohr and B.R.Mottelson, *Nuclear Structure*, Vol.2 (Benjamin, New York, 1975)
[20] A.Bohr and B.R.Mottelson, Dan. Mat.-Fys. Medd. 30 (1955) no.1

Varna2015 IOP Publishing
Journal of Physics: Conference Series **724** (2016) 012018 doi:10.1088/1742-6596/724/1/012018

Beyond-mean-field corrections within the second random-phase approximation

M. Grasso
Institut de Physique Nucléaire, IN2P3-CNRS, Université Paris-Sud, F-91406 Orsay Cedex, France
E-mail: grasso@ipno.in2p3.fr

D. Gambacurta
Extreme Light Infrastructure - Nuclear Physics (ELI-NP), Horia Hulubei National Institute for Physics and Nuclear Engineering, 30 Reactorului Street, RO-077125 Magurele, Jud. Ilfov, Romania

J. Engel
Department of Physics and Astronomy, University of North Carolina, Chapel Hill, North Carolina 27516-3255, USA

Abstract. A subtraction procedure, introduced to overcome double-counting problems in beyond-mean-field theories, is used in the second random-phase approximation (SRPA). Double-counting problems arise in the energy-density functional framework in all cases where effective interactions tailored at leading order are used for higher-order calculations, such as those done in the SRPA model. It was recently shown that this subtraction procedure also guarantees that the stability condition related to the Thouless theorem is verified in extended RPA models. We discuss applications of the subtraction procedure, introduced within the SRPA model, to the nucleus ^{16}O. The application of the subtraction procedure leads to: (i) stable results that are weakly cutoff dependent; (ii) a considerable upwards correction of the SRPA spectra (which were systematically shifted downwards by several MeV with respect to RPA spectra, in all previous calculations). With this important implementation of the model, many applications may be foreseen to analyze the genuine impact of 2 particle-2 hole configurations (without any cutoff dependences and anomalous shifts) on the excitation spectra of medium-mass and heavy nuclei.

1. Introduction

Several aspects of the nuclear many–body problem can be well described within the energy–density functional (EDF) theory at the mean–field level. However, a precise description of nuclear phenomena requires in some cases to go beyond this independent–particle picture. A special care has to be paid then to the fact that the starting point is not a bare many–body Hamiltonian but an energy functional of the density that is generated, in most cases, by a density–dependent effective interaction.

In general, mean–field–based models within the EDF theory correspond to the leading order of the many–body perturbative problem, that is described by the Dyson equation [1]. A way to go beyond this level of approximation is to truncate the perturbative expansion at the next order, by including also the second–order contribution. By making this, however, an obvious risk of double counting exists. Since the parameters of the effective interaction are adjusted at the leading order, that is, with mean–field calculations, the strategy followed within the mean–field–based EDF is very similar to that governing the so–called density–functional theory, currently used in chemistry and solid–state physics: all the physics is supposed to be well described at the leading order and this is the reason why parameters are adjusted at this order in the EDF framework. When correlations are explicitly included in the theoretical model (for instance, when the computation is carried out at an higher order beyond the mean–field level), those correlations which are already implicitly contained in the numerical values of the parameters can then be obviously double counted. This problem may be however rigorously handled in the framework of the EDF and some solutions have been proposed. One of them is a subtraction procedure tailored for those beyond–mean–field cases where the mass operator acquires an energy dependence. This procedure was first introduced to cure double–counting problems in the framework of particle–vibration models [2] and then formulated so to be used in extended random–phase approximation (RPA) models, such as the second RPA (SRPA) [3].

2. SRPA and the subtraction method

2.1. Formalism of SRPA

The SRPA model is a natural extension of the RPA model, where the 2 particle–2 hole (2p2h) elementary configurations are added to the RPA 1 particle–1 hole (1p1h) excitations. The excitation operators are then written as linear superpositions of particle–hole, hole–particle, 2 particle–2 hole, and 2 hole–2 particle elementary excitations, where particle and hole single–particle states are defined with respect to the Hartree–Fock (HF) ground state.

The SRPA equations are very similar to the RPA equations and are written in a compact form as

$$\begin{pmatrix} \mathcal{A} & \mathcal{B} \\ -\mathcal{B}^* & -\mathcal{A}^* \end{pmatrix} \begin{pmatrix} \mathcal{X}^\nu \\ \mathcal{Y}^\nu \end{pmatrix} = \omega_\nu \begin{pmatrix} \mathcal{X}^\nu \\ \mathcal{Y}^\nu \end{pmatrix}, \tag{1}$$

where:

$$\mathcal{A} = \begin{pmatrix} A_{11'} & A_{12} \\ A_{21} & A_{22'} \end{pmatrix}, \mathcal{B} = \begin{pmatrix} B_{11'} & B_{12} \\ B_{21} & B_{22'} \end{pmatrix},$$

$$\mathcal{X}^\nu = \begin{pmatrix} X_1^\nu \\ X_2^\nu \end{pmatrix}, \quad \mathcal{Y}^\nu = \begin{pmatrix} Y_1^\nu \\ Y_2^\nu \end{pmatrix}.$$

The '11' part of the matrices represents the 1p1h sector, that corresponds to the RPA sub–matrices. New matrix elements enter into play in the '12' and '22' sectors and describe the 1p1h–2p2h and 2p2h–2p2h couplings, respectively. The 1p1h–2p2h coupling, in particular, is responsible for providing a physical description of the fragmentation and of the spreading width of the excitation modes, that cannot be done within a simple mean–field model.

The first large–scale SRPA calculations were performed only quite recently [4, 5, 6, 7, 8, 9], without any cut or simplification in the matrices to diagonalize and with large values of the energy cutoff. These calculations have allowed us to fully address for the first time some specific issues related to the SRPA model such as, for instance, the proper derivation of the new rearrangement terms generated by the density dependence of the force in the beyond–RPA matrix elements [7].

When traditional effective interactions are used for these calculations, the already mentioned double–counting problems arise and they should be handled. Moreover, an ultraviolet divergence occurs and has to be regularized when a zero–range force is used. This can be seen in all the calculations that were performed with effective interactions such as Skyrme or Gogny forces in Refs. [6, 7, 8, 9]. There are two types of results that have been obtained: (i) in the case of giant resonances and, in general, in those cases where the excitations are already well described by the RPA, a strong downwards shift in energy of the response function was found with respect to the RPA. This strong and anomalous shift was a very surprising and unexpected result. In most cases, it produces strength distributions that are in less good agreement with the experimental data than the RPA. One would expect however that the SRPA, that contains the RPA as a simpler case, should provide similar results as the RPA in those cases where such model works well, that is, in those cases where the 1p1h configurations are the dominant configurations. (ii) For states that are mostly multiparticle–multihole states, the SRPA provides a better description than the standard RPA, which is very poor in describing such excitations.

In all cases, there is a strong cutoff dependence generated by the ultraviolet divergence associated to the zero–range of the used effective force.

Recently, it was pointed out that the standard SRPA does not satisfy the stability condition related to the Thouless theorem [10]. This was pointed out by Papakonstantinou in Ref. [11]. Such drawback of the SRPA and a procedure to overcome it will be discussed in the next Subsection.

2.2. Subtraction method

The formal properties of the SRPA model in its standard formulation, and all its advantages and drawbacks, were analyzed recently in Ref. [11]. In particular, a specific drawback of the model related to the stability condition (Thouless theorem) [10] was discussed. A possible direction to overcome this formal limitation of the standard SRPA model was suggested and consists in using a correlated ground state in the calculations instead of the HF state. This means that the so–called quasiboson approximation, also used in the RPA model, should be avoided in the SRPA model to ensure that the stability condition is satisfied. In a less recent work, Tselyaev suggested an alternative direction based on a subtraction procedure [3], that he had previously introduced to cure double–counting problems in models based on the particle–vibration coupling [2]. In Ref. [3], Tselyaev demonstrated that the use of such subtraction method ensures not only the cancelation of any double counting in extended RPA models such as the SRPA, but also the validity of the stability condition in those models.

The SRPA equations showed above may be written, after some manipulations, as RPA–type equations (that is, with only the '11' sector) where, however, the matrix elements acquire an explicit energy dependence. The cancelation of double counting is guaranteed when one imposes that the SRPA energy–dependent matrix calculated at zero energy is equal to the RPA matrix. This may be obtained by subtracting, to the energy–dependent contributions in the SRPA matrix elements, the same contributions calculated at zero energy. What Tselyaev calls the static correlations, that are already included in the model in the numerical values of the parameters of the effective interactions, are in this way counted only once. The genuine dynamic correlations, that this beyond–mean–field model introduces, are thus taken into account in a clean and proper way. As Tselyaev showed, it turns out that such procedure also allows one to demonstrate the validity of the Thouless theorem in the SRPA.

Recently, this subtraction method was applied within the SRPA and it was found that the anomalous huge downwards shift in energy that the standard SRPA produces with respect to the RPA is not obtained when the subtraction procedure is used [12]. This result indicates that such shift is not related to the enriched physical description of the excitation modes provided by the SRPA, but to the drawback of the model related to the violation of the stability condition. Moreover, the ultraviolet divergences are also canceled by the same subtraction. A very interesting remark on this method is the following: it appears clear that the subtraction acts mostly on the elementary 1p1h excitation energies by renormalizing them. This explains why the effects of the subtraction are mostly visible on those excitations that have a strong 1p1h nature and are very weakly mixed with multiparticle–multihole configurations. The subtraction actually acts as a corrective term that is added to the '11' part of the matrix. This corrective term is calculated as a sum of terms with a cutoff in the 2p2h space. A coherent calculation is done if this cutoff is chosen equal to the 2p2h cutoff used to compute the matrix elements.

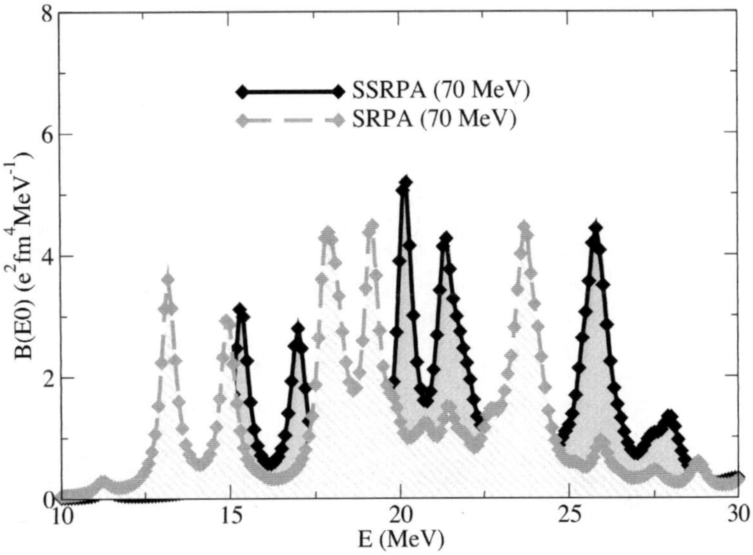

Figure 1. Isoscalar monopole response calculated for the nucleus ^{16}O with the Skyrme interaction SGII. The yellow area and orange dashed line refer to a standard SRPA calculation with a cutoff of 70 Mev in the 2p2h configurations. The cyan area and solid black line correspond to a subtracted SRPA calculation with the same cutoff of 70 MeV both in the matrices and in the corrective term for the 2p2h configurations.

On the other side, multiparticle–multihole excitations (that were already well described by the standard SRPA model) are practically not affected by the subtraction.

An illustration of the found results is shown in Fig. 1, where the isoscalar monopole response calculated for the nucleus ^{16}O with the Skyrme interaction SGII is presented. The yellow area and orange line refer to a standard SRPA calculation with a cutoff of 70 Mev in the 2p2h configurations. The cyan area and black solid line correspond to subtracted SRPA (SSRPA) calculations with the same cutoff of 70 MeV in the matrices and in the subtractive term for the 2p2h configurations.

One observes that the subtraction produces an upwards shift of the energies that goes in the good direction to cure the anomalous downwards shift generated by the standard SRPA. This goes in the correct direction also to provide a better comparison with the experimental results. The agreement is improved with respect to both the standard SRPA and the RPA. For example, for the isoscalar quadrupole response of the nucleus ^{16}O, the subtracted SRPA model provides a centroid energy and a width of 20.21 and 4.05 MeV, respectively. The RPA model leads to 20.73 and 2.42 MeV, respectively. The corresponding experimental values [13] are 19.76 and 5.11 MeV, respectvely. One observes that the energy of the centroid is slightly decreased by the subtracted SRPA with respect to the RPA and is closer to the experimental value. The width is remarkably increased by the SRPA and this is indeed the most relevant improvement expected from the SRPA model, with respect to the RPA model, in the description of giant resonances.

3. Conclusions

A subtraction procedure proposed by Tselyaev for extended RPA models [3] was recently applied to the SRPA model to remove double–counting problems, to ensure the validity of the stability condition, and to cancel all the ultraviolet divergences associated to the zero range of the used effective interaction.

An application to the nucleus ^{16}O was discussed and a comparison with the RPA spectrum and the experimental measurements was shown for the isoscalar quadrupole resonance. The subtracted SRPA model leads to a centroid energy which is in better agreement with the experimental value compared to the RPA centroid. The most important improvement provided by the SRPA is found however in the description of the spreading width of the excitation mode, which is remarkably increased by the SRPA with respect to the RPA. The great advantage of the subtracted SRPA, with respect to the standard SRPA, is not only the suppression of the anomalous downwards shift of the energies that the standard SRPA produces with respect to the RPA, but also the fact that the predictions are fully cutoff independent because all the ultraviolet divergences are removed by the subtraction.

References

[1] Fetter A L and Walecka J D *Quantum Theory of Many–Particle Systems* (McGraw-Hill, New York)
[2] Tselyaev V I 2007 *Phys. Rev.* **C 75** 024306
[3] Tselyaev V I 2013 *Phys. Rev.* **C 88** 054301
[4] Papakonstantinou P and Roth R 2009 *Phys. Lett.* **B 671** 356
[5] Papakonstantinou P and Roth R 2010 *Phys. Rev.* **C 81** 024317
[6] Gambacurta D Grasso M and Catara F 2010 *Phys. Rev.* **C 81** 054312
[7] Gambacurta D Grasso M and Catara F 2011 *J. Phys.* **G 38** 035103
[8] Gambacurta D Grasso M and Catara F 2011 *Phys. Rev.* **C 84** 034301
[9] Gambacurta D Grasso M De Donno V Co' G and Catara F 2012 *Phys. Rev.* **C 86** 021304(R)
[10] Ring P and Schuck P *The Nuclear Many–Body Problem* (Springer-Verlag, New York)
[11] Papakonstantinou P 2014 *Phys. Rev.* **C 90** 024305
[12] Gambacurta D Grasso M and Engel J *Phys. Rev..* **C 92** 034303
[13] Lui Y W Clark H L and Youngblood D H 2001 *Phys. Rev.* **C 64** 064308

The $A = 8$ isotriplet in Fermionic Molecular Dynamics

KR Henninger

GSI Helmholtzzentrum für Schwerionenforschung GmbH, Planckstrasse 1, 64291 Darmstadt, Germany

E-mail: k.henninger@gsi.de

Abstract. Nuclear structure of the $A = 8$ isotriplet is investigated in the Fermionic Molecular Dynamics (FMD) model. All three nuclei have importance to astrophysics, and exhibit clustering or haloes. FMD uses a wave-packet basis and is well-suited for modelling such structures. For a multiconfiguration treatment we construct the many-body Hilbert space from antisymmetrised angular-momentum projected 8-particle states. First results show formation of a proton halo in ^{8}B, and reasonable reproduction of the T=1 states in ^{8}Be.

1. Introduction

Determination of stellar nucleosynthetic reaction rates may rely on input from nuclear structure calculations: experimentally, these reactions are very low-yield at astrophysical energies, giving large uncertainties and necessitating theoretical input [1]. Weakly-bound nuclei appear as intermediates in these reactions, meaning that calculation of these reaction rates often requires modeling of weakly-bound nuclear structure. Modeling such is, however, challenging, due to proximity to the continuum, which requires that one access both scattering states and compact ("shell-model like") configurations in one's chosen model, ideally in a consistent way. This is particularly relevant to the asymptotic part of the wave function (*i.e.* at distances large compared to the nuclear radius), where the scattering solutions contribute strongly ("coupling to the continuum"). Asymptotic structure may influence reaction rate heavily [2], making accurate modeling thereof an important goal.

Fermionic Molecular Dynamics (FMD) [3] is a microscopic nuclear model whose chief benefit is in accessing clustering and shell-model like states in a consistent way, due to its flexible basis (see *e.g.* [4]). The aim of this study is to model the nuclei ^{8}Li, ^{8}Be and ^{8}B in the FMD model, with a view to astrophysical applications.

Decay of ^{8}B to ^{8}Be gives rise to almost the entire high-energy solar-neutrino flux [5], making ^{8}B structure important both for determining the solar core temperature [5] and reducing uncertainty in the Standard Solar Model, where the S factor (S_{17}) for ^{7}Be(p,γ)^{8}B is imprecisely known [6, 7]. Other features in this isotriplet include the two-alpha cluster structure of the ^{8}Be ground state, which is challenging to model, especially if modeling the higher-lying, compact T=1 states consistently. ^{8}Li is of interest as the "mirror" of ^{8}B: phenomena like Thomas-Ehrman shift [8–10] are worth a revisit in mirror pairs where one nucleus has a proton halo.

2. Fermionic Molecular Dynamics

Starting with protons and neutrons as basic nuclear constituents, we choose intrinsic many-body configurations $|Q\rangle$ that are Slater determinants of single-particle states $|q\rangle$, or [11]:

$$|Q\rangle = \hat{\mathcal{A}}\left\{|q_1\rangle \otimes ... \otimes |q_A\rangle\right\} \tag{1}$$

where $\hat{\mathcal{A}}$ is the antisymmetrisation operator. The single-particle states corresponding to the protons and neutrons are:

$$|q\rangle = \sum_i |a_i, \vec{b}_i\rangle \otimes |\chi_i^{\uparrow}, \chi_i^{\downarrow}\rangle \otimes |\xi_i\rangle c_i, \tag{2}$$

where a superposition of wave-packets $|a_i, \vec{b}_i\rangle$ in the spatial part aids description of extended distributions (*e.g.* haloes). Parameters $\vec{b}$ relate to the mean position and mean momentum of the wave-packets and parameters a to their width [12]. The ket $|\chi^{\uparrow}, \chi^{\downarrow}\rangle$ is the most general spinor. The isospin part $|\xi\rangle$ defines a particle as a proton or neutron, allowing no superpositions of the two.

In phase space, the nucleon wave functions are Gaussian wavepackets, *e.g.* in co-ordinate space we have:

$$\langle\vec{x}|q\rangle = \exp\left\{-\frac{(\vec{x}-\vec{b})^2}{2a}\right\}. \tag{3}$$

It is necessary to project out of intrinsic states $|Q\rangle$ the eigenstates of angular momentum and parity, via projection operators [12]:

$$\begin{aligned} \hat{P}^{\pi} &= \frac{1}{2}(1+\pi\hat{\Pi}) \\ \hat{P}^{J}_{MK} &= \frac{2J+1}{8\pi^2}\int d\Omega\, D^{J}_{MK}(\Omega)^{*}\, \hat{R}(\Omega), \end{aligned}$$

where $D^J_{MK}(\alpha,\beta,\gamma)$ is the Wigner D-matrix, $\hat{R}(\alpha,\beta,\gamma)$ and $\hat{\Pi}$ are rotation and parity operators and $\Omega = \{\alpha,\beta,\gamma\}$ symbolises the Euler angles.

To obtain a good approximation to the nuclear ground state, the parameters $\{q_\nu\}$ describing the intrinsic basis states $|Q\rangle$ are evaluated by requiring [4]:

$$\min_{\{q_\nu\}} \quad \frac{\langle Q|\hat{H}-\hat{T}_{cm}|Q\rangle}{\langle Q|Q\rangle}, \tag{4}$$

where including the operator $\hat{T}_{cm}$ ensures that centre-of-mass kinetic energy is subtracted out. Using Eq.(4) is called Projection after Variation (PAV), as the variational principle is applied and then projection operators act on the minimised state. It is more accurate to vary energy of the projected state (Variation After Propection (VAP)), as then it is certain the projected state has minimum energy [4].

To improve the description, additional basis states can be generated by performing minimisation (Eq.(4)) subject to various constraints. These constraints could be on *e.g.* radii, moments or deformation parameters [4]; allowing inclusion of many different configurations in the basis thus created. The Hamiltonian may then be diagonalised in this basis. One may also create basis states in which one explicitly imposes that the nucleus is composed of 2 or 3 clusters [4], which may then be added to the constrained VAP basis to improve the description of the asymptotic region.

2.1. Interactions for FMD
FMD requires the Hamiltonian to be in operator form, so short range correlations are treated via the Unitary Correlation Operator Method (UCOM) (see refs [13–15]).

The UCOM involves applying a unitary correlation operator $\hat{C}$ to the many-body state, to incorporate the short-range correlations. This gives rise to an effective Hamiltonian $\hat{H}_{eff}$, as [14]:

$$\langle\Psi|\hat{C}^{\dagger}\hat{H}\hat{C}|\Psi\rangle = \langle\tilde{\Psi}|\hat{H}|\tilde{\Psi}\rangle$$
$$= \langle\Psi|\hat{H}_{eff}|\Psi\rangle,$$

where $\hat{H}_{eff} = \hat{C}^{\dagger}\hat{H}\hat{C}$ and $\hat{C}|\Psi\rangle = |\tilde{\Psi}\rangle$.

This transformation of the Hamiltonian induces many-body terms, *i.e.* [14]:

$$\hat{H}_{eff} = \hat{H}^{[1]}_{eff} + \hat{H}^{[2]}_{eff} + \hat{H}^{[3]}_{eff} + \cdots , \tag{5}$$

where the number in square brackets indicates number of bodies involved. The estimated contribution of the three- and higher-body terms turns out to be very small compared to the two-body contribution [13]. The transformed Hamiltonian up to two-body is therefore what one calls the UCOM Hamiltonian. One compensates for "missing" three-body and higher terms by adjusting the strength factor of the spin-orbit term, according to what best reproduces the nuclear observables. In these calculations, we use a UCOM-transformed Argonne $v18$ (AV18) interaction [16], as No-Core Shell Model calculations using this interaction show it to give good descriptions of light p-shell nuclei [14].

3. Calculations
A VAP basis set was created for each of the three nuclei (^{8}Li, ^{8}Be and ^{8}B), using constraints on matter radius. Using the same constraint for all three nuclei ensures that one may compare results. The maximum basis size was chosen by adding states to the basis, diagonalising, and looking for convergence behaviour in energies of the low-lying states (see Fig. 1).

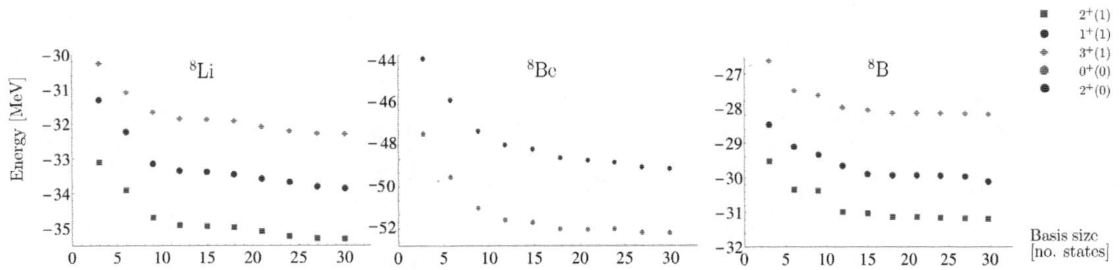

Figure 1: Energies of the low-lying states of the $A = 8$ nuclei, plotted with basis size, indicating convergence behaviour as basis size is increased. States are labelled with spin, parity and isospin as in the key.

In order to compare results for the whole isomultiplet, it was decided to add the isospin partner states into each basis. These are obtained for the ^{8}Li-^{8}B mirror pair via rotations in isospin space, or

$$|\tilde{Q}\rangle = exp\left\{i\pi\hat{T}_2\right\}|Q\rangle = exp\left\{i\pi\sum_j \hat{t}_2(j)\right\}\left(\mathcal{A}\left\{|t(1),t_3(1)\rangle\otimes\ldots\otimes|t(A),t_3(A)\rangle\right\}\right), \tag{6}$$

where $|t(j), t_3(j)\rangle$ represents the isospin state of the jth nucleon, $|\tilde{Q}\rangle$ is a state in ^{8}B and $|Q\rangle$ is a state in ^{8}Li, say. For transforming states in ^{8}Li or ^{8}B to states in ^{8}Be, formally one applies the isospin raising and lowering operators. In practise, one successively changes each of the five protons to a neutron in each radius-constrained ^{8}B basis state $|Q\rangle$ (or *vice versa*); giving 5 new ^{8}Be basis states for each $|Q\rangle$ of ^{8}B or ^{8}Li. The two methods are equivalent: With isospin raising and lowering operators, one would obtain a linear combination of these aforementioned states, but we will reach such a linear combination when diagonalizing the Hamiltonian in this basis.

Besides creation and selection of basis states, one also has to consider the strength factor of the spin-orbit term, as this incorporates the three- and higher-body terms in the interaction. We initially perform all calculations using a spin-orbit strength factor of 2, and then used calculated transition-strengths and energies as an indicator for adjusting the strength. In the case of weakly-bound nuclei with spatially-extended structures, (such as ^{8}B), one may need less spin-orbit strength. In ^{8}Be, the T=1 states are more compact than the (α-cluster based) T=0 states, and will thus be affected more strongly by three- and higher-body forces. We thus try larger values of the spin-orbit strength factor when treating ^{8}Be. Results and selected strength factors are discussed below (Section 4).

4. Results

4.1. Observables

For all three nuclei, one obtains energy-levels that compare quite favourably to experiment (Fig. 2), and especially to the calculated thresholds, in all three cases. The final basis set used for each nucleus is summarised in Table 3. The calculated thresholds are ground-state energies of the relevant core nuclei, calculated in radius-constrained FMD basis sets of comparable dimension to those for the nuclei of interest. In the case of ^{8}Li, the ^{7}Li+n threshold was calculated (making ^{7}Li the core nucleus), for ^{8}Be, the two α particle threshold, and for ^{8}B the ^{7}Be+p threshold.

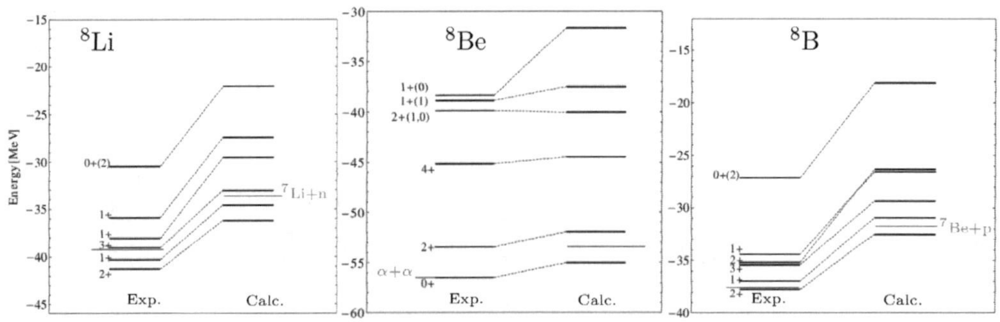

Figure 2: Experimental and calculated level-schemes for nuclei of the A=8 isotriplet. Basis sets used are described in Table 3. Note the different y-axes for energy. The thin horizontal lines indicate experimental and calculated thresholds for the decay process indicated (red labels).

Table 1: Transition strengths for some intense transitions in the A=8 nuclei. Measured values from [17–19]. "Exp." and "Calc." indicate experimental and calculated energies. Numbers in round braces indicate the strength of the spin-orbit term. Transition strengths were calculated in the "full" basis (supplied in Table 3).

	States	Energies [MeV]				Transition strengths			
		Exp.	Calc. (1.5)	Calc. (2.0)	Calc. (3.0)	Exp.	Calc.(1.5)	Calc. (2.0)	Calc. (3.0)
^{8}Li	$1^+ \to 2^+$	0.980	1.261	1.622	2.541	5.00(16) μ_N^2	5.080 μ_N^2	5.317 μ_N^2	5.825 μ_N^2
^{8}Be	$2^+ \to 0^+$	3.03	3.387	3.367	3.117	25(8) e^{2}fm^4	21.175 e^{2}fm^4	18.515 e^{2}fm^4	12.281 e^{2}fm^4
^{8}B	$1^+ \to 2^+$	0.77	1.207	1.570	2.521	4.71(21) μ_N^2	4.267 μ_N^2	4.385 μ_N^2	4.915 μ_N^2

Table 2: Calculated rms radii for ground states of the A=8 isotriplet. References for the experimental values are in square brackets. Radii were calculated in the "full" basis (as in Table 3). Numbers in round braces indicate the strength of the spin-orbit term.

		R_{matter} [fm]	R_p [fm]	R_n [fm]	R_{charge} [fm]
^{8}Li	Exp.	2.37(2) [20]	-	-	2.29(8) fm [21]
	Calc. (1.5)	2.315	2.118	2.430	2.255
	Calc. (2.0)	2.243	2.070	2.341	2.211
^{8}Be	Calc.(2.0)	2.404	2.408	2.400	2.545
	Calc.(3.0)	2.278	2.282	2.274	2.427
^{8}B	Exp.	2.38(4) [20]	-	-	-
	Calc.(1.5)	2.367	2.492	2.142	2.634
	Calc.(2.0)	2.282	2.389	2.091	2.537

Table 3: Details of basis sets.

Nucleus:	Basis states:
^{8}Li	45 basis states: 27 radius-constrained VAP states for ^{8}Li (9 each for $J^\pi = 1^+, 2^+, 3^+$), 11 cluster states (^{7}Li+n) and 18 isospin-partner states from ^{8}B (radius-constrained VAP states projected on $J^\pi = 1^+, 2^+, 3^+$).
^{8}Be	260 basis states: 24 radius-constrained VAP states for ^{8}Li (8 each for $J^\pi = 0^+, 2^+, 4^+$), 11 cluster states (2-α) and 225 isospin-partner states from ^{8}B and ^{8}Li (radius-constrained VAP states projected = J^π of $1^+, 2^+, 3^+$).
^{8}B	45 basis states: 18 radius-constrained VAP states for ^{8}B (6 each for $J^\pi = 1^+, 2^+, 3^+$), 11 cluster states (^{7}Be+p) and 27 isospin-partner states from ^{8}Li (radius-constrained VAP states projected on $J^\pi = 1^+, 2^+, 3^+$).

4.2. Spin-orbit strength factor

As discussed, we choose a spin-orbit strength factor that best incorporates higher-body terms. We deduce from reproduction of observables (Tables 1-2) that spin-orbit strength factor 1.5 is preferable for ^{8}Li, and factor 2 for ^{8}B (radii are better-reproduced with factor 1.5, but transitions and energies with factor 2). Spin-orbit strength strongly affects energy of ^{8}Be T=1 states (Fig. 3). The T=1 states are well-reproduced when using a spin-orbit strength factor 3.

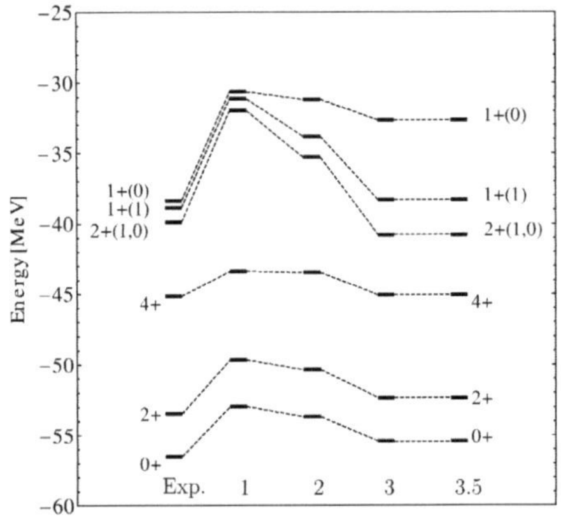

Figure 3: Level-schemes for ^{8}Be, comparing experimental and calculated energies when strength factor of the spin-orbit term is varied. Numbers below each band indicate the strength factor of the spin-orbit term.

4.3. Comparison to No-Core Shell Model

Our results are compared to current many-body theoretical results from the No-Core Shell Model (NCSM) in Fig. 4. The NCSM calculations of [19] use an interaction obtained from chiral effective field theory (χ-EFT). We compare to the energies they obtain using a χ-EFTH interaction with up to two body terms included because the UCOM-transformed AV18 is truncated at two-body, as discussed. In order to compare to their calculations using two-body forces, we performed calculations using the UCOM-transformed AV18 with an $L \cdot S$ term with strength factor 1. Our calculations compare relatively well with theirs, especially as regards binding relative to threshold. The model-space of [19] is an $8\hbar\Omega$ space.

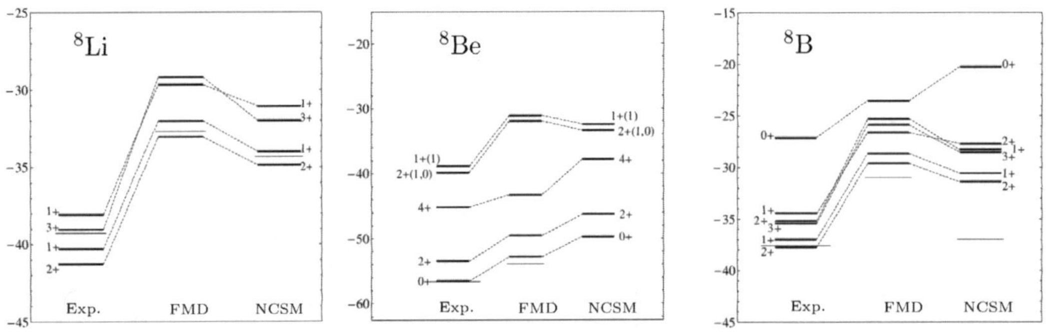

Figure 4: Our results for the A=8 isotriplet (FMD), compared to NCSM results of [19] (NCSM). Both are compared to the experimental levels (Exp). The FMD results shown here were calculated using a spin-orbit term with strength factor 1, as the NCSM energy-levels shown were calculated using a χ-EFT interaction truncated at the two-body level. Note the different y-axes for energy. The thin horizontal lines indicate experimental and calculated thresholds, which are the same as those in Fig. 2. The NCSM results were calculated in an $8\hbar\omega$ model-space.

5. Conclusion and Summary

The results obtained here show that our FMD calculations reproduce energies, transition-strengths and radii of the A=8 isotriplet nuclei favourably, and compare well with the latest theoretical results. Our results show formation of a proton halo in ^{8}B (a calculated proton radius of 2.389 fm, neutron radius of 2.091 fm and matter radius of 2.282 fm).We also achieve reasonable reproduction of the T=1 states in ^{8}Be; which indicates FMD is able to reproduce both clustering and shell-model like states consistently.

Acknowlegements

The author thanks Prof. Dr Hans Feldmeier and Dr Thomas Neff for extensive and expert guidance throughout the work.

References

[1] C.E. Rolfs and W.S. Rodney. *Cauldrons in the Cosmos: Nuclear Astrophysics.* The University of Chicago Press, Chicago, 1988. Pp 150, 156, 493.
[2] K. Riisager and A.S. Jensen. *The radius of ^{8}B and Solar Neutrinos.* Phys. Lett. ***B***, **301**:6, 1993.
[3] H. Feldmeier. *Fermionic Molecular Dynamics.* Nucl. Phys., **A515**:147, 1990.
[4] T. Neff and H. Feldmeier. Cluster Structure in the Fermionic Molecular Dynamics approach. In M. Brenner, editor, *Cluster Structure of Atomic Nuclei*, pages 67–94. Research Signpost publishing, 2010.
[5] R. E. Tribble A. M. Mukhamedzhanov H.M. Xu, C. A. Gagliardi and N. K. Timofeyuk. *Overall Normalization of the Astrophysical S-factor and the Nuclear Vertex Constant for ^{7}Be(p,γ)^{8}B Reactions.* Phys. Rev. Lett, 73:2027, 1994.
[6] A. Csòtò. *Off-Shell Effects in the Energy-Dependence of the ^{7}Be(p,γ)^{8}B Astrophysical S-factor.* Phys. Lett. ***B***, **394**:247, 1997.
[7] P.D. Parker W.C. Haxton and C.E. Rolfs. *Solar Hydrogen burning and neutrinos.* Nucl. Phys., **A777**:226, 2006.
[8] R.G. Thomas. *An Analysis of the Energy Levels of the Mirror Nuclei, ^{13}C and ^{13}N.* Phys. Rev., **88**:1109, 1952.
[9] J. B. Ehrman. *On The Displacement of Corresponding Energy Levels of ^{13}C and ^{13}N.* Phys. Rev., **81**:412, 1951.
[10] JA Nolen and JP Schiffer. *Coulomb Energies.* Ann. Rev. Nucl. Sci., **19**:472, 1969.
[11] H. Feldmeier and J. Schnack. *Fermionic Molecular Dynamics.* Prog. Part. Nucl. Phys., **39**:343, 1997.
[12] T. Neff and H. Feldmeier. *Clustering and Other Exotic Phenomena in Nuclei.* Eur. Phys. J., **156**:69, 2008.
[13] T. Neff H. Feldmeier, R. Roth and J. Schnack. *A Unitary Correlation Operator Method.* Nucl. Phys., **A632**:61, 1998.
[14] T. Neff R. Roth and H. Feldmeier. *Nuclear Structure in the Framework of the Unitary Correlation Operator Method.* Prog. Part. Nucl. Phys., **65**:50, 2010.
[15] T. Neff and H. Feldmeier. *Tensor Correlations in the Unitary Correlation Operator Method.* Nucl. Phys. ***A***, **713**:311, 2003.
[16] R. Schiavilla R. B. Wiringa, B.G.J. Stokes. *Accurate Nucleon-Nucleon Potential with Charge-Independence Breaking.* Phys. Rev ***C***, **51**:38, 1995.
[17] J.L.Godwin D.J. Millener J.E. Purcell C.G.Sheu D.R. Tilley, H.Kelley and H.R.Weller. *Energy Levels of Light Nuclei A = 8,9,10.* Nucl. Phys. , **A745**:155, 2004 (revised 2014).
[18] D.R. Chakrabarty V. Nanal E.T. Mirgule A. Mitra V.M. Datar, S. Kumar and H.H. Oza. *Direct Observation of the 4+-to-2$^+$ Gamma Transition in ^{8}Be.* Phys. Rev. Lett., **94**:122502, 2005.
[19] J.P. Vary P. Maris and P. Navratil. *Structure of A=7-8 nuclei with two- plus three-nucleon interactions from Chiral EFT.* Phys. Rev. C, **87**:014327, 2013.
[20] T. Suzuki A. Ozawa and I. Tanihata. *Nuclear Size and Related Topics .* Nucl. Phys. **A**, **693**:32, 2001.
[21] A. Dax S. Götte R. Kirchner H.-J. Kluge Th. Kühl R. Sanchez A. Wojtaszek B. A. Bushaw G. W. F. Drake Z.-C. Yan G. Ewald, W. Nörtershäuser and C. Zimmermann. *Nuclear Charge Radii of ^{8,9}Li determined by Laser Spectroscopy.* Phys. Rev. Lett., **93**:113002, 2004.

Charged-current inclusive neutrino cross sections in the SuperScaling model

M.V. Ivanov[1,2], G.D. Megias[3], R. González-Jiménez[4], O. Moreno[5], M.B. Barbaro[6], J.A. Caballero[3], T.W. Donnelly[5], A.N. Antonov[1], E. Moya de Guerra[2], and J.M. Udías[2]

[1] Institute for Nuclear Research and Nuclear Energy, Bulgarian Academy of Sciences, Sofia 1784, Bulgaria

[2] Grupo de Física Nuclear, Departamento de Física Atómica, Molecular y Nuclear, Facultad de Ciencias Físicas, Universidad Complutense de Madrid, Madrid E-28040, Spain

[3] Departamento de Física Atómica, Molecular y Nuclear, Universidad de Sevilla, 41080 Sevilla, Spain

[4] Department of Physics and Astronomy, Ghent University, Proeftuinstraat 86, B-9000 Gent, Belgium

[5] Center for Theoretical Physics, Laboratory for Nuclear Science and Department of Physics, Massachusetts Institute of Technology, Cambridge, Massachusetts 02139, USA

[6] Dipartimento di Fisica, Università di Torino and INFN, Sezione di Torino, Via P. Giuria 1, 10125 Torino, Italy

E-mail: martin.inrne@gmail.com

Abstract. SuperScaling model (SuSA) predictions to neutrino-induced charged-current π^+ production in the Δ-resonance region are explored under MiniBooNE experimental conditions. The SuSA charged-current π^+ results are in good agreement with data on neutrino flux-averaged double-differential cross sections. The SuSA model for quasielastic scattering and its extension to the pion production region are used for predictions of charged-current inclusive neutrino-nucleus cross sections. The contribution of two-particle-two-hole vector meson-exchange current excitations is also considered within a fully relativistic model. Results are compared with the T2K experimental data for inclusive scattering.

1. Introduction

The analysis of neutrino oscillations is at present one of the main research topics in Physics. This explains the huge activity in the field and the numerous experiments that have been proposed in several facilities and covering a very wide range in energy. For the first time, experimental data on neutrino interactions are being measured with a high statistics and small error bands. As a consequence, a great variety of studies on the theoretical side have also appeared trying to describe the data. These include analyses from low-to-intermediate energies where the nuclear effects may play an important role, to very high energies where the nucleonic and/or subnucleonic degrees of freedom are the principal ingredients. In all cases, the understanding of neutrino properties is a key ingredient in elementary particle physics.

In most neutrino experiments, the interactions of the neutrinos occur with nucleons bound in nuclei. Model predictions for these reactions involve many different effects such as nuclear correlations, interactions in the final state, possible modification of the nucleon properties inside the nuclear medium, that presently cannot be computed in an unambiguous and precise way. This is particularly true for the channels where neutrino interactions take place by means of excitation of a nucleon resonance and ulterior production of mesons. The data on neutrino-induced charged-current (CC) charged pion production cross sections on mineral oil recently released by the MiniBooNE collaboration [1] provide an unprecedented opportunity to carry out a systematic study of double differential cross section of the processes: $\nu_\mu p \to \mu^- p \pi^+$ and $\nu_\mu n \to \mu^- n \pi^+$ averaged over the neutrino flux.

New measurements of inclusive charged-current (CC) neutrino-nucleus scattering cross sections, where only the outgoing lepton is detected, have been recently performed by the T2K [2] collaboration. For neutrino energies around 1 GeV (T2K) the main contributions to the cross sections are associated with quasielastic (QE) scattering and one pion (1π) production. In the present work we evaluate the CC neutrino inclusive cross sections within the SuperScaling approach (SuSA), introduced in [3] to describe neutrino-nucleus scattering by using electron scattering data instead of relying on specific nuclear models. Since meson-exchange currents (MEC) are known to violate superscaling, their contribution must be added to the superscaling result: this will be accomplished by using a parametrization [4] of the relativistic calculation of [5]. The parametrization is necessary in order to reduce the computation time, since the exact calculation involves 7-dimensional integrals.

2. The SuSA model and charged-current neutrino-nucleus cross section

2.1. The SuSA scaling function in the QE- and Δ-region

One way of avoiding model-dependencies is to use the nuclear response to other leptonic probes, such as electrons, under similar conditions to the neutrino experiments. The analyses of the world data on inclusive electron-nucleus scattering [6, 7] confirmed the observation of superscaling and thus justified the extraction of a universal nuclear response to be also used for weak interacting probes. The present study uses the results found from studies superscaling (based on ψ-scaling variable) obtained from analyses of inclusive electron scattering data. The latter consists in constructing a "superscaling function" $f(\psi)$ obtained by removing the single-nucleon content from the double-differential cross section and plotting it versus a scaling variable $\psi(q,\omega)$. Scaling of the first kind of scaling function [*i.e.*, no explicit q dependence of $f(\psi)$] can be seen at excitation energies below the QE peak. Scaling of second kind [*i.e.*, no dependence of $f(\psi)$ on the mass number] turns out to be excellent in the same region. When scaling of both the first and second type occurs, one says that superscaling has taken place. However, while there is a number of theoretical models that exhibit superscaling, such as for instance the relativistic Fermi gas (RFG) [8, 9], the nuclear response departs from the one derived from the experimental data. This showed the necessity to consider more complex dynamical pictures of finite nuclear systems – beyond the RFG – in order to describe the nuclear response at intermediate energies. SuSA predictions are based on the phenomenological superscaling function extracted from the world data on quasielastic electron scattering [10]. In figure 1 we present the quasielastic SuSA scaling function (phenomenological parametrization of the experimental data taken from [3]) in comparison with the RFG scaling function: $f^{\rm QE}_{\rm RFG}(\psi_{\rm QE}) = \frac{3}{4}(1-\psi^2_{\rm QE})\theta(1-\psi^2_{\rm QE})$, where $\psi_{\rm QE}$ is the QE scaling variable.

The SuSA model has been extended to the Δ-resonance region [3] where the response of the nuclear system proceeds through excitation of internal nucleonic degrees of freedom. Indeed, a non-quasielastic cross section for the excitation region in which nucleon excitations, particularly the Δ, play a major role was obtained by subtracting from the data QE-equivalent cross sections given by SuSA [11, 12]. This procedure has been possible due to the large amount of available high-quality data of inelastic electron scattering cross sections on ^{12}C, including also separate information on the longitudinal and transverse responses, the latter containing important contributions introduced by effects beyond the impulse approximation (non-nucleonic).

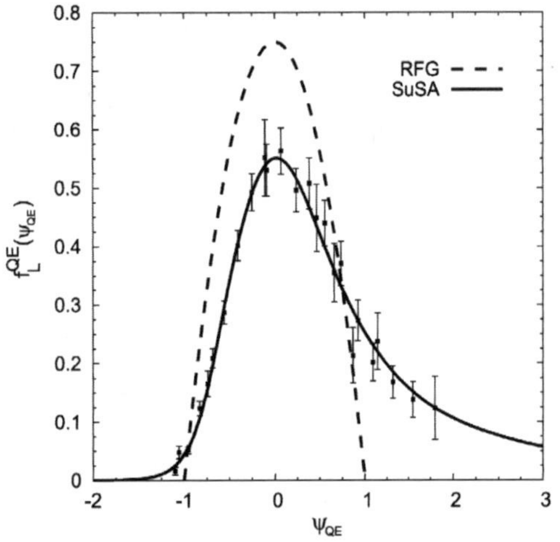

Figure 1. Averaged experimental $f_L^{\rm QE}(\psi_{\rm QE})$ versus $\psi_{\rm QE}$ in the quasielastic region together with a phenomenological parametrization of the data (solid line). The dashed line shows the scaling function in the RFG model.

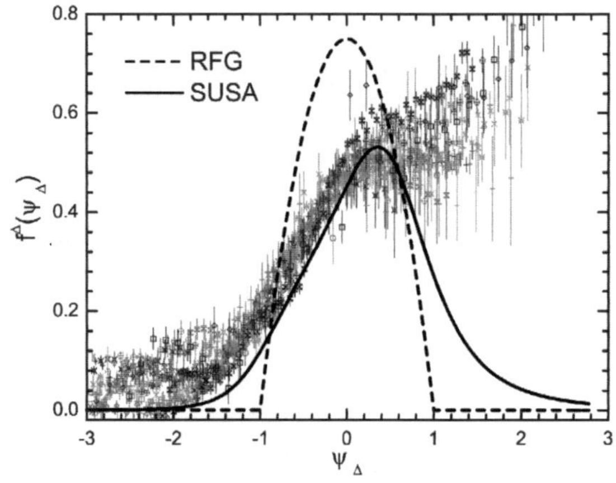

Figure 2. The SuSA scaling function in the Δ-region $f^{\Delta}(\psi_\Delta)$ (solid line) extracted from the world data on electron scattering [3]. The dashed line shows the scaling function $f^{\Delta}_{\rm RFG}(\psi_\Delta)$ in the RFG model.

We have extended the analysis to CC pion production cross-section measured at MiniBooNE [13], that from the theoretical point of view can be seen as more challenging. For instance, Δ properties in the nuclear medium, as well as both coherent and incoherent pion production for the nucleus should be considered in any theoretical approach, while in the SuSA procedure they are included phenomenologically extracted from the electron scattering data. All what is assumed within SuSA approach is the nuclear response to be factorized into a single-nucleon part and a 'nuclear function' accounting for the overall interaction among nucleons. As mentioned before, the SuSA assumptions have been tested against a great deal of electron-nucleus scattering data with fair success. The factorization assumption allows one to apply the same nuclear responses derived from electron scattering to neutrino-induced reactions, with a mere use of the adequate single-nucleon terms for this case. To show the importance of nuclear interaction effects as predicted within SuSA, as a reference, we also show results obtained within the RFG, with no interactions among nucleons, for which the scaling function in the Δ-domain is simply given as $f^{\Delta}_{\rm RFG}(\psi_\Delta) = \frac{3}{4}(1-\psi^2_\Delta)\theta(1-\psi^2_\Delta)$ with ψ_Δ the dimensionless scaling variable extracted from the RFG analysis that incorporates the typical momentum scale for the selected nucleus [3, 14]. In figure 2 we compare the Δ-region SuSA [3] and RFG scaling functions, which we use in our study.

2.2. Charged-current neutrino-nucleus cross section

We follow the description of the formalism concerning the CC (anti)neutrino-nucleus cross section given in Ref. [3]. The CC neutrino cross section in the target laboratory frame is given in the form:

$$\left[\frac{d^2\sigma}{d\Omega dk'}\right]_\chi \equiv \sigma_0 \mathcal{F}^2_\chi, \tag{1}$$

where $\chi = +$ for neutrino-induced reaction (*e.g.*, $\nu_\ell + n \to \ell^- + p$, where $\ell = e, \mu, \tau$) and $\chi = -$ for antineutrino-induced reactions (*e.g.*, $\overline{\nu}_\ell + p \to \ell^+ + n$),

$$\sigma_0 \equiv \frac{(G\cos\theta_c)^2}{2\pi^2}[k'\cos\tilde{\theta}/2]^2, \tag{2}$$

$G = 1.16639 \times 10^{-5}$ GeV^{-2} being the Fermi constant, θ_c being the Cabibbo angle ($\cos\theta_c = 0.9741$), and

$$\tan^2\tilde{\theta}/2 \equiv \frac{|Q|^2}{v_0}, \tag{3}$$

$$v_0 \equiv (\epsilon + \epsilon')^2 - q^2 = 4\epsilon\epsilon' - |Q|^2. \tag{4}$$

In Eqs. (1)-(4) Ω, k' and ϵ' are the scattering angle, momentum and energy of the outgoing lepton. The quantitiy $\mathcal{F}_\chi^2$ in Eq. (1) depends on the nuclear structure and it is presented in Ref. [3] as a generalized Rosenbluth decomposition having charge-charge, charge-longitudinal, longitudinal-longitudinal, and two types of transverse responses (R's). These nuclear response functions are expressed in terms of the nuclear tensor $W^{\mu\nu}$ in the QE (as well as in the Δ-region) by means of its relationships with the scaling functions from the particular model used.

2.3. π^+ production in the MiniBooNE experiment

We present RFG and SuSA predictions for the double-differential cross section for CC neutrino-induced π^+ production (from Δ resonance region) on CH_2 averaged over the neutrino flux $\Phi(\epsilon_\nu)$, namely

$$\frac{d^2\sigma}{dT_\mu d\cos\theta} = \frac{1}{\Phi_{\text{tot}}}\int\left[\frac{d^2\sigma}{dT_\mu d\cos\theta}\right]_{\epsilon_\nu}\Phi(\epsilon_\nu)d\epsilon_\nu, \tag{5}$$

where T_μ and θ are correspondingly the kinetic energy and scattering angle of the outgoing muon, ϵ_ν is the neutrino energy and Φ_{tot} is the total integrated ν_μ flux factor for the MiniBooNE experiment [1]. The nuclear response functions in Δ-region are expressed in terms of the nuclear tensor $W^{\mu\nu}$ in the corresponding region. The basic expressions used to calculate the single-nucleon cross sections are given in [3]. These involve the leptonic and hadronic tensors as well as the response and structure functions for single nucleons. A convenient parametrization of the single-nucleon $W^+n \to \Delta^+$ vertex is given in terms of eight form-factors: four vector ($C^V_{3,4,5,6}$) and four axial ($C^A_{3,4,5,6}$) ones. Vector form factors have been determined from the analysis of photo and electro-production data, mostly on a deuteron target. Among the axial form factors, the most important contribution comes from C^A_5. The factor C^A_6, whose contribution to the differential cross section vanishes for massless leptons, can be related to C^A_5 by PCAC. Since there are no other theoretical constraints for $C^A_{3,4,5}(q^2)$, they have to be fitted to data. We use two different parameterizations: the one given in [15] where deuteron effects were evaluated (authors estimated that the latter reduce the cross section by 10%), denoted as "PR1", and the one from [16], called "PR2".

With these ingredients, we evaluate the cross section for CC Δ^{++} and Δ^+ production on proton and neutron, respectively. Once produced, the Δ decays into πN pairs. For the amplitudes $\mathcal{A}$ of pion production the following isospin decomposition applies: $\mathcal{A}(\nu_l p \to l^- p\pi^+) = \mathcal{A}_3$, $\mathcal{A}(\nu_l n \to l^- n\pi^+) = \frac{1}{3}\mathcal{A}_3 + \frac{2\sqrt{2}}{3}\mathcal{A}_1$, $\mathcal{A}(\nu_l n \to l^- p\pi^0) = -\frac{\sqrt{2}}{3}\mathcal{A}_3 + \frac{2}{3}\mathcal{A}_1$, with $\mathcal{A}_3$ being the amplitude for the isospin 3/2 state of the πN system, predominantly Δ, and $\mathcal{A}_1$ the amplitude for the isospin 1/2 state that is not considered here.

The double-differential cross section for CC neutrino-induced π^+ production averaged over the neutrino energy flux as a function of the muon kinetic energy T_μ is presented in figure 3. Each panel corresponds to a bin of $\cos\theta$. PR1 and PR2 parametrizations have been considered. Results with the PR1 parameterization are about 5% higher, that is a measure of the degree

Journal of Physics: Conference Series **724** (2016) 012020 doi:10.1088/1742-6596/724/1/012020

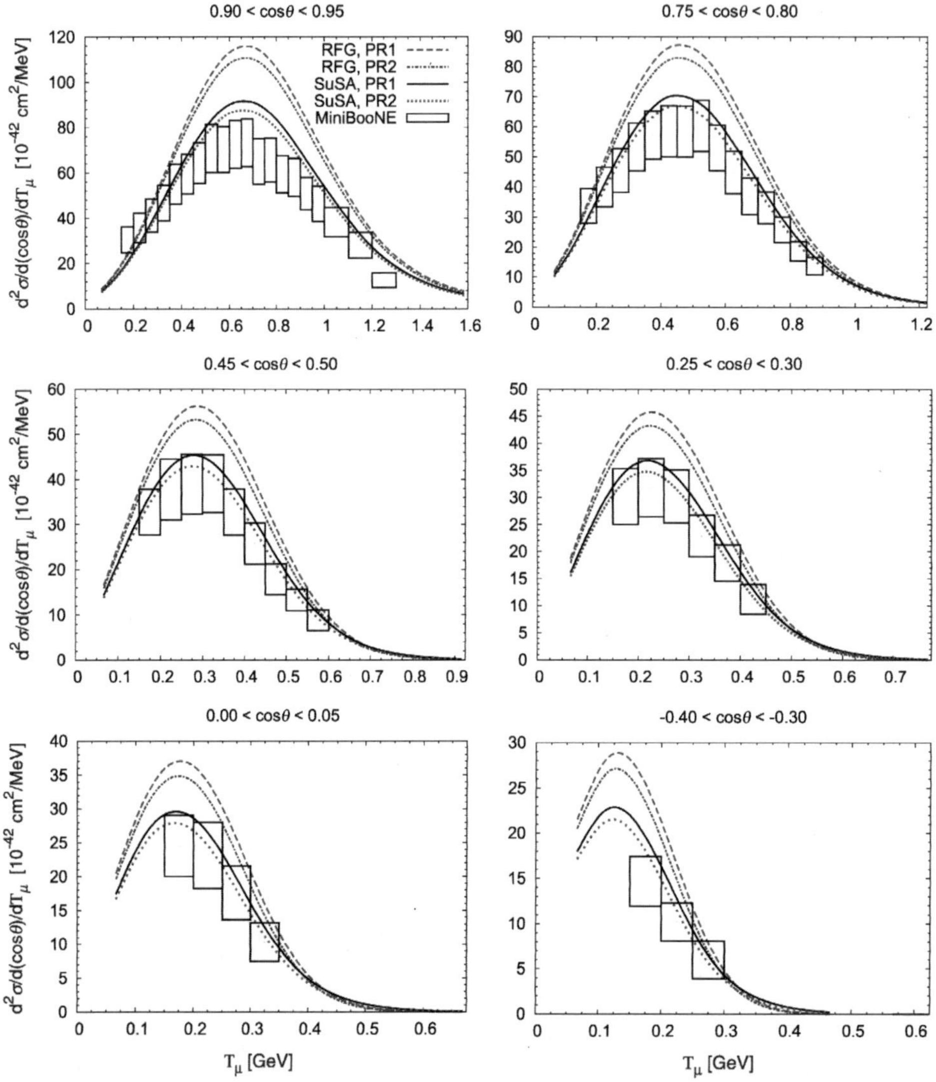

Figure 3. (Color online) The double-differential cross section averaged over the neutrino energy flux as a function of the muon kinetic energy T_μ obtained by SuSA and RFG Δ-region scaling functions. In each subfigure the results have been averaged over the corresponding angular bin of $\cos\theta$. For vector and axial form-factors two parameterizations, "PR1" [15] and "PR2" [16], are used.

of uncertainty that we expect from the choice of the single-nucleon response for this reaction. We compare the predictions of SuSA and RFG with the MiniBooNE data [1]. Here we show that SuSA predictions are in good agreement with the MiniBooNE experimental data for π^+ cross-section in the case of the flux averaged data.

2.4. Inclusive charged-current neutrino cross sections in the T2K experiment

In figure 4 we show the CC inclusive $\nu_\mu-^{12}$C double-differential cross section per nucleon versus the muon momentum, p_μ, for different angular bins, folded with the T2K flux. The QE curve (dashed line) corresponds to the results obtained using SuSA scaling function in the QE-region [3]

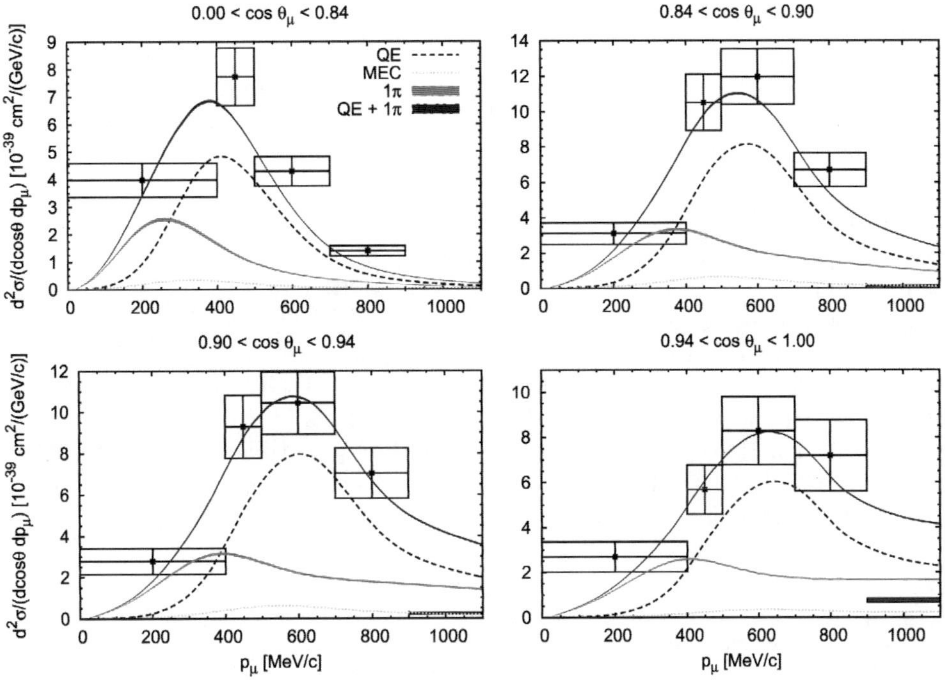

Figure 4. The CC inclusive T2K flux-folded ν_μ-^{12}C double-differential cross section per nucleon evaluated using SuSA scaling functions in the QE region [QE] and in the Δ-region [1π] is displayed as a function of the muon momentum for different bins in the muon angle. The separate contributions of the QE, 1π, and vector 2p2h MEC are displayed. The data are from [2].

and Pauli blocking effects in the scaling function introduced in Ref. [17]. The standard value of the axial mass $M_A = 1.03$ GeV is used in the QE calculations. The resonant pion production curve (1π) is derived with the SuSA scaling function in the Δ-region $f^{\Delta}(\psi_\Delta)$ (figure 2). The band corresponds to the two different parametrizations, PR1 and PR2, described in the previous Section.

As first noticed in [3] and studied in more depth in [12], sizeable deviations (10 – 15%) from scaling are observed in the region where the QE and Δ responses overlap. In this region effects stemming from correlations and 2p2h MEC can play an important role and they cannot be reproduced by models that assume impulsive, quasifree scattering on bound nucleons. Therefore these effects must be added to the QE and Δ scaling functions. In figure 4 the MEC curve (dotted line) corresponds to the fully relativistic calculation of 2p2h excitations induced by pionic vector two-body current of [5] and parameterized in [4].

We observe that the model yields good agreement with the T2K data. The main contribution in the cross section comes from the QE and pion production mechanisms. On the contrary, MEC play a minor role at these kinematics. It should be noted however that the present model does not include the axial two-body current.

3. Conclusions

The SuSA approach provides neutrino-nucleus cross section predictions, based on the observed nuclear response to electron projectile and the universal character of the scaling function. We show that SuSA predictions are in good agreement with the MiniBooNE experimental data for pionic cross section in the case of the flux averaged data. We conclude that the idea of the

SuSA approach for the QE- and Δ-regions (extracted from electron scattering experiments), when being extended to neutrino processes, proves to be successful in describing ν_μ inclusive charged-current cross sections. The model also includes 2p2h excitations induced by vector meson-exchange currents carried by the pion. Axial-vector MEC contributions have yet to be included, that might be expected to improve the agreement with the T2K data.

Acknowledgments
This work was partially supported by INFN under project MANYBODY, by Spanish DGI and FEDER funds (FIS2011-28738-C02-01, FPA2013-41267), by the Junta de Andalucia, by the Spanish Consolider-Ingenio 2000 program CPAN (CSD2007-00042), by the Campus of Excellence International (CEI) of Moncloa project (Madrid) and Andalucia Tech, by the Bulgarian National Science Fund under contracts Nos. DFNI-T02/19 and DFNI-E02/6, by the Office of Nuclear Physics of the U.S. Department of Energy under Grant Contract Number DE-FG02-94ER40818 (T.W.D.) and by the European Union 7th FP Marie Curie IOF project ELECTROWEAK (O.M.). R.G.J. acknowledges financial help from the Interuniversity Attraction Poles Programme initiated by the Belgian Science Policy Office.

References

[1] Aguilar-Arevalo A A *et al.* (MiniBooNE Collaboration) 2011 *Phys. Rev. D* **83** 052007
[2] Abe K *et al.* (T2K Collaboration) 2013 *Phys. Rev. D* **87** 092003
[3] Amaro J E *et al.* 2005 *Phys. Rev. C* **71** 015501
[4] Megias G D *et al.* 2015 *Phys. Rev. D* **91** 073004
[5] De Pace A, Nardi M, Alberico W M, Donnelly T W and Molinari A 2003 *Nucl. Phys. A* **726** 303
[6] Donnelly T W and Sick I 1999 *Phys. Rev. Lett.* **82** 3212
[7] Donnelly T W and Sick I 1999 *Phys. Rev. C* **60** 065502
[8] Alberico W M *et al.* 1988 *Phys. Rev. C* **38** 1801
[9] Barbaro M B, Cenni R, Pace A D, Donnelly T W and Molinari A 1998 *Nucl. Phys. A* **643** 137
[10] Jourdan J 1996 *Nucl. Phys. A* **603** 117
[11] Barbaro M B, Caballero J A, Donnelly T W, and Maieron C 2004 *Phys. Rev. C* **69** 035502
[12] Maieron C *et al.* 2009 *Phys. Rev. C* **80** 035504
[13] Ivanov M V *et al.* 2012 *Phys. Lett. B* **711** 178
[14] Maieron C, Donnelly T W and Sick I 2002 *Phys. Rev. C* **65** 025502
[15] Alvarez-Ruso L, Singh S K and Vacas M J V 1999 *Phys. Rev. C* **59** 3386
[16] Paschos E A, Yu J Y and Sakuda M 2004 *Phys. Rev. D* **69** 014013
[17] Megias G D *et al.* 2014 *Phys. Rev. D* **89** 093002

Manifestation of cluster effects in collective octupole and superdeformed states of heavy nuclei.

T.M.Shneidman, G.G.Adamian, N.V.Antonenko, R.V.Jolos

Joint Institute for Nuclear Research, 141980 Dubna, Russia

E-mail: jolos@theor.jinr.ru

Abstract. The effects of reflection-asymmetric deformation on the properties of the low-lying negative-parity collective states and superdeformed states of heavy nuclei are analyzed basing on dinuclear model. The results of consideration of the alternating parity bands in actinides and the superdeformed bands in ^{60}Zn, Pb and Hg isotopes are discussed.

1. Introduction

There are several approaches to treat the collective motion related to reflection-asymmetric degrees of freedom. one of them is based on the assumption that the most suitable conditions for manifestation of softness with respect to octupole vibrations or even octupole deformation are realized in nuclei with a single particle level scheme in which two single particle levels with $\Delta l = \Delta j = 3$ are very close together and close to the Fermi surface. One of them is located below and the other – above Fermi surface [1]. These two single particle states are connected by a strong E3 transition matrix element. This effect is stronger if this situation is realized in both proton and neutron systems.

At the same time the low-lying negative parity states are known in light nuclei, for instance, in ^{20}Ne where $\Delta l = \Delta j = 3$ pair of the single particle levels is characterized by a too large excitation energy. It is commonly assumed that octupole collectivity in light nuclei is related to the α-clustering [2,3]. Several Nilsson-Strutinsky-type calculations for light nuclei demonstrate that nuclear configurations corresponding to the minima of the potential energy are characterized by density profiles clearly demonstrating cluster structure.

It was also suggested in the framework of the Interacting Boson Model that the collective low-lying negative-parity states in heavy nuclei like Ra and Th isotopes are related to manifestation of α-clustering in these nuclei [4]. Later it was indicated on the correlations between the α-decay hindrance factor and the excitation energy of the lowest 1^- state in heavy nuclei [5].

2. Dinuclear system concept

At the end of 60th deep inelastic collisions (DIC) of heavy ions have been discovered [6,7]. The characteristic feature of these reactions is large variation of mass (charge) distribution. To interpret the experimental data on DIC it was assumed that during the collision a configuration of two touching nuclei (clusters) which keep their individuality is formed. This configuration was named by V.V.Volkov [8] as a dinuclear system. Such system has two main degrees of freedom which govern its dynamics: the relative motion between the nuclei and a multinucleon transfer leading to dependence of dynamics on mass and charge asymmetries. A multinucleon transfer is described by the mass and charge asymmetry coordinates

$$\eta = \frac{A_1 - A_2}{A_1 + A_2} \tag{1}$$

and

$$\eta_Z = \frac{Z_1 - Z_2}{Z_1 + Z_2} \tag{2}$$

These coordinates can be assumed as continuous or discrete quantities. For $\eta = \eta_Z = 0$ we have a symmetric clusterization with two equal nuclei, and if η approaches the values ± 1 or if A_1 or A_2 is equal to zero a mononucleus is formed. These coordinates have been originally introduced by W.Greiner [9] and applied for description of mass distribution in fission [10]. In [11] it was suggested to apply dinuclear system concept to description of the reflection-asymmetric low-lying collective states.

3. Ground state alternating parity bands. Parity splitting

The degrees of freedom choosen to characterize a dinuclear system are related to description of rotation of a dinuclear system as a whole, and a transfer of nucleons between the fragments. The hamiltonian of the model can be presented in the form

$$H = -\frac{\hbar^2}{2}\frac{d}{d\eta}\frac{1}{B(\eta)}\frac{d}{d\eta} + U(\eta, I) \tag{3}$$

The method of calculation of the inertia coefficient $B(\eta)$ has been suggested in [12]. Our calculations have shown that $B(\eta)$ is a smooth function of the mass number A. As a consequence, we take nearly the same value of $B(\eta) = 20 \times 10^4 m \cdot fm^2$ for almost all considered actinide nuclei. Here m is the nucleon mass. The potential energy $U(\eta, I)$ is determined by the biding energies $(B_{1,2})$ of nuclei forming dinuclear system and their nuclear V_N and Coulomb V_{Coul} interaction, and by the rotational energy of a dinuclear system as a whole V_{rot}

$$U(\eta, I) = B_1(\eta) + B_2(\eta) - B + V(R = R_m, \eta, I), \tag{4}$$

$$V(R, \eta, I) = V_N(R, \eta) + V_{Coul}(R, \eta) + V_{rot}(R, \eta, I). \tag{5}$$

Above R is a distance between the centers of mass of nuclei forming dinuclear system and R_m is a position of the minimum of the radial potential energy.

It was further shown that among different cluster configurations the only system which gives a significant contribution to the formation of the low-lying states is the α-cluster configuration. The energies of the Li-cluster configurations are about 15 MeV larger than the binding energies of the considered mononuclei. Therefore, for small excitations only oscillations in η are of interest which lie in the vicinity of $|\eta|$=1, i.e., only cluster configurations up to Li clusters need to be considered.

It is seen from the expression of the Hamiltonian (1) that concrete information about a nucleus under consideration is presented only by the binding energies of nuclei forming dinuclear system. Since in all cases α-particle is presented as a light fragment of a dinuclear system a specific information which vary from nucleus to nucleus is a binding energy of a heavy fragment of a system. However, a binding energy of the heavy fragment is given by the experimental data and, therefore, in dinuclear model evolution of the properties of the low-lying negative parity states is determined by the experimental data on biding energies. This circumstances give us a possibility to check a correctness of the basic concept of dinuclear model in its application to description of the properties of the low-lying collective negative-parity states. It was especially interesting to check how well dinuclear model describes evolution of the excitation energy of the lowest 1^- state with the mass number in the chains of the isotopes of some actinide nuclei where this excitation energy varies significantly.

The results of application of dinuclear system model to description of the excitation energies of the lowest 1^- states ($E(1_1^-)$) in Ra and Th isotopes were quite successful. The model describes quite well evolution of the values of $E(1_1^-)$ with the mass number. For instance, the experimental values of the excitation energies of the 1_1^- states in Th isotopes decrease from 714 keV in ^{232}Th to 230 keV in ^{226}Th and then keep the values 251 keV and 250 keV in ^{224}Th and ^{222}Th, respectively. The calculated values are: 693 keV in ^{232}Th, 254 keV in ^{226}Th, 204 keV in ^{224}Th and 195 keV in ^{222}Th. The excitation energies of the negative-parity states with higher angular momenta have been also calculated and the results obtained demonstrate a good description of the experimental data.

The results of calculations of the wave functions of the states of the ground state alternaing parity bands have shown that the weight of the α-cluster component in the wave functions of these states increases with angular momentum in the light Ra and Th isotopes from approximately 5% at I=1 to 40 % at around I=20. In contrast, in ^{232}Th, which has the largest energy of the 1_1^- state the weight of the α-cluster component at $I \approx 20$ does not exceed 15%. The value of the weight of the α-cluster component at I=1 equal to 5% is in a correspondence with the spectroscopic factor for α-decay of the light Ra and Th isotopes. The results obtained also show that the weight of the α-cluster component in the wave functions of negative-parity states is higher than in the neighboring positive-parity states. This fact is manifested in the staggering of the weight of the α-cluster component with smooth increase of the angular momentum.

We have also calculated the E1 transition matrix elements. The results obtained show significant increase of the value of $\langle I||Q(E1)||I+1\rangle$ with angular momentum. This is a manifestation of an increase with angular momentum in the wave function of the weight of the α-cluster component compare to the mononucleus component. Since Z to N ratio in α-particle is quite different from the Z to N ratio in the heavy cluster $(A-4, Z-2)$ the relative motion of the α-particle and a rest of a nucleus creates strong electric dipole transitional moment. This fact is important for analysis of predictions of different models describing octupole motion in heavy nuclei at high angular momenta.

4. Cluster effects in the structure of the superdeformed band in ^{60}Zn

In many light nuclei clustering is a very prominent feature in a large number of states. Typical examples are ^{16}O and ^{20}Ne where $^{12}C+\alpha$ and $^{16}O+\alpha$ structures are particularly stable. It is interesting that clustering in ^{20}Ne is realized in the ground state band, but in ^{16}O cluster structure has the excited band built on 6.06 MeV intruder state. In heavier nuclei α-cluster structures have been predicted to be stable in ^{40}Ca and ^{44}Ti. In the pairs of nuclei ^{16}O–^{20}Ne and ^{40}Ca–^{44}Ti one of the partners is a spherical double magic nucleus and the other is a double magic nucleus plus α-particle. The next pair of this kind is ^{56}Ni–^{60}Zn. In ^{56}Ni the excited deformed rotational band is known up to $I^\pi = 12^+$. In ^{60}Zn the threshold for the α-decay is only 2.7 MeV higher than the ground state. Therefore it is quite possible that the ground state band of ^{60}Zn contains the α-cluster component. However, in ^{60}Zn there is also a superdeformed band. At angular momenta higher than 12 only E2 transitions between the states of the superdeformed band have been seen. In the angular momentum region I=8–12 the superdeformed band decay into the states of the ground state band. Decay of the superdeformed rotational bands into normally deformed or spherical states is one of the interesting nuclear structure phenomena.

In the superdeformed band of ^{60}Zn the moment of inertia depending on I takes the values

(692-795)$m \cdot fm^2$, where m is the nucleon mass. These values are close to the sticking moment of inertia of the ^{52}Fe+^{8}Be cluster configuration which is 750$m \cdot fm^2$. We mention also that the threshold energy for decay of ^{60}Zn into ^{52}Fe+^{8}Be (10.8 MeV) and ^{48}Cr+^{12}C (11.2 MeV) are close to the extrapolated values of the superdeformed band head, which is approximately 7.5 MeV. Thus, it is quite possible that the superdeformed band in ^{60}Zn corresponds to the Be cluster configuration. Our calculations for ^{60}Zn have shown that dinuclear configuration with α-cluster as a light cluster has a potential energy smaller than the energy of the mononucleus at $|\eta|$=1. With respect to the ground state energy the potential energy at the α-minimum is –4.5 MeV. Next important minima correspond to ^{8}Be and ^{12}C cluster configurations with the following values of the potential energy at the minima with respect to the energy of the mononucleus: 5.1 MeV and 9.0 MeV, respectively. At the values of η corresponding to Li- and B-clusters potential enrgy has maxima.

We have calculated [13] the branching ratios of the E2 ΔI=2 intensities of the γ-transitions. In the experiment for the 18^+_{sd}, 16^+_{sd} and 14^+_{sd} states only decay into the superdeformed states have been observed. The calculated branching ratios for these transitions are

$$\frac{I(18^+_{sd} \rightarrow 16^+_{gs})}{I(18^+_{sd} \rightarrow 16^+_{sd})} = 0.02, \quad (6)$$

$$\frac{I(16^+_{sd} \rightarrow 14^+_{gs})}{I(16^+_{sd} \rightarrow 14^+_{sd})} = 0.07, \quad (7)$$

$$\frac{I(14^+_{sd} \rightarrow 12^+_{gs})}{I(14^+_{sd} \rightarrow 12^+_{sd})} = 0.18. \quad (8)$$

For the ratio $I(12^+_{sd} \rightarrow 10^+_{gs})/I(12^+_{sd} \rightarrow 10^+_{sd})$ the calculated value is 0.42. For the ratio $I(10^+_{sd} \rightarrow 8^+_{gs})/I(10^+_{sd} \rightarrow 8^+_{sd})$ the calculated value is 0.63. The calculated value of the $I(8^+_{sd} \rightarrow 6^+_{gs})$ is 0.19 W.u. which is large enough to explain an absence of the $8^+_{sd} \rightarrow 6^+_{sd}$ transition.

5. Decay out of superdeformed bands in the mass region $A \approx$190 within dinuclear model approach

Over two hundred superdeformed bands have been investigated in different mass regions of the nuclide chart. While the rotational transitions between the superdeformed states are easy to detect with modern Ge arrays, it is hard to localize the superdeformed bands in excitation energy, angular momentum and parity and to link them to the normal deformed bands. This is because of the remarkable feature of the superdeformed states: the intraband $E2$ rotational transitions follow the band down with practically constant intensity and drop sharply at some angular momentum. This phenomenon is referred to as the decay out of the superdeformed band.

Analysis of the potential energy of dinuclear system has shown that the important minima of the potential energy are: mononucleus, α, ^{8}Be and ^{12}C cluster configurations. Dinuclear system with α-cluster has a potential energy at I=0 which is close to the energy of a mononucleus at $|\eta|$=1. As a consequence, the states of the normal deformed bands have a significant contribution of the α-cluster component. The states of the lowest and excited superdeformed bands are described mainly as ^{8}Be and ^{12}C cluster configurations, respectively.

The wave functions of the normal deformed and superdeformed states are well separated at low angular momenta. Only a very small tail of the wave functions of the lowest superdeformed states in the normal deformed minimum supplies a small coupling of the superdeformed state with a complex spectrum of the compound states in the normal deformed minimum. Since different cluster configurations have different moments of inertia, the potential energy depends on the angular momentum of the system. This is an origin of the angular momentum dependence of the mixing between the superdeformed and the normal deformed states. At large angular momenta the energies of the superdeformed states are lower than the energies of the neighboring collective normal deformed statesconnected to the mass-asymmetry degree of freedom. With decreasing angular momentum the energy interval between these states decreases and at some angular momentum I between 6 and 14 the collective normal deformed state become energetically lower than the superdeformed state with the same angular momentum. In all nuclei belonging to the $A \approx 190$ region considered in the framework of dinuclear model the calculated angular momentum is close to the observed angular momentum at which decay out happens. Thus, the decay out occurs through the normal deformed doorway state. In dinuclear model this normal deformed state is the excited vibrational state corresponding to the motion in mass asymmetry degree of freedom. At high excitation energy in the normal deformed potential well, the normal deformed doorway state is spread among the sea of the dense compound states. This spreading leads to a large width of a doorway state. So, statistical mixing with highly excited normal deformed states is one of the reason for decay out of the superdeformed band. The sudden decay out takes place near a crossing of the superdeformed band with the nearest neighboring excited collective normal deformed band where a weight of the normal deformed doorway state increases. Even at the near band crossing point the normal deformed admixture in the superdeformed state is relatively small but decay out occurs due to the large width of doorway state with respect to the width of the superdeformed state. The maximal normal deformed admixture of the superdeformed states were found to be in a range of a few percent [14].

6. Conclusions

We have shown that there are correlations between an increase of collectivity of octupole mode and manifestation of α-clustering. Dinuclear model and driving potential calculated following the rules of the model qualitatively explain evolution of the characteristics of the low-lying octupole states with the atomic mass and charge numbers.

Properties of the superdeformed bands are also qualitatively described in the framework of dinuclear model.

6.1. Acknowledgments

We are grateful to Profs. P. von Brentano, W. Scheid and H. Lenske for collaboration. This work was supported in parts by the DFG (Germany) under Contract Br 799/25-1, by the RFBR (Russia) under grant No. 13-02-00080 and by the Heisenberg-Landau program.

7. References

[1] Butler P and Nazarewicz W 1996 *Rev. Mod. Phys.* **68** 349
[2] Rae W D M 1988 *Int. J. Mod. Phys.* A **3** 1343
[3] Freer M and Merchant A C 997 *J. Phys.* G **23** 261
[4] Iachello F and Jackson A D 1982 *Phys. Lett.* B **108** 151
[5] Sheline R K and Riley M A 2000 *Phys. Rev.* C **61** 057301
[6] Wilczynski J, Volkov V V, Dezovski P 1967 *Yadernaya Fizika* **5** 942
[7] Gridnev G F, Volkov V V, Wilczynski J 1970 *Nucl. Phys.* A **142** 385
[8] Volkov V V 1986 *Izv. AN SSSR ser. fiz.* **50** 879
[9] Fink H J, Maruhn J, Scheid W and Greiner W 1974 *Z. Phys.* **268** 321
[10] Maruhn J and Greiner W 1974 *Phys. Rev. Lett.* **32** 548
[11] Shneidman T M, Adamian G G, Antonenko N V, Jolos R V 2003 *Phys. Rev.* C **67** 014313

[12] Adamian G G, Antonenko N V, Jolos R V 1995 *Nucl. Phys.* A **584** 205
[13] Adamian G G, Antonenko N V, Jolos R V, Palchikov Yu V, and Scheid W 2003 *Phys. Rev.* C **67** 054303
[14] Adamian G G, Antonenko N V, Jolos R V, Palchikov Yu V, Scheid W, and Shneidman T M 2004 *Phys. Rev.* C **69** 054310

Double-neutron capture reaction and natural abundance of ^{183}W, ^{195}Pt and ^{199}Hg isotopes

Karamian S. A., Aksenov N. V. and Bozhikov G. A.

Joint Institute for Nuclear Research, Dubna, 141980, Moscow region, Russia

e-mail: karamian@jinr.ru

Abstract. There are much data on neutron cross sections over the chart of nuclides for stable isotopes and not as much for the radioactive ones. Double neutron capture experiments could be fruitful to provide more data. Time-integrated mean flux of slow neutrons reaches the value of $2.3{\cdot}10^{12}$ n/cm^2 s at the irradiation port near the active zone of the IBR-2 pulsed reactor of JINR. This is enough to detect the double neutron capture products by the activation method. A high capture cross section is obtained in the present experiment for intermediate radioactive ^{182}Ta and ^{194}Ir target nuclides. Together with the known data for ^{198}Au, these values may prove an essential role of double neutron capture process for nucleosynthesis of ^{183}W, ^{195}Pt and ^{199}Hg isotopes at stellar conditions.

Keywords: neutron flux, radioactive products, double capture, isotope abundance

1. Introduction

In a double neutron capture process the relatively short-lived ^{194}Ir (19.28 *h*) is present as an intermediate target nucleus. The ^{195}Pt could be found among the final products of activation due to the following process: ^{194}Ir (*n*, γ)^{195m}Ir $\beta^- \rightarrow$ ^{195m}Pt (13/2$^+$). The latter nuclide is convenient for detection being 4.01 *d*–lived and the main channel (40%) in the decay of 3.67 *h*–lived ^{195m}Ir (11/2$^-$) isomer. The ground state of ^{195}Ir (3/2$^+$) (2.29 *h)* is produced with a large yield in the reaction, but it decays directly to the ground state of the stable ^{195}Pt, and not to the isomer [1]. Such properties provide an advantageous option for production of ^{195m}Pt through the double neutron capture reaction. In the case of ^{nat}Ta target, the consequent two-neutron capture leads to ^{183}Ta ($T_{1/2}$ = 5.1 *d*) via the intermediate 114.4 *d*–lived ^{182}Ta isotope. In the present work the ^{195m}Pt and ^{183}Ta activities were detected by the γ-spectroscopy method after irradiation of ^{193}Ir (98.5% enriched) and ^{nat}Ta targets at the IBR-2 reactor. The method of ^{195m}Pt production by double neutron capture reaction was stressed in [2] as useful for clinical applications.

In fact, there are recommended neutron cross section values given in [3 – 6]. However, there are not enough data available to describe the formation of ^{195m}Pt, which puts the most significant restriction on calculation of the yield. Only two of six important values are known for thermal cross section plus one value for the resonance integral. It would be impossible to evaluate the yield of ^{195m}Pt from such initial data, even in the case when the well-developed computer program is available. The results of [2] were obviously obtained using theoretically estimated cross sections, though it is known that the cross sections, the resonance integral values and isomer-to-ground state ratios cannot not be predicted in theory. One may hope for fast progress in neutron data, but the experiment on observation of ^{195m}Pt in neutron irradiations seems more direct and fruitful. The flux supplied by the IBR-2 reactor is enough to detect the activation of double neutron capture products.

Initially, a test experiment [7] has been performed using relatively low neutron flux generated by the MT-25 electron accelerator at FLNR, JINR. The flux about 108 n/cm^2 s was not enough to observe the products of double neutron capture, but major products of Ir isotopes activation were successfully detected and the corresponding σ_{th} and I_γ values were measured. The results are given in Table 1. One of the important conclusions is that the most efficient way to obtain ^{195m}Pt is through the process ^{194g}Ir (*n*, γ) ^{195m}Ir $\beta^- \rightarrow$ ^{195m}Pt. The alternative branch with the population of ^{194m}Ir (171 *d*) at the first step of neutron capture is inefficient because of the low cross section measured for this high-spin product (see Table 1). A large spin-difference $\Delta I \approx 9$ between the initial ^{193}Ir (3/2+) and the final ^{194m}Ir (10 or 11) nuclides suppresses the yield of the product in agreement with the systematics [8] and the results obtained for the Hf isomers [9]. At the second step of neutron capture, the

^{195m}Ir product has a twice-lower spin $11/2^-$ and may be successfully populated in (n, γ) reaction, it then decays in 40% events to ^{195m}Pt. Unfortunately, the ground state of ^{195}Ir decays only to the ground state of ^{195}Pt and there is no branch to the ^{195m}Pt isomer. Therefore, the only efficient sequence of processes leading to the ^{195m}Pt isomer is obvious.

Table 1. Measured cross sections for the slow-neutron capture reactions by Ir isotopes. Comparison to the data of [3 – 6] is given in [7].

Target	Product	$T_{1/2}$	σ_{th}, barn	I_γ, barn
^{191}Ir(3/2$^+$)	^{192g}Ir(4$^+$)	73.83 *d*	550±65	3560±150
	^{192m1}Ir(1$^-$)	1.5 *min*		
	^{192m2}Ir(11$^-$)	241 *y*	–	–
^{193}Ir(3/2$^+$)	^{194g}Ir(1$^-$)	19.28 *h*	98±12	1550±90
	^{194m1}Ir(4$^+$)	32 *ms*		
	^{194m2}Ir(10 or 11)	171 *d*	(1.0±2.8)10^{-3}	0.15±0.05

2. The ^{195m}Pt yield due to double neutron capture (experiment)

The numbers of radioactive atoms accumulated after irradiation time t as a result of single and double neutron capture N_1 and N_2, respectively may be obtained by the solution of linear differential equations taking into account the accumulation and exponential decay law. The following notations are used below: the decay constants λ_1 and λ_2 for the products with mass numbers $(A_t + 1)$ and $(A_t + 2)$, where $\lambda = \ln2/T_{1/2}$, A_t is the mass number of the stable target. Obviously, the yield of products is proportional to the number of target atoms $N0$ and is determined by the flux F of neutrons per cm^2s. Let us assume that only thermal neutrons are involved and their cross sections are σ_1 and σ_2. At moderate neutron flux we may neglect the target material exhausting as well as burning-up of the $(A_t + 2)$ product due to the capture of the third neutron.
If necessary (at high fluxes), a factor of burning-up for the $(A_t + 1)$ product may be introduced replacing λ_1 with $(\lambda_1 + \sigma_2 \cdot F)$. The resonance neutron contribution is described by similar equations substituting the resonance integral I_γ instead of σ. However, the resonance neutron flux Fr and the Westscott parameter for the activation product must be specified.

$$N_1(t) = \frac{N_0\sigma_1 F}{\lambda_1}(1-e^{-\lambda_1 t}) \; ; \qquad (1)$$

$$N_2(t) = \frac{N_0\sigma_1\sigma_2 F^2}{\lambda_1\lambda_2}\left[1-\frac{\lambda_1 e^{-\lambda_2 t}}{(\lambda_1-\lambda_2)}+\frac{\lambda_2 e^{-\lambda_1 t}}{(\lambda_1-\lambda_2)}\right]. \qquad (2)$$

As mentioned above, the accumulation of ^{195m}Pt (4.01 *d*) proceeds through the radioactive β^- decay of 3.67 *h*-lived ^{195m}Ir. This means that the longer-lived product is formed after the decay of the short-lived predecessor. The half-life 3.67 *h* of ^{195m}Ir is much shorter than a typical irradiation time comparable to the half-life of a product. Therefore, it is logical to assume that ^{195m}Ir is transformed to ^{195m}Pt with no time delay and the parameter λ_2 corresponds to the decay of ^{195m}Pt (4.01 *d*). Definitely, the population efficiency $k = 0.40$ for the final product must also be introduced into Eq. (2) as a reducing factor. The cross section of ^{194g}Ir activation at the first step is known, but the branch leading to the 195mIr isomer at the second step remained uncertain until now. In the present experiment, the corresponding cross section and the resonance integral were successfully determined at IBR-2 using the fluxes about $2.3 \cdot 10^{12}$ and $2.0 \cdot 10^{11}$ n/cm^2s for thermal and resonance neutrons, respectively.

The method of Cd-difference was applied when two enriched 193Ir(98.5%) targets of 20 mg weight each were exposed at the vertical channel of the IBR-2 reactor, FLNP, JINR. The targets with and without Cd shielding were irradiated during the 17 – d reactor run. Metal foils of Ta served as spectators. The Ir samples were dissolved by electrochemical method for consequent isolation of the Pt fraction applying the chromatography. Gamma spectroscopy with HP Ge detector was used for the activity measurements. The dissolving yield was calibrated by the 192Ir activity (present due to the 191Ir admixture), while the Pt isolation method was tested elsewhere. Finally, the gamma lines of ^{195m}Pt decay were measured with a good statistical accuracy, and the production process ^{194}Ir(n, γ)^{195m}Ir → ^{195m}Pt is characterized by the following values: σ_{th} = 5150 b and I_γ =295 b including the reduction factor due to the β– decay branch.
The cross sections determined now are enough to evaluate the activity yield at the high neutron flux about $2.5 \cdot 10^{15}$ n/cm^2s, as at the Oak Ridge reactor. In calculations one must take into account that the N_1 intermediate product is partially exhausted due to the second neutron capture. Then, the equilibrium activity of ^{195m}Pt may reach 1.0 Ci per mg of the ^{193}Ir target material (more details are given below). This value of activity satisfies the requirements for production of great specific-activity solutions necessary for radiotherapy applications. Of course, proper technical tools and methods must be developed for the chemical processing of intense β,γ-ray sources.

3. Nuclear and astrophysical consequences

A high cross section is obtained for the double-neutron capture process ^{193}Ir (n,γ)^{194}Ir (n, γ)^{195m}Ir (3.67 h) → ^{195m}Pt. Taking into account the efficiency of the β^- decay branch leading to ^{195m}Pt, one immediately deduces the values of σ_{th} = 12900 barns and I_γ = 740 b characterizing the constituent ^{194}Ir(n, γ)^{195m}Ir(11/2$^-$) reaction. The decay branch of low-spin ^{195g}Ir(3/2$^+$) to ^{195m}Pt(13/2$^+$) was not observed being negligible [1]. In reaction the cross section for the 2.29 h-lived products must exceed by an order of magnitude the observed one for the population of isomeric ^{195m}Ir(11/2$^-$) state as follows from the typical isomer-to-ground state ratios depending on the spin for (n, γ) products [9]. Therefore, a total capture cross section for the short-lived ^{194}Ir(19.28 h) target nuclide must be extremely high (105 barns) unlike the cross section assumed in [10]. This is a surprising result in itself.
In addition, a high cross section of 10^5 barns must result in the great effect of exhausting for the intermediate ^{194}Ir nuclide with the suppression of the final ^{195m}Pt product yield. In particular, for calculation one has to substitute in Eq. (2) the value $0.40 \cdot \sigma_2^m$ instead of σ_2 and (λ_1 + $\sigma_2^t F$) instead of λ_1, where σ_2^m corresponds to the population branch of ^{195m}Ir after neutron capture and σ_2^t to a total cross section of neutron capture by ^{194}Ir. Within the indicated choice of values the yield of ^{195m}Pt is calculated as a function of the irradiation time t and neutron flux F. From Figure 1 it can be seen that the saturation with time is reached earlier, than in 2 $T_{1/2}$, and at lower values of N_2. In addition, almost linear dependence of the yield on the flux is observed, instead of the flux square F^2 function predicted assuming a negligible burn-up effect.

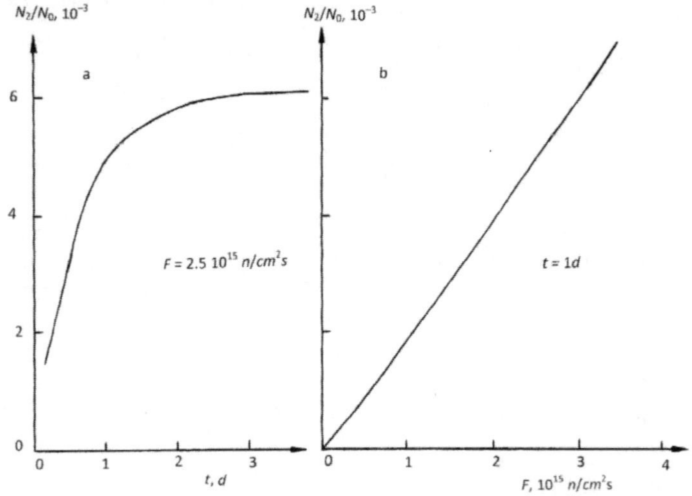

Figure 1. Calculated number of produced ^{195m}Pt nuclei as a function of time (a) and neutron flux (b).

Experimentally observed neutron cross section deduced here for radioactive ^{194}Ir target is comparable to the highest thermal cross sections known over the nuclide chart and it requires an appropriate interpretation, probably, due to a strong compound resonance exactly near the neutron binding energy in ^{195}Ir. Both *m* and *g* products of ^{195}Ir reach the ^{195}Pt ground state after decay, and the known abundance of stable ^{195}Pt isotope comprises the production through the double-neutron capture by ^{193}Ir.
In the same experiment the spectator ^{nat}Ta targets were also irradiated and the second-step ^{182}Ta$(n, \gamma)^{183}$Ta reaction demonstrated values of σ_{th}= 25300 *b* and I_γ= 16600 *b* substantially exceeding the tabular data [3]. Meanwhile, the cross section 47000 *b* was reported for ^{182}Ta in the publication [11] not specifying σ_{th} and I_γ. The numerical values given here (except the estimate 10^5 *b* for ^{194}Ir) were obtained with the standard deviation about 10% including the errors due to the calibration and recalculations. A high value of σ_{th}= 25100 *b* was obtained in [3] for the neutron capture by radioactive ^{198}Au with production of ^{199}Au and then ^{199}Hg after β^- decay. Measured cross sections for radioactive odd-odd nuclides, such as ^{182}Ta, ^{194}Ir, and ^{198}Au are given in Table 2. The natural abundances of ^{183}W, ^{195}Pt, and ^{199}Hg nuclides are correspondingly 14.3, 33.8, 16.9%, and may include a contribution from the double neutron capture at stellar nucleosynthesis. In general, the double neutron capture way differs from the standard *s*- and *r*-processes. The second neutron capture occurs prior the β– decay of the first capture product (unlike the *s*-process), while the capture of the third and further neutrons is improbable (unlike the *r*-process).

Table 2. Parameters of the thermal and resonance-neutron capture reactions measured in the present work for radioactive odd-odd isotopes of ^{182}Ta, ^{194g}Ir and the known data [3] for ^{198}Au.

(n, γ) reaction	Neutron number	E^*, MeV		σ_{th}, *b*	I_γ, *b*
^{182}Ta→^{183}Ta	109→110	6.934		25300	16600
^{194}Ir→^{195m}Ir	117→118	7.232	*m*	12900	740
			total	≈10^5	≈10^4
^{198}Au→^{199}Au	119→120	7.584		25100	–

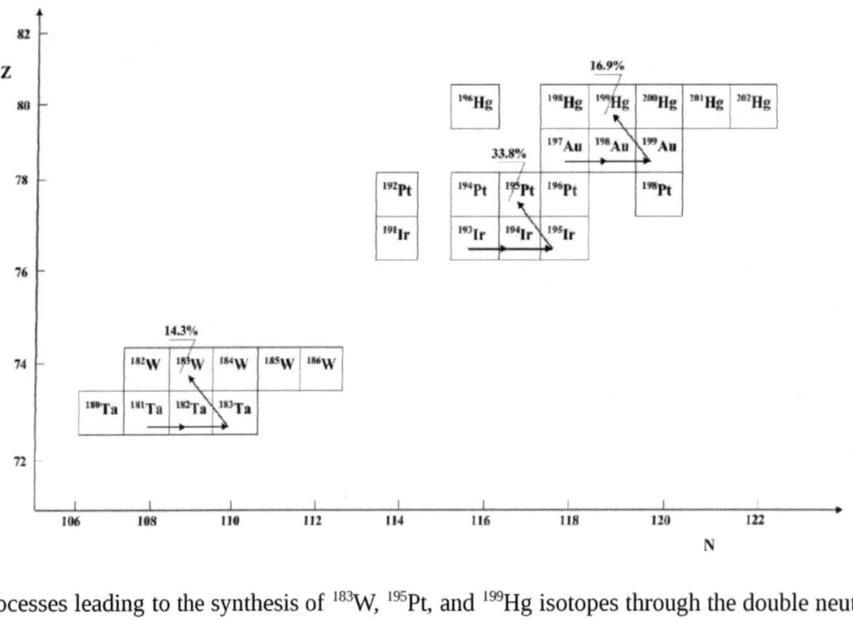

Figure 2. Processes leading to the synthesis of ^{183}W, ^{195}Pt, and ^{199}Hg isotopes through the double neutron capture reaction in addition to the standard *s*-process way.

Concerning the nucleosynthesis one must realize that the room-temperature conditions differ strongly from the typical ones in the Universe. A great variety of conditions exists in the Universe, but neutrons are generated within dense and hot sites. Produced neutrons are moderated to reach the temperature of the surrounding matter. The temperature value $T \geq 10^6$ K approximately corresponds to the neutron energy $E_n > 100$ eV. The room-temperature thermal cross sections determined in the laboratory conditions are not applicable to the evaluation of nucleosynthesis at stellar conditions. Even the resonance integral for heavy nuclei results mostly from the neutron energies (1-10) eV which are lower than the thermal energies in Universe.

Our conclusion about the productive role of the double neutron capture seems doubtful, but this verdict is not absolute. Indeed, the Maxwell distribution for neutron kinetic energies involves the asymptotic $W(E) \sim E^{1/2}$ at $E \to 0$. At the same time, the neutron absorption cross section in geometrical approximation is expressed with the factor $\sigma \sim E^{-1/2}$. Therefore, the reaction yield in the general case remains almost constant at low energies. The high thermal cross section at laboratory conditions may arise only due to the presence of a strong resonance at the neutron energy about zero. This resonance changes the asymptotics of σ providing a strong enhancement and the significant reaction yield even despite a high temperature of the moderator in stellar conditions. Our measured cross sections for ^{182}Ta and ^{194}Ir together with the known data [3] for ^{198}Au are just the cases of strong enhancement of cross sections by the resonance at $E \to 0$. The possibility of synthesis of ^{183}W, ^{195}Pt, and ^{199}Hg isotopes is illustrated in Figure 2.

4. Summary

Production and chemical isolation of ^{195m}Pt isomeric activity is of interest for radiotherapy of patients. The efficiency of double neutron capture reaction for accumulation of ^{195m}Pt is now proved by the experiment on irradiation of the ^{193}Ir enriched target at the IBR-2 reactor. A high cross section is obtained for the neutron capture reaction by radioactive ^{194g}Ir nuclide. The ^{nat}Ta targets were also irradiated over the experiment as spectators and they demonstrated a high cross section for the second neutron capture by radioactive ^{182}Ta. The cross sections of neutron capture by the odd-odd radioactive targets, such as ^{194g}Ir, ^{182}Ta, and ^{198}Au, exceeding 10^4 b are of importance for understanding within the nuclear reaction theory. On the other hand, the natural isotope abundances are influenced due to the observed high probability of the double-neutron capture process, in particular, for the production of ^{183}W, ^{195}Pt, and ^{199}Hg isotopes.

References

[1] Zhou Chunmei 1999 Nucl. Data Sheets 86 645; Huang Xialong and Kang Mengxiao 2014 *Nucl. Data Sheets* **121** 395

[2] Russ) Knapp F F Jr, Mirzadeh S, Beets A L and M. Du 2005 *J. Radioanal. Nucl. Chem.* **263** 503

[3] Mughabghab S F 2006 Atlas of Neutron Resonances 5th Edition (Brookhaven: NNDC); Mughabghab S F (2006) *Neutron Cross Sections* **1** Part B (New York: Academic)

[4] 2012 *Evaluated Data Library* ENDF/B-VII//BNL-98403-2012-JA (Brookhaven)

[5] Magil J, Pfennig G and Galy G 2006 *Karlsruher Nuclidkarte 7th Edition* (Karlsruhe)

[6] Horiguchi T, Tachibana T, Koura H, Katakura J 2000 *Chart of Nuclides* (Ibaraki: Nucl. Data Center)

[7] Karamian S A, Aksenov N V, Albin Y A, Belov A G, Bozhikov G A, Dmitriev S N and Starodub G Y 2014 *Bull. Rus. Acad. Sciences. Phys.* **78** 367

[8] Karamian S A 2013 *Phys. Atom. Nucl.* **76** 1437

[9] Karamian S A and Carroll J J 2006 *High Energy Density Phys.* **2** 48

[10] Calvin G G, Börner H G, Geltenbort P, Hoyler F, Kerr S A, Schrekenbach K and Cizewski J A 1987 *Nucl. Phys.* A **465** 240

[11] Der Mateosian E 1955 *Phys. Rev.* **97** 1023

Search for mixed-symmetry states in ^{212}Po

D. Kocheva[1], G. Rainovski[1], J. Jolie[2], N. Pietralla[3], C. Stahl[3], P. Petkov[4], A. Blazhev[2], A. Hennig[2], A. Astier[5], Th. Braunroth[2], L. Cortes[3], A. Dewald[2], M. Djongolov[1], C. Fransen[2], K. Gladnishki[1], V. Karayonchev[2], J. Litzinger[2], C. Müller-Gatermann[2], M. Scheck[3], Ph. Scholz[2], C. Schramm[2], P. Tölle[2], V. Werner[2], W. Witt[2], D. Wölk[2] and P. Van Isacker[6]

[1] Faculty of Physics, St. Kliment Ohridski University of Sofia, 1164 Sofia, Bulgaria
[2] Institut für Kernphysik, Universität zu Köln, D-50937 Köln, Germany
[3] Institut für Kernphysik, Technische Universität Darmstadt, D-64289 Darmstadt, Germany
[4] Bulgarian Academy of Sciences, Institute for Nuclear Research and Nuclear Energy, 1784 Sofia, Bulgaria
[5] CSNSM, IN2P3/CNRS and Universite Paris-Sud, F-91405 Orsay Campus, France
[6] Grand Accélérateur National d'Ions Lourds, CEA/DSM-CNRS/IN2P3, BP 55027, F-14076 Caen Cedex 5, France

E-mail: dkocheva@phys.uni-sofia.bg

Abstract. In this work we present an experiment dedicated to searching for quarupole-collective isovector valence-shell excitation – the states with so-called mixed proton-neutron symmetry (MSSs), in the nucleus ^{212}Po. The states of interest were populated and studied by an α-transfer reaction. The experiment provides indication for existence of one-phonon MSS in the nucleus ^{212}Po which is the first experimentally identified MSS in the region around double magic nucleus ^{208}Pb.

1. Introduction

States with mixed proton-neutron symmetry (MSSs) are the lowest-lying isovector excitations in the valence shell [1] for vibrational nuclei. These states have been defined in the framework of the Interacting Boson Model with proton-neutron degrees of freedom (IBM-2), proposed by Arima and Iachello[2, 3]. The states' properties simultaneously depend on the nuclear quadrupole collectivity, the underlying single particle structure, and the proton-neutron balance of the wave function. Mixed-symmetry states can be identified experimentally by their unique decay to the low-lying fully-symmetry states (FSSs) [4]. This comprises a major challenge since it requires full spectroscopic infomation, i.e. the spin-parities of these non-yrast states, the lifetimes, the branching ratio and the multipole mixing ratio have to be determined. The one-phonon mixed-symmetry state decays to the 2^+_1 by a strong $M1$-transition. The best examples of MSSs are found in the regions with atomic mass around A≈90 [5] and A≈130 [6, 7, 8, 9, 10, 11]. The available information on MSSs in vibrational nuclei shows that good candidates for MSSs reside around double magic nuclei. Thus, the main goal of the present study is the identification of MSS in the vicinity of double magic nucleus ^{208}Pb.

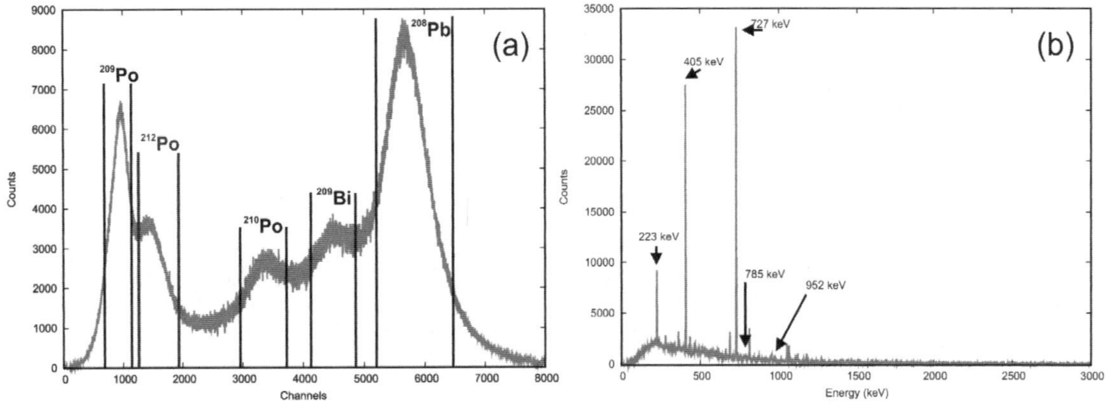

Figure 1. (a) The projection of the particle-γ matrix (at 142°). The vertical lines represent groups of particles found to be in coincidences with the γ-rays from the indicated nuclei. (b) The γ-ray spectrum in coincidences with the region indicated as "^{212}Po" in panel (a).

2. Identification of the one-phonon 2^+ mixed-symmetry state of ^{212}Po

The experiment was performed at the FN Tandem facility of the University of Cologne, Germany. The excited states of interest were populated using the transfer reaction ^{208}Pb(^{12}C,^{8}Be)^{212}Po. The beam was accelerated to 62 MeV which is about 2 MeV below the Coulomb barrier. The target was a self-supporting 10 mg/cm^2 thick Pb foil enriched up to 99.14% with the isotope ^{208}Pb. The reaction took place in the reaction chamber of the Cologne coincidence plunger device [12]. To detect the light reaction fragments six solar cells (10 mm × 10 mm) were used. The array of solar cells was mounted at backward angles with respect to the beam axis, covering an annular space between 116.8° and 167.2°. The solar cells were placed at a distance of about 15 mm between their centers and the target.

The emitted gamma-rays were registered by 12 HPGe detectors mounted outside the plunger chamber in three rings at, on average, 12 cm from the target. Five detectors were positioned at backward angles (142.3° with respect to the beam axes). Six detectors were placed at forward angles (35°) and a single detector was positioned at 0°. The master trigger was set for at least one particle (signal in the solar cells) and one γ-ray (signal in the HPGe detector) or at least two γ-rays. The particle-γ coincidence data was sorted in three matrices with respect to the angular position of the HPGe detectors.

In Fig. 1(a) is shown the projection of the particle-γ matrix. Note that the actual counts correspond to the fragments (ejectiles) from the transfer reactions. For instance, the γ-spectrum of ^{212}Po (shown in Fig. 1(b)) is in coincidence with ^{8}Be (or 2 α-particles). This spectrum is dominated by the 727-keV, the 405-keV and the 223-keV lines which are the transitions from the first three yrast states in ^{212}Po [13]. Besides some contaminants from ^{211}Po all other γ lines in the spectrum in Fig. 1(b) also originated from the decay of excited states in ^{212}Po.

A partial level scheme of the states in ^{212}Po populated in the present experiment was constructed on the basis of the $\gamma-\gamma$ coincidence relationships. The level scheme is shown in Fig. 2. The levels are observed in another α-transfer reaction, namely ^{208}Pb(^{18}O,^{14}C)^{212}Po [13] but we have also populated two non-yrast 2^+ states at excitation energies at 1512 keV and at 1679 keV, respectively [14]. Taking into account the branching ratio and the multipole mixing ratio of these two levels(+0.09(3) and +0.65(50) [14], respectively) these states show a predominant $M1$-transition to the 2^+_1 state via the 785-keV and the 952-keV, respectively [18]. That makes these two 2^+ states potential candidates for the one-phonon MSS in ^{212}Po. The only missing piece of experimental information that is needed for identification of these states

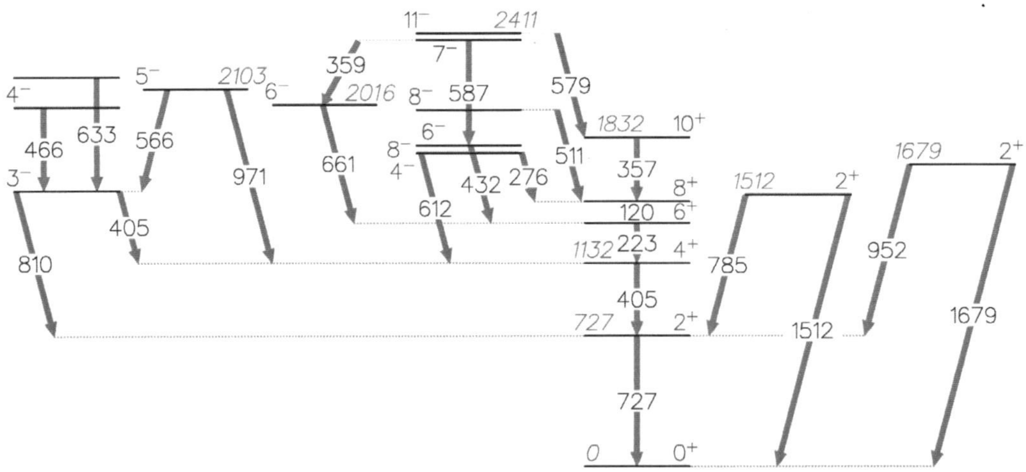

Figure 2. Partial level scheme of ^{212}Po obtained in the present work. For the spin and parity qantum numbers we have adopted the values reported in the previous studies [13, 14, 15, 16, 17]. The energy of the transitions are given in keV. The candidates for the one-phonon MSS are placed on the right of the yrast states.

as MSSs are short lifetimes. The determination of these lifetimes is possible by means of the Doppler Shift Attenuation Method (DSAM) (cf. Ref. [19] and references therein) due to the well pronounced Doppler shapes in both lines of interest.

For extracting the lifetimes two parallel analyses were used. The first one, labelled here Analysis I, is based on a Monte Carlo (MC) simulation by means of a modified [20, 21] version of the program DESASTOP [22] in order to describe the slowing down of the recoiling nuclei. The electronic stopping powers are taken from the Northcliffe and Schilling tables [23] with corrections for the atomic structure of the medium, as discussed in Ref. [24]. An empirical reduction of $f_n = 0.7$ was applied [25] to downscale the nuclear stopping power predicted by the theory of Lindhard, Scharff, and Schitt [26]. In the second analysis, labelled here Analysis II, an integrated software named APCAD (Analysis Program for Continuous Angle DSAM) is used [27]. In APCAD, the slowing down process is simulated by GEANT4 [28] as the electronic stopping process is modelled in the same way as in Analysis I. On the other side, APCAD adopts a simpler approach to modelling the nuclear stopping process, compared to the completely discrete approach used in Analysis I; in Analysis II the angular straggling due to nuclear collisions is modelled discretely by means of MC simulation while the corresponding energy loss is considered to emerge as a result from a continuous process and the nuclear stopping powers were taken from SRIM2013 [29] and reduced by 40%. Both analyses take into account the response of the HPGe detectors, the experimental geometry and the restrictions on the reaction kinematics imposed by the solar cells array. The feeding histories of the levels of interest were determined by the $\gamma{-}\gamma$ coincidence data. Slow feeding was introduced and fitted in the analyses only if the analysed transitions were observed in coincidence with transitions from higher-lying states. In the opposite case, only very fast feeding which can be associated with direct population of the levels of interest, was considered. Under these assumptions both analyses produced similar results for the level at 2016 keV in ^{212}Po (see Table 1).

For extracting the lifetime from the lineshape of the 661.3-keV transition a feeding history similar to the one used in Ref. [13] (58% fast feeding and 42% slow feeding) is used (see Table 1). The lifetimes of the 2_2^+ state at 1512 keV and 2_3^+ state at 1679 keV were extracted from the lineshapes of the transitions $E_\gamma = 785$ keV and $E_\gamma = 952$ keV, respectively. Both of these

Table 1. The results for the lifetime of the 6^- state at 2016 keV (see Fig. 2) extracted by both parallel analyses.

	Ref. analysis [17]	Analysis 1	Analysis 2
$\tau(ps)$	0.49(16)	0.50(4)	0.47(3)

Table 2. Properties of the 2_2^+ and the 2_3^+ states of ^{212}Po. The results for the lifetimes and the transitions strengths are preliminary.

$E_{level}(keV)$	J^π	$\tau(ps)$ [0]	$E_\gamma(keV)$	I_γ, Ref. [18]	α, Ref. [18]	J^π_{final}	δ, Ref. [14]	Transition strength [1] $B(E2)$ in e^2fm^4; $B(M1)$ in μ_N^2
1512	2_2^+	0.67(5)	1512.7	26(3)		0_1^+		B(E2)=32(4)
			785.4	100(1)	0.0408(2)	2_1^+	0.09(3)	B(M1)=0.133(9)
								B(E2)=25(16)
1679	2_3^+	0.68(2)	1679.7	35(8)		0_1^+		B(E2)=23(5)
			952.1	100(19)	0.020(5)	2_1^+	0.65(50)	B(M1)=0.049(17)
								B(E2)=331(306)

[0] From the present experiment. Preliminarily results.
[1] From the present experiment. Preliminarily results.

transitions are in coincidence with 727-keV transition, only. As a result only fast feeding was considered. The preliminary results for the lifetimes together with the available spectroscopic information and the resulting transition strengths are summarized in Table 2. The 2_2^+ state in ^{212}Po at 1512 keV excitation energy decays with a sizeable $M1$-transition to the 2_1^+ state. This allows us to conclude that the 2_2^+ state in ^{212}Po is, at least, a fragment of the one-phonon MSS.

3. The spectrum of ^{212}Po from a simple perspective

The nucleus ^{212}Po has two protons and two neutrons outside the core of ^{208}Pb. In the simplest approach the neutrons are in the $2g_{9/2}$ orbital and the protons – in $1h_{9/2}$. States with two neutrons in the $2g_{9/2}$ orbital and two protons in the $1h_{9/2}$ orbital can be described in the basis

$$|(2g_{9/2})^2 J_\nu, (1h_{9/2})^2 J_\pi; J\rangle \equiv |J_\nu J_\pi J\rangle \tag{1}$$

The neutron-neutron and proton-proton interactions are diagonal in this basis which is mixed by the neutron-proton interaction

$$\langle J_\nu J_\pi J|\hat{H}|J'_\nu J'_\pi J'\rangle = (V^{J_\nu}_{\nu\nu} + V^{J_\pi}_{\pi\pi})\delta_{J_\nu J'_\nu}\delta_{J_\pi J'_\pi} + \tag{2}$$
$$+4\sqrt{(2J_\nu+1)(2J_\pi+1)(2J'_\nu+1)(2J'_\pi+1)} \times \sum_R (2R+1) \begin{bmatrix} j_\nu & j_\pi & J_\pi & J_\nu & \\ & R & j_\pi & J & j_\nu \\ j_\nu & j_\pi & J'_\pi & J'_\nu & \end{bmatrix} V^R_{\nu\pi}$$

where $V^{J_\nu}_{\nu\nu}$, $V^{J_\pi}_{\pi\pi}$ and $V^R_{\nu\pi}$ are the neutron-neutron, proton-proton and neutron-proton interaction matrix elements, respectively, and the symbol in square brackets is a $12j$ coefficient [30]. These three interactions are taken from the energy spectra of the neighbouring nuclei with neutrons and protons in the required shell configurations. In particular, in the single-shell approximation

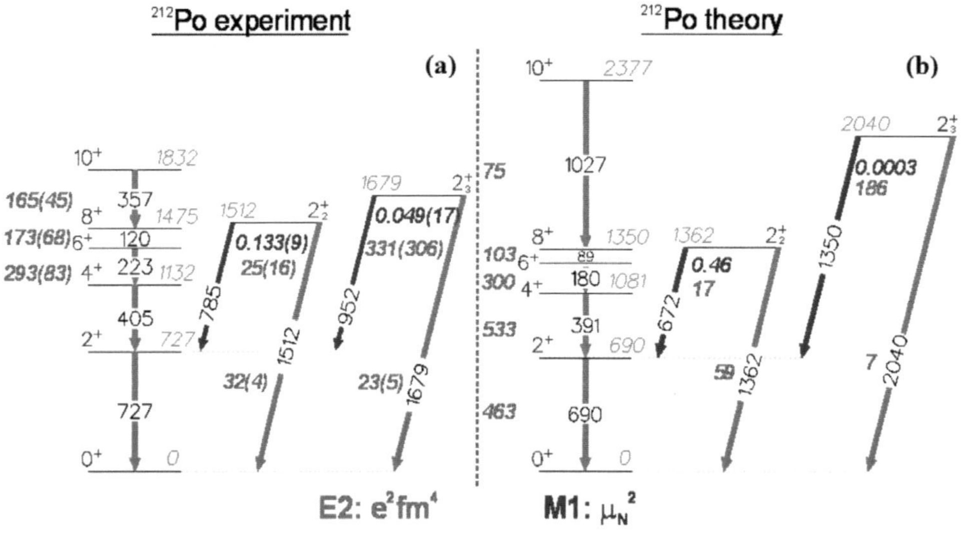

Figure 3. (a) The part of the expertimental spectrum of ^{212}Po representing the yrast states and the levels of interest. (b) The calculated (for details see the text) yrast states and the two lowest 2^+ energy levels of ^{212}Po.

^{210}Pb corresponds to two neutrons in the $2g_{9/2}$ orbital and ^{210}Po to two protons in $1h_{9/2}$. Both of these nuclei display seniority-like spectra. The interaction between a neutron in the $2g_{9/2}$ orbital and a proton in the $1h_{9/2}$ orbital can be determined by the energy spectrum of ^{210}Bi.

In the single-shell approximation the $M1$ operator can be determined by the magnetic moments of the ground states of ^{209}Pb and ^{209}Bi – $\mu(9/2_1^+) = -1.4735\ \mu_N$ and $\mu(9/2_1^-) = +4.1103\ \mu_N$. This yields an effective neutron g factor of $g_\nu = -0.33\ \mu_N$ and an effective proton g factor of $g_\pi = +0.91\ \mu_N$. The $M1$ operator can be written as

$$\hat{T}_\mu(M1) = \sqrt{\frac{3}{4\pi}}\sqrt{\frac{j(j+1)(2j+1)}{3}}[g_\nu(\nu_j^+ \times \tilde{\nu}_j)_\mu^{(1)} + g_\pi(\pi_j^+ \times \tilde{\pi}_j)_\mu^{(1)}], \tag{3}$$

where ν_j^+ and π_j^+ are the neutron and proton create operators in the j orbital (here $j = j_\nu = j_\pi = 9/2$) and $\tilde{\nu}_{jm} = (-)^{j+m}\nu_{j-m}$ $(\tilde{\pi}_{jm} = (-)^{j+m}\pi_{j-m})$.

The $E2$ operator can be determined by the nuclei ^{210}Pb and ^{210}Po due to the pure two-nucleon configurations in 8_1^+ and 6_1^+ states in these nuclei. Therefore, the effective charges of $e_\nu = 0.88$ and $e_\pi = 1.11$ are derived from the transition $B(E2; 8^+ \to 6^+)$. The form of the $E2$ operator is

$$\hat{T}_\mu(E2) = -\sqrt{\frac{(2j-1)(2j+1)(2j+3)}{64\pi j(j+1)}}(N+\frac{3}{2})b^2[e_\nu(\nu_j^+ \times \tilde{\nu}_j)_\mu^{(2)} + e_\pi(\pi_j^+ \times \tilde{\pi}_j)_\mu^{(2)}], \tag{4}$$

where N is the major oscillation quantum number and b is the length parameter of the oscillator, $b \approx 1.0A^{1/6}$ fm. With this information an empirical description of the nucleus ^{212}Po is possible. The results from the calculations are compared to the experimental ones in Fig. 3.

Both schemes in Fig. 3(a) and 3(b) show a qualitative agreement, except for the 10_1^+ and the 2_3^+ states. The similar behaviour is valid for transition strengths, too.

Of particular interest is to examine the wave functions' structures of the 2^+ states. There are twelve different ways to couple all possible J_ν with all possible J_π to total angular momentum $J = 2$. The two lowest energy levels can be described by the following wave functions:

$$\begin{aligned} |2_1^+\rangle &= 0.448|J_\nu = 0, J_\pi = 2, J = 2\rangle + 0.819|J_\nu = 2, J_\pi = 0, J = 2\rangle + \ldots \\ |2_2^+\rangle &= 0.813|J_\nu = 0, J_\pi = 2, J = 2\rangle - 0.517|J_\nu = 2, J_\pi = 0, J = 2\rangle + \ldots \end{aligned} \tag{5}$$

In both cases the two components define a good fraction of the entire state (87% and 93%, respectively) and define almost mutually orthogonal states. Therefore, the first excited two 2^+ states can be analyzed in terms of the components of proton and neutron S ($J = 0$) and D ($J = 2$) pairs. The opposite signs in both wave functions indicate that the second 2^+ state is of isovector nature, in agreement with the experimentally observed sizeable $M1$ transition between the 2_2^+ and the 2_1^+.

The third 2^+ state has an entirely different structure. Its wave function can be written as

$$\begin{aligned} |2_3^+\rangle =& 0.772|J_\nu = 2, J_\pi = 2, J = 2\rangle + 0.401|J_\nu = 4, J_\pi = 4, J = 2\rangle + \\ & 0.307|J_\nu = 6, J_\pi = 6, J = 2\rangle + 0.242|J_\nu = 8, J_\pi = 8, J = 2\rangle + \ldots \end{aligned} \tag{6}$$

This state cannot be described in IBM-2 since it invokes pairs of all multipolarities, not only S and D.

4. Conclusions

In summary, we have presented an experiment dedicated to searching of the one-phonon mixed-symmetry state in ^{212}Po which is in the vicinity of the double magic nucleus ^{208}Pb. The preliminary results from this investigation provide indication for the existence of this state that is the first experimentally identified one-phonon MSS in the region with atomic mass A≈208.

Acknowledgments

D.K. acknowledges the support from SF of University of Sofia under contract №152/2015. G.R. acknowledges the support from the Alexander von Humboldt foundation. This work was supported by the partnership agreement between the University of Cologne and University of Sofia, by the DAAD German-Bulgarian exchange program under grants No. PPP57082997 and No. DNTS/01/05/2014, by the DFG under grant Pi393/2-3, and by the BMBF under grants 05P12RDCI8 and 05P15RDCIA.

References

[1] N. Lo Iudice and F. Palumbo, Phys. Rev. Lett. **41**, 1532 (1978).
[2] F. Iachello, Phys. Rev. Lett. **53**, 1427 (1984).
[3] F. Iachello and A. Arima, *The interacting boson model* (Cambridge University Press, Cambridge, 1987).
[4] N. Pietralla, P. von Brentano, A.F. Lisetskiy, Prog. Part. Nucl. Phys. **60**, 225 (2008) and the references therein.
[5] K. Heyde, P. von Neumann-Cosel, A. Richter, Rev. Mod. Phys. **82**, 2366 (2010).
[6] G. Rainovski *et al.*, Phys. Rev. Lett. **96**, 122501 (2006).
[7] T. Ahn *et al.*, Phys. Lett. **B679**, 19 (2009).
[8] L. Coquard *et al.*, Phys. Rev. C **82**, 024317 (2010).
[9] K.A. Gladnishki *et al.*, Phys. Rev. C **82**, 037302 (2010).
[10] M. Danchev *et al.*, Phys. Rev. C **84**, 061306(R) (2011).
[11] T. Ahn *et al.*, Phys. Rev. C **86**, 014303 (2012).
[12] A.Dewald, O.Möller, P.Petkov, Prog. Part. Nucl. Phys. **67**, 786 (2012).
[13] A. Astier, P. Petkov, M.-G. Porquet, D. S. Delion, and P. Schuck, Phys. Rev. Lett. **104**, 042701 (2010); A. Astier, P. Petkov, M.-G. Porquet, D. S. Delion, and P. Schuck, Eur. Phys. J. A **46**, 165 (2010).

[14] B. Bengtson *et al.*, Nucl. Phys. A **378**, 1 (1982).
[15] A.R. Poletti *et al.*, Nucl. Phys. A **473**, 595 (1987).
[16] Zs. Podolýak *et al.*, Nucl. Instrum. Methods Phys. Res. A **511**, 354 (2003).
[17] A.B. Garnsworthy *et al.*, J. Phys. G **31**, S1851 (2005).
[18] E. Browne, Nuclear Data Sheets **104**, 427 (2005).
[19] T. K. Alexander and J. S. Forster, Adv. Nucl. Phys. **10**, 197 (1978).
[20] P. Petkov *et al.*, Nucl. Phys. A **640**, 293 (1998).
[21] P. Petkov *et al.*, Nucl. Instrum. Methods Phys. Res. A **431**, 208 (1999).
[22] G. Winter, ZfK. Rossendorf Report ZfK-497, 1983; G. Winter, Nucl. Instrum. Methods **214**, 537 (1983).
[23] L. C. Northcliffe and R. F. Schilling, Nucl. Data Sect. **7**, 233 (1970).
[24] J. F. Ziegler and J. P. Biersack, in *Treatise on Heavy Ion Science*, edited by D. A. Bromley (Plenum Press, New York, 1985), Vol. 6, p. 95.
[25] J. Keinonen, AIP Conf. Proc. **125**, 557 (1985).
[26] J. Lindhard, M. Scharff, and H. E. Schitt, Kgl. Dan. Vid. Selsk. Mat. Fys. Medd. **33**, 14 (1963).
[27] C. Stahl, PhD thesis, TU Darmstadt (2015).
[28] S. Agostinelli *et al.*, Nucl. Instrum. Methods Phys. Res. A **506**, 250 (2003).
[29] J.F. Ziegler, M.D. Ziegler, J.P. Biersack, Nucl. Instr Meth., B **268**, 1823 (2010).
[30] A.P. Yutsis, I.B. Levinson and V.V. Vanagas, *The theory of angular momentum* (Israel Program for Scientific Translations, Jerusalem, 1962).

Search for heavy lithium isotopes $^{10\text{-}12}$Li in stopped pion absorption reactions

B.A. Chernyshev, Yu. B. Gurov, L.Yu. Korotkova, S.V. Lapushkin,
R.V. Pritula, V.G. Sandukovskiy
korotkovalara@gmail.com
National Research Nuclear University "MEPhI",115409 Moscow, Russia

The experimental search for the production of heavy lithium isotopes $^{10\text{-}12}$Li was carried out in the stopped π^--mesons absorption reactions on the targets ^{11}B, ^{12,14}C. Study of ^{10}Li was performed in the correlation and inclusive measurements of reactions ^{12}C(π^-,pp)X, ^{14}C(π^-,pt)X, ^{14}C(π^-,dd)X and ^{11}B(π^-,p)X. Three highly excited states of ^{10}Li with the resonance energies E_r = 6.1 ± 0.1 MeV, 7.8 ± 0.1 MeV and 10.1 ± 0.1 MeV were identified together with the well-known levels. The resonance parameters of three low-lying excited states of ^{11}Li were determined in the ^{14}C(π^-,pd)X reaction. In the ^{14}C(π^-,pp)X reaction ^{12}Li excited state with E_r = 4.0 ± 0.2 MeV and an indication on a level with $E_r \approx 7$ MeV were found.

1. Introduction

Interest in the study of the heavy lithium isotopes is due to extreme properties of these nuclei. Branch of the nuclear physics related to the research of exotic nuclear systems close to the drip line began to develop after the discovery of abnormally large radius of the nucleus ^{11}Li in the radioactive ion beams experiments [1, 2].

To explain this phenomenon a model based on the hypothesis of a neutron halo of ^{11}Li was introduced. In this model ^{11}Li is described as ^{9}Li core surrounded by two neutrons. A treatment of ^{11}Li as a three-body system requires the information on the parameters of the two-body subsystems: n + n and n + ^{9}Li. The properties of latter subsystem are directly related to the parameters of ^{10}Li low-lying states, whose states are nucleon-unstable ones [3]. Besides, the importance of this information for understanding the structure of ^{11}Li, the interest is due to the fact that ^{10}Li ground state has an abnormal parity, which is the direct evidence of the appearance of new nucleon magic numbers, instead of traditional [4]. Note that ^{11}Li is the so-called Borromean nucleus, which means that it is a coupled system, consisting of three items, where each pair is unbound.

^{11}Li is the heaviest nucleon-stable isotope with Z = 3, but the properties of the heavier isotopes of lithium are also important in terms of studying the drip line boundary area for light nuclei and for the development of theoretical models to describe these systems. ^{12}Li isotope which is virtual state in the system ^{11}Li + n was recently discovered [5-7] and the question of its structure is still open.

Experimental studies of the level structure of heavy lithium isotopes ^{10,11}Li are in progress for a quite long time, but still there are many open questions. The ^{10}Li ground state is a virtual s-wave state with the scattering length a ~ -25 fm [5, 8-9] (refs. to earlier works are presented in [10]). In the effective mass spectrum this state appears as a peak near the E_r = 0 MeV (E_r – energy relative to the ^{9}Li + n threshold). Information on the first excited state which is p-wave resonance is more uncertain: E_r is in the range $0.3 \le E_r \le 0.7$ MeV [5, 8-10]. Data on highly excited states are very limited [10-12].

Experimental studies of the ^{11}Li excitation spectrum were carried out in a large number of works [6, 9, 14-22]. In all the papers except [16] the ^{11}Li level was observed at the excited energy 0.6 MeV < E_x < 1.3 MeV with different widths 0.26 MeV< Γ < 0.75 MeV [19, 22]. High-energy excited levels were observed only in [9, 16, 18], several excited states with the energies 2.5 MeV < E_x < 11.3 MeV were found. Thus the problem of the precise measurements of ^{11}Li excited levels and their widths remains unsolved.

The first experimental evidence of the possible existence of the resonant state of ^{12}Li was presented in the work [23]. Preliminary data obtained in stopped pion absorption ^{14}C(π^-,pp)X indicated on the existence of the level of ^{12}Li with E_r = 4.9 ± 1.5 MeV (E_r – energy relative to the ^{11}Li + n threshold). New evidences on the ^{12}Li production have been appeared recently [5 - 7]. Authors observed the virtual s-wave state with a scattering length a ~ -13.7 ± 1.6 fm [5, 7]. Two excited levels with the resonance energies E_{r1} = 250 ± 20 keV and E_{r2} = 555 ± 20 keV were also found in [7].

Thus, the experimental information on the level structure of heavy lithium isotopes is quite limited and inconsistent especially for ^{12}Li. In this case new data are needed, it is also very important for the development of theoretical models. Our measurements were carried out at low energy pion channel of the Los Alamos Meson Physics Facility (LAMPF) with a two-arm multilayer semiconductor spectrometer [26]. The spectrometer and experimental technique are described more detail in [25, 26].

2. Results

2.1. Spectroscopy of ^{10}Li isotope

2.1.1. Inclusive measurements on ^{11}B target

The missing mass spectra for ^{11}B(π^-,p)X and ^{12}C(π^-,pp)X reactions are shown on Fig. 1. The sum of neutron and ^{9}Li masses is taken as a reference point.

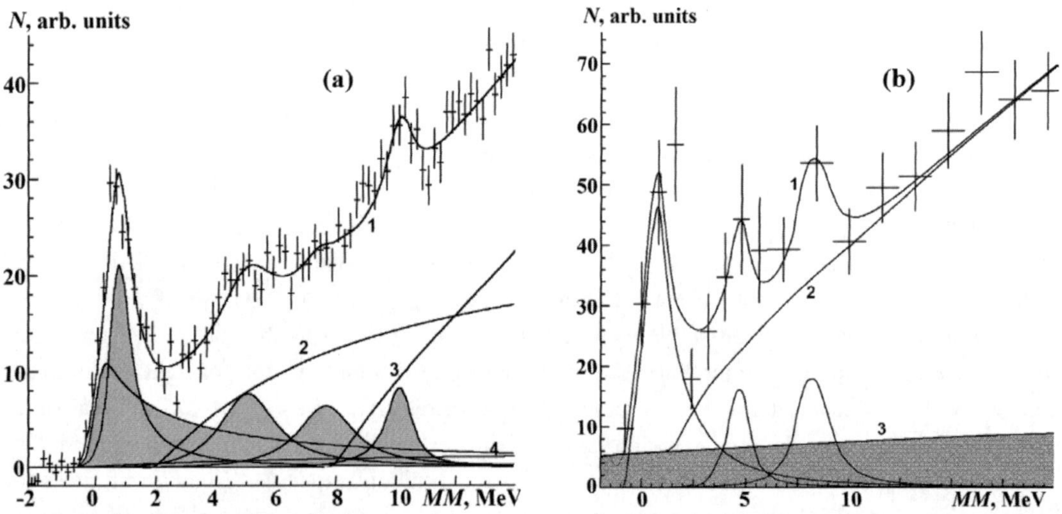

Fig. 1. The *MM* spectra for ^{11}B(π^-,p)X (a) and ^{12}C(π^-,pp)X (b) reactions. Dots with error bars are the experimental data. Curves: 1 is the total fit. (a): 2 and 3 are the phase-space distributions for ^{9}Lipn and ^{7}Hept respectively, 4 is the background coincidences. Shaded areas are the Breit-Wigner distributions for the two-particle channels with the formation of the ^{10}Li. (b): 2 is the sum of the phase-space distributions for many-body channels, 3 is the background coincidences.

In order to identify the states of ^{10}Li and determine their parameters, we used the least square method in the fitting of the experimental spectrum by the sum of Breit–Wigner shaped resonances, n-particle phase-space distributions ($n \geq 3$) and the background coincidences. The ground state and the first excited state were described with the Breit-Wigner threshold functions for s-wave and p-wave states respectively.

The best description of the peak close to zero in the *MM* spectrum on Fig. 1a can be achieved by introducing the ground state and the first excited state with the resonance energy $E_r = 0.6 \pm 0.1$ MeV with $\Gamma \approx 1$ MeV. Also in the experimental spectrum we can see the level with $E_r = 4.8 \pm 0.1$ MeV and $\Gamma \approx 2$ MeV. Highly excited states of the ^{10}Li with $E_r = 7.7 \pm 0.1$ MeV with $\Gamma \approx 2$ MeV and 10.1 ± 0.1 MeV with $\Gamma \approx 0.5$ MeV were observed for the first time.

3.1.2. Correlation measurements on ^{12}C target

Three-body reaction channel with the formation of ^{10}Li was also studied in the correlation measurements. The search for the production of the ^{10}Li isotope was performed in the ^{12}C(π^-,pp)X reaction. Measured *MM* spectrum is shown on Fig. 1b. Here we consider the phase-space distribution with the number of particles in the final state not less than four.

A statistically accurate description of the experimental spectrum can be achieved without taking into account the ^{10}Li ground state. One of the possible explanations of this fact is the mechanism of the π^--meson absorption reaction with the formation of two fast protons [11].

The peaks due to the two highly excited states of ^{10}Li are clearly seen. The resonance energy of the first of these states $E_r = 4.7 \pm 0.1$ MeV is close to $E_r = 4.64 \pm 0.10$ MeV obtained in [7], however, in our measurements the level is more than 3 times wider. The broad state with $E_r = 7.9 \pm 0.2$ MeV and $\Gamma \approx 2$ MeV was observed for the first time.

3.1.3. Correlation measurements on ^{14}C target

The *MM* spectra obtained in the correlation measurements of dd- and pt-pairs on the radioactive target ^{14}C are shown on Fig. 2. The contribution of the impurity of ^{12}C is already subtracted. The ^{12}C impurity background was determined using the results from the correlation measurements of dd- and pt-pairs on the ^{12}C target. The *MM* spectra for the reactions ^{12}C(π^-,dd)*X* and ^{12}C(π^-,pt)*X* were normalized to the relative impurity contribution (23%) and was subtracted from the experimental spectra. The small background in the region of negative *MM* is due to several causes: the statistical errors of the subtraction procedure, background coincidences and the contribution of uncontrolled impurities.

A statistically satisfactory description of the experimental dd-spectrum (Fig. 2a) can be achieved by introducing two excited states of ^{10}Li with the following resonance parameters (E_r, Γ): (0.8 ± 0.1 MeV, ≈ 0.5 MeV) and (6.1 ± 0.1 MeV, ≈ 1 MeV). It is seen that near $MM \approx 0$ MeV there are some events which may be associated with the formation of a virtual *s*-wave ground state of ^{10}Li. However the relatively low statistics is not sufficient to allow strict conclusions about its parameters.

The spectrum *MM* obtained in the correlation measurements pt-pairs on radioactive target ^{14}C was considered in a manner similar to the case of dd-pairs. The spectrum obtained after the subtraction the contribution of the impurity for the reaction ^{14}C(π^-,pt)*X* is shown on Fig. 2b. A statistically satisfactory description of the experimental spectrum can be achieved by introducing three excited states of ^{10}Li with the following resonance parameters (E_r, Γ): (0.7 ± 0.1 MeV, ≈ 0.2 MeV), (2.2 ± 0.1 MeV, ≈ 1 MeV) and (6.1 ± 0.1 MeV, ≈ 2 MeV). As for ^{12}C(π^-,dd)*X* reaction, an evidence for the ^{10}Li ground state is absent.

The results for the ^{10}Li level structure obtained in all reaction channels are not in the contradiction. The lack of the level with $E_r = 2.2$ MeV in the measurements of ^{14}C(π^-, dd)X is apparently due to the differences in the mechanisms of the (π^-, dd) and (π^-, pt) reactions. Absorption of π^- on ^{3}He and ^{4}He clusters plays an important role in light nuclei [11]. Therefore the (π^-, pt) reaction can occur by the charge exchange (n, p) after the absorption $\pi^- + {}^4\text{He} \rightarrow \text{n} + \text{t}$. In turn the ($\pi^-$, dd) reaction can occur by pick-up (n, d) after the absorption $\pi^- + {}^3\text{He} \rightarrow \text{n} + \text{d}$.

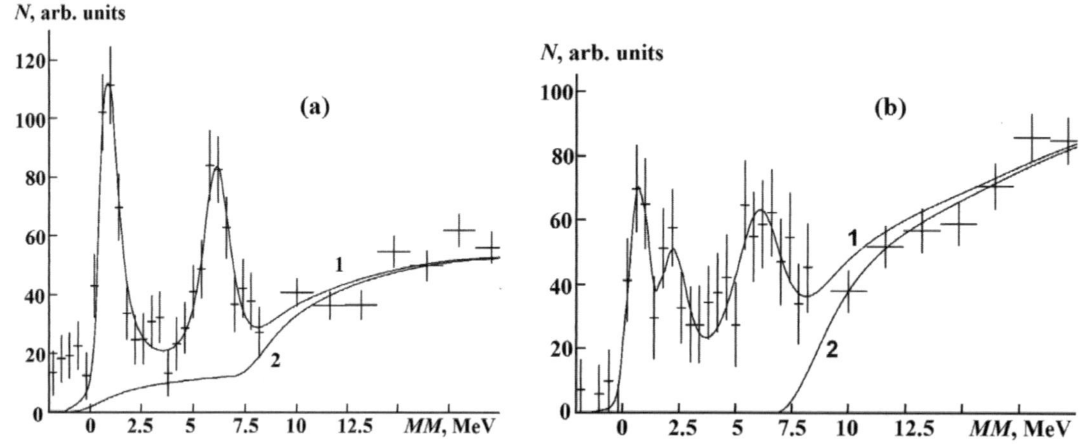

Fig. 2. The *MM* spectra for the $^{14}C(\pi^-,dd)X$ (a) and $^{14}C(\pi^-,pt)X$ (b) reactions. Dots with error bars are the experimental data obtained after subtracting ^{12}C background. Curves: 1 is the total fit, 2 is the sum of the phase-space distributions for many-body channels.

In our measurements the resonance energy of the ^{10}Li first excited state (average weighted value is 0.7 ± 0.1 MeV) exceeds the value observed in recent studies: ≈ 0.4 MeV [6], 0.510 ± 0.044MeV [4] and 0.566 ± 0.014MeV [5]. The reason for this discrepancy is unclear. However, the parameters of the first excited state, obtained in inclusive measurements on the target ^{11}B, are in good agreement with [3].

We do not observe a large number of narrow levels with E_r > 1 MeV as in [7]. Also, these states are not observed in [4] and [5], where the energy spectrum of the ^{10}Li was measured up to 3 MeV and 4 MeV respectively. However, we should like to point out an agreement between the resonance parameters (E_r = 2.35 ± 0.10 MeV, Γ = 1.2 ± 0.4 MeV) obtained in [7] and our results for the second level observed in the $^{14}C(\pi^-, pt)^{10}Li$ reaction. So the experimental situation with the level structure of ^{10}Li is quite ambiguous, which requires new experimental information.

3.2. Spectroscopy of ^{11}Li isotope

The spectrum of the *MM* obtained in the correlation measurements of pd-pairs on the radioactive target ^{14}C after impurity subtraction is shown on Fig. 3. The ^{11}Li mass is taken as reference point. The contribution of the background caused by ^{12}C impurity was determined using the results from correlation measurements dd-pairs on the ^{12}C target [6].

A statistically satisfactory description of the experimental spectrum can be achieved by introducing three excited states of ^{11}Li with the following resonance parameters (E_x, Γ): (0.9 ± 0.1, ≈ 0.3 MeV), (2.3 ± 0.3, ≈ 0.7 MeV) and (3.9 ± 0.3, < 0.2 MeV).

Our results for the energy and width of the first excited level are close to the data obtained for fragmentation reactions in [9, 15, 17-20, 22]. The E_x variation from 0.6 MeV to 1.3 MeV in these papers was probably caused by data analysis ambiguity. The width observed by us was in agreement with a result Γ = 0.26 ± 0.24 MeV that was recently obtained in [9]. These values are considerably smaller than the results of the microscopic model [28], in which it was predicted that the peak with Γ = 0.75 ± 0.60 MeV observed in [19] is the sum of two ^{11}Li resonance states.

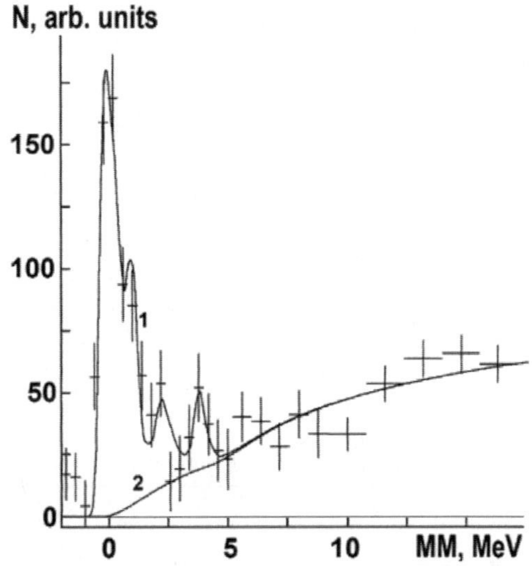

Fig.3. The *MM* spectrum for the reaction $^{14}C(\pi^-,pd)X$. Dots with error bars were obtained after subtracting of ^{12}C background. The solid lines are: 1 – the fit and 2 – the sum of phase-space distributions.

The energy E_x = 2.3 ± 0.3 MeV of the second excited state coincides within the experimental error with results obtained in heavy ion reactions E_x = 2.47 ± 0.07 MeV [16] and fragmentation reaction E_x = 2.45 ± 0.27 MeV [9], however, the widths observed in these reactions are appreciably larger, i.e. 1.2 ± 0.2 MeV and 2.91 ± 0.72 MeV respectively. The narrow state with E_x = 3.9 ± 0.3 MeV had not been observed in other works yet.

3.3. Spectroscopy of ^{12}Li isotope

The spectrum of the *MM* obtained in the correlation measurements of pp-pairs on the radioactive target ^{14}C after the impurity subtraction is shown on Fig. 4. The sum of neutron and ^{11}Li masses is taken as an initial point. The results of measurements for the reaction $^{12}C(\pi^-,pp)X$ were presented in [26].

The *MM* spectrum shown on Fig. 4 contains the pronounced peak caused by the ^{12}Li state with the resonance parameters E_r = 4.0 ± 0.2 MeV and $\Gamma \approx 1$ MeV. Note that the subtraction of the ^{12}C impurity results in a significant contribution to errors.

In our data near $MM \approx 0$ MeV there is no indication on the existence of the virtual s-wave state observed in [5, 7]. One cause of the suppression of this state might be the appreciable selectivity of the level populations in the $A(\pi^-,pp)X$ reaction [29]. Note that certain amplification is observed in the effective mass spectrum of the ^{11}Li+n system, measured in [5] at E_r > 3 MeV. However, the presented data are restricted to the value E_r = 4 MeV, which does not allow us to attribute this amplification to the ^{12}Li state we observed. Even smaller range of the E_r (< 2 MeV) was investigated in [7]. The problem of the ^{12}Li level structure thus remains open.

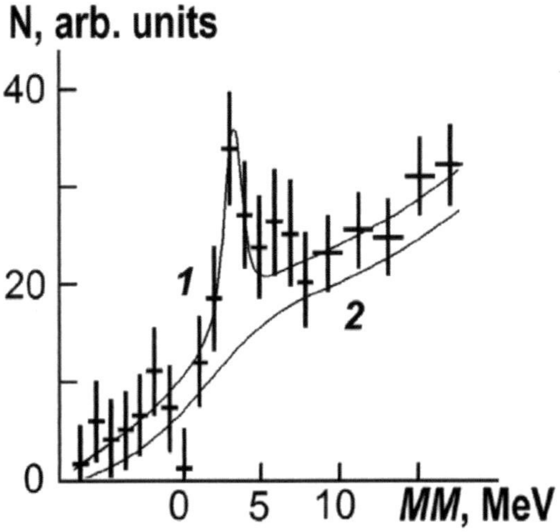

Fig.4. The *MM* spectrum for the reaction $^{14}C(\pi^-,pp)X$. Dots with error bars were obtained after subtracting of ^{12}C background. The solid lines are the fit and the sum of phase-space distributions.

4. Conclusion

The production of $^{10\text{-}12}Li$ heavy lithium isotopes was studied in the reactions of stopped pion absorption on ^{11}B and $^{12,14}C$ nuclei. The results of our measurements are in reasonable agreement with the results of other studies for the reliably defined states of $^{10,11}Li$. The exception is the ^{10}Li ground state. It is similarly in the missing mass spectrum of the reaction $^{14}C(\pi^-,pp)X$: we do not obtained statistically significant indications on ^{12}Li state at $E_r \cong 0$ MeV, which was observed in the fragmentation reactions of radioactive beams in [5, 7]. At the same time in our measurements ^{10}Li states with $E_r = 6.1$ MeV, 7.8 MeV and 10.1 MeV, ^{11}Li state with $E_x = 3.9$ MeV and ^{12}Li state with $E_r = 4.0 \pm 0.2$ MeV were first identified.

This research was supported by the grant of Russian Ministry of Education and Science № 1729.

References

1. Tanihata I. et al. 1985 *Phys. Rev. Lett.* **55** 2676.
2. Tanihata I. et al. 1988 *Phys. Lett. B* **206** 592.
3. Audi G. et al. 2003 *Nucl. Phys. A* **729** 3.
4. Jonson B. 2004 *Phys. Rep* **389** 1.
5. Aksyutina Yu. et al. 2008 *Phys. Lett. B* **666** 430.
6. Gurov Yu.B. et al. 2010 *Bull. of RAS.: Phys* **74** 469.
7. Hall C.C. et al. 2010 *Phys. Rev. C* **81** 021302(R).
8. Jeppesen H.B. et al. 2006 *Phys. Lett. B.* **642** 449.
9. Simon H. et al. 2007 *Nucl. Phys. A* **791** 267.
10. Tilley D.R. et al. 2004 *Nucl. Phys. A* **745** 155.
11. Bohlen H.G. al. 1999 *Prog. Part. Nucl. Phys* **42** 17.
12. Kobayashi T. et al. 1997 *Nucl. Phys. A* **616** 223c.
13. Roger T. et al. 2009 *Phys. Rev. C* **79** 031603(R).
14. Kobayashi T. 1992 *Nucl. Phys. A* **538** 343c.

15. Sackett D. et al. 1993 *Phys. Rev. C* **48** 118.
16. Bohlen H.G. et al. 1995 *Z. Phys* **351** 7.
17. Shimoura S. et al. 1995 *Phys. Lett. B* **348** 29.
18. Korsheninnikov A.A. et al. 1996 *Phys. Rev. C* **53** 537(R).
19. Korsheninnikov A.A. et al. 1997 *Phys. Rev. Lett.* **78** 2317.
20. Zinser M. et al. 1997 *Nucl. Phys. A* **619** 151.
21. Gurov Yu.B. et al. 1998 *Phys. Rev. Lett.* **81** 4325.
22. Nakamura T. et al. 2006 *Phys. Rev. Lett.* **96** 252502.
23. Gornov M.G. 1996 *PANIC - 1996 Abstracts, Sweden. Uppsala: Un. Press* 216.
24. Backenenstoss G. 1970 *Ann. Rev. Nucl. Sci.* **20** 467.
25. Gurov Yu.B. et al. 2009 *Phys. of Part. and Nucl.* **40** 1063.
26. Gornov M.G. et al. 2000 *NIM in Phys. Res. A* **446** 461.
27. Gurov Yu.B. et al. 2010 *Bull. of RAS.: Phys.* **75** 491.
28. Ershov S.N. et al. 2004 *Phys. Rev. C* **70** 054608.
29. Gurov Yu.B. et al. 2006 *Phys. of Atom. Nucl.* **69** 1448.

High precision electron scattering off ^{96}Zr

C. Kremer, S. Aslanidou, S. Bassauer, A. Krugmann, N. Pietralla, V. Ponomarev, M. Singer, P. von Neumann-Cosel, and M. Zweidinger

Institut für Kernphysik, Technische Universität Darmstadt, D-64289 Darmstadt, Germany

E-mail: `ckremer@ikp.tu-darmstadt.de`

Abstract. We report on our recent electron scattering experiment off ^{96}Zr. The B(E2; $0_1^+ \to 2_2^+$) transition strength is extracted in a relative analysis and allows the determination of all decay strengths of the 2_2^+ state. The strong collectivity of the $2_2^+ \to 0_2^+$ transition establishes the 0_2^+ state and the states build on top of it as an isolated deformed band.

1. Introduction

The Zirconium isotopic chain is of considerable interest from the point of view of nuclear structure physics. It is located at the subshell closure $Z = 40$ ($2\,p_{1/2}$ filled) and contains both, the shell closure at $N = 50$ ($1\,g_{9/2}$ filled) and the subshell closure at $N = 56$ ($2\,d_{5/2}$ filled). The nuclei ^{90}Zr and ^{96}Zr, located at the neutron subshell closures, show the typical indications of double magic nuclei, e.g. a high-lying 2_1^+ state. Two different nuclear structure phenomena in the zirconium isotopes are of special interest. First, there is the concept of states with mixed proton neutron symmetry. Within the zirconium isotopic chain such states have been identified in ^{92}Zr [1] and ^{94}Zr [2] and it is unclear if and how they persist in the neutron rich zirconium isotopes. Second, there is a smooth shape phase transition from a spherical ground state in $^{90-98}$Zr to a rotational deformed one in neutron rich ^{102}Zr [3]. The nucleus ^{96}Zr is of special interest, because neither a low-lying 2^+ mixed symmetric state nor a deformed 0^+ state have been identified. In order to clarify the low-lying structure of this nucleus additional experimental data is needed. Especially the unknown decay strength of the 2_2^+ state would help to answer the question whether this state is a mixed symmetry state or a rotational excitation build on top of the 0_2^+ state, which is the first excited state of ^{96}Zr. As the multipole mixing ratios and branching ratios of the decays of the 2_2^+ state are known, all decay strengths can be determined by measuring the B(E2; $0_1^+ \to 2_2^+$) value in electron scattering.

2. Experiment

The electron scattering experiment on ^{96}Zr was conducted during different experimental beam times in December 2014, March 2015, and June 2015 at the Lintott spectrometer using the electron beam provided by the **S**uperconducting-**DA**rmstadt-**LIN**ear-**AC**celerator (S-DALINAC). Data were taken at scattering angles of 81°, 93°, 117°, and 141°. The measurement at 117° was conducted with a beam energy of 69 MeV while the other measurements were performed at 43 MeV. The covered momentum transfer (q) values are $q = 0.59\,\text{fm}^{-1}$, $q = 0.40\,\text{fm}^{-1}$, $q = 0.31\,\text{fm}^{-1}$, and $q = 0.28\,\text{fm}^{-1}$. Intensities of the beam ranged from 500 nA up to 2.5 μA and were limited by the dead time of the data acquisition system. The target used was a 2 cm by 3 cm self-supporting zirconium foil of thickness 10 $^{mg}/_{cm^2}$. It was enriched in ^{96}Zr up to 57.36 % and also contained ^{92}Zr (27.2 %), ^{90}Zr (9.2 %), ^{94}Zr (4.3 %), and ^{91}Zr (2.0 %). The resolution of the obtained spectra ranged from 12.3 keV (FWHM) to 17.5 keV (FWHM). For the measurement at 141° the obtained data are shown in Fig. 1. In addition,

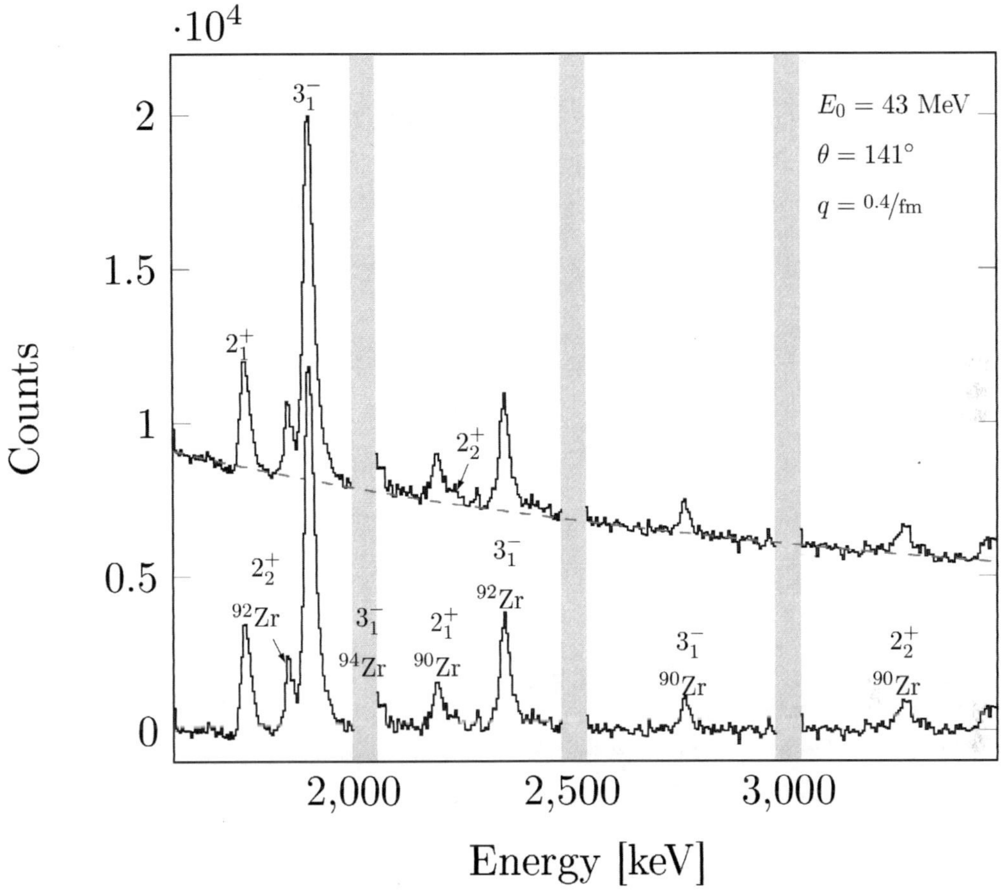

Figure 1. Summed and efficiency corrected experimental data for $\theta = 141°$ and $E_0 = 43$ MeV. The upper histogram shows the data including contributions from the elastic line. The lower histogram shows the same data with the radiative tail of the elastic line (red, dashed) subtracted. If a transition does not stem from ^{96}Zr its origin has been stated. Grey areas correspond to inactive segments of the detector system.

several measurements were taken to calibrate the relative efficiency of the detectors. These measurements were ^{208}Pb(e, e'), ^{12}C(e, e'), and ^{196}Pt(e, e') at sufficiently high excitation energies ($\approx$ 20 MeV to $\approx$ 35 MeV) to be above the giant dipole resonance of the respective nucleus.

3. Data analysis and results

In preparation of the analysis, the experimental raw data, consisting of many single runs, are efficiency corrected, energy calibrated, and summed up. Then, the elastic background is removed (cf. Fig. 1) and the number of detected counts A_i for each excited state i can be determined by a fitting procedure. The extracted peak areas can be used to determine the transition strength of the $0^+_1 \to 2^+_2$ transition relative to that of the $0^+_1 \to 2^+_1$ transition using Plane Wave Born

Approximation (PWBA) [4]. The necessary Coulomb corrections, to account for the distortion of the electron wave functions by the nucleus, are expected to cancel in this relative analysis to better than 1 % over the momentum transfer range of interest. Employing the PWBA formalism, using Siegert's theorem, and expanding the transition strengths $B(\mathrm{C}\,\lambda, q)$ in powers of the momentum transfer (q) and transition radius (R_{tr}) yields a relation of experimentally determined peak areas to the ratio of $B(\mathrm{E2})$ values:

$$R_F(q)\sqrt{\frac{A_2}{A_1}} \approx \sqrt{\frac{B(\mathrm{E2};\,2_2^+\to 0_1^+,\,k_2)}{B(\mathrm{E2};\,2_1^+\to 0_1^+,\,k_1)}}\cdot\left(\frac{1-\frac{q_2^2}{14}\,(R_{tr,1}+\Delta R)^2+\frac{q_2^4}{504}\,(R_{tr,1}+\Delta R)^4}{1-\frac{q_1^2}{14}\,(R_{tr,1})^2+\frac{q_1^4}{504}\,(R_{tr,1})^4}\right), \tag{1}$$

where $R_F(q)$ denotes a ratio of kinematical factors, $R_{tr,1}$ is the transition radius of the transition to the 2_1^+ state, $\Delta R = R_{tr,2} - R_{tr,1}$, and k_1 and k_2 denote the photon points for the excitations of 2_1^+ and 2_2^+ states. A χ^2 minimization of Eq. 1 with respect to the square root of the $B(\mathrm{E2})$ ratio and the difference in transition radii ΔR yields

$$T(2_2^+,\,2_1^+) = \sqrt{\frac{B(\mathrm{E2};\,2_2^+\to 0_1^+)}{B(\mathrm{E2};\,2_1^+\to 0_1^+)}} = 0.335^{+0.050}_{-0.040}\,,\ \text{and} \tag{2}$$

$$\Delta R = \left(-0.22^{+0.87}_{-0.92}\right)\ \mathrm{fm}. \tag{3}$$

The transition radius $R_{tr,1}$ used is 5.38 fm, which is suggested by a QRPA calculation [5]. The

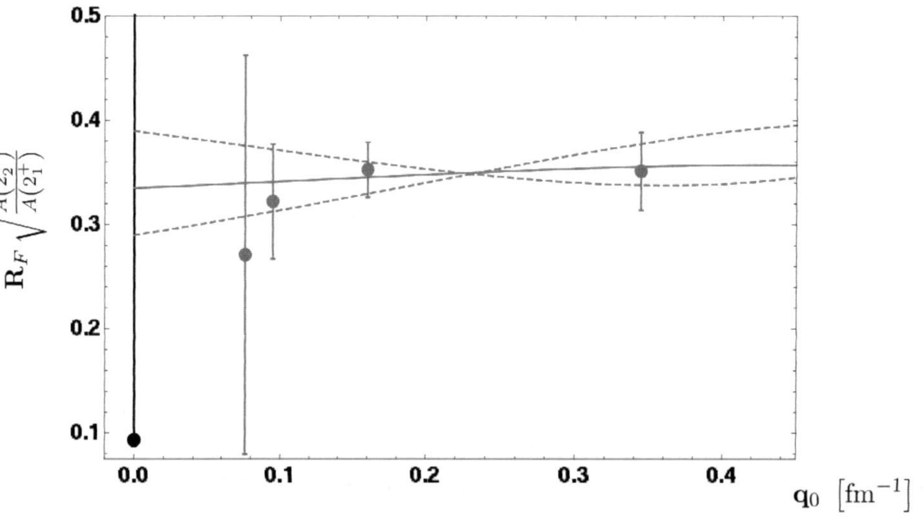

Figure 2. The solid red line shows the best fit of Eq. 1 to the experimental data (blue). The dashed lines represent the $1\,\sigma$ uncertainties with respect to $T(2_2^+,\,2_1^+)$. The previously known experimental limit [6] is shown in black.

best fit is represented by a solid red line in Fig. 2 whereas the dashed lines represent the solutions that define the $1\,\sigma$ uncertainties for $T(2_2^+,\,2_1^+)$. The extracted value of $T(2_2^+,\,2_1^+)$ is largely independent to the choice of $R_{tr,1}$ (at least up to ± 1 fm). It is evident that this experiment is unable to determine the value for ΔR with high precision. The extracted value is consistent with zero. The situation for the transitions strengths is different. Combining Eq. 2 with the literature value $B(\mathrm{E2};\,2_1^+\to 0_1^+) = 2.3\pm 0.3$ W.u. [7] yields $B(\mathrm{E2};\,2_2^+\to 0_1^+) = 0.26^{+0.09}_{-0.07}$ W.u.. Together with the known multipole mixing ratios and the known branching rations [7] it is then possible to determine all the transition strengths for decays of the 2_2^+ state (see Tab. 1). Using the new decay data results in a lifetime of $\tau = \left(0.34^{+0.9}_{-0.7}\right)$ ps or equivalently a half-life of $T_{1/2} = \left(0.24^{+0.6}_{-0.5}\right)$ ps for the 2_2^+ state of ^{96}Zr.

Table 1. Transition strengths for decays of the 2_2^+ state of ^{96}Zr obtained from experimental data. The $B(\text{M2};\, 2_2^+ \rightarrow 3_1^+)$ is consistent with zero. All data except transition strengths are taken from Ref. [7].

J_f^π	E_γ [keV]	I_γ	Multipolarity	δ	Transition strength
3_1^-	328.75(3)	14(1)	E1+M2	−0.02(5)	$B(\text{E1}) = \left(27.6^{+9.9}_{-7.8}\right) \cdot 10^{-3}$ W.u.
2_1^+	475.33(1)	57(1)	M1+E2	$-0.09^{+0.01}_{-0.02}$	$B(\text{M1}) = \left(0.14^{+0.05}_{-0.04}\right)\ \mu_N^2$
					$B(\text{E2}) = \left(2.7^{+1.3}_{-1.2}\right)$ W.u.
0_2^+	644.18(6)	28(1)	E2		$B(\text{E2}) = \left(35.6^{+12.5}_{-9.8}\right)$ W.u.
0_1^+	2225.93(5)	100(5)	E2		$B(\text{E2}) = \left(0.26^{+0.09}_{-0.07}\right)$ W.u.

4. Interpretation

The strong $B(\text{E2};\, 2_2^+ \rightarrow 0_2^+)$ hints at a common underlying structure (deformed) of the 0_2^+ and 2_2^+ states, which, as indicated by the weak $B(\text{E2};\, 2_2^+ \rightarrow 0_1^+)$, is different from the structure of the spherical ground-state band. It is instructive to look at the data via a simple mixing calculation. Assuming that the experimentally observed states are mixtures of deformed and spherical structures their wave functions can be written as:

$$\begin{aligned} \mid 0_1^+\rangle &= \alpha \mid 0_{\text{sph}}^+\rangle + \beta \mid 0_{\text{def}}^+\rangle, && (4)\\ \mid 0_2^+\rangle &= -\beta \mid 0_{\text{sph}}^+\rangle + \alpha \mid 0_{\text{def}}^+\rangle, && (5)\\ \mid 2_1^+\rangle &= \gamma \mid 2_{\text{sph}}^+\rangle + \delta \mid 2_{\text{def}}^+\rangle, \text{ and} && (6)\\ \mid 2_2^+\rangle &= -\delta \mid 2_{\text{sph}}^+\rangle + \gamma \mid 2_{\text{def}}^+\rangle. && (7) \end{aligned}$$

Using the experimentally observed energies of these states and the observed transition strengths $B(\text{E2};\, 2_1^+ \rightarrow 0_1^+)$ and $B(\text{E2};\, 2_2^+ \rightarrow 0_2^+)$ as input data, the mixing amplitudes can be computed under the assumption that the mixing matrix element V_{mix} is identical for the mixing of the 0^+ and 2^+ states. Carrying out the calculation yields

$$\alpha^2 = 99.8\,\% \quad \beta^2 = 0.2\,\% \qquad \gamma^2 = 97.5\,\% \quad \delta^2 = 2.5\,\%, \tag{8}$$

which shows that the states are barely mixed and the ground-state band is almost purely spherical, whereas the band on top of the 0_2^+ state corresponds almost exclusively to a deformed configuration. The mixing matrix element amounts to $V_{mix} = 76$ keV and the quadrupole deformation parameter $\beta_{2,\text{def}}$ of the deformed band is estimated to be

$$\beta_{2,\text{def}} = \frac{4\,\pi}{3\,Z\,R_0^2} \cdot \left(\frac{B(\text{E2};\, 0_{\text{def}}^+ \rightarrow 2_{\text{def}}^+)}{e^2}\right)^{\frac{1}{2}} \approx 0.11, \tag{9}$$

where $R_0 = 1.2\ \text{fm} \cdot A^{1/3}$ has been used.

5. Conclusion

Using electron scattering it was possible to deduce the transition strengths of the decays of the 2_2^+ state of ^{96}Zr . Its strong decay to the 0_2^+ state and weak decay to the ground state establishes the 0_2^+ state, and the band build on top of it, as a deformed structure, which is clearly separated from the spherical ground-state band.

6. Acknowledgments

The authors thank the S-DALINAC accelerator crew for delivering an excellent electron beam. This work was supported by the Deutsche Forschungsgemeinschaft under grant No. SFB 634.

Journal of Physics: Conference Series **724** (2016) 012025 doi:10.1088/1742-6596/724/1/012025

References

[1] V. Werner et al., Phys. Lett. B **550**, 140 (2002).
[2] E. Elhami et al., Phys. Rev. C. **75**, 011301 (2007).
[3] H. Mach et al., Phys. Lett. B **230**, 21 (1989).
[4] A. Scheikh Obeid et al., Phys. Rev. C **87**, 014337 (2013).
[5] Soloviev, V. G. Theory of Atomic Nuclei: Quasiparticles and Phonons (Institute of Physics Publishing, Bristol, 1992)
[6] H. Mach et al., Phys. Rev. C **42**, R811 (1990).
[7] Evaluated Nuclear Structure Data Files, http://www.nndc.bnl.gov/ensdf/ (accessed August 2, 2015).

Journal of Physics: Conference Series **724** (2016) 012026 doi:10.1088/1742-6596/724/1/012026

Nuclear reactions in the storage ring ESR with EXL

Th Kröll[1], M von Schmid[1], J C Zamora[1], S Bagchi[2], S Bönig[1], M Csatlós[3], I Dillmann[4], C Dimopoulou[4], P Egelhof[4], V Eremin[5], T Furuno[6], H Geissel[4], R Gernhäuser[7], M N Harakeh[2], A L Hartig[1], S Ilieva[1], N Kalantar-Nayestanaki[2], O Kiselev[4], H Kollmus[4], C Kozhuharov[4], A Krasznahorkay[3], M Kuilman[2], S Litvinov[4], Yu A Litvinov[4], M Mahjour-Shafiei[2,8], M Mutterer[4], D Nagae[9], M A Najafi[2], C Nociforo[4], F Nolden[4], U Popp[4], C Rigollet[2], S Roy[2], C Scheidenberger[4], M Steck[4], B Streicher[2,4], L Stuhl[3], M Thürauf[1], T Uesaka[10], H Weick[4], J S Winfield[4], D Winters[4], P J Woods[11], T Yamaguchi[12], K Yue[1,4,13], J Zenihiro[10] for the EXL collaboration[14]

[1]Institut für Kernphysik, TU Darmstadt, Germany
[2]KVI-CART, University of Groningen, The Netherlands
[3]Institute for Nuclear Research, MTA-Atomki, Debrecen, Hungary
[4]GSI Helmholtzzentrum für Schwerionenforschung GmbH, Darmstadt, Germany
[5]Ioffe Physical-Technical Institute, St. Petersburg, Russia
[6]Kyoto University, Japan
[7]Physik-Department E12, TU München, Germany
[8]University of Tehran, Iran
[9]University of Tsukuba, Japan
[10]RIKEN Nishina Center, Japan
[11]University of Edinburgh, UK
[12]Saitama University, Japan
[13]Institute of Modern Physics, Lanzhou, China

E-mail: `tkroell@ikp.tu-darmstadt.de`

Abstract.
The EXL project aims to study nuclear structure by direct reactions in inverse kinematics at the storage rings of the future FAIR facility. In this contribution, we present the status of the project: the technical implementation at the ESR at GSI and preliminary results of the EXL campaign in 2012, the first using also a radioactive beam.

1. Introduction

EXL (EXotic nuclei studied in Light-ion induced reactions at storage rings) is part of NUSTAR (NUclear STructure, Astrophysics and Reactions) at FAIR, the future Facility for Antiproton and Ion Research, which is currently under construction at the site of the GSI Helmholtzzentrum für Schwerionenforschung in Darmstadt [1]. EXL aims for the investigation of nuclear structure with direct reactions in inverse kinematics on an internal target at storage rings [2]. The programme is mainly focussed on reactions with very low momentum transfers and therefore largely complementary to R^3B [3].

[14] http://www.rug.nl/kvi/Research/hnp/Research/EXL/index

The physics menu of EXL comprises:

- elastic and inelastic scattering
- capture, transfer and charge-exchange reactions
- knockout and quasi-free scattering

EXL aims for kinematically complete measurements which requires a large dynamic range and angular coverage. In the final implementation, this will be achieved by an onion-like design with different layers of Si detectors (ESPA: EXL Silicon Particle Array) to measure target-like recoils. This will be surrounded by a CsI(Tl) calorimeter and spectrometer (EGPA: EXL Gamma and Particle Array) to measure photons and high-energy charged particles punching through ESPA. In addition, detectors for beam-like ejectiles and neutrons will be placed downstream of the ESPA/EGPA setup centred around the target.

The experiments presented in this contribution have been performed with two prototype detectors, which have been built to demonstrate the feasibility of the EXL programme. They will be described later in detail.

2. Technical implementation at ESR

The existing storage ring ESR at GSI provides a unique opportunity to perform part of the programme already. The ESR has a circumference of 108.4 m and a maximal magnetic rigidity of 10 Tm. After electron cooling, the momentum spread is in the order of $\delta p/p \approx 10^{-5}$ [4]. The operation of a storage ring requires an ultra-high vacuum (UHV) better than 10^{-10} mbar which imposes a major challenge on experiments performed in the ring.

In our experiments, the stored beam impinged on an internal gas-jet target of H_2 or He [5]. Although the energy of the stored beams is high in the order of a few 100 MeV/u, scattering in inverse kinematics at low momentum transfer implies that the recoiling target-like particles have very low energy in the laboratory system. Figure 2 shows as example the elastic scattering of ^{56}Ni on protons in inverse kinematics at 400 MeV/u.

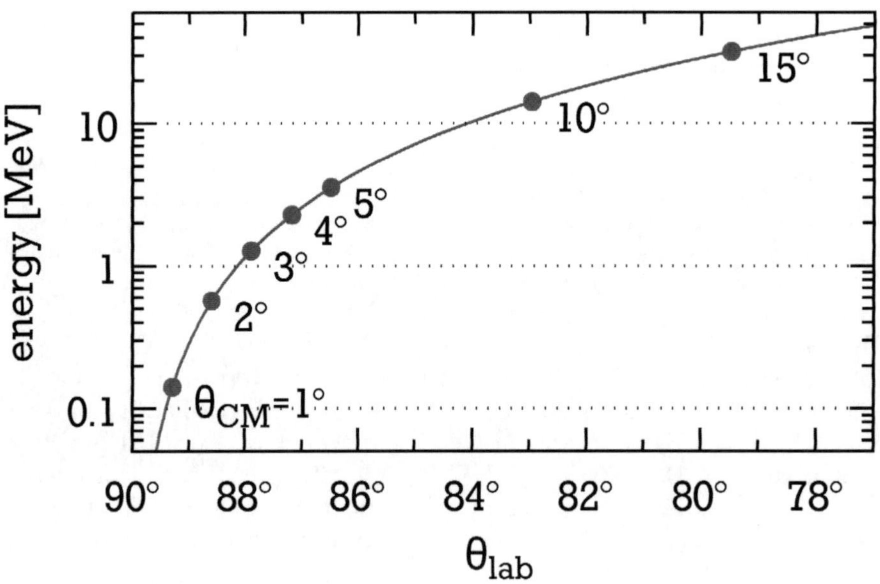

Figure 1. Kinetic energy of the recoiling protons in the laboratory system following the ^{56}Ni(p,p) reaction at 400 MeV/u (adapted from [6])

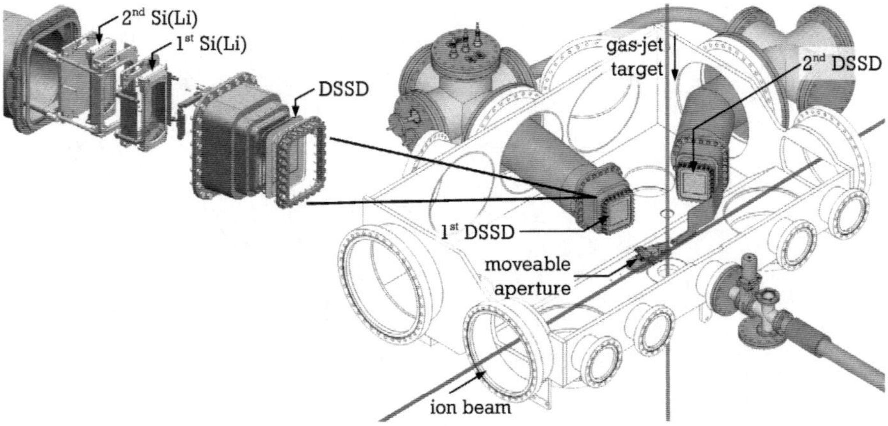

Figure 2. The two detectors, the movable aperture (see text) and indications of the beam direction and the gas-jet target (drawings by M. Lindemulder, adapted from [6])

Therefore, the main challenge was the development of window-less highly-segmented Si strip detectors which can be operated under UHV conditions [7, 8]. In our concept, these detectors serve as active windows separating the UHV of the storage ring and an auxilliary vacuum in a so-called pocket which accommodates the non-bakeable parts of the system. In order to achieve the UHV conditions, the DSSDs were mounted on ceramic AlN boards, a material which is non-outgassing and bakeable. This concept was implemented, tested in the laboratory and, finally, two detectors were built.

Figure 2 shows the experimental setup used in our experiments with the two detectors mounted in the scattering chamber which was placed in the ESR. The first detector, labelled "1st DSSD" in Figure 2, consisted of a 6×6 cm^2 DSSD (Double-sided Si Strip Detector) of 285 μm thickness with 128×64 segments followed by two 8×4 cm^2 Si(Li) detectors of 6.5 mm thickness with 4×2 segments. The Si(Li)s were equipped with a cooling system to keep their temperature low even during the baking of the pocket. This detector was centred at 80.5° in the laboratory system. The second detector, labelled "2nd DSSD" in Figure 2, consisted only of a DSSD of the same type and was centred at 32.5°.

3. First physics experiments

In 2012, we successfully performed an experiment (E105) with stable ^{58}Ni as well as radioactive ^{56}Ni beams [9]. Before our experiment, we had the opportunity to take data with a stable ^{20}Ne beam from another experiment (E087) as well. The beams hit the internal gas-jet target (H_2 or He) and the respective target recoils were measured by our newly developed UHV compatible Si detector setup described above.

The first physics goal was to deduce the nuclear matter distribution of ^{56}Ni from elastic proton scattering at 400 MeV/u [6, 10]. The radioactive ^{56}Ni was produced by projectile fragmentation of a primary ^{58}Ni beam in the FRagment Separator (FRS) of GSI. Approximately $3 \cdot 10^6$ particles were stored in the ESR. As every spill from the FRS contained only about $4 \cdot 10^4$ ^{56}Ni ions, several injections were needed, the so-called beam stacking [11]. The average lifetime of the stored beam, limited by collisions with the residual gas, was about 1.5 h, before the ring had to be refilled.

The small target density of the H_2 gas-jet target of $3 \cdot 10^{13}$ cm^{-2} and the available beam intensity of the stored beam were compensated by the revolution frequency of about 2 MHz to obtain a luminosity of about $2 \cdot 10^{26}$ cm^{-2}s^{-1} which allowed us to perform the experiment.

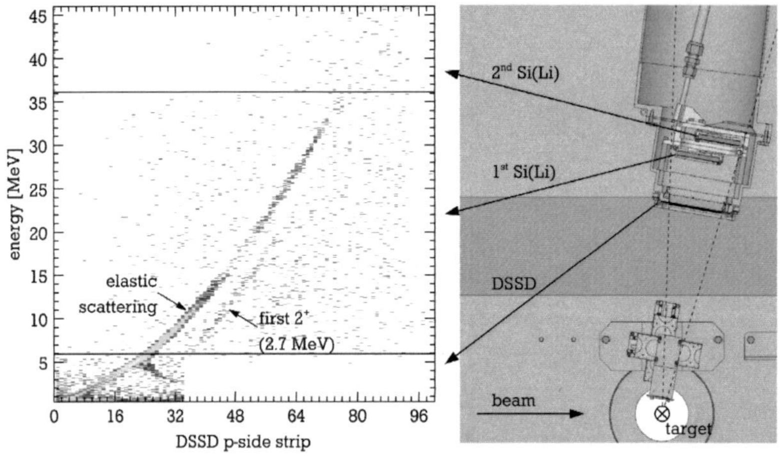

Figure 3. Reconstructed laboratory energies of the recoiling protons from the reaction ^{56}Ni(p,p) at 400 MeV/u using a 1 mm slit aperture (adapted from [6])

The achievable angular distribution is limited by the size of the beam ($\sigma \approx 0.6$ mm) and, in our case the most severe limitation, by the size of the target (radius about 3.2 mm). In order to separate the inelastic scattering to the first 2^+ state in ^{56}Ni at 2.7 MeV from the elastic scattering a movable aperture with a 1 mm slit between the target and the detector was introduced. Figure 3 shows the energy spectrum obtained with this setup. The energy of the protons was reconstructed by summing up the energy deposits in the DSSD and the subsequent Si(Li)s. Protons with energies above 6 MeV punch through the DSSD and above 36 MeV also through the first Si(Li).

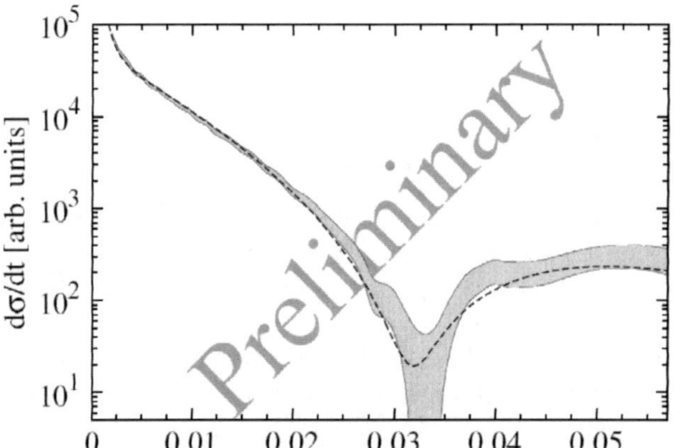

Figure 4. Experimental cross section of ^{56}Ni(p,p) at 400 MeV/u (the grey area between the solid lines represents the $\pm 1\sigma$ accuracy of the experimental cross section in bins corresponding to the vertical detector strips of the first DSSD converted into momentum transfer) and a fit to the data using Glauber multiple-scattering theory (dashed line) (adapted from [6])

The extracted differential cross section was fitted assuming a symmetrised Fermi function to describe the matter distribution and applying a Glauber multiple-scattering theory (Figure 4).

In order to estimate the effect originating from the a-priori assumption of this particular shape of the matter distribution, also a Sum-of-Gaussians (SOG) analysis has been performed. For consistency, all steps of the analysis have been analogously performed with data obtained under the same experimental conditions with a stable ^{58}Ni beam for which the matter radius is well known [12].

A preliminary RMS matter radius of 3.7-3.8 fm could be extracted which is well within the range of predictions by most of theory calculations [13, 14] whereas older predictions with smaller values are already excluded [15]. This experiment was the *first successfully observed nuclear reaction with a stored radioactive beam.*

The second part of the experimental campaign was the first study of isoscalar giant resonances populated in inelastic α scattering on stored nuclei [16, 17]. Of particular interest was the isoscalar giant monopole resonance, the excitation energy of which is related to the nuclear incompressibility, an important quantity in the equation of state for nuclear matter. This quantity is also of high interest in nuclear astrophysics to describe neutron stars [18].

As isoscalar giant resonance can be selectively excited using isoscalar probes, for this experiment we used a He gas-jet target (mainly the isoscalar isotope ^{4}He). The density is an order of magnitude lower compared to hydrogen, therefore this experiment has been performed using a stable ^{58}Ni beam. The giant monopole resonance is the dominant channel at very forward angles in the centre-of-mass system. In the laboratory system, the corresponding inelastically scattered α particles are detected at forward laboratory angles. For this experiment, the second detector centred at 32.5° was important (see Figure 2). The cross section increases with beam energy, however, the kinetic energy of the inelastically scattered α particles decreases. As compromise, an energy of 100 MeV/u was chosen resulting in recoil energies of a few 100 keV. Again, the use of a window-less detector was mandatory.

Figure 5 shows the measured energy spectra for two different scattering angles in the laboratory system. In the analysis, the δ-electrons seen in the low-energy part of the spectra turned out to be a challenge as their energies in forward direction are only slightly lower than those of the recoiling α particles. This interpretation of the spectra has been confirmed by extensive GEANT4 simulations.

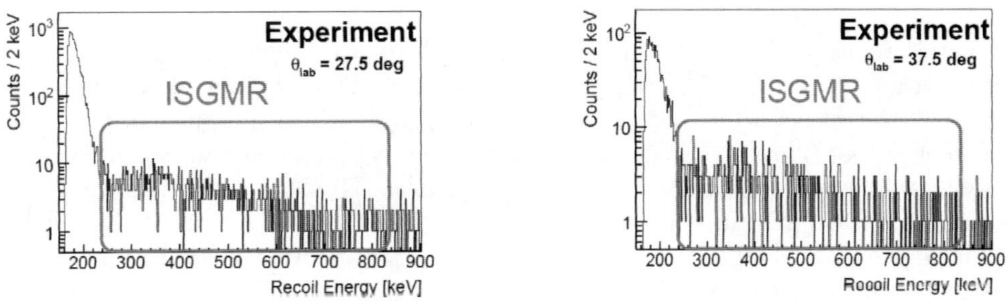

Figure 5. Experimental α spectra measured at two scattering angles in the laboratory system. The region corresponding to the excitation of the isoscalar giant monopole resonance is indicated, the peak at low energies is due to δ-electrons (adapted from [17])

For the double-differential cross section a multipole-decomposition analysis has been performed. The dominant contribution to the double-differential cross section for the inelastic α scattering at angles in the centre-of-mass system $\vartheta_{CM} < 1.5°$ was identified to be the excitation of the isoscalar giant monopole resonance. Figure 6 shows the double-differential cross section in the centre-of-mass system at an average scattering angle of $\bar{\Theta}_{CM} = 1°$. The most important contributions, the isoscalar monopole (ISGMR) and quadrupole (ISGQR) giant resonances, are indicated. The preliminary results for the position, width and strength of the ISGMR in ^{58}Ni are in very good agreement with previous experiments performed in normal kinematics.

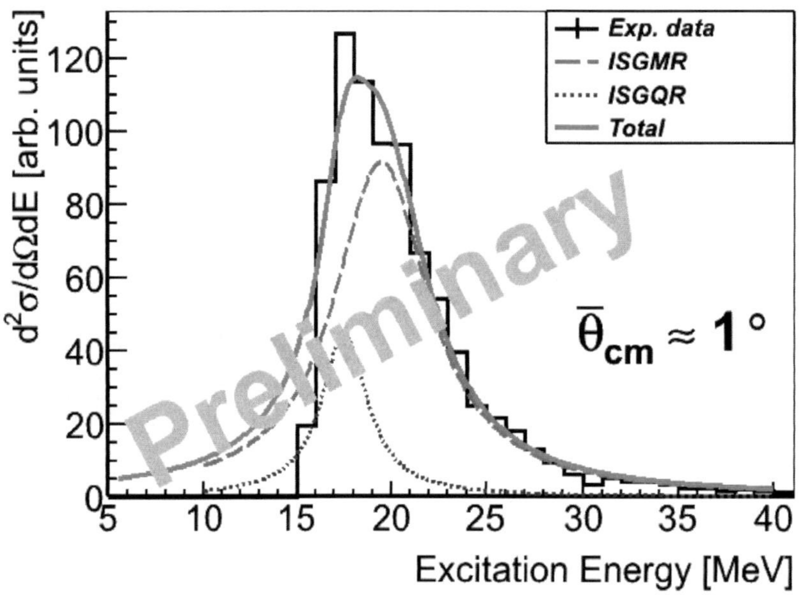

Figure 6. Double-differential cross section for ^{58}Ni(α,α') at 100 MeV/u transformed to the centre-of-mass system. From the results of the multipole decomposition the monopole and quadrupole terms are indicated (adapted from [17])

During the preparations of the experiment with the detectors already mounted at the ESR, it was possible to use a stable ^{20}Ne beam at 50 MeV/u with the hydrogen target. Besides the elastic scattering, also the ^{20}Ne(p,d)^{19}Ne one-neutron transfer reaction was observed [17]. Although the kinetic energies of the deuterons were too high to be stopped in the DSSD, the measured energy losses clearly showed the kinematic curve with a maximum angle $\vartheta < 35°$ in the laboratory system which is characteristic for this reaction. The obtained energy resolution was not sufficient to resolve the transfer to the ground state and the low-lying excited states in ^{19}Ne, however, a multipole decomposition of the differential cross section identified the $L = 0$ and $L = 2$ as the dominant contributions. The extracted spectroscopic factors using the Adiabatic Distorted Wave Approximation (ADWA) approach are well described by shell model calculations

4. Summary and outlook towards FAIR

The first experimental campaign has successfully demonstrated the feasibility of the EXL concept. As major breakthrough, the use of highly-segmented Si detectors as active vacuum windows separating the UHV of the storage ring from an auxilliary vacuum has been proven to work for the detection of low-energy recoils.

For three of the major reaction types comprised in the physics programme of EXL - elastic and inelastic scattering as well as nucleon transfer reactions - first physics results were obtained. The measurement of the matter radius of ^{56}Ni with elastic proton scattering was *the first nuclear reaction performed with a stored radioactive ion beam and an internal target.*

An upgraded detector setup covering a larger solid angle is under design which will be used for further reaction experiments at the ESR. EXL is a very versatile setup and its use at FAIR is considered for experiments with low-energy (at the CRYRING), medium-energy (at the ESR) and high-energy beams up to 5 GeV/u (at the HESR).

Acknowledgments

This work was supported by German BMBF (06DA9040I, 05P12RDFN8 and 05P15RDFN1), the European Commission within the Seventh Framework Programme through IA-ENSAR (contract no. RII3-CT-2010-262010), the Hungarian OTKA Foundation No.K106035, the HGF through the Helmholtz-CAS Joint Research Group HCJRG-108, HIC for FAIR, GSIRUG/ KVI collaboration agreement and TU Darmstadt-GSI cooperation contract.

References

[1] Gutbrod H H et al. (Eds.) 2006 *FAIR Baseline Technical Report* (ISBN-3-9811298-0-6)
[2] *http://www.rug.nl/kvi-cart/research/hnp/research/exl/*
[3] *https://www.gsi.de/r3b*
[4] Litvinov Yu A et al. 2013 *Nucl. Instr. and Meth.* B **317** 603
[5] Reich H et al. 1997 *Nucl. Phys.* A **626** 417
[6] von Schmid M 2015 *Doctoral Thesis* (TU Darmstadt)
[7] Streicher B et al. 2011 *Nucl. Instr. and Meth.* A **654** 604
[8] Mutterer M et al. 2015 *Physica Scripta* (in press)
[9] Egelhof P et al. 2015 *JPS Conf. Proc.* **6** 020049
[10] von Schmid M et al. 2014 *EPJ web conf.* **66** 03093
[11] Nolden F et al. 2013 *Proceedings of IPAC2013* 91
[12] Alkhazov G D et al. 2011 *Int. J. Mod. Phys.* E **20** 583
[13] Lenske H and Kienle P 2007 *Phys. Lett.* B **647** 82
[14] Antonov A N et al. 2005 *Phys. Rev.* C **72** 044307
[15] Tarbutton R M and Davies K T R 1968 *Nucl. Phys.* A **120** 1
[16] Zamora J C et al. 2015 *Physica Scripta* (in press)
[17] Zamora J C 2015 *Doctoral Thesis* (TU Darmstadt)
[18] Lattimer J M 2012 *Annu. Rev. Nucl. Part. Sci.* **62** 485

Journal of Physics: Conference Series **724** (2016) 012027 doi:10.1088/1742-6596/724/1/012027

Magnetic Moments of States in ^{110}Sn.

G J Kumbartzki

Department of Physics and Astronomy, Rutgers University, New Brunswick, NJ, USA

E-mail: kum@physics.rutgers.edu

Abstract. The semi-magic Sn isotopes with Z = 50 are the subject of extensive experimental and theoretical studies. The measured $B(E2)$ values to the 2_1^+ states for the neutron-deficient side of the isotope chain suggest enhanced collectivity when fewer particles are available if the proton shell is not broken. Magnetic moments which are sensitive to proton and neutron contributions to the wave functions of the states could provide critical and relevant information. Magnetic moments were previously measured only for the even stable and a few neutron-rich unstable Sn isotopes. A measurement of the g factors of excited states in ^{110}Sn using the transient field technique was performed at the 88-Inch Cyclotron at the LBNL in Berkeley. The ^{110}Sn nuclei were produced via an α-particle transfer to ^{106}Cd.

The reduced transition probabilities, $B(E2; 0_{gs}^+ \rightarrow 2_1^+)$, to the first excited 2^+ states in the Sn isotope chain on the neutron-deficient side exhibit a remarkable deviation from the shell model expectation. A recent compilation of the results can be found in Ref. [1]. In a simple shell-model picture the $B(E2)$ values are expected to have their maximum at midshell and fall off towards the closed shells in a parabolic shape. This picture holds for the upper half of the Sn isotope chain. But, on the neutron-deficient side of the chain the measured $B(E2)$ values are larger, indicating collectivity, as fewer and fewer particles are available in the pure neutron valence space. Unless, as strongly suggested by Bader *et al.* [2], "enhanced proton excitations across $Z = 50$" occur "as $N = Z$ is approached".

The $B(E2)$ measurements of the neutron-deficient Sn isotopes are difficult and the results scatter across various experiments and have large errors. The Coulomb-excitation measurements suffer from low-intensity and often very high-energy beams. Many experiments therefore are relative measurements using "known" $B(E2)$'s obtained under "similar" conditions.

Magnetic moments of the unstable neutron-deficient Sn isotopes have not been previously measured. They, potentially, can provide valuable information on the proton-neutron contributions to the wave functions of the exited Sn states, given the well known large intrinsic g factors with opposite signs for neutrons and protons.

Magnetic moments reported for the 2_1^+ states in the stable even Sn isotopes [3-5] and the neutron-rich unstable ^{126,128}Sn isotopes [6,7] are shown in Figure 1. The trend is from positive values at ^{112}Sn to negative values for the heavier neutron-rich isotopes, confirming the dominant role of neutrons. Most measurements used the transient field technique (TF) and some the recoil-in-vacuum method (RIV).

The measurements on unstable isotopes depend on the availability of pure isotopic beams. The light neutron-deficient isotopes can be produced by beam fragmentation. Beams of ^{124}Xe at

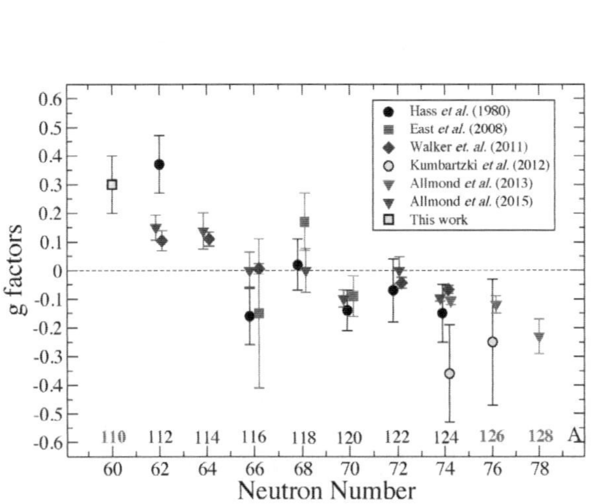

Figure 1. Compilation of measured $g(2_1^+)$ factors of the Sn isotope chain.

hundreds of MeV/nucleon are smashed into a thick Be target. From the beam fragments the Sn isotopes of interest are selected. Because of their high velocities these isotopes are not suitable for g-factor measurements by either the TF or RIV methods. More suitable are the beams produced at REX-ISOLDE at CERN. Besides technical problems, like contaminant radioactivities in the beam, the low beam intensities of about 10^{-6}p/s impose severe experimental limits.

In this TF experiment the ^{110}Sn ($T_{1/2}$ = 4.1h) was produced by a transfer reaction of an α particle from ^{12}C to ^{106}Cd. An isotopically pure beam of ^{106}Cd was accelerated to 410 MeV by the Berkeley 88-inch cyclotron. The traditional multilayer target consisted of 0.636 mg/cm^2 C deposited on subsequent layers of Gd (8.34 mg/cm^2), Ta (1.1 mg/cm^2) and Cu (5.40 mg/cm^2). The α-transfer reaction in inverse kinematics was successfully used in many TF experiments, see for example Ref. [8] and references therein. The hallmark of this method is the accessibility of some radioactive nuclei produced in conditions suitable for a TF experiment; the nuclei should be excited, spin-aligned and forward-scattered with sufficient velocity to traverse the polarized ferromagnetic foil (Figure 2).

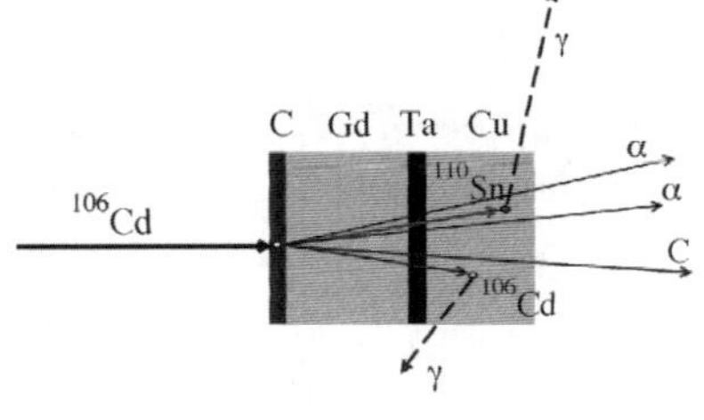

Figure 2. Multilayer target. The excited and spin aligned probe ions ^{110}Sn and ^{106}Cd, which are created in the first layer, traverse the Gd, where their spins precess in the dynamic magnetic TF. The target is designed to stop the probe ions in the Cu layer, a field-free environment. The nuclei deexcite in the Cu layer without further spin rotation.

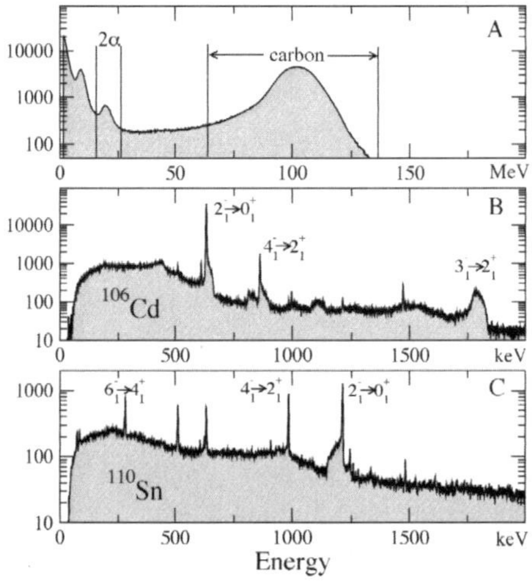

Figure 3. Particle and gated γ spectra. Panel A: Coincident particle spectrum with areas indicating the detection of the two α particles hitting the detector simultaneously and the carbon recoils. Panel B: Gated on the carbon peak showing only γ transitions in ^{106}Cd for a clover detector at 60°. Panel C: ^{110}Sn spectrum obtained by gating on the 2α particle peak in a detector at 120°.The prominent line shapes in some γ lines are due to decay in flight and can be used to determine the lifetimes of the states by the Doppler Shift Attenuation Method.

In the α-transfer reaction ^{12}C(^{106}Cd,^{8}Be)^{110}Sn the residual ^{8}Be nuclei break up into two α particles moving forward in the beam direction. This reaction competes with the Coulomb excitation of the projectiles, which is usually used to measure magnetic moments. This simultaneous measurement of g factors provides an ideal check of the TF experiment. The α-pickup has its maximum yield near the Coulomb barrier of the projectile and ^{12}C which can be considered a three α-cluster nucleus. Compared to Coulomb excitation of the beam projectiles, the α-transfer reaction has a much lower yield, comes with a reduced spin alignment and the population of higher states in the nuclei of interest making feeding corrections possibly necessary.

Typical particle and γ spectra are shown in Figure 3. A description of the detector arrangement and data acquisition can be found in Ref. [8]. Requiring a particle-γ coincidence provides not only cleaner spectra but also selects an ensemble of aligned nuclei which decay with a characteristic angular distribution for each state of interest. The spin rotation in the transient magnetic field causes a rotation of the angular correlation pattern. This rotation is measured as a rate change in the stationary γ detectors when the polarizing field at the target is reversed. The precession effect $\epsilon = (\rho - 1)/(\rho + 1)$ is calculated from quadruple ratios involving four detectors, with the same quadrant angle θ_γ:

$$\rho = \sqrt{\rho_{1,4}/\rho_{2,3}} \text{ , where } \rho_{\mathrm{i,j}} = \sqrt{(\mathrm{N_i^\uparrow N_j^\downarrow})/(\mathrm{N_i^\downarrow N_j^\uparrow})} \ . \tag{1}$$

N↑ and N↓ are the integrated γ-peak rates for the two magnetic field directions. The polarizing field of 0.07 T at the target was reversed every 150 s. This field should not be confused with the transient hyperfine field which is 10^5 times stronger. The logarithmic slope, $S(\theta_\gamma) = \frac{1}{W(\theta_\gamma)} \cdot \frac{dW(\theta_\gamma)}{d\theta}$, of the angular correlation is needed to transform the rate change at

Table 1. Preliminary results from the α-transfer reaction measured simultaneously with the Coulomb excitation of ^{106}Cd.

Nucleus	E_x(keV)	I^π	Transition	τ(ps)	$\|S(67°\|$	g
^{110}Sn	1212.0	2_1^+	$2_1^+ \to 0_1^+$	0.69(6)	0.411(55)	+0.31(11)
	2197.0	4_1^+	$4_1^+ \to 2_1^+$	> 4.0	0.463(55)	+0.07(23)
	2477.7	6_1^+	$6_1^+ \to 4_1^+$	$8.1(4) \cdot 10^3$	0.549(126)	+0.05(22)
^{106}Cd	632.7	2_1^+	$2_1^+ \to 0_1^+$	10.5(12))	1.76(3)	+0.407(17)
	1493.8	4_1^+	$4_1^+ \to 2_1^+$	1.26(15)	0.663(24)	+0.24(5)

the detector positions into the rotation angle $\Delta\theta = \epsilon/S(\theta_\gamma)$. The particle-$\gamma$ angular correlation, using the usual notation,

$$W(\theta) = 1 + A_2^{exp} Q_2 P_2(cos\theta) + A_4^{exp} Q_4 P_4(cos\theta) \quad (2)$$

has been determined from anisotropy ratios obtained from the same precession data in the individual clover segments. The g factor follows from equation (3)

$$g = \frac{\Delta\theta}{\frac{\mu_N}{\hbar} \cdot \int_{t_{in}}^{t_{out}} B_{TF}(v(t), Z) \cdot e^{-t/\tau} dt} \quad (3)$$

with B_{TF} the transient field acting between t_{in} and t_{out}, the time spent in the ferromagnetic layer.

The g-factor results of the initial data analysis are shown in Table 1. For ^{110}Sn only coincident γ spectra were used when both α particles were recorded in the particle detector. The g factor of the long-lived 6_1^+ state was previously measured to a high precision [9], $g = +0.012(3)$. The lifetime of the 4_1^+ state is unknown but judging from the lineshape of the transition, its lifetime is long enough not to affect the TF measurement. Both states feed into the 2_1^+ state contributing to the stopped component in the line shape. Therefore, to avoid the feeding corrections only the tail of the 1212 keV line was used in the precession analysis of the 2_1^+ state.

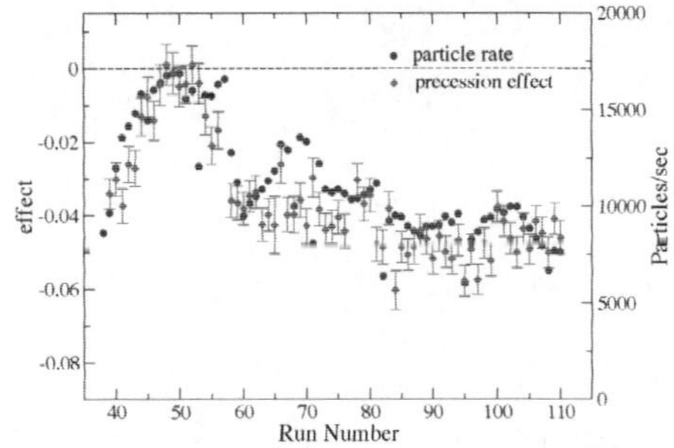

Figure 4. Precession effect of the 2_1^+ state in ^{106}Cd and average total particle rate versus run number. In this experiment the observed effect is negative. A lower beam correlates with a larger effect.

The observed $g(^{110}\text{Sn}; 2_1^+) = +0.31(11)$ is in line with empirical single-particle g factors calculated for neutrons as discussed in Ref. [4]. A break-up of the $Z = 50$ shell would result in a large positive g factor, $g > 1$.

The published $g(^{106}\text{Cd}; 2_1^+) = 0.393(31)$ value in Ref. [10] was reproduced in a measurement at 400 MeV and used as "monitor" of the target magnetization throughout the run period. As a byproduct the g factor of the 4_1^+ state in ^{106}Cd was measured for the first time.

The temperature of the target spot varies with the beam load and that can affect the magnetization and ultimately the strength of the magnetic field parametrized as

$$B_{TF} = 96.7MZ^{1.1}(v/v_0)^{0.45} \tag{4}$$

where M is the magnetization of the ferromagnetic layer.

In Figure 4 the influence of beam heating on the precession effect ϵ is demonstrated. Shown is the correlation of the beam intensity (average particle rate) and the observed precession effect for each run. When the particle rate exceeded 15000 p/s the measured effect was zero. Even at particle rates below 10000 p/s the deduced magnetization from the known g factor of ^{106}Cd was reduced from the offline-measured value of $M = 0.1800$ T. This example clearly shows the importance of controlling the beam load at the target. The magnetization derived from the ^{106}Cd data was applied in the g factor calculation for the ^{110}Sn.

Acknowledgments
Special thanks to my close collaborators N. Benczer-Koller from Rutgers University and K.-H. Speidel fron the University of Bonn (Germany). The help of the Berkeley 88-inch cyclotron scientists and staff was essential to the success of the experiment. The work was supported in part by the U.S. National Science Foundation and the U.S. Department of Energy.

References

[1] P. Doornenbal *et al.* 2014 *Phys. Rev.* C **90** 061302
[2] V.M. Bader *et al.* 2013 *Phys. Rev.* C **88** 051301(R)
[3] M. Hass *et al.* 1980 *Phys. Rev.* C **22** 97
[4] M.C. East *et al.* 2008 *Phys. Lett.* B **665** 147
[5] J. Walker *et al.* 2011 *Phys. Rev.* C **84** 014319
[6] G.J. Kumbartzki *et al.* 2012 *Phys. Rev.* C **86** 034319
[7] J.M. Allmond *et al.* 2013 *Phys. Rev.* C **87** 054325
[8] G.J. Kumbartzki *et al.* 2014 *Phys. Rev.* C **89** 064305
[9] D.A. Volkov *et al.* 1989 Izv.Akad.Nauk SSSR, Ser.Fiz.53, 2188
[10] S.K. Chamoli *et al.* 2011 *Phys. Rev.* C **83** 054318

DSAM lifetime measurements for the chiral bands in ^{194}Tl

P L Masiteng[1,2,3], A A Pasternak[4], E A Lawrie[1], O Shirinda[1], J J Lawrie[1], R A Bark[1], S P Bvumbi[1,2,3], N Y Kheswa[1], R Lindsay[2], E O Lieder[1], R M Lieder[1], T E Madiba[1,2], S M Mullins[1], S H T Murray[1], J Ndayishimye[1,5], S S Ntshangase[1,6], P Papka[5,1] and J F Sharpey-Schafer [1,2]

[1] iThemba LABS, National Research Foundation, PO Box 722, 7129 Somerset West, South Africa
[2] University of the Western Cape, Private Bag X17, 7535 Bellville, South Africa
[3] University of Johannesburg, PO Box 524, Auckland Park, South Africa
[4] A.F. Ioffe Physical-Technical Institute, 194021 St.-Petersburg, Russia
[5] Department of Physics, University of Stellenbosch, Private Bag X1, 7602 Matieland, South Africa
[6] Department of Physics, University of Cape Town, Private Bag, 7701 Rondebosch, South Africa

E-mail: elena@tlabs.ac.za

Abstract. When a left handed and a right-handed nuclear system form in angular momentum space, a pair of nearly degenerate rotational bands is observed. To identify chiral symmetry most important is to establish near-degeneracy not only in excitation energies of the partner bands, but also in their intra-band and inter-band B(M1) and B(E2) transition probabilities. This needs dedicated lifetime measurements. Such measurements were performed for four bands of ^{194}Tl. Two of these have very close near-degeneracy and form a prime candidate for best chiral pair. The lifetime measurements confirm the excellent near-degeneracy in this chiral pair.

1. Chiral symmetry in nuclei.

Chiral symmetry in nuclei forms in angular momentum space [1]. It occurs when a system, which is made of angular momenta, shows well defined handedness, i.e. the mirror image of the system does not coincide with the original. A simple chiral system is made of three mutually orthogonal angular momenta, which can be arranged either in a left-handed or in a right-handed system. A nucleus with ideal chiral symmetry shows a pair of degenerate rotational bands. In real nuclei, however, the partner bands may show considerable divergence from degeneracy [2,3]. Many chiral partners were observed to date in the mass regions of 80, 100, 150, and 190. A review of the experimental data can be found in Ref. [4]. More recent results are published in Refs. [5-14]. The observed chiral bands show certain similarities, but not true degeneracy in all their properties. It is, however, expected that pairs with closer near-degeneracy should correspond to purer chiral symmetry. Thus, search for chiral pairs with good near-degeneracy is being carried out in many laboratories all over the world for the last couple of decades.

2. Chiral pair with excellent near-degeneracy in ^{194}Tl.

At iThemba LABS in South Africa the heavier Tl isotopes were investigated searching for chirality. In particular a pair of rotational bands with very good near-degeneracy was discovered in ^{194}Tl [8]. This experiment used a thin ^{181}Ta target, ^{18}O beam at 91 and 93 MeV, and the AFRODITE array, comprising of 8 Compton suppressed clover detectors [15,16]. The data analysis yielded more than a 100 new transitions, placed in several new bands [9].

The pair of negative parity bands, shown in Fig.1, was proposed as a chiral pair [8]. The two bands are assigned a $\pi h_{9/2} \times \nu i_{13/2}^{-1}$ configuration at low spins. The proton lies at the bottom of the $h_{9/2}$ shell and has particle nature, thus its angular momentum tends to align along the short nuclear axis. The neutron occupies an orbital at the upper part of the $i_{13/2}$ shell, it has hole nature and its angular momentum aligns along the long nuclear axis.

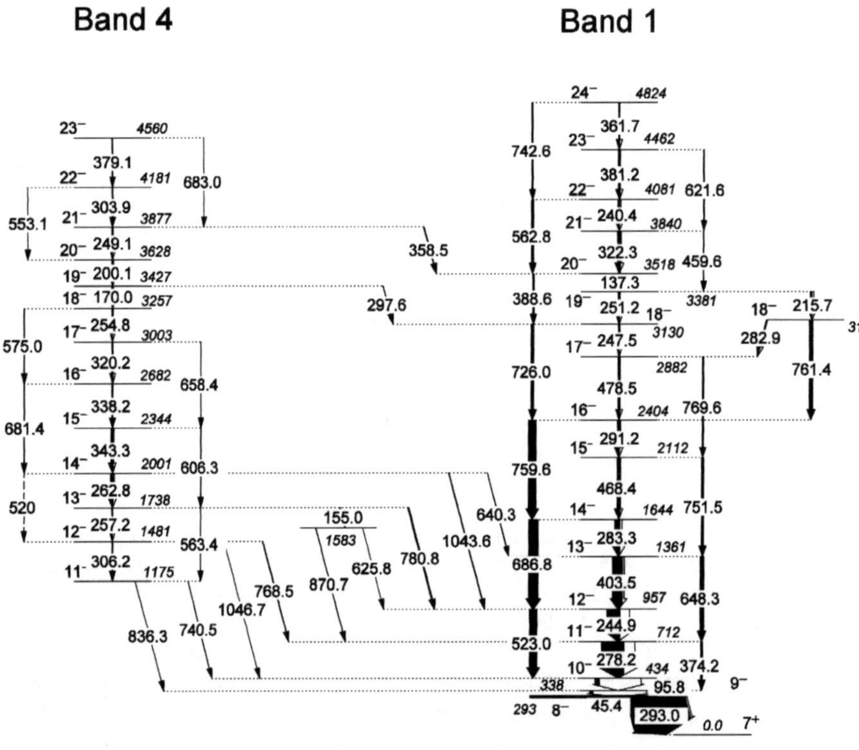

Figure 1. The pair of negative parity bands in ^{194}Tl, proposed as chiral partners. From Ref. [8].

In addition, the calculations performed with the Cranked Nilsson Strutinsky code suggest that ^{194}Tl has a triaxial shape, with a rotational angular momentum that is oriented along the intermediate axis [8]. Therefore the three angular momenta, those of the odd proton, the odd neutron and the rotation are oriented predominantly along the three nuclear axes, and should thus form a chiral system. The pair of observed rotational bands, see Fig. 1, shows some similarity at low spins. For instance the difference in the excitation energies of the two partners decreases from $\Delta E \sim 463$ keV at $I = 11$ to $\Delta E \sim 127$ keV at $I = 18$. The near-degeneracy, however, improves further at high spins.

In the spin range of 16 - 19 ħ a backbend occurs. A pair of $i_{13/2}$ neutrons is broken and their angular momenta are aligned along the long nuclear axis. Thus, above $I \sim 19$ ħ the partner bands are associated with a $\pi h_{9/2} \times \nu i_{13/2}^{-3}$ configuration [8]. This is the first 4-quasiparticle chiral pair observed to date. It is also the only case in which the chiral symmetry persists through a backbend. Still more remarkable is the excellent near-degeneracy observed for this pair at high spins. A comparison with other excellent chiral pairs is shown in Fig. 2. It is obvious that the pair in ^{194}Tl exhibits particularly close near-degeneracy in all properties, such as excitation energies, alignments, and ratios of B(M1)/B(E2) transition probabilities. It is, thus, a prime candidate for best chiral pair found to date.

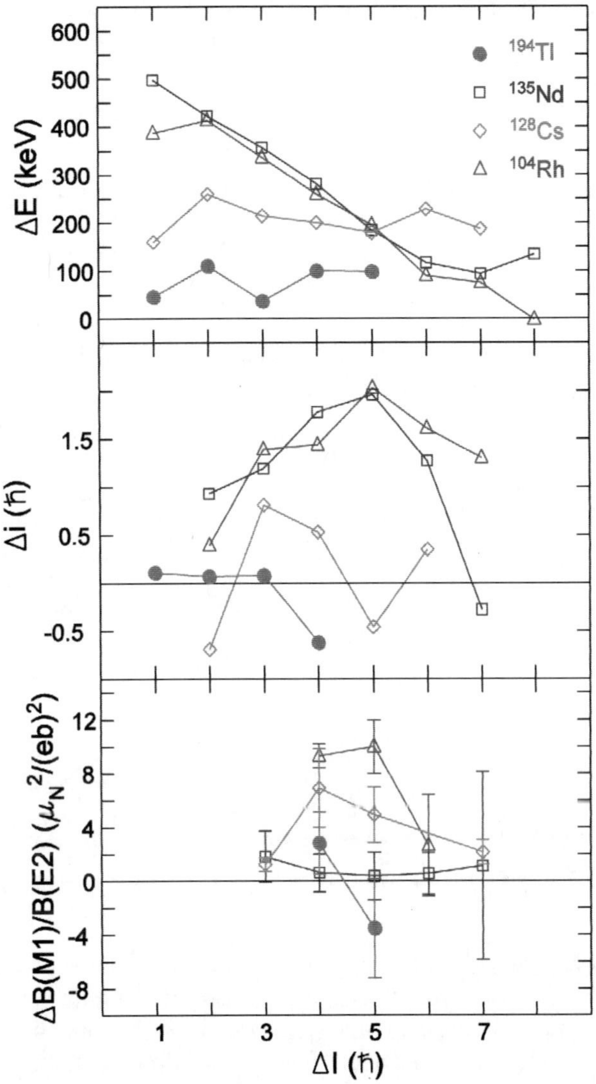

Figure 2. The difference in the excitation energies ΔE, alignments Δi, and ratios of the reduced transition probabilities ΔB(M1)/B(E2) as a function of the spin I with respect to the band head spin for the chiral pairs in ^{194}Tl (the 4-quasiparticle pair), ^{135}Nd, ^{128}Cs and ^{104}Rh. From Ref. [8].

To test whether this pair exhibits good near-degeneracy for the B(M1) and B(E2) reduced transition probabilities, we have performed lifetime measurements in ^{194}Tl using the Doppler Shift Attenuation Method (DSAM). Such measurements are crucial, since as previously shown [17], good energy near-degeneracy alone does not confirm unambiguously that a pair of bands represents a good chiral system.

3. DSAM lifetime measurements in ^{194}Tl.

The experiment was performed at iThemba LABS, South Africa. The high spin states in ^{194}Tl were populated in the ^{181}Ta(^{18}O,5n) reaction at a beam energy of 91 MeV. The target was a 1 mg/cm^2 foil of

^{181}Ta on which a thick layer of Bi was evaporated. The emitted γ rays were detected with the AFRODITE array, consisting of 9 Compton suppressed clover detectors and 6 LEPS detectors. Four clovers were arranged at 45° and four clovers were placed at 135° with respect to the beam direction. The remaining detectors were mounted at 90°. The γ rays were emitted while the ^{194}Tl nuclei were slowing down in the Bi backing. Thus, the γ-ray peaks detected at 45° and 135° showed Doppler broadening, which was then analyzed to extract the lifetime of the nuclear states.

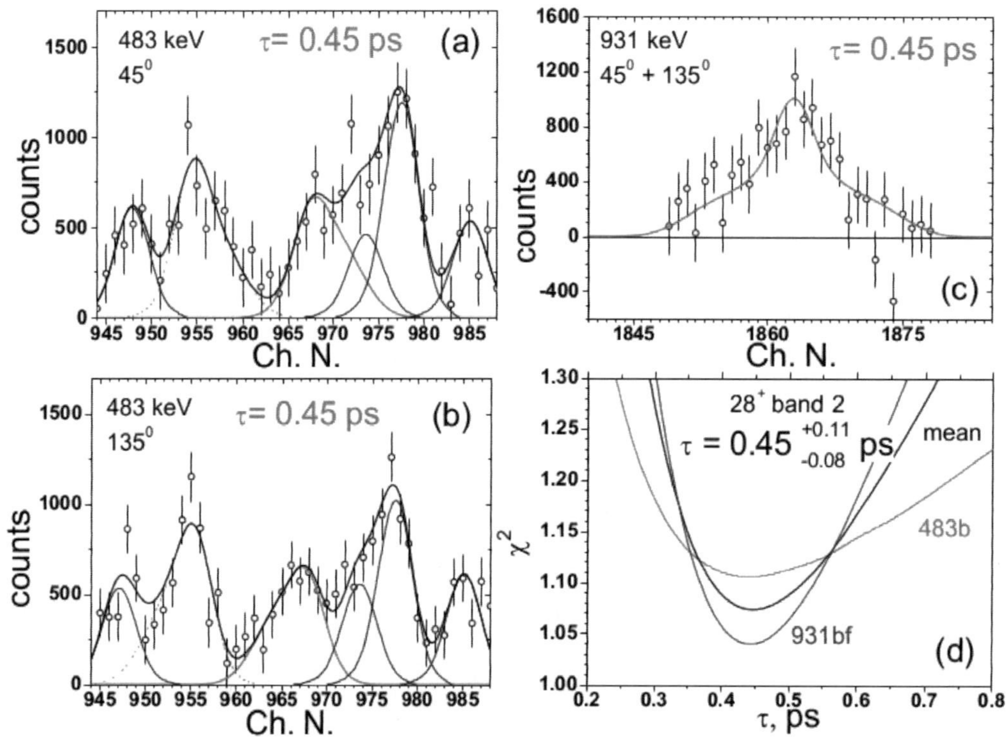

Figure 3. Analysis of the Doppler broadening to extract the lifetime of the 28^+ level in ^{194}Tl. Gated spectra at forward, (a), and backward, (b), angles are shown illustrating the Doppler broadening of the 483 keV peak, shown in red. The spectrum in (c) shows the sum of the forward and backward angles spectra for the 931 keV transition. In panel (d) the χ^2 functions from the analysis of the spectra in panels (b) and (c) are shown.

Data were sorted into background subtracted gated spectra at forward (45°) and backward (135°) angles. The DSAM analysis was performed using the programs COMPA, GAMMA, and SHAPE [18-22], which have been previously applied with the AFRODITE array [11,23]. Monte-Carlo methods were employed to simulate the entry state and the side feeding cascades to the level of interest. Lifetimes were determined step by step starting from the highest-energy level of a band. An example of DSAM analysis is shown in Fig. 3 for the 28^+ level, which decays by a 483 and a 931 keV transitions. One can notice the difference in the broadening of the low-energy part of the medium-energy 483 keV peak, shown in red in panels (a) and (b). The Doppler broadening is much larger for the higher-energy 931 keV peak, which appears symmetric in the summed spectrum of forward and backward angles, see panel (c) of Fig. 3. The analysis of the χ^2 functions for the spectra in panels (b) and (c) is shown in panel (d) and illustrates the extraction of the lifetime of the 28^+ level. Lifetimes were measured for several levels in four rotational bands in ^{194}Tl.

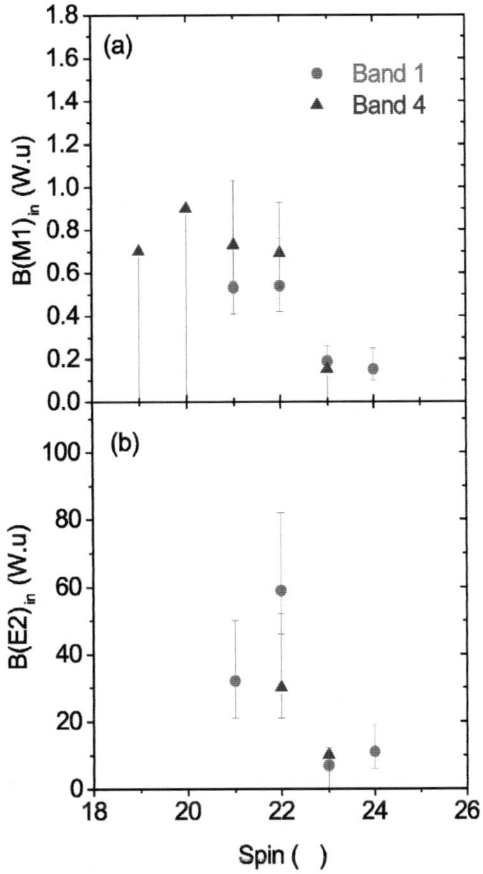

Figure 4. The measured B(M1) and B(E2) in-band transition rates in the yrast (Band 1) and side (Band 4) bands of the chiral pair in ^{194}Tl.

The extracted B(M1) and B(E2) reduced transition probabilities for the states of the chiral pair are shown in Fig. 4. Both bands exhibit very similar values for the in-band transition probabilities. This confirms that the pair shows excellent near-degeneracy, including in the B(M1) and B(E2) rates. It indicates that the bands in ^{194}Tl are (among) the best known chiral symmetry partners.

We would like to thank the crew of the iThemba LABS separated sector cyclotron for the delivery of the oxygen beam. This work is based upon research supported by the National Research Foundation, South Africa, with grants UID 76632 and 93531, and by an International Science and Technology Agreement of South Africa and Russia with grants UID 75248 and RFBR11-02-93963-SA_a.

References

[1] Frauendorf S and Meng J 1997 *Nucl. Phys.* A **617** 131
[2] Lawrie E A and Shirinda O 2010 *Phys. Lett.* B **689**, 66
[3] Shirinda O and Lawrie E A 2012 *Eur. Phys. J.* A **48** 118
[4] Koike T 2010 *Nucl. Phys.* A **834** 36c
[5] Grodner E *et al.* 2011 *Phys. Lett.* B **703** 46
[6] Wang S Y *et al.* 2011 *Phys. Lett.* B **703** 30
[7] Ayangeakaa A D *et al.* 2013 *Phys. Rev. Lett.* **110**, 172504

[8] Masiteng P L *et al.* 2013 *Phys. Lett.* B **719** 83
[9] Masiteng P L *et al.* 2014 *Eur. Phys. J.* A **50** 119
[10] Kuti I *et al*. 2014 *Phys. Rev. Lett.* **113** 032501
[11] Lieder E O *et al.* 2014 *Phys. Rev. Lett.* **112** 202502
[12] Rather N *et al.*, 2014 *Phys. Rev. Lett.* **112** 202503
[13] Tonev D *et al.* 2014 *Phys. Rev. Lett.* **112** 052501
[14] Yavahchova M, presented at the XXI International School on Nuclear Physocs, Neutron Physics and applications, Varna 2015, submitted to J. Phys. G.
[15] Newman R T *et al.* 1998 *Balkan Phys. Lett.*, **Special Issue**, 182
[16] Sharpey-Schafer J F 2004 *Nucl. Phys. News* **14** 5
[17] Tonev D *et al.* 2006 *Phys. Rev. Lett.* **96** 052501
[18] Lemberg I Kh, Pasternak A A, Modern Methods of Nuclear Spectroscopy (Nauka, Leningrad, 1985).
[19] Grodner E *et al.* 2006 *Eur. Phys. J.* A **27** 325
[20] Srebrny J *et al.*, 2001 *Nucl. Phys. A* **683** 21
[21] Lieder R M *et al.*, 2004 *Eur. Phys. J.* A **21** 37
[22] Lieder E O *et al.*, 2008 *Eur. Phys. J.* A **35** 135
[23] Lieder E O *et al.*, 2009 *Nucl. Instr. and Meth.* A **607** 591

Journal of Physics: Conference Series **724** (2016) 012029 doi:10.1088/1742-6596/724/1/012029

Simultaneous occurrence of distinct symmetries in nuclei

A. Leviatan

Racah Institute of Physics, The Hebrew University, Jerusalem 91904, Israel

E-mail: ami@phys.huji.ac.il

Abstract. We show that distinct emergent symmetries, such as partial dynamical symmetry and quasi dynamical symmetry, can occur simultaneously in the same or different eigenstates of the Hamiltonian. Implications for nuclear spectroscopy in the rare-earth region and for first-order quantum phase transitions between spherical and deformed shapes, are considered.

1. Introduction

Symmetries play an important role in the understanding of complex systems. However, complete dynamical symmetries (DS) are manifest in only a few nuclei. Generic Hamiltonians involve interaction terms with competing incompatible symmetries, which break the DS. More often one finds that the assumed symmetry is not obeyed uniformly, *i.e.*, is fulfilled by some of the states but not by others. The need to address such situations, but still preserve important symmetry remnants, has led to the introduction of partial dynamical symmetry (PDS) [1]. For the latter, only part of the eigenspectrum observes a symmetry and retains good quantum numbers and/or complete solvability. Various types of PDS were proposed and extensive tests have provided empirical evidence for their relevance to a broad range of nuclei, *e.g.*, in the rare-earth [2–4], platinum [5, 6] and actinide [7] regions, in light nuclei [8] and in semi-magic nuclei [9]. In parallel, the notion of quasi dynamical symmetry (QDS) was introduced and discussed in the context of nuclear models [10, 11]. It expresses the tendency of a Hamiltonian to exhibit characteristic properties of the closest DS, for a certain range of its parameters. This "apparent" (but broken) symmetry, is due to a coherent mixing of representations in selected states, which share a common intrinsic structure.

The concept of PDS reflects the purity of selected states, as opposed to QDS which reflects a coherent mixing. Up to now, these two symmetry concepts have been considered to be unrelated. In the present contribution, we establish a linkage between them and show that both can occur simultaneously in the same or in different eigenstates of the Hamiltonian. Implications for the spectroscopy of rare-earth nuclei [12] and for shape-phase transitions [13–15] are considered.

2. The interacting boson model: test ground for competing symmetries

The interacting boson model (IBM) [16] provides a convenient framework for exploring the role of symmetries in nuclei. It has been widely used to describe quadrupole collective states in nuclei in terms of N monopole ($s^{\dagger}$) and quadrupole ($d^{\dagger}$) bosons, representing valence nucleon pairs. The model has U(6) as a spectrum generating algebra and exhibits three DS limits associated

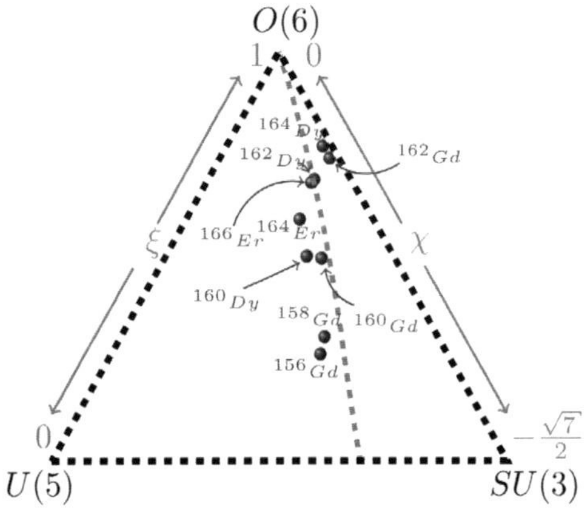

Figure 1. The ECQF symmetry triangle with the position of the rare-earth nuclei of Table 1 indicated by bullets. The calculated (for $N = 60$) red dashed line, correspond to a region of an approximate ground-state O(6) symmetry, exemplifying an O(6)-PDS, as discussed in Section 3. From [20].

with the following chains of nested subalgebras,

$$\begin{aligned} &U(6) \supset U(5) \supset O(5) \supset O(3) && |N,\, n_d,\, \tau,\, n_\Delta,\, L\rangle && \text{(1a)}\\ &U(6) \supset SU(3) \supset O(3) && |N,\, (\lambda,\mu),\, K,\, L\rangle && \text{(1b)}\\ &U(6) \supset O(6) \supset O(5) \supset O(3) && |N,\, \sigma,\, \tau,\, n_\Delta, L\rangle\ . && \text{(1c)} \end{aligned}$$

The indicated basis states are classified by quantum numbers which are the labels of irreducible representation (irreps) of the algebras in each chain. These solvable limits correspond to known benchmarks of the geometric description of nuclei, involving vibrational [U(5)], γ-soft [O(6)], and rotational [SU(3)] types of dynamics. This identification is consistent with the geometric visualization of the model. The latter is obtained by an energy surface, $E(\beta,\gamma)$, defined by the expectation value of the Hamiltonian in the coherent (intrinsic) state [17],

$$|\beta,\gamma;N\rangle \;=\; (N!)^{-1/2}(b_c^\dagger)^N\,|0\,\rangle\ , \tag{2}$$

where, $b_c^\dagger \propto \beta\cos\gamma d_0^\dagger + \beta\sin\gamma(d_2^\dagger + d_{-2}^\dagger)/\sqrt{2} + s^\dagger$. Here (β,γ) are quadrupole shape parameters whose values, $(\beta_{\rm eq},\gamma_{\rm eq})$, at the global minimum of $E(\beta,\gamma)$ define the equilibrium shape for a given Hamiltonian. For one- and two-body interactions, the energy surface is of the form

$$E(\beta,\gamma) \propto (1+\beta^2)^{-2}\beta^2\left[a - b\beta\cos 3\gamma + c\beta^2\right]\ . \tag{3}$$

The shape can be spherical ($\beta = 0$) or deformed ($\beta > 0$) with $\gamma = 0$ (prolate), $\gamma = \pi/3$ (oblate), or γ-independent. The equilibrium deformations associated with the DS limits are $\beta_{\rm eq} = 0$ for U(5), ($\beta_{\rm eq} = \sqrt{2}, \gamma_{\rm eq} = 0$) for SU(3) and ($\beta_{\rm eq} = 1, \gamma_{\rm eq}$ arbitrary) for O(6).

The extended consistent-Q formalism (ECQF) [18] of the IBM, uses the same quadrupole operator, $\hat{Q}^\chi = d^\dagger s + s^\dagger\tilde{d} + \chi\,(d^\dagger\tilde{d})^{(2)}$, in the E2 transition operator and in the Hamiltonian,

$$\hat{H}_{\rm ECQF} = \omega\left[(1-\xi)\,\hat{n}_d - \frac{\xi}{4N}\,\hat{Q}^\chi\cdot\hat{Q}^\chi\right]. \tag{4}$$

Here $\hat{n}_d$ is the d-boson number operator, $\tilde{d}_m = (-)^m d_{-m}$ and the dot implies a scalar product. ξ and χ are the sole structural parameters of the model since ω is a scaling factor. The parameter ranges $0 \leq \xi \leq 1$ and $-\frac{\sqrt{7}}{2} \leq \chi \leq 0$ interpolate between the U(5), O(6) and SU(3) DS limits, which are reached for $(\xi,\chi) = (0,\chi)$, $(1,0)$, and $(1,-\frac{\sqrt{7}}{2})$, respectively. It is customary to represent the parameter space by a symmetry triangle [19], shown in Fig. 1, whose vertices correspond to these limits. The ECQF has been used extensively for describing nuclei [21] and it was found that the vast majority of nuclei are best described by ECQF parameters in the interior of the triangle, away from any DS limit. In what follows we examine the O(6) symmetry properties of ground-band states in such nuclei, in the rare-earth region, using $\hat{H}_{\rm ECQF}$ (4).

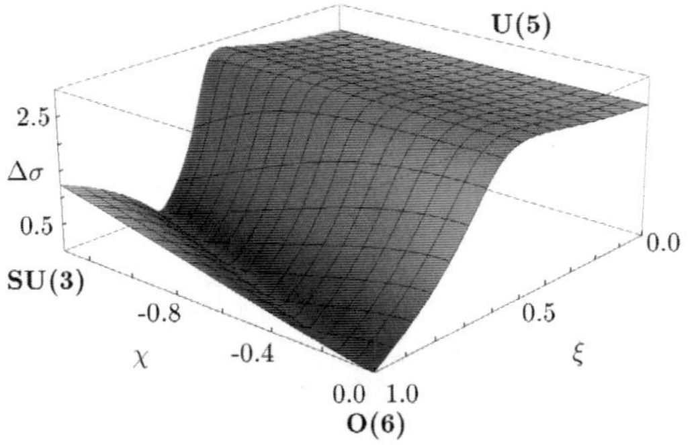

Figure 2. Ground-state ($L=0_1^+$) fluctuations $\Delta\sigma_0$ (5) for $\hat{H}_{\mathrm{ECQF}}$ (4) with $N = 14$. The fluctuations vanish at the O(6) DS limit, saturate towards the U(5) DS limit, and are of the order 10^{-2} in the valley. Adapted from [12].

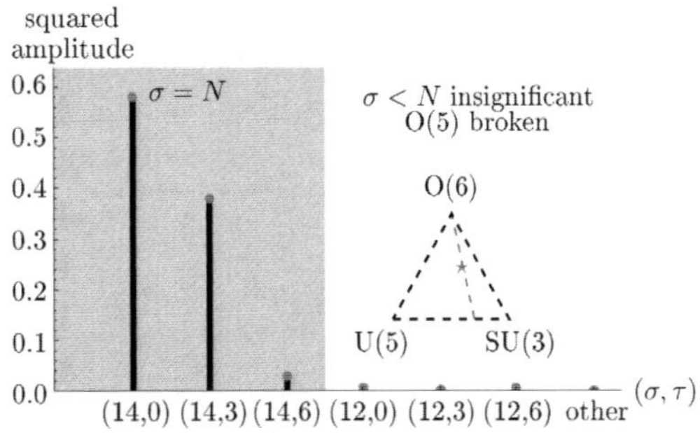

Figure 3. Squared amplitudes C_i^2 in the expansion in the O(6) basis (1c), of the $L=0_1^+$ ground state of the ECQF Hamiltonian (4) with parameters $\xi = 0.84$, $\chi = -0.53$, $N = 14$, appropriate for ^{160}Gd. Adapted from [12].

3. Simultaneous occurrence of O(6)-PDS and SU(3)-QDS in rotational nuclei

Given an eigenstate $|L\rangle$, with angular momentum L, of the ECQF Hamiltonian (4), its expansion in the O(6) DS basis reads $|L\rangle = \sum_i C_i\,|N, \sigma_i, \tau_i, L\rangle$. The degree of O(6) symmetry of the state $|L\rangle$ is inferred from the fluctuations in σ, calculated as [12],

$$\Delta\sigma_L = \sqrt{\sum_i C_i^2\,\sigma_i^2 - \left(\sum_i C_i^2\,\sigma_i\right)^2}\,. \tag{5}$$

If $|L\rangle$ carries an exact O(6) quantum number, σ fluctuations are zero, $\Delta\sigma_L = 0$. If $|L\rangle$ contains basis states with different O(6) quantum numbers, then $\Delta\sigma_L > 0$, indicating that the O(6) symmetry is broken. Note that $\Delta\sigma_L$ also vanishes for a state with a mixture of components with the same σ but different O(5) quantum numbers τ, corresponding to a state $|L\rangle$ with good O(6) but mixed O(5) character. $\Delta\sigma_L$ has the same physical content as wave-function entropy.

The fluctuations $\Delta\sigma_L$ can now be examined for the entire parameter space of the ECQF Hamiltonian (4). Results of this calculation for the ground state, $|L = 0_1^+\rangle$, with $N = 14$, are shown in Fig. 2. At the O(6) DS limit ($\xi = 1$, $\chi = 0$), $\Delta\sigma_0 \equiv \Delta\sigma_{L=0_1}$ vanishes per construction whereas it is greater than zero for all other parameter pairs. Towards the U(5) DS limit ($\xi = 0$), the fluctuations reach a saturation value of $\Delta\sigma_0 \approx 2.47$. At the SU(3) DS limit ($\xi = 1$, $\chi = -\frac{\sqrt{7}}{2}$) the fluctuations are $\Delta\sigma_0 \approx 1.25$. Surprisingly, one recognizes a valley of almost vanishing $\Delta\sigma_0$ values, two orders of magnitude lower than at saturation. This region (depicted by a red dashed arc in the triangle of Fig. 1), represents a parameter range of the IBM, outside the O(6) DS limit, where the ground-state wave function exhibits an exceptionally high degree of purity with respect to the O(6) quantum number σ.

The ground-state wave functions in the valley of low $\Delta\sigma_0$, can be expanded in the O(6)-DS basis (1c). At the O(6) DS limit only one O(6) basis state, with $\sigma = N$ and $\tau = 0$ contributes, while outside this limit the wave function consists of multiple O(6) basis states. Investigation of the wave function for parameter combinations inside the valley reveals an overwhelming dominance of the O(6) basis states with $\sigma = N$. This is seen in Fig. 3 for the ground-state wave function of $\hat{H}_{\mathrm{ECQF}}$ (4), with parameter values that apply to the nucleus ^{160}Gd. The $\sigma = N$ states comprise more than 99% of the ground-state wave function at the bottom of the valley and their dominance causes $\Delta\sigma_0$ to be small. At the same time, the O(5) symmetry is broken, as basis states with different quantum number τ contribute significantly to the wave

Table 1. Calculated σ fluctuations $\Delta\sigma_L$, Eq. (5), for rare earth nuclei in the vicinity of the identified region of approximate ground-state-O(6) symmetry [12]. Also shown are the fraction $f^{(L)}_{\sigma=\mathrm{N}}$ of O(6) basis states with $\sigma = N$ contained in the $L = 0, 2, 4$ states, members of the ground band. The structure parameters ξ and χ of the ECQF Hamiltonian (4) employed, are taken from [21].

Nucleus	N	ξ	χ	$\Delta\sigma_0$	$f^{(0)}_{\sigma=\mathrm{N}}$	$\Delta\sigma_2$	$f^{(2)}_{\sigma=\mathrm{N}}$	$\Delta\sigma_4$	$f^{(4)}_{\sigma=\mathrm{N}}$
^{156}Gd	12	0.72	-0.86	0.46	95.3%	0.43	95.8%	0.38	96.6%
^{158}Gd	13	0.75	-0.80	0.35	97.2%	0.33	97.5%	0.30	97.9%
^{160}Gd	14	0.84	-0.53	0.19	99.1%	0.19	99.2%	0.17	99.3%
^{162}Gd	15	0.98	-0.30	0.17	99.3%	0.17	99.3%	0.16	99.3%
^{160}Dy	14	0.81	-0.49	0.44	96.2%	0.39	96.4%	0.36	96.8%
^{162}Dy	15	0.92	-0.31	0.07	99.9%	0.07	99.9%	0.06	99.9%
^{164}Dy	16	0.98	-0.26	0.13	99.6%	0.13	99.6%	0.13	99.6%
^{164}Er	14	0.84	-0.37	0.39	96.5%	0.37	96.7%	0.35	97.1%
^{166}Er	15	0.91	-0.31	0.12	99.7%	0.11	99.7%	0.10	99.7%

function. Consequently, the valley can be identified as an entire region in the symmetry triangle with an approximate O(6)-PDS, which means that some of the eigenstates exhibit some of the symmetries in the chain (1c). Outside this valley the ground state is a mixture of several σ values and $\Delta\sigma_0$ increases.

Detailed ECQF fits for energies and electromagnetic transitions of rare-earth nuclei [21], allow one to relate the structure of collective nuclei to the parameter space of the ECQF Hamiltonian (4). From the extracted (ξ, χ) parameters one can calculate the fluctuations $\Delta\sigma_L$ and the fractions $f_{\sigma=N}$ of squared $\sigma = N$ amplitude. Nuclei with $\Delta\sigma_0 < 0.5$ and $f_{\sigma=N} > 95\%$ in the ground-state ($L = 0^+_1$) are listed in Table 1. These quantities are also calculated for yrast states with $L > 0$ and exhibit similar values in each nucleus. It is evident that a large set of rotational rare earth nuclei are located in the valley of small σ fluctuations. They can be identified as candidate nuclei with an approximate O(6)-PDS not only for the ground state, but also for the members of the band built on top of it.

The experimental spectrum of a representative nucleus from Table 1, ^{160}Gd, along with its ECQF description (4), is shown in Fig. 4(a). Figures 4(b) and 4(c) show the decomposition into O(6) and SU(3) basis states, respectively, for yrast states with $L = 0, 2, 4$. It is evident that the SU(3) symmetry is broken, as significant contributions of basis states with different SU(3) quantum numbers (λ, μ) occur. It is also clear from Fig. 4(c) that this mixing occurs in a coherent manner with similar patterns for the different members of the ground-state band. Such coherent mixing is the hallmark of SU(3) QDS [11]. On the other hand, as seen in Fig. 4, the same yrast states with $L = 0, 2, 4$ are almost entirely composed out of O(6) basis states with $\sigma = N = 14$ which implies small fluctuations $\Delta\sigma_L$ (5) and the preservation of O(6) symmetry (but with broken O(5) symmetry) in the ground-state band. Thus an empirically-manifested link is established between SU(3) QDS and O(6) PDS.

The simultaneous occurrence of SU(3)-QDS and O(6)-PDS for members of the ground band, signals the existence of a single intrinsic state with good O(6) character. Such an intrinsic state is the condensate of Eq. (2) with $(\beta = 1, \gamma = 0)$, that has $\sigma = N$. The states projected from it keep exact O(6) symmetry ($\sigma = N$) but break the O(5) symmetry (mixed τ) and have a high overlap with the yrast eigenstates of $\hat{H}_{\mathrm{ECQF}}$ (more than 99% for the ground states $L = 0^+_1$). This suggests that the indicated intrinsic state provides a good approximation, in a variational sense, to the ground band of $\hat{H}_{\mathrm{ECQF}}$ along the valley of low-$\Delta\sigma_0$. The two extremum equations for $E(\beta, \gamma)$, Eq. (3), $\partial E/\partial\beta = \partial E/\partial\gamma = 0$, have $\beta = 1$ and $\gamma = 0$ as a solution,

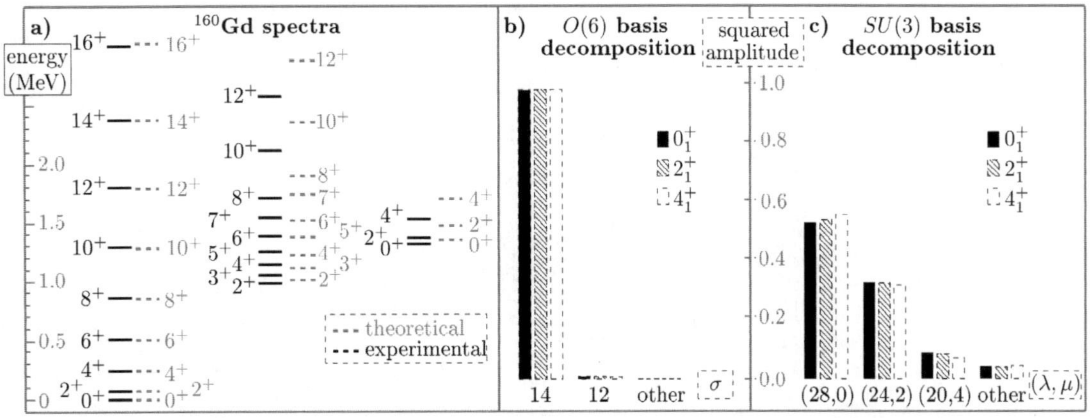

Figure 4. a) The experimental spectrum of ^{160}Gd compared with the IBM calculation using the ECQF Hamiltonian (4) with parameters $\xi = 0.84$ and $\chi = -0.53$ taken from Ref. [21]. b) The O(6) decomposition in σ components of yrast states with $L = 0, 2, 4$. c) The SU(3) decomposition in (λ, μ) components of the same yrast states. Adapted from [12].

provided $b = 2c$. For large N, the energy surface coefficients of $\hat{H}_{\rm ECQF}$ are $b = -\omega\xi\sqrt{\frac{2}{7}}\chi/N$ and $c = \omega\left[1 - \xi - \xi\chi^2/14\right]/N$. Thus, in the valley of low $\Delta\sigma_0$ the desired condition, $b = 2c$, fixes ξ to be $\xi = 1/[1 - \chi/\sqrt{14} + \chi^2/14]$. This relation predicts the location of the region of approximate ground-state O(6) symmetry for large N very precisely [12]. These results demonstrate that coherent mixing of one symmetry (QDS) can result in the purity of a quantum number associated with partial conservation of a different, incompatible symmetry (PDS).

4. Simultaneous occurrence of U(5)-PDS and SU(3)-QDS in shape-coexistence

PDS (partial purity) and QDS (coherent mixing) of distinct symmetries can occur simultaneously also in different sets of eigenstates of a given Hamiltonian. Such a situation is encountered, for example, in a first-order quantum phase transition (QPT) involving shape coexistence. Focusing on the intrinsic dynamics [22] at the critical-point such QPT between spherical [U(5)] and deformed [SU(3)] shapes, the relevant IBM Hamiltonian annihilates the corresponding condensates (2) with $\beta = 0$ and $(\beta = \sqrt{2}, \gamma = 0)$, and can be transcribed in the form [23, 24]

$$\hat{H}_{\rm cri} = h_2\, P^\dagger_2 \cdot \tilde{P}_2 \quad , \quad P^\dagger_{2m} = 2d^\dagger_m s^\dagger + \sqrt{7}\,(d^\dagger\, d^\dagger)^{(2)}_m \quad , \quad \tilde{P}_{2m} = (-)^m P_{2,-m}\ . \tag{6}$$

$\hat{H}_{\rm cri}$ mixes terms from different DS chains of the IBM, hence is non-integrable. The corresponding classical Hamiltonian, obtained by Glauber coherent states, has a Landau potential with two degenerate spherical and prolate-deformed minima, as shown in Fig. 5. A detailed classical analysis [13–15] reveals a robustly regular dynamics in the region of the deformed minimum and a change with energy from regular to chaotic dynamics in the region of the spherical minimum. The mixed but well separated dynamics persists even at energies far exceeding the barrier height. A quantum analysis is based on Peres lattices [25], which are constructed by plotting the expectation values $x_i \equiv \sqrt{2\langle i|\hat{n}_d|i\rangle/N}$ of the d-boson number operator, versus the energy $E_i = \langle i|\hat{H}|i\rangle$ of the Hamiltonian eigenstates $|i\rangle$. The lattices $\{x_i, E_i\}$ corresponding to regular dynamics display an ordered pattern, while chaotic dynamics leads to disordered meshes of points [25]. The quantity x_i is related to the coordinate x in the classical potential, hence the indicated lattices can distinguish regular from irregular states and associate them with a given region in phase space.

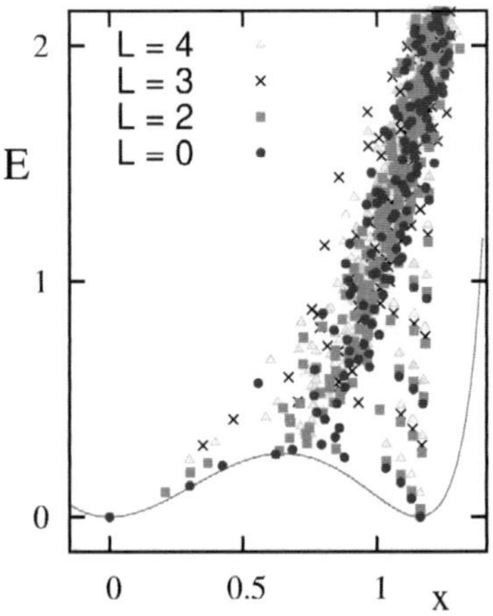

Figure 5. Peres lattices $\{x_i, E_i\}$ for $L=0,2,3,4$ and $N=50$ eigenstates of $\hat{H}_{\rm cri}$, Eq. (6), overlayed on the classical potential. Here $x_i \equiv \sqrt{2\langle L_i|\hat{n}_d|L_i\rangle/N}$. Adapted from [15].

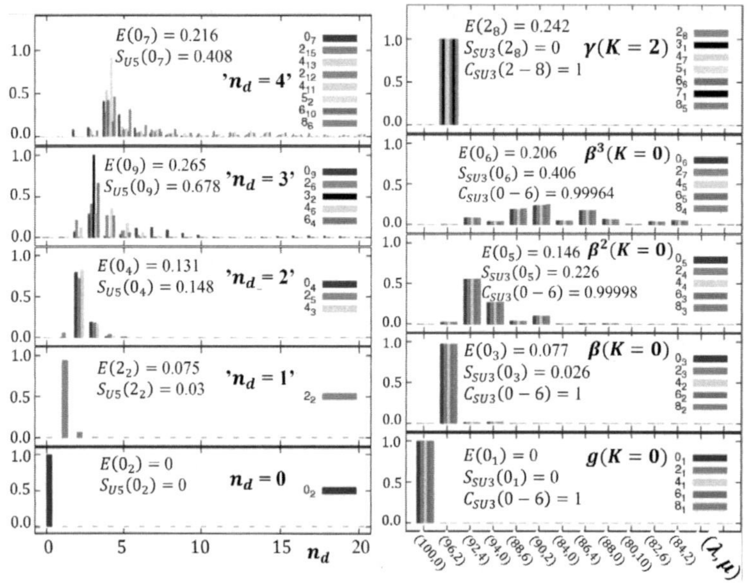

Figure 6. Left column: U(5) n_d-probability distribution for spherical type of eigenstates of $\hat{H}_{\rm cri}$, Eq. (6), arranged in U(5)-like n_d-multiplets. Right column: SU(3) (λ,μ)-probability distribution for deformed type of eigenstates arranged in rotational K-bands. Shannon entropies $S_{\rm U5}(L_i)\approx 0$ and $S_{\rm SU3}(L_i)\approx 0$ signal U(5)-PDS and SU(3)-PDS, respectively. SU(3) Pearson correlator $C_{\rm SU3}(0\text{–}6)\approx 1$, signals SU(3)-QDS. Adapted from [15].

The Peres lattices for eigenstates of $\hat{H}_{\rm cri}$ (6), with $L=0,2,3,4$, are shown in Fig. 5, overlayed on the classical potential. They disclose regular sequences of states localized within and above the deformed well. They are comprised of rotational states with $L=0,2,4,\ldots$ forming regular $K=0$ bands and sequences $L=2,3,4,\ldots$ forming $K=2$ bands. Additional K-bands (not shown in Fig. 5), corresponding to multiple β and γ vibrations about the deformed shape, can also be identified. Such ordered band-structures persist to energies above the barrier and are not present in the disordered (chaotic) portions of the Peres lattice. At low-energy, in the vicinity of the spherical well, one can also detect multiplets of states with $L=0$, $L=2$ and $L=0,2,4$, typical of quadrupole excitations with $n_d=0,1,2$, of a spherical shape.

The nature of the surviving regular sequences of selected states, is revealed in a symmetry analysis of their wave functions. The left column of Fig. 6 shows the U(5) n_d-probabilities, $P^{(L_i)}_{n_d}$, for eigenstates of $\hat{H}_{\rm cri}$ (6), selected on the basis of having the largest components with $n_d=0,1,2,3,4$, within the given L spectra. The states are arranged into panels labeled by 'n_d' to conform with the structure of the n_d-multiplets of the U(5) DS limit. In particular, the zero-energy $L=0^+_2$ state is seen to be a pure $n_d=0$ state, with vanishing U(5) Shannon entropy, $S_{\rm U5}=0$. It is a solvable eigenstate of $\hat{H}_{\rm cri}$, exemplifying U(5)-PDS [24]. The state 2^+_2 has a pronounced $n_d=1$ component (96%) and the states ($L=0^+_4$, 2^+_5, 4^+_3) in the third panel, have a pronounced $n_d=2$ component and a low value of $S_{\rm U5}<0.15$. All the above states with '$n_d\leq 2$' have a dominant single n_d component, and hence qualify as 'spherical' type of states. They are the left-most states in the Peres lattices of Fig. 5. In contrast, the states in the panels '$n_d=3$' and '$n_d=4$' of Fig. 6, are significantly fragmented. A notable exception is the $L=3^+_2$ state, which is a solvable U(5)-PDS eigenstate with $n_d=3$ [24]. The existence in the spectrum of specific spherical-type of states with either $P^{(L)}_{n_d}=1$ [$S_{\rm U5}(L)=0$] or $P^{(L)}_{n_d}\approx 1$ [$S_{\rm U5}(L)\approx 0$], exemplifies the presence of an exact or approximate U(5) PDS at the critical-point.

The states considered in the right column of Fig. 6 have a different character. They belong to the five lowest regular sequences seen in the Peres lattices of Fig. 5, in the region $x \geq 1$. They have a broad n_d-distribution, hence are qualified as 'deformed'-type of states, forming rotational bands: $g(K=0)$, $\beta(K=0)$, $\beta^2(K=0)$, $\beta^3(K=0)$ and $\gamma(K=2)$. The ground $g(K=0)$ and $\gamma(K=2)$ bands are pure [SU(3) (λ,μ)-distribution, $P^{(L)}_{(\lambda,\mu)} = 1$, and SU(3) Shannon entropy, $S_{\rm SU3}(L) = 0$] with $(\lambda,\mu) = (2N,0)$ and $(2N-4,2)$ SU(3) character, respectively. These are solvable bands of $\hat{H}_{\rm cri}$ and exemplify SU(3)-PDS [24]. The non-solvable K-bands are mixed with respect to SU(3) in a coherent, L-independent manner, hence exemplify SU(3)-QDS. As expected, the Pearson correlator [26] is $C_{\rm SU3}(0_i\!-\!6) \approx 1$, for these regular K-bands. The above results demonstrate that PDS and QDS can characterize the remaining regularity in a system, amidst a complicated (at time chaotic) environment of other states.

5. Concluding remarks

Both PDS and QDS do not arise from invariance properties of the Hamiltonian, hence can occur simultaneously in atomic nuclei. As shown, the existence of a region of almost exact ground-state-band O(6) symmetry outside the O(6) DS limit of the IBM, can be understood in terms of an approximate O(6) PDS. The same wave functions display coherent (L-independent) mixing of SU(3) irreps and hence comply with the conditions of an SU(3) QDS. Many rare-earth nuclei do exhibit this linkage. Both types of emergent symmetries can characterize persisting regular patterns in nuclei, *e.g.*, U(5)-like and SU(3)-like multiplets, amidst a complicated environment of other states, a situation encountered in a quantum shape-phase transition.

The work reported in Section 3, was done in collaboration with C. Kremer, J. Beller, N. Pietralla, R. Trippel (Darmstadt), G. Rainovski (Sofia), P. Van Isacker (GANIL) and in Section 4, with M. Macek (Yale), and is supported by the Israel Science Foundation.

References

[1] For a review see, Leviatan A 2011 *Prog. Part. Nucl. Phys.* **66** 93
[2] Leviatan A 1996 *Phys. Rev. Lett.* **77** 818; Leviatan A and Sinai I 1999 *Phys. Rev. C* **60** 061301
[3] Leviatan A, García-Ramos J E and Van Isacker P 2013 *Phys. Rev. C* **87** 021302(R)
[4] Casten R F, Cakirli R B, Blaum K, and Couture A 2014 *Phys. Rev. Lett.* **113** 112501
[5] García-Ramos J E, Leviatan A and Van Isacker P 2009 *Phys. Rev. Lett.* **102** 112502
[6] Van Isacker P, Jolie J, Thomas T and Leviatan A 2015 *Phys. Rev. C* **92** 011301(R)
[7] Couture A, Casten R F, and Cakirli R B 2015 *Phys. Rev. C* **91** 014312
[8] Escher J and Leviatan A 2000 *Phys. Rev. Lett.* **84** 1866; *ibid* 2002 *Phys. Rev. C* **65** 054309
[9] Van Isacker P and Heinze S 2008 *Phys. Rev. Lett.* **100** 052501; *ibid* 2014 *Ann. Phys. (NY)* **349** 73
[10] Bahri C and Rowe D J 2000 *Nucl. Phys. A* **662** 125
[11] Rowe D J 2004 *Nucl. Phys. A* **745** 47; Rosensteel G and Rowe D J 2005 *Nucl. Phys. A* **759** 92
[12] Kremer C, Beller J, Leviatan A, Pietralla N, Rainovski G, Trippel R, and Van Isacker P 2014 *Phys. Rev. C* **89** 041302(R); *ibid* 2015 *Phys. Rev. C* **92** 039902(E)
[13] Macek M and Leviatan A 2011 *Phys. Rev. C* **84** 041302(R)
[14] Leviatan A and Macek M 2012 *Phys. Lett. B* **714** 110
[15] Macek M and Leviatan A 2014 *Ann. Phys. (NY)* **351** 302
[16] Iachello F and Arima A 1987 *The Interacting Boson Model* (Cambridge: Cambridge Univ. Press)
[17] Ginocchio J N and Kirson M W 1980 *Phys. Rev. Lett.* **44** 1744
[18] Warner D D and Casten R F 1983 *Phys. Rev. C* **28** 1798
[19] Casten R F and Warner D D 1983 *Prog. Part. Nucl. Phys.* **9** 311
[20] Kremer C 2015 private communication
[21] McCutchan E A, Zamfir N V and Casten R F 2004 *Phys. Rev. C* **69** 064306
[22] Leviatan A 1987 *Ann. Phys. (NY)* **179** 201; Leviatan A and Kirson M W 1990 *Ann. Phys. (NY)* **201** 13
[23] Leviatan A 2006 *Phys. Rev. C* **74** 051301(R)
[24] Leviatan A 2007 *Phys. Rev. Lett.* **98** 242502
[25] Peres A 1984 *Phys. Rev. Lett.* **53** 1711
[26] Macek M, Dobeš J and Cejnar P 2010 *Phys. Rev. C* **82** 014308

Are the nuclei beyond ^{132}Sn very exotic?

R. Lozeva[1], H. Naïdja[1,2,3], F. Nowacki[1], A. Odahara[4] and C.-B.Moon[5] for the NP1112-RIBF87 collaboration[6]

[1] IPHC, CNRS, IN2P3, University of Strasbourg, 67037 Strasbourg, Cedex 2, France
[2] GSI Helmholtzzentrum für Schwerionenforschung GmbH, 64291 Darmstadt, Germany
[3] University of Constantine, 25000 Constantine, Algeria
[4] Department of Physics, Osaka University, Osaka 560-0043 Toyonaka, Japan
[5] Department of Display Engineering, Hoseo University, Chung-Nam 336-795, Korea
[6] RIKEN Nishina Center, Wako-shi, Saitama 351-0198, Japan

E-mail: radomira.lozeva@iphc.cnrs.fr

Abstract. The term exotic nucleus is used for nuclei that have different from normal behavior. However, it turns out that the term normal is valid only for nuclei close to stability and more particularly for regions close to double-shell closures. As long as one drives away in the neutron-rich nuclei, especially at intermediate mass number, interplay between normal single-particle and many collective particle-hole excitations compete. In some cases with the addition of neutrons, these may turn to evolve as a skin, acting against the core nucleus that may also influence its shell evolution. Knowledge of these nuclear ingredients is especially interesting beyond the doubly-magic ^{132}Sn, however a little is known on how the excitations modes develop with the addition of both protons and neutrons. Especially for the Sb nuclei, where one gradually increases these valence particles, the orbital evolution and its impact on exoticness is very intriguing. Experimental studies were conducted on several such isotopes using isomer and β-decay spectroscopy at RIBF within EURICA. In particular, new data on ^{140}Sb and ^{136}Sb are examined and investigated in the framework of shell model calculations.

1. Introduction

One particular region on nuclei receives a special attention from both experiment and theory since several decades. These are the chain of Sn nuclei, known to have superfluid characteristics, owing to the constant behavior of their 2^+ and 4^+ excitation energies all over the whole isotopic chain. Moreover, experimental data confirms the well developed Z=50 and N=82 magicity, contrary to some suggestions for a shell quenching in the neutron-rich Sn isotopes due to a neutron excess and a reduction in the spin-orbit force [1]. With increasing the proton particles or the neutron holes from the doubly-magic ^{132}Sn, a different behavior can be seen, particularly related to the $\pi g_{7/2}$-$\pi d_{5/2}$ orbitals. For example, with increasing Z in the semi-magic N=82 isotones, the 2^+, 4^+ excitation energies exhibit a hump at Z=64, indicating a sub-shell closure of the $\pi g_{7/2}$-$\pi d_{5/2}$ orbitals [2].

Following the systematics of the low-lying states in the odd-even Sn isotopes below $N = 82$, compared to the even-odd N=82 isotones, one can trace that the difference in their spectra with respect to the even-even nuclei is rather large. Their evolution shows that the levels are compressed at N, Z=64, manifesting a clear decoupling of $\pi d_{5/2}$ and $\pi g_{7/2}$ orbitals with respect to the other three. It is therefore believed (as observed in the lowered energies of the $1/2^+$, $3/2^+$

and $11/2^-$ states in $^{142}_{65}Tb_{77}$) that the gap between the proton $d_{5/2}$, $g_{7/2}$ and the $d_{3/2}$, $s_{1/2}$ and $h_{11/2}$ orbitals is responsible for the rise of the 2^+, 4^+ excitation energies in $^{146}_{64}Gd_{82}$ [2, 3, 4]. On the contrary for the Sn isotopes, the $1/2^+$, $3/2^+$ and $11/2^-$ decrease fast to mix up with the $5/2^+$, $7/2^+$ in $^{111}_{50}Sn_{61}$ [2, 3].

2. Physics motivation

The orbital evolution beyond ^{132}Sn is expected to be similar to the one below N=82 and Z=50. On one side, the interaction of the $\pi d_{5/2}$ and $\pi g_{7/2}$ orbitals with the $\nu h_{11/2}$ orbital shall persist due to the same parity of $\nu f_{7/2}$, resulting in repulsion of $\pi d_{5/2}$ and $\pi g_{7/2}$. However, experimental data manifest the opposite behavior with a fast drop for ^{133}Sb [5] and ^{135}Sb [6] in the observed position of the $5/2^+$ states with respect to the $7/2^+$ g.s. Interestingly, this effect was also detected for the I nuclei beyond ^{132}Sn where also drop was observed for the measured $5/2^+$ states. Several explanations were given in this context, from decrease of $\pi d_{5/2}$ and $\pi g_{7/2}$ spacing needed in shell model (SM) calculations in order to explain the experimental data in ^{135}Sb and ^{136}I [7, 8, 9], to an additional seniority mixing in the wave functions of these states [10]. According to some of the calculations, the distance between $\pi d_{5/2}$ and $\pi g_{7/2}$ decreases with both N and Z [11]. Thus, one may expect their crossing for the more exotic nuclei. The evolution of these particular proton orbitals was suggested to be related to the development of a neutron skin in the neutron-rich isotopes beyond ^{132}Sn [12]. For example, according to Mean field (MF) calculations, using SKX Skryme interaction [13], a decreased spacing was predicted with the increase of Z. Suggestion about the same proton orbital evolution with N was given using HFB-D1S calculations [14], with a possible crossing at N=90. They indicated, that the radii of the systems and the spin-orbit interaction, known to be concentrated mainly at the nuclear surface, play an important role in such crossing, even though some additional $\pi\nu$ interactions also may give a contribution. It is challenging to find how such expectations would influence the excited schemes in these very exotic nuclei and could these be directly related to the formation of skin.

The evolution of the neutron orbitals beyond ^{132}Sn is also intriguing. With the increase of the neutrons, using empirical SM calculations [11], an opening of a sub-shell gap (of about 2 MeV) was suggested at N=90 between the $\nu p_{3/2}$ and $\nu f_{7/2}$ orbitals. At the end of the $\nu f_{7/2}$ shell (^{140}Sn) similar effect was independently proposed in the MF calculations using the Cooper-pair formalism [15]. Investigating the effect on the transition rates of $(f^n_{7/2})_{6^+}$ isomers, where n equals the number of valence neutrons in $^{134-138}$Sn, it was shown in [16] that one would expect apparent differences for these nuclei which were not seen. Therefore, the expected gap would be inconsistent with the isomer data [17], similarly to the purely theoretical conclusion of [18].

With the increase of both π and ν, anomalies were observed also in Te 2^+ excitation energies, dropping fast beyond N=82, similarly to the odd-proton Sb. In SM calculations [19], it was pointed out that increasing the neutron weight in the lowest collective 2^+ states, strong $p-n$ exchange asymmetry is developing with the addition of π between the Te and Xe isotopes. Along, strong polarisation effects were proposed also for the odd-proton I nuclei, as unexplained isomerism was observed in ^{138}I and exchange in g.s. spin/parity was detected between ^{138}I and ^{140}I [20]. Therefore, the interplay of several effects would influence the structure of the isotopes beyond ^{132}Sn and it is an experimental challenge to measure them and provide crucial ingredients to their understanding and the development of theory.

3. Experimental techniques

The experimental studies were performed using an isomer and β-decay measurement technique at the RIBF facility at RIKEN [21] in the framework of the EURICA project [22, 23]. The system represented one of the most powerful to date combinations of high-intensity uranium beams and a highly efficient γ-ray detector system. It enabled new and detailed spectroscopic studies from light to heavy very-neutron rich nuclei carried out within several experimental campaigns. In this case, an in-flight $^{238}U^{86+}$ fission at 345 MeV/u was initiated on ^{9}Be target with a thickness of 2.9 mm. The beam intensity varied between 1-5 pnA during the measurement time. The nuclei of interest were transported and selected by the BigRIPS and ZeroDegree spectrometers [24] and identified using the standard $\Delta E - ToF - B\rho$ method and a system of tracking and identification detectors at the focal planes [25]. The ions were further implanted in the wide-range active stopper for β and ion detection, WAS3ABi, comprising of a stack of five (1 mm thick) Double-Sided Silicon-Strip detectors (DSSSD) detectors with a segmentation of 60 $\times$ 40 strips (1-mm pitch) [22, 23].

For the nuclei produced in an isomeric state, with lifetime long enough to survive in-flight decay through the spectrometers (of the order of 600 ns), delayed γ rays were observed after their implantation. These were detected using the EURICA (4π) array surrounding the stopper (in a close-packed geometry), assembled from twelve Ge Cluster detectors (of seven HPGe crystals). Ion-γ coincidences (for isomeric or β-decay studies) were acquired for different time windows after the implantation using either a standard TDC for short-(μs) or DGF for a longer- (tens of μs) time ranges. Other experimental details can be found in [26, 27, 28, 29].

4. The Sb isotopes

In these studies we populated five exotic Sb isotopes between ^{136}Sb and ^{140}Sb [26], thus systematic studies could be performed. The data was in addition compared to earlier experiments on ^{134}Sb and ^{136}Sb at the beginning of the same $\nu f_{7/2}$ shell [2]. Low- and intermediate-energy states in these Sb nuclei were described in analogy with ^{210}Bi [30, 31, 8] and ^{212}Bi [32, 33], considered in the evolution of the equivalent $\pi g_{7/2}\nu f_{7/2}$ and $\pi h_{9/2}\nu g_{9/2}$ proton-neutron multiplets in the two different neutron-rich mass-regions. No data on the evolution of this resemblance for the more exotic species was, however, available up to date.

4.1. Earlier observations

A long-lived (10.07(5) s [34]) β-decaying isomer was reported in ^{134}Sb ($T_{1/2}$(g.s) = 0.75(7) s [35]), assigned as a 7^- level [36]. It was placed at 279 keV [8] together with other yrast states observed in an earlier fission fragment study [30]. The low-spin structure of ^{134}Sb was assigned from several β-decay studies: [37] suggested 0^- g.s. followed by a 1^- level 13 keV higher in energy, in contrast to the 1^- level proposed earlier at about 300 keV [35]. Spins 2^-, 3^- and 4^- were assigned to the other observed states below 1 MeV and reasonably explained with SM and Kuo-Herling (KH) interaction KH5082 [37]. However, according to some of the calculations (using scaled and unscaled KH208 and KH5082 for $p-n$) in [8] also 1^- g.s. should be considered, although they could describe the data of [35] using CD Bonn (for $n-n$). The long-lived state was attributed to the lowest member of the multiplet (0^- to 7^-), whose wave function was dominated (by 90% or more) with the $\pi g_{7/2}\nu f_{7/2}$ configuration, while the isomeric state was interpreted to the maximum-aligned configuration.

An isomer of the same $p-n$ excitation with suggested $\pi g_{7/2}\nu f^3_{7/2}$ configuration was observed also in ^{136}Sb [38, 39, 40]. A spin/parity (6^-) was suggested due to its appearance as an yrast trap, in contrast to the maximum-aligned configuration trap 7^- as in ^{134}Sb. For ^{136}Sb, in ^{238}U projectile fission at relativistic energies only one isomeric γ-ray of 173 keV was detected [38] with $T_{1/2}$=565(50) ns. In a later work [39], the ^{136}Sb isomer was populated using (thermal) neutron-induced ^{241}Pu fission together with other A=136 isobars and detected using γ and conversion-electron spectroscopy with $T_{1/2}$=480(100) ns. In that case, except the 173.0 keV transition, another low-energy γ-ray of 53.4(3) keV was identified to belong to the isomeric decay. The conversion electron (Si) spectrum in coincidence with 173.0 (Ge) line suggested another candidate, calculated to be a 51.4(5) keV γ transition, which was consistent with their analysis of more than two transitions with $M1-E2$ multipolarity. The unobserved transition was set as the isomeric (followed by the cascade of the other two transitions) in the level scheme, constructed in accordance to SM calculations [32, 39]. In a more recent ^{238}U fission run, however, no sign of this low-energy line was seen, reporting only the two γ rays of 53.9 keV and 173.1 keV and $T_{1/2}$=570(5) ns [40].

The g.s. of ^{136}Sb was suggested to have a spin/parity of 1^- from β-decay data [41], mainly in analogy to ^{212}Bi (with g.s. of 1^-). It was questioned in some later theoretical works, supporting 2^- assignment, as such a scenario allowed them to better describe experimental data in the neighboring ^{134,135}Sb, as well as the N=84 isotones [44, 9]. No spectroscopy results are currently known for the isotopes beyond ^{136}Sb with very few outcomes from first β-decay and g.s. lifetime measurement on ^{137}Sb, performed in [12, 42].

4.2. New experimental observations

In the current studies, for ^{136}Sb, we detected three (mutually-coincident) delayed γ-rays with $T_{1/2}$= 489(40) ns. Two of them, the 53.4 keV and the 172.5 keV transitions, were already observed and a third, unobserved, transition was placed in the level scheme [39]. In our case instead, we identified another low-energy line of 43.4 keV, to which we assigned $M1$ multipolarity using the intensity ratios between all observed transitions and theoretical conversion coefficients from [43]. With our data set, we suggested a different multipolarity for the 53.4 keV line of $E2$ type, while for the 172.5 keV γ-ray the known $E2$ multipolarity seemed consistent with our work. Further experimental details on these new observations are given elsewhere [27]. Taking into account all four studies available on this isomer [27, 38, 39, 40] , the adopted $T_{1/2}$ amounts to 533(40) ns.

We rearranged the level scheme with the new information by placing the observed 53.4 keV as isomeric transition and by changing the energy of the first excited state to 43.4 keV. Comparing our results to SM calculations using the KH interaction [29], we confirmed the earlier suggestions for the isomer spin/parity of (6^-). The level scheme is presented in comparison to ^{134}Sb in Fig.1.

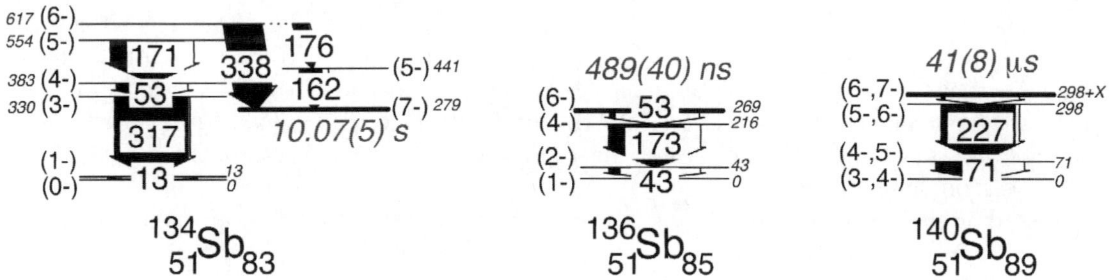

Figure 1. Isomeric level schemes in the neutron-rich Sb isotopes with A=134−140.

Furthermore, with these studies we were able to provide first spectroscopic data on ^{140}Sb [29]. We detected two (mutually-coincident) delayed γ transitions of 70.9(8) and 227.3(5) keV with $T_{1/2}$ = 41(8) μs. Using intensity analysis and conversion coefficients, we assigned their order in the level scheme as shown in Fig.1. However, according to our estimates another low-energy line shall be responsible for the isomeric lifetime, with an energy below our detection threshold. Both $M1$ and $E2$ multipolarity are possible for this transition [29]. In addition, according to our β-decay ^{140}Sb→^{140}Te analysis [45]), for the g.s. of ^{140}Sb both (3^- and 4^-) options are probable. Thus, for the higher-lying states in the level scheme we proposed two respective possibilities, in accordance to the available experimental information. As a consequence of that, for the isomeric state both (6^-) and (7^-) spin/parities are probable.

4.3. Interpretation and discussion

Despite the origin of its excited states, the level scheme of ^{136}Sb is certainly much different in comparison to the one in ^{134}Sb (see Fig.1) due to the extra neutrons. Their isomeric states are also suggested to be with $\pi g^1_{7/2} \otimes \nu f^1_{7/2}$ in ^{134}Sb and $\pi g^1_{7/2} \otimes \nu f^3_{7/2}$ in ^{136}Sb configurations.

As described in [27], we performed SM calculations using ^{132}Sn as a core nucleus and the whole valence space beyond. The yrast schemes for all of the three isotopes are shown in Fig.2 and compared to the experimental data. For these calculations the KH interaction was employed, shown already to be successful in the description of the ^{134}Sb nucleus [37], and the neutron-rich N=82-84 isotones near ^{132}Sn [44]. The results for ^{134}Sb (Fig.2(a)) are compared to the those for ^{136}Sb (Fig.2(b)) and ^{140}Sb (Fig.2(c)). A modification of the KH interaction [29] was employed for the better description of ^{140}Sb (Fig.2(d)). As it can be seen, the yrast spin traps for both ^{134,136}Sb are well reproduced with the original version of the KH interaction, however the ^{140}Sb seemed to be difficult as only traps for the g.s. could be expected with largely compressed levels in variance to the observed by the experiment. In order to tune the modified KH interaction, we used experimental information from the neighboring isotopes [45, 46], including [17]. We may note that the energy compression, especially for the 6^- and 7^- levels, may be further reduced, resulting in an increased spacing with respect to the lower-lying states. This is, however, at the expense of the g.s. information for the neighboring ^{139}Sb and ^{141}Sb nuclei. In this case it was set to the g.s. of ^{139}Sb according to our experimental information [46], in order to constrain the observable. In the modified version of the KH interaction, we inspected yrast (I exited) and yrare (II excited) states. As no spin-traps are expected (see Fig.2(d)), except for the (2^- and 4^-) candidates for the g.s., we suggested the possibility of occupying both yrast and yrare states in the isomeric decay. We note that such a scenario would be consistent with the observed isomeric ratios that we determined for ^{136,140}Sb [27, 29].

Concerning ^{136}Sb, theoretically, the 2^- state can be expected at less than a keV distance from the 1^- state, thus one may clearly have both possibilities for g.s. spin/parity. Although our data [27] would more be consistent with the 1^- assignment due to the order of the transitions and their multipolarity, a firm measurement would be extremely valuable for fixing one of them with a larger certitude. Concerning ^{140}Sb, theoretically, the 2^- state can be expected as a g.s. for this nucleus, closely followed by a 4^- state. Experimentally, such scenario would be in variance with the observed transitions and lifetimes of the populated states [29], therefore the 4^- suggestion would be more appropriate.

According to the calculations, the g.s. and the first excited states in ^{136}Sb are characterized by dominant $\nu f^3_{7/2} \otimes \pi g^1_{7/2}$ component, which contributes to 66% of the wave function, in comparison to about 90% in ^{134}Sb. The $\nu f^{-1}_{7/2} \otimes \pi g^1_{7/2}$ component in ^{140}Sb amounts to about 60% for both the 6^- and 7^- states, according to the calculations. Therefore, one can conclude that the same isomeric structure is kept all over the whole $\nu f_{7/2}$ shell.

In order to search for isomeric candidates in ^{140}Sb we examined in detail the configurations of all excited states. We found, for example, that all of the low-lying states contained the strongly occupied $\pi d_{5/2}$ orbital, while the occupation of $\pi g_{7/2}$ was sizable only for the first 6^- and 7^- states. In contrast, the occupation of the $\pi d_{5/2}$ orbital was expected to be very weak in the lighter Sb isotopes as e.g. in ^{136}Sb. Based on detailed analysis of transition rates and orbital occupations [29], we interpreted the isomeric state to arise most probably due the change of the configuration. It shall result from the crossing of the $\pi d_{5/2}$ and $\pi g_{7/2}$ orbitals at this extreme neutron number and provides valuable information about their evolution in the very neutron-rich Sb isotopes. Our conclusion correlates also with the predictions for crossing between these

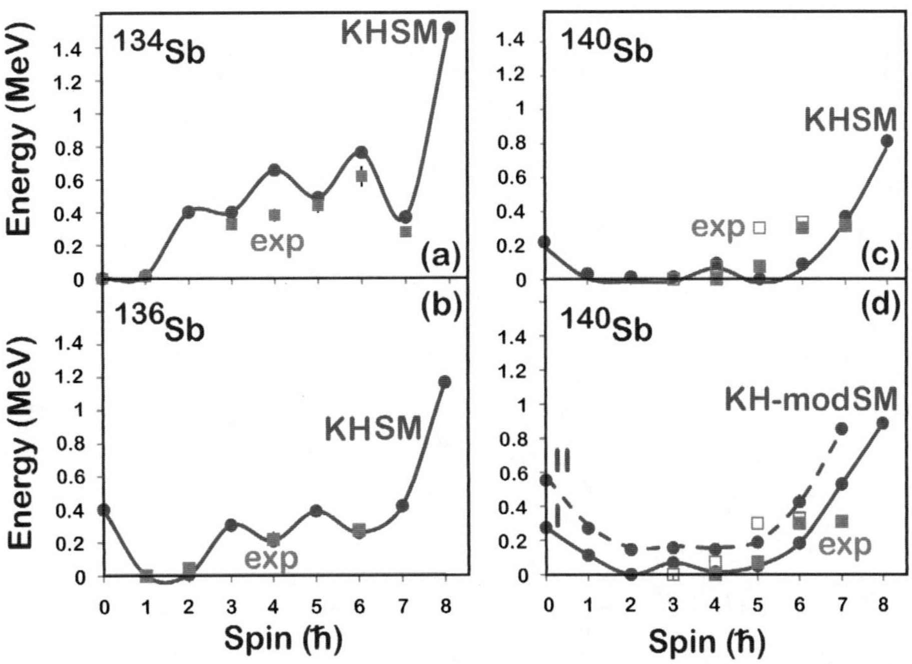

Figure 2. Yrast schemes in ^{134}Sb (a) and ^{136}Sb (b) compared to yrast (I excited) and yrare (II excited) states in ^{140}Sb (c,d) are shown by full and dashed lines, respectively. SM calculations using KH interaction (a,b,c) and modified KH interaction (d) are given for comparison with the experiment. The experimental data for ^{140}Sb, separating the two possible options for spin/parities (see text), are shown with empty and full squares.

π orbitals at N=90 [14], that we experimentally evidence at N=89. Thus, we may conclude, that the large amount of neutrons are strongly influencing the proton shell evolution beyond ^{132}Sn which may effect the development of a neutron skin [12, 47].

5. Summary and Outlook

Among the very neutron-rich isotopes beyond ^{132}Sn which we populated and analyzed (subject to several forthcoming articles), we observed five Sb isotopes, with ^{140}Sb being the most n-rich known up to date. Therefore, the study provides extension of the spectroscopy knowledge at extreme neutron number and probes the evolution of the excited states in these nuclei. The level scheme of the known (6^-) isomer in ^{136}Sb was revised with new experimental information and a new isomeric state was detected in ^{140}Sb with a most probable spin/parity of (6^-). While for ^{136}Sb we confirmed the appearance of an yrast isomer, for ^{138}Sb, we did not detect an isomeric state in the μs lifetime range. The isomer we found again in ^{140}Sb was observed with a much longer μs lifetime, and both yrast and yrare states were suggested to be populated in its decay. Interpretation of the new information was performed in the SM framework using realistic interaction, suggesting that the configuration of the isomeric state in ^{140}Sb is similar to that in the lighter ^{134,136}Sb nuclei. It results from the coupling $\pi g^{1}_{7/2}\nu f^{-1}_{7/2}$ and possibly appears due to the interchanging position of the $\pi d_{5/2}$ and $\pi g_{7/2}$ orbitals at N=89. The results from the coupling of two additional protons to the system is a subject to another forthcoming article, where interestingly we observe another isomeric state at the same N=89.

Acknowledgements

For this work we acknowledge the support staff of the RIBF at RIKEN Nishina Center accelerator complex for providing stable and high-intensity beams to our experiment, the EUROBALL Owners Committee for the loan of the Ge detectors, the PreSpec Collaboration for the readout electronics of the Cluster detectors, the Rare Isotope Science Project, the Ministry of Science, ICT and MSIP and NRF of Korea for part of the WAS3ABi detectors, the Japanese Government fund KAKENHI (No. 25247045), the FR-JP LIA support and the support from Helmholtz association through the Nuclear Astrophysics Virtual Institute, NAVI (No. VH-VI-417).

References

[1] M.M. Sharma et al, *Phys. Rev.* C **61**, 054306 (2000).
[2] NNDC database (http://www.nndc.bnl.gov/). Accessed 2015-09-07.
[3] A. Gniady et al, APS/123-QED (2009) unpublished.
[4] M. Ogawa et al, *Phys. Rev. Lett.* **41**, 289 (1978).
[5] A. Korgul et al, *Eur. Phys. J.* A **32**, 25 (2007).
[6] H. Mach et al, *Act. Phys. Pol.* B **38** Vol. 4, 1213 (2007).
[7] W. Urban et al, *Eur. Phys. J.* A **27**, 257 (2006).
[8] J. Shergur et al, *Phys. Rev.* C **72**, 024305 (2005).
[9] S. Sarkar et al, *Eur. Phys. J.* A **21**, 61 (2004).
[10] L. Coraggio *Phys. Rev.* C **87**, 034309 (2013).
[11] S. Sarkar et al, *Phys. Rev.* C **81**, 064328 (2010).
[12] J. Shergur et al, *Phys. Rev.* C **65**, 034313 (2002).
[13] B.A. Brown et al, *Phys. Rev. Lett.* **58**, 220 (1998).
[14] M.G. Porquet et al, *Eur. Phys. J.* A **25**, 319 (2005).
[15] H. Shimoyama and M. Matsuo, *Phys. Rev.* C **88**, 054308 (2013).
[16] H. Naidja et al, *Journ. Phys. Conf. Ser.* **580**, 012030 (2015).
[17] G. Simpson et al, *Phys. Rev. Lett.* **113**, 132502 (2014).
[18] A. Covello et al, *Journ. Phys. Conf. Ser.* **267**, 012019 (2011).
[19] D. Bianco et al, *Phys. Rev.* C **88**, 024303 (2013).
[20] A. Lindroth et al, *Phys. Rev. Lett.* **82**, 4783, (1999).
[21] T. Ohnishi et al, *J. Phys. Soc. Jpn.* **79**, 073201 (2010).
[22] S. Nishimura et al, *Prog. Theor. Exp. Phys.* **2012**, 03C006 (2012).
[23] P.A. Söderström et al, *Nucl. Instrum. Meth. Phys. Res.* B **317**, 649 (2013).
[24] T. Kubo et al, *Prog. Theor. Exp. Phys.* **2012**, 03C003 (2012).
[25] N. Fukuda et al, *Nucl. Instrum. Meth. Phys. Res.* B **317**, 323 (2013).
[26] R. Lozeva et al, *RIKEN Accel. Prog. Rep.* **47**, 4 (2014).
[27] R. Lozeva et al, *Phys. Rev.* C **92**, 024304 (2015).
[28] P. Lee et al, *Phys. Rev.* C **92**, 044320 (2015).
[29] R. Lozeva et al, *Phys. Rev.* C, in press.
[30] W. Urban et al, *Eur. Phys. J.* A **5**, 239 (1999).
[31] B. Fornal et al, *Phys. Rev.* C **63**, 024322 (2001).
[32] L. Corraggio et al, *Phys. Rev.* C **80**, 021305(R) (2009).
[33] A. Covello et al, *Jour. Phys. G: Conf. Ser.* **205**, 012004 (2010).
[34] G. Rudstam et al, *At. Dat. Nucl. Data Tabl.* **53**, 1 (1993).
[35] B. Fogelberg et al, *Phys. Rev.* C **41**, R1890 (1990).
[36] J. P. Omtvedt et al, *Phys. Rev. Lett.* **75**, 3090 (1995).
[37] A. Korgul et al, *Eur. Phys. J.* A **15**, 181 (2002).
[38] M. Mineva et al, *Eur. Phys. J.* A **11**, 9 (2001).
[39] G. Simpson et al, *Phys. Rev.* C **76**, 041303(R) (2007).
[40] D. Kameda et al, *Phys. Rev.* C **86**, 054319 (2012).
[41] P. Hoff et al, *Phys. Rev.* C **56**, 2865 (1997).
[42] O. Arndt et al, *Phys. Rev.* C **84**, 061307(R) (2011).
[43] T. Kibedi et al, *Nucl. Instrum. Meth. Phys. Res.* A **589**, 202 (2008).
[44] S. Sarkar et al, *J. Phys. IAC* **53** (6), 1 (1999).
[45] P. Lee et al, *Phys. Rev*, in preparation.
[46] R. Lozeva et al, *Phys. Rev*, in preparation.
[47] P. Sarriguren et al, *Phys. Rev.* C **76**, 044322 (2007).

Cluster Structure of ^{9}Be from ^{3}He+^{9}Be Reaction

S.M. Lukyanov[1], M.N. Harakeh[2], M.A. Naumenko[1], Yi Xu[3], W.H. Trzaska[4], V. Burjan[3], V. Kroha[3], J. Mrazek[3], V. Glagolev[3], Š. Piskoř[3], E.I. Voskoboynik[1], S.V. Khlebnikov[5], Yu.E. Penionzhkevich[1,6], N.K. Skobelev[1], Yu.G. Sobolev[1], G.P. Tyurin[3], K. Kuterbekov[7] and Yu. Tuleushev[8]

[1] Flerov Laboratory of Nuclear Reactions, Dubna, Russian Federation
[2] KVI-CART, University of Groningen, Groningen, Netherlands
[3] Nuclear Physics Institute, Řež, Czech Republic
[4] Department of Physics, University of Jyväskylä, Jyväskylä, Finland
[5] Khlopin Institute, St. Petersburg, Russian Federation
[6] National Research Nuclear University "MEPhI", Moscow, Russian Federation
[7] Eurasian Gumilev University, Astana, Kazakhstan
[8] Nuclear Physics Institute, Almaty, Kazakhstan

E-mail: lukyan@nrmail.jinr.ru

Abstract. The study of inelastic scattering and multi-nucleon transfer reactions was performed by bombarding a ^{9}Be target with a ^{3}He beam at an incident energy of 30 MeV. Angular distributions for ^{9}Be(^{3}He,^{3}He)^{9}Be, ^{9}Be(^{3}He,^{4}He)^{8}Be, ^{9}Be(^{3}He,^{7}Be)^{5}He, ^{9}Be(^{3}He,^{6}Li)^{6}Li and ^{9}Be(^{3}He,^{7}Li)^{5}Li reaction channels were measured. Cross sections for channels leading to unbound ^{5}He$_{g.s.}$, ^{5}Li$_{g.s.}$ and ^{8}Be systems were obtained from singles measurements where the relationship between the energy and the scattering angle of the observed stable ejectile is constrained by two-body kinematics. Information on the cluster structure of ^{9}Be was obtained from the transfer channels. It was concluded that cluster transfer is an important mechanism in the investigated nuclear reactions. In the present work an attempt was made to estimate the relative strengths of the interesting ^{8}Be+n and ^{5}He+α cluster configurations in ^{9}Be. The branching ratios have been determined confirming that the ^{5}He+α configuration plays an important role. The configuration of ^{9}Be consisting of two bound helium clusters ^{3}He+^{6}He is significantly suppressed, whereas the two-body configurations ^{8}Be+n and ^{5}He+α including unbound ^{8}Be and ^{5}He are found more probable.

1. Introduction

In recent years the study of light radioactive nuclei [1, 2] has intensified due to the significant progress made with radioactive beam facilities. It has led to a decrease of interest in the study of light stable nuclei such as ^{6,7}Li and ^{9}Be. It has been shown that in light nuclei the nucleons tend to group into clusters, the relative motion of which defines to a large extent the properties of these nuclei. Consequently, the cluster structure of their ground as well as low-lying excited states became the focus of theoretical and experimental studies. For example, ^{6}Li and ^{7}Li nuclei are both well described by the two-body cluster models ($\alpha + d$ and $\alpha + t$, respectively).

Due to its Borromean structure, a special attention has been focused on the ^{9}Be nucleus, which could break up directly to α and a neutron, or via one of two unstable intermediate nuclei: ^{8}Be or ^{5}He. The breakup of ^{9}Be via ^{8}Be$_{g.s.}$ has been measured for many of the low-lying excited states of ^{9}Be [3, 4]. However, the breakup branching via the first-excited 2^+ state of ^{8}Be and via ^{5}He remained uncertain.

In Ref. [5], the structure of ^{9}Be was discussed in the frame of isobaric analogue states of ^{9}B. Charity et al. [5] found that the decay of ^{9}B has the dominant branch α+^{5}Li implying that "the corresponding mirror state in ^{9}Be would be expected to decay through the mirror channel α+^{5}He, instead of through the n+^{8}Be(2^+) channel The mirror state of ^{9}Be at 2.429 MeV is also reported to decay by α emission". In this case it would decay to the unstable ground state of ^{5}He producing the n+2α final state. On the contrary, another experimental work [4] claimed that this state decays almost exclusively by n emission to the unstable first-excited state of ^{8}Be.

The excited states of ^{9}Be have been populated in various ways including among others β-decay [6]. Most experiments confirm that the 2.429 MeV state has a branching ratio to the ^{8}Be$_{g.s.}$+n channel of only ~7% [3], but could not determine whether the remaining strength was in the ^{8}Be(2^+)+n or the ^{5}He+α channel. However, it was reported in Ref. [6] that the ratio 2:1 could be assigned for the two channels, respectively.

The results of Refs. [3, 4, 5] along with the qualitative breakup data discussed above suggest the necessity of obtaining quantitative branching-ratio data for the low-lying states in the Borromean nucleus ^{9}Be.

The present experiment was designed to study the breakup of ^{9}Be in the attempt to determine the contribution of ^{5}He+α and n+^{8}Be channels. Our data are based on inclusive measurements, whereas the experimental results of Refs. [3, 4] were obtained in exclusive measurements. Another purpose of the present study was the attempt to find the cluster structure (for instance, ^{5}He) and how the cluster structure influences the nuclear reaction mechanism.

The angular distributions of the ^{9}Be(^{3}He,^{7}Be)^{5}He and ^{9}Be(^{3}He,^{6}Li)^{6}Li reaction channels were scrupulously measured at the energy of 60 MeV for the transitions to the ground states of ^{5}He and ^{6}Li by Rudchik et al. [8]. The experimental data were analysed using the coupled reaction channels (CRC) model including one- and two-step cluster transfer and cluster spectroscopic amplitudes calculated in the framework of the translation-invariant shell model. It was concluded that the cluster transfer is unimportant: in the ^{9}Be(^{3}He,^{7}Be)^{5}He reaction channel the α-transfer dominates only at small angles while the transfer of two neutrons dominates at large angles. Nevertheless, the conclusion of Ref. [8] in general supports the idea of cluster transfer.

The second motivation for studying again the ^{3}He+^{9}Be reaction was the attempt to find evidence for the cluster structure of these nuclear systems by the analysis of the reaction products.

2. Experimental procedure

The ^{3}He+^{9}Be experiments were performed at the K130 Cyclotron facility of the Accelerator Laboratory of the Physics Department of Jyväskylä University and at the Nuclear Physics Institute (NPI), Řež, Czech Republic. The ^{3}He beam energy was 30 MeV. The average beam current during the experiments was maintained at 10 nA. The self-supporting Be target was prepared from a 99% pure thin foil of beryllium. The target thickness was 12 μm. Peaks due to carbon and oxygen contaminations were not observed in the energy spectra.

To measure (in)elastically scattered ions four Si-Si(Li) telescopes each consisting of ΔE_0, ΔE and E_r detectors with thicknesses 10 μm, 100 μm and 3 mm, respectively, were used. Each telescope was mounted at a distance of ~45 cm from the target. Particle identification was based on the measurements of energy-loss ΔE and residual energy E_r (the so-called ΔE-E method). The telescopes were mounted on rotating supports, which allowed to obtain data from θ_{lab} = 20° to θ_{lab} = 107° in steps of 1-2°. The two-dimensional plots energy loss ΔE vs. residual energy E_r and ΔE_0 vs. ΔE are shown in Fig. 1a and Fig. 1b, respectively. The excellent energy resolutions of both ΔE and E_r detectors allowed unambiguous identification of A and Z of each product. The panel (a) is mainly related to the transfer from the projectile to the target,

Journal of Physics: Conference Series **724** (2016) 012031 doi:10.1088/1742-6596/724/1/012031

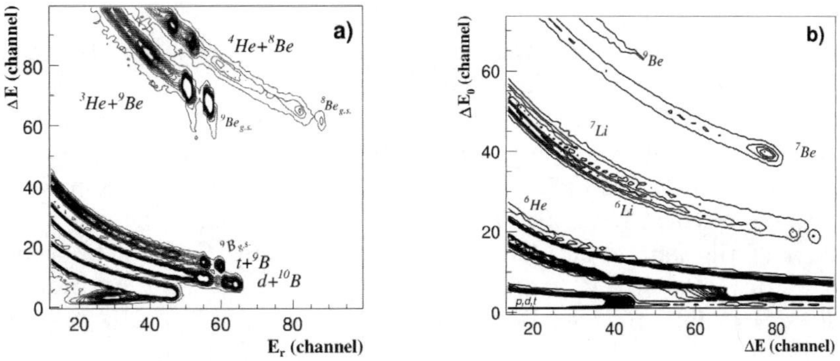

Figure 1. Particle identification plots for the products of the ^{3}He+^{9}Be reaction: a) p, d, t and ^{3,4}He; b) ^{6}He, ^{6,7}Li and ^{7,9}Be. ΔE is the energy loss and E_r is the residual energy. The loci for the products are indicated.

whereas the panel (b) is completely related to the transfer from the target to the projectile.

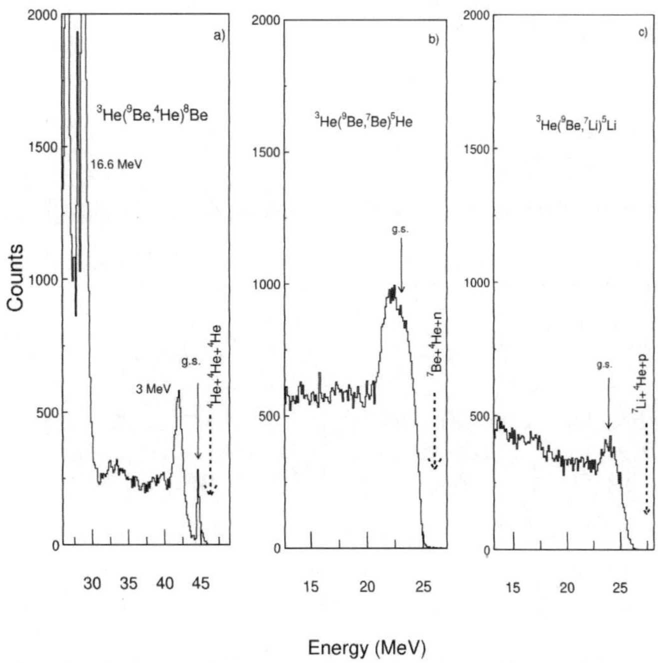

Figure 2. Spectra of the total deposited energy for the detected ^{4}He (a), ^{7}Be (b) and ^{7}Li (b), measured at 18° in the laboratory system for the reaction ^{3}He(30 MeV)+^{9}Be.

Fig. 2 shows experimental spectra of the total deposited energy for the detected ^{4}He, ^{7}Be and ^{7}Li, measured at 18° in the laboratory system for the reaction ^{3}He(30 MeV)+^{9}Be. The plotted total energies were calculated as the sum of the calibrated energy losses ΔEs and the residual energy E_r. The ground and the most populated excited states of ^{8}Be, ^{5}He and ^{5}Li as well as the ground states populated in the reaction channels ^{9}Be(^{3}He,^{4}He)^{8}Be, ^{9}Be(^{3}He,^{7}Be)^{5}He and ^{9}Be(^{3}He,^{7}Li)^{5}Li were unambiguously identified.

Since ^{8}Be, ^{5}He, ^{5}Li and are unbound nuclei and decay into ^{4}He+^{4}He, ^{4}He+n and ^{4}He+p, respectively, we calculated the positions in the measured spectra (Fig. 2) of the maximum energies of ^{4}He, ^{7}Be and ^{7}Li, assuming either two bodies or three bodies in the exit channels. Calculations were performed with the aid of the NRV server [9]. The solid arrows show the maximum values of the kinetic energies of the detected ^{4}He, ^{7}Be and^{7}Li, which correspond in the case of ^{4}He, ^{7}Be and^{7}Li, to the two-body character of the reaction in the exit channels:^{9}Be(^{3}He,^{4}He)^{8}Be, ^{9}Be(^{3}He,^{7}Be)^{5}He and ^{9}Be(^{3}He,^{7}Li)^{5}Li, respectively. The maximum values of the kinetic energies in the case of three-body final states in the reaction channels ^{9}Be(^{3}He,^{4}He+^{4}He)^{4}He, ^{9}Be(^{3}He,^{7}Be)^{4}He+n and ^{9}Be(^{3}He,^{7}Li)^{4}He+p are shown by the dashed arrows. As can be seen from Fig. 2, the local maxima at the g.s. are close to the values corresponding to the two-body exit channels in all studied reaction channels. This indicates that the two-body reaction channels dominate leading to the unbound ^{8}Be, ^{5}He and ^{5}Li in their ground states. Therefore, it may certainly be concluded that one-step multi-nucleon transfer is the main reaction mechanism leading to the binary exit channels.

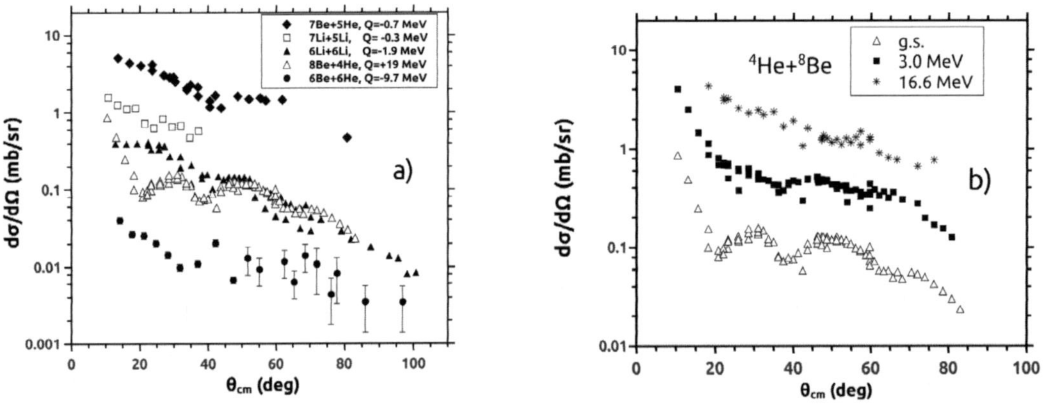

Figure 3. Angular distributions for a) ^{3}He + ^{9}Be → ^{5}He + ^{7}Be (♦), ^{3}He + ^{9}Be → ^{5}Li + ^{7}Li$_{g.s.}$ (□), ^{3}He + ^{9}Be → ^{6}Li + ^{6}Li$_{g.s.}$ (▲), ^{3}He + ^{9}Be → ^{4}He + ^{8}Be$_{g.s.}$ (△), ^{3}He + ^{9}Be → ^{6}He + ^{6}Be$_{g.s.}$ (•); b) ^{3}He + ^{9}Be → ^{4}He + ^{8}Be$_{g.s.}$ (△), 3.0 MeV (■) and 16.6 MeV (∗) excited states.

3. Results

3.1. Elastic and inelastic scattering

The measured angular distributions for the reaction channels ^{3}He + ^{9}Be → ^{5}He + ^{7}Be (♦), ^{3}He + ^{9}Be → ^{5}Li + ^{7}Li$_{g.s.}$ (□), ^{3}He + ^{9}Be → ^{6}Li + ^{6}Li$_{g.s.}$ (▲), ^{3}He + ^{9}Be → ^{4}He + ^{8}Be$_{g.s.}$ (△), ^{3}He + ^{9}Be → ^{6}He + ^{6}Be$_{g.s.}$ (•) are shown in Fig. 3a.

First of all, one may notice that the cross section for the case of ^{5}He+^{7}Be (♦) is much larger than for the other reaction channels shown in Fig. 3.

The ^{7}Be energy spectrum (Fig. 2d) indicates the large probability of transitions to the ground state of the ^{5}He nucleus in conjunction with the 3^- ground state and the $1/2^-$ first-excited (0.429 MeV) state of ^{7}Be (unresolved in the present experiment). Therefore, the angular distribution for the ^{5}He+^{7}Be (♦) channel corresponds to the ground state and the first-excited state of ^{7}Be.

As it was mentioned, this exit channel of the reaction has the largest cross section among the channels shown in Fig. 3a. The calculated Q-values of the investigated channels are given in the inset of Fig. 3a. It may be seen that the behaviour of the cross sections does not strictly follow the famous Q-reaction systematics. For instance, the measured differential cross section for the channel ^{4}He+^{8}Be$_{g.s.}$ with the maximum value of $Q = +19$ MeV is much less than for the channel ^{5}He+^{7}Be ($Q = -0.7$ MeV). The lowest cross section was observed for ^{6}He+^{6}Be$_{g.s.}$ ($Q = -9.7$ MeV).

Fig. 3b shows the angular distributions of the differential cross sections for the ^{3}He(30 MeV)+^{9}Be $\rightarrow$ ^{4}He+^{8}Be reaction populating the *g.s.*, and the 3.0 MeV and around 16.6 MeV excited states. It seems that the high positive Q-value of the reaction (18.9 MeV) for this channel allows to excite very high-lying levels in the unbound ^{8}Be nucleus. It is clearly visible from Fig. 3b that in the reaction ^{3}He+^{9}Be $\rightarrow$ ^{4}He+^{8}Be the cross section for 16.6 MeV level dominates over the other reaction channels.

The experimental data for ^{9}Be+^{3}He elastic scattering were fitted. The obtained OM potential parameters are listed in Table 1 in comparison with the parameters of Ref. [8]. The results of fitting are shown in Figs. 4 and 5 together with the experimental data.

The difference between the obtained parameters and those of Ref. [8] is probably due to the different projectile energies: our data were obtained at 30 MeV and the data of Rudchik et al. [8] were obtained at 60 MeV. For instance, the depths of the real and imaginary parts are smaller at the lower incident energy, whereas the radii are larger than in Ref. [8].

The cross section for the transfer channel ^{9}Be(^{3}He,^{4}He)^{8}Be$_{g.s.}$ was fitted and the OM potential parameters for the exit channel ^{4}He+^{8}Be were obtained. In the fitting process, the OM potential parameters for the entrance channel ^{9}Be+^{3}He obtained in the first step were used.

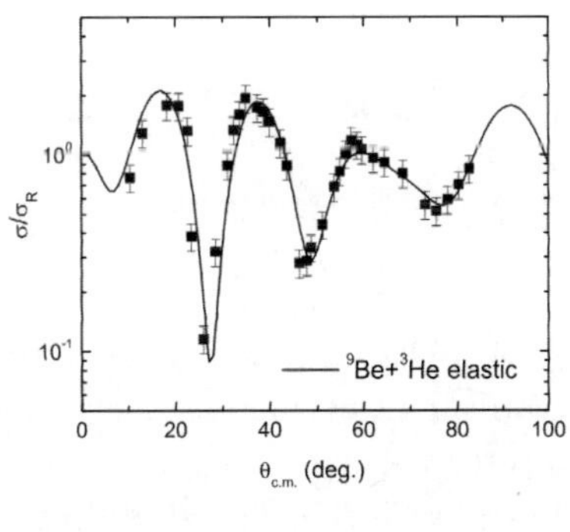

Figure 4. The ratio of the measured elastic scattering cross section to the Rutherford cross section for ^{3}He+^{9}Be at the incident energy 30 MeV in comparison with the OM fit.

The cross section for the transfer reactions ^{9}Be(^{3}He,^{7}Be)^{5}He was fitted simultaneously and the OM potential parameters for the exit channel ^{5}He+^{7}Be were obtained (Table 1). In the fitting process, the OM potential parameters for the entrance channel ^{9}Be+^{3}He obtained in the first step were used. In the calculations of the total cross section both transition amplitudes corresponding to the alpha-transfer mechanism (the ^{9}Be(^{3}He,^{7}Be)^{5}He channel) and the $2n$-transfer mechanism (the ^{9}Be(^{3}He,^{5}He)^{7}Be channel) have been taken into account. All calculations and fitting have been performed using the FRESCO code [10] in the framework of the CC method.

By fitting the experimental angular distributions, the attempt was made to determine which reaction mechanisms are the most important for the measured distributions. The results are shown in Fig. 5 for the following channels: a) ^{3}He + ^{9}Be $\rightarrow$ ^{4}He + ^{8}Be$_{g.s.}$, b) ^{3}He + ^{9}Be $\rightarrow$ ^{5}He + ^{7}Be and c) ^{3}He + ^{9}Be $\rightarrow$ ^{6}Li + ^{6}Li$_{g.s.}$.

Table 1. OM potential parameters used within the optical model and CC approaches for the reaction ^{3}He+^{9}Be in comparison with the parameters of Ref. [8].

reaction channel	V_0 [MeV]	r_v [fm]	a_v [fm]	W_0 [MeV]	r_w [fm]	a_w [fm]	V_{so} [MeV]	r_{so} [fm]	a_{so} [fm]
^{3}He+^{9}Be	108.5	1.123	0.54	15.69	1.15	0.855	13.63	1.83	0.4
^{3}He+^{9}Be [8]	143.4	1.02	1.10	38.3	1.405	1.17			
α+^{8}Be	90.98	1.382	0.404	12.36	1.01	0.4	–	–	–
^{5}He+^{7}Be	60.88	1.25	0.65	2.8	1.25	0.65	8.93	1.28	0.65
^{5}He+^{7}Be [8]	122.5	0.9768	0.61	51.71	0.914	1.178			
^{6}Li+^{6}Li	110.95	1.307	0.621	2.48	1.25	0.65	1.05	1.25	0.65

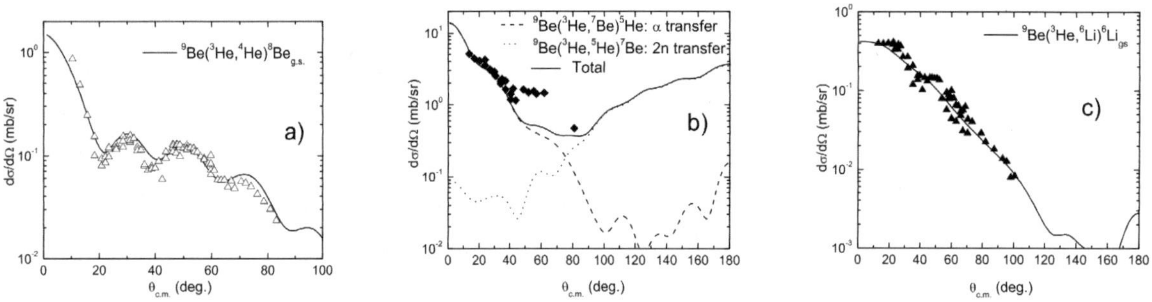

Figure 5. The angular distributions for a) ^{3}He + ^{9}Be $\rightarrow$ ^{4}He + ^{8}Be$_{g.s.}$ ($\triangle$), b) ^{3}He + ^{9}Be $\rightarrow$ ^{5}He + ^{7}Be ($\blacklozenge$) and c) ^{3}He + ^{9}Be $\rightarrow$ ^{6}Li + ^{6}Li$_{g.s.}$ ($\blacktriangle$). The curves are the results of the optical-model and the coupled-channel calculations (see text).

It is clear that in the case of ^{3}He+^{9}Be $\rightarrow$ ^{7}Be+^{5}He reaction channel the α-transfer (long-dashed curve) dominates at forward angles, whereas the transfer of the two neutrons (short-dashed curve) dominates at backward angles.

The fit for ^{3}He+^{9}Be $\rightarrow$ ^{6}Li+^{6}Li (solid curve) was obtained assuming t transfer from the target nucleus to the projectile. A good agreement between the experimental data and the calculations is observed. A similar situation may be expected for the case of ^{3}He+^{9}Be $\rightarrow$ ^{5}Li+^{7}Li where d transfer dominates. All the values of the spectroscopic amplitudes are taken from [8].

As could be seen from Fig. 5, the main processes describing the angular distributions of the exit channels ^{4}He+^{8}Be$_{g.s.}$, ^{5}He+^{7}Be and ^{6}Li+^{6}Li$_{g.s.}$ are, respectively, the neutron, α and t transfer from the target nucleus to the projectile.

3.2. Cluster transfer in ^{3}He+^{9}Be reaction

Due to the Borromean structure of ^{9}Be the breakup of this nucleus may, in principle, proceed directly into two α particles plus a neutron or via one of the unstable intermediate nuclei: ^{8}Be or ^{5}He. The attempt was made to get some insights into n+^{8}Be, ^{4}He+^{5}He, d+^{7}Li and t+^{6}Li cluster configurations inside of ^{9}Be and especially to estimate the relative strengths of the most interesting ^{8}Be+n and ^{5}He+α cluster configurations in ^{9}Be.

Based on the angular distributions shown in Fig. 3, one may estimate the probability of cluster configurations in ^{9}Be. To obtain the partion probabilities for the observed channels, the angular distributions for each excited state (shown in Fig.3) were integrated over the solid angle. The probability was calculated as weighted integrals among the channels listed above. The obtained values of the partion probabilities are given in Table 2. The ratio of ^{8}Be+n to ^{5}He+α is 2.7, which is close to the ratio 2:1 given in Ref. [6].

Table 2. Probability of cluster configurations in the ^{9}Be nucleus

Exit channel	^{9}Be cluster configuration	Partion probabilities (%)
^{4}He + ^{8}Be	n + ^{8}Be	≤68.7 ±10
^{5}He + ^{7}Be	α + ^{5}He	≤25.1 ±5
^{6}Li + ^{6}Li$_{g.s.}$	t + ^{6}Li	≥3.3 ±2
^{5}Li + ^{7}Li$_{g.s.}$	d + ^{7}Li	≥2.7 ±2
^{6}Be + ^{6}He$_{g.s.}$	^{3}He + ^{6}He	≥0.2 ±0.7

Acknowledgments
We would like to thank the JYFL Accelerator Laboratory and NPI (Řež) for giving us the opportunity to perform this study as well as the cyclotron staff of both institutes for the excellent beam quality. This work was supported in part by the Russian Foundation for Basic Research (project numbers: 13-02-00533 and 14-02-91053), the CANAM (IPN ASCR) and by the grants to JINR (Dubna) from the Czech Republic, the Republic of Poland and the mobility grant from the Academy of Finland.

References
[1] Yu E Penionzhkevich, S M Lukyanov, R A Astabatyan, N A Demekhina, M P Ivanov, R Kalpakchieva, A A Kulko, E R Markaryan, V A Maslov, Yu A Muzychka, R V Revenko, N K Skobelev, V I Smirnov, and Yu G Sobolev, *J. Phys. G* **36**, 025104 (2009).
[2] N K Skobelev, N A Demekhina, R Kalpakchieva, A A Kulko, S M Lukyanov, Yu A Muzychka, Yu E Penionzhkevich, and T V Chuvilskaya, *Phys. Part. Nucl. Lett.* **6**, 208 (2009).
[3] T A D Brown, P Papka, B R Fulton, D L Watson, S P Fox, D Groombridge, M Freer, N M Clarke, N I Ashwood, N Curtis, V Ziman, P McEwan, S Ahmed, W N Catford, D Mahboub, C N Timis, T D Baldwin, D C Weisser, *Phys. Rev. C* **76**, 054605 (2007).
[4] P Papka, T A D Brown, B R Fulton, D L Watson, S P Fox, D Groombridge, M Freer, N M Clarke, N I Ashwood, N Curtis, V Ziman, P McEwan, S Ahmed, W N Catford, D Mahboub, C N Timis, T D Baldwin, D C Weisser, *Phys. Rev. C* **75**, 045803 (2007).
[5] R J Charity, T D Wiser, K Mercurio, R Shane, L G Sobotka, A H Wuosmaa, A Banu, L Trache, and R E Tribble, *Phys. Rev. C* **80**, 024306 (2009).
[6] G Nyman, R E Azuma, P G Hansen, B Jonson, P O Larsson, S Mattsson, A Richter, K Riisager, O Tengblad, K Wilhelmsen, and the ISOLDE Collaboration *et al.*, *Nucl. Phys. A* **510**, 189 (1990).
[7] S M Lukyanov, A S Denikin, E I Voskoboynik, S V Khlebnikov, M N Harakeh, V A Maslov, Yu E Penionzhkevich, Yu G Sobolev, W H Trzaska, G P Tyurin and K A Kuterbekov, *J. Phys. G* **41**, 035103 (2014).
[8] A T Rudchik, E I Koshhchy, A Budzanovsky, R Siudak, A Szczurek, I Skwirczynska, Yu Mashkarov, L Glowalska, J Turkiewicz, I Zalyubovsky, V Ziman, N Burtebaev, A Duyselbaev, V Adodin, *Nucl. Phys. A* **609**, 147 (1996).
[9] V Zagrebaev, A Denikin and A Alekseev, DWBA for nucleon transfer reactions http://nrv.jinr.ru/nrv/webnrv/transfer/ (2009), Nuclear Reaction Video Project.
[10] I Thompson, http://www.fresco.org.uk/, http://www.ianthompson.org/surrey/.
[11] F C Young, and A R Knudson, *Nucl. Phys. A* **184**, 563 (1972).

Nuclear charge radii as signature for structural changes

I Angeli[1] and K Marinova[2]

[1]Institute of Experimental Physics, University of Debrecen, H-4010 Debrecen Pf. 105, Hungary
[2]Joint Institute for Nuclear Research, 141980 Dubna, Moscow Region, Russia

E-mail: marinova@jinr.ru

Abstract. The correlation of nuclear charge radii with other ground and excited state nuclear observables is considered. An empirical approach is used to deal with a large amount of experimental information, which is properly handled to obtain interesting correlations among different observables as one moves away from the line of stability. Especially the appearance of new magic numbers and/or disappearance of traditional ones as well as the onset of deformation in the region of light nuclei ($A < 30$) are discussed.

PACS: 21.10.Ft; 21.10.-k; 29.87+g; 29.85.Fj

1. Introduction

Recently, a large table of experimental charge radii R that contains more than 900 isotopes has been published [1]. It is obtained by applying appropriately chosen statistical methods for analysis of two types of experimental data: *i)* radii changes determined from optical and K_αX-ray isotope shifts and *ii)* absolute radii $R = \langle r^2\rangle^{1/2}$ measured by muonic spectra and electronic scattering experiments. The optical isotope shifts are the main source of information on mean square radii changes $\delta\langle r^2\rangle$. Due to modern laser spectroscopic methods with their high precision and sensitivity, isotope shifts away from the line of stability are accessible. This determines the extension of the radii tabulation towards the neutron numbers far from stability.

The table of experimental charge radii $R = R(N,Z)$ can be considered from different aspects. Those include the influence of the magic nucleon numbers, of deformation and of the odd-even staggering [2]. Also a comparison of the radii trend with other nuclear observables becomes accessible. This is possible not only in isotopic but also in isotonic dependences of R over the whole nuclide chart. Remarkable correlations between nuclear charge radii and other nuclear observables have been recognized in their isotopic evolution [3] for $A > 100$ where sufficient data exist to make meaningful conclusion. The authors proposed this as a simple approach that can be exploited to predict trends of behaviour in more exotic systems. In [4], our knowledge is extended over the broad region of nuclear chart by investigating the correlations of R and other nuclear observables that elucidate nuclear structure and its evolution with Z.

When necessary also the evolution with N was additionally considered. Special attention is paid to the region of light nuclei where structural variations are better seen. The abrupt changes in the behaviour of charge radii have been compared to similar abrupt behaviour of other nuclear observables.

This paper presents brief highlights of new single Z or double (Z, N) magic numbers in the region of light nuclei determined in [4]. The same empirical approach is used as well to show other nuclei in this region for which the structural changes – subshell closure and onset of deformation - are reflected in a correlated behaviour of the discussed nuclear observables.

2. Observables discussed

We will search for the correlations between the nuclear charge radii and the following nuclear observables (see [3, 4]): the energies, $E_1(2^+)$, of the first excited 2^{+1} state; the ratios of the energies of the first 4^{+1} and 2^{+1} states - $R_{4/2} = E_1(4^+)/E_1(2^+)$; the reduced transition probabilities $B(E2)\uparrow = B(E2; 2^{+1} \rightarrow 0^{+1})$ and—in some cases—the effective two nucleon shell gap Δ_p (Δ_n) expressed as the difference of the two nucleon separation energy (see [4] and [5]). All these exhibit familiar behaviour at closed shells and at the onset of deformation. The excitation energy $E_1(2^+)$ sensitively reflects the structure of even-even nuclei, peaks at magic numbers, drops at the onset of deformation and thus leads to clear patterns over the whole nuclide chart. The ratio $R_{4/2}$ is an indicator of nuclear structure (although not its unambiguous determinant) that takes values < 2 for nuclei in the closed vicinity of magic numbers, $R_{4/2} \sim 2.0$–2.4 in vibrational nuclei and $R_{4/2} \sim 3.33$ in well-deformed rotors. The $B(E2$: $2^{+1} \rightarrow 0^{+1}$) values have a similar behaviour as $R_{4/2}$. The effective two-neutron and two-proton shell gap goes toward maximum at shell closures and vanishes in deformed nuclei. In the case of light nuclei, a peak of Δ_p (Δ_n) for $Z = N$ nuclei is dominant due to the Wigner effect that makes the $Z = N$ nuclei more tightly bound than their neighbours. The evolution of radii with neutron or proton numbers is well known and discussed in a large amount of papers. At magic nucleon numbers, $R(N)$ and $R(Z)$-curves show characteristic kinks: the slope of radii curve goes to saturation before and increases steep after a magic number. The shape transition regions are characterized with a sharp rise of R. In each case, sources of empirical information are the most recent databases (see references [1, 6-7].

The examples considered in [3, 4] and below (section 3.2) probe a variety of structural features that present a unified view of the structural evolution and, that can be applied to the study of exotic nuclei.

3. Nuclei with $Z \leq 16$

3.1. Summary of earlier empirical results

The detailed analysis performed in [3, 4] shows an excellent agreement among the evolution of nuclear charge radii, $E_1(2^+)$, $R_{4/2}$, $B(E2)\uparrow$ and Δ_p (Δ_n) for the major proton and neutron magic numbers 50 and 82. This gives an unambiguous experimental confirmation that charge radii can be considered as an additional signature for shell closures in the case of nuclei with $A > 100$. The approach has been exploited also in the regions of lower proton magic numbers 28 and 20 [4]. For some light nuclei around (Z, N) = 20, 28 and below an abnormal behaviour was detected. It is shown that for low Z the similarity with the corresponding pictures around the traditional proton magic numbers 50 and 82 is too poor. It turns out that the abrupt changes of the magic number signatures which occurs in heavy nuclei, are either absent in light nuclei or somewhat displaced from traditional magic numbers. This is due to two reasons: *i*) the lack of sufficient information in the region of light nuclei and *ii*) the appearance of exotic properties.

The comparison of the (Z, N)-radii evolution with the other shell closure signatures gives an evidence that also for light nuclei R can be considered as a reliable indicator for structural changes. Unambiguous experimental confirmation is obtained [4] that $Z = 6$ (8) and 14 (16) are magic-like for several N values: $Z = 6$ is magic for $N = 8$; $Z = 8$ is magic not only for $N = 8$ but also for $N = 6$; $Z = 14$ is magic for $N = 16$, Z

= 14 and may be $Z = 16$ are magic for $N = 20$ [4]. Strong evidences are presented that ^{34}Si and the mirror nuclei ^{14}C–^{14}O are new double magic-like nuclei. The predicted double-magicity of ^{34}S is also discussed but no unambiguous evidence is found.

Special attention is paid to the mirror nuclei ^{30}Si and ^{30}S. Those are very interesting due to the conflict between the theoretical predictions of [9, 10] and empirical rule of the so called "j-j" coupling [11]. The latter claims a double magicity of both nuclei. According to our previous work [4] the double magicity of ^{30}Si and ^{30}S appears to be as yet questionable contrary to the statement of [11]. This may be a consequence of the insertion of additional observable to the usually considered $E_1(2^+)$ and $B(E2)\uparrow$ and/or the use of more recent experimental data.

3.2. Light nuclei around Z = 10

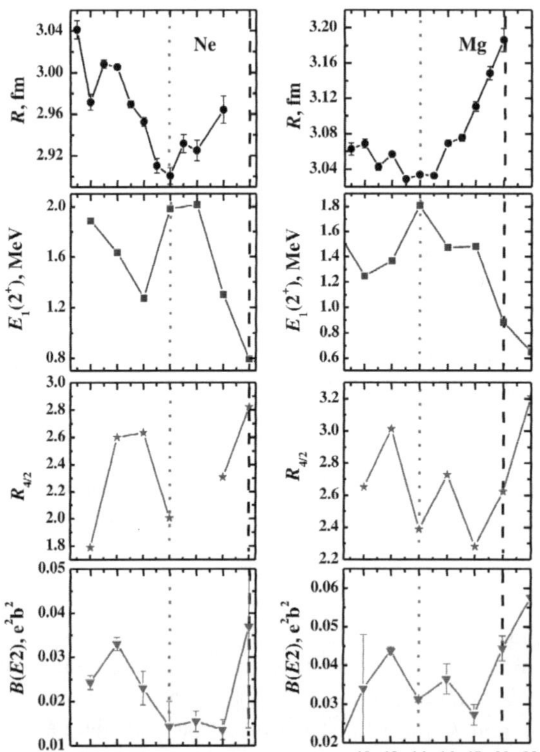

Figure 1. Isotopic sequences of Ne and Mg: comparison of charge radii development with $E_1(2^+)$, $R_{4/2}$ and $B(E2)\uparrow$. Except charge radii development all curves refer to even-even nuclei. Error bars are shown only if larger than the symbols. The dashed vertical lines indicate the traditional magic number $N = 20$; the short dashed lines mark the anomalies at $N = 14$. These presentation illustrates the breakdown of $N = 20$ shell gap for Ne and Mg and the appearance of structural changes around $N = 14$ or $N = 16$.

Usually, rapid nuclear structure changes appear in localized regions of the nuclear chart. Most of the theoretical predictions and experimental data refer to light nuclei from neutron drip line [15]. Concerning the nuclei around $Z = 10$ different theoretical predictions exist: e.g. according to [16] $N = 14$ is magic for $Z = 7$-10, while [17] predicted magicity of $N = 16$ for $Z = 10$-18. We test the situation from an experimental point of view using the correlation between nuclear charge radii and the observables pointed above. The charge radii evolution in the Ne [12] , Na [13], and Mg[14] isotopic chains, spanning the neutron v*sd* shell and beyond, provide excellent examples for considering different types of abrupt changes in nuclear structure.

In the vicinity of the traditional neutron shell closure $N = 20$ the radii curves do not show the expected saturation but increase steeply signaling the weakening of the $N = 20$ magic number. This is confirmed by the evolutions of the other observables that are as well typical for an onset of deformation: an abnormal low energy of the 2^{+1} state, much larger $B(E2)\uparrow$ and $R_{4/2}$ values than that expected from a shell closure. The decreased value of the effective two neutron shell gap toward $N = 20$ ($\Delta_n = 2.2$ MeV in Ne and $\Delta_n =$ 1.4 MeV in Mg) is also in favour of the disappearance of the traditional neutron shell closure. As explained in [18, 19], this is caused by a deformed intruder configuration formed by promoting two neutrons from the *sd* shell across the $N = 20$ shell gap into the *fp* shell.

The development of the Ne and Mg radii, as well as of Na (see [13]), around a minimum at $N = 14$ (figure 1) holds the characteristic features of a subshell closure. For Ne the trends of radii and $B(E2)\uparrow$ are in agreement, showing a minimum at $N = 14$; $R_{4/2}$ is not informative due to the lack of data at $N = 16$ [6]. The local "plateau" between $N = 14$ and $N = 16$ of a strongly enhanced $E_1(2^+)$ correlates with the "plateau" of increased Δ_n values ($\Delta_n \sim 4.4$ MeV) suggesting a competition between $N = 14$ (closed 1v$d_{5/2}$-subshell) and $N = 16$ (closed 2v$s_{1/2}$ subshell) as candidates for magic-like neutron number for $Z = 10$.

In the case of Mg, the agreement between the behaviour of all observables presented in figure 1 proposes more definitely the persistence of shell closure at $N = 14$: the minimum of charge radii at $N = 14$ (if one neglect the odd-even staggering) agrees with the local maximum of $E_1(2^+)$ and the local minima of $R_{4/2}$ and $B(E2)\uparrow$. However, effective two neutron shell gap exhibits a well-defined maximum of 4.9 MeV at $N = 16$ instead at $N = 14$.

4. Outlook

The examples presented in [4] and above give an experimental evidence for new magic-like numbers, for disappearance of some traditional magic numbers and for unexpected onset of deformation. The interpretation concerning $E_1(2^+)$ and $B(E2)\uparrow$ is not in itself new as namely these observables are considered in most papers. What we try to show is that sharp changes in nuclear charge radii can be an additional signature for sub-shell effects. The main focus is on light nuclei, where effects related to the shell or subshell changes are studied. Signatures of shell closure, as well as abnormal behaviour or deviations of expected one are explored and, as a result 1) new candidates for magic-like numbers are presented and 2) disappearance of other traditional magic numbers is predicted. The results confirm that a proper arrangement of raw empirical information can lead to the finding of hidden properties that show up once correlations among various observables are established. Such investigations provide important insight into the shell evolution and pose a stringent benchmark for theory.

Although the correlations point out to the existence (disappearance) of new (traditional) magic-like numbers, in some cases adding new observables to the usually discussed $E_1(2^+)$ and $B(E2)\uparrow$ leads to discrepancies between different empirical data (see [4] and section 3.2). This happens even though some of the observables are related, e.g. $E_1(2^+)$ and $B(E2)\uparrow$ (see e.g. figure 4b an figure 5a of Ref. [4]) as well as R and $B(E2)\uparrow$ through the quadrupole deformation. The reason is simply that different observables may differ widely in sensitivity to the specific physics, thus remaining in many cases open questions. For example, it is important to be cleared whether only the sudden change of $E(2^+)$ toward a maximum can be considered as a signature for shell closure or it serves only as a benchmark for structural changes.

What would happen when the number of observables considered increases, e.g. may their growth switch the aspect considered over other equally important features of the nucleus?

The problem is widely discussed in the literature (see e.g. [20, 21], but no unambiguous answer is found. Though, it seems that the sudden change of a single observable does not give in itself a sufficient reason to draw convincing conclusion without further experimental evidence or proper theoretical background. This emphasizes the importance of the empirical approach applied. It lies primarily in new regions of exotic nuclei far from stability where the experimental information is very poor, usually limited to mass and energy level data. Examining the correlation among those signatures, on which information is available, it would be possible to draw implications on other observables. But this also presents a challenge because of the minimal data that will often be available, especially at the extremes of accessibility. "Therefore, just as it is essential to construct the facilities and instruments to access exotic nuclei, it is also critical to develop new approaches to extract the most physics from minimal information" [20].

References

[1] Angeli I and Marinova K 2013 *Atom. Data Nucl. Data Tables* **99** 69
[2] Angeli I, Gangrsky Yu P, Marinova K *et al* 2009 *J. Phys. G: Nucl. Part. Phys.* **36** 085102
[3] Cakirli R B, Casten R F and Blaum K 2010 *Phys. Rev.* C **82** 061306(R)
[4] Angeli I and Marinova K 2015 *J. Phys. G: Nucl. Part. Phys.* **42**, 055108
[5] Lunny D, Pearson J M and Thibault C 2003 *Rev. Mod. Phys.* **75** 1021-1082
[6] Nuclear Levels and Gamma Search (http://www.nndc.bnl.gov/nudat2/indx_adopted.jsp), version 07.2015
[7] Pritychenko B, Birch M, Singh B and Horoi M arXiv:1312.5975v3 [nucl-th] 12 Sep 2014
[8] Audi G, Kondev F G, Wang M *et al* 2012 *Chin. Phys.* C **36** 1157 http://amdc.in2p3.fr/bulletins/filel.html , version 2013
[9] Otsuka T, Fujimoto R, Utsuno Y *et al* 2001 *Phys. Rev. Lett.* **87** 082502
[10] Brown B A 2001 *Prog. Part. Nucl. Phys.* **47** (2001)517
[11] Boboshin I N, Varlamov V V, Ishkhanov B S and Romanovskii A E 2007 *Bull. Russ. Acad. Sci.: Phys.* **71** 325
[12] Marinova K, Geithner W, Kowalska M *et al* 2011 *Phys. Rev.* C **84** 034313.
[13] Huber G, Touchard F , Buettgenbach S *et al* 1978 *Phys. Rev.* C **18** 2342.
[14] Yordanov D T, Bissell M L, Blaum K *et al* 2012 *Phys. Rev. Lett.* **108** 042504
[15] Jonson B 2004 *Phys. Rep.* **389** 1
[16] Samanta C and Adhikari A 2002 *Phys. Rev.* C **65** 037301
[17] Obertelli A *et al.* 2005 *Phys. Rev.* C **71** 024304
[18] Wildenthal B H and Chung W1980 *Phys. Rev.* C **22** 2260
[19] Warburton E K, Becker J A and Brown B A 1990, *Phys. Rev.* C **41** 1447
[20] Alexandra Gade 2014 *Nuclear Physics News* **23** 10-16
[21] Casten R F 2015 *J. Phys. G: Nucl. Part. Phys.* **42** 034029

STUDY OF EXOTIC WEAKLY BOUND NUCLEI USING MAGNETIC ANALYZER MAVR

V.A.Maslov[1,2], V.I.Kazacha[1], I.V.Kolesov[1], S.M.Lukyanov[1], V.N.Melnikov[1], N.F.Osipov[1], Yu.E.Penionzhkevich[1,3], N.K.Skobelev[1], Yu.G.Sobolev[1,2], E.I.Voskoboinik[1]
[1]Joint Institute for Nuclear Research, Dubna, Moscow region, Russia
[2]NRC Kurchatov Institute, Moscow , Russia
[3]Federal Research Nuclear University "MEPhI", Moscow, Russia

E-mail: maslov_vova@mail.ru

Abstract. A project of the high-resolution magnetic analyzer MAVR is proposed. The analyzer will comprise new magnetic optical and detecting systems for separation and identification of reaction products in a wide range of masses (5–150) and charges (1–60). The magnetic optical system consists of the MSP–144 magnet and a doublet of quadrupole lenses. This will allow the solid angle of the spectrometer to be increased by an order of magnitude up to 30 msr. The magnetic analyzer will have a high momentum resolution (10^{-4}) and high focal-plane dispersion (1.9 m). It will allow products of nuclear reactions at energies up to 30 MeV/nucleon to be detected with the charge resolution ~1/60. Implementation of the project is divided into two stages: conversion of the magnetic analyzer proper and construction of the nuclear reaction products identification system. The MULTI detecting system is being developed for the MAVR magnetic analyzer to allow detection of nuclear reaction products and their identification by charge Q, atomic number Z, and mass A with a high absolute accuracy. The identification will be performed by measuring the energy loss (ΔE), time of flight (TOF), and total kinetic energy (TKE) of reaction products. The particle trajectories in the analyzer will also be determined using the drift chamber developed jointly with GANIL. The MAVR analyzer will operate in both primary beams of heavy ions and beams of radioactive nuclei produced by the U400 - U400M acceleration complex. It will also be used for measuring energy spectra of nuclear reaction products and as an energy monochromator.

1. Introduction

One of the scientific research fields at the Flerov Laboratory of Nuclear Reactions (FLNR), JINR, is studying the mechanism of nuclear reactions in beams of accelerated stable and radioactive nuclei. Another research direction is the study of properties of nuclei near the nucleon drip line. Experiments are carried out at the U400 and U400M cyclotrons of FLNR and in cooperation with the research centers GANIL (France), RIKEN (Japan), and Cyclotron Laboratory (Juväskylä, Finland). The experimental procedure involves magnetic spectrometers and separators that provide high purity separation of reaction products from the primary beam and high resolution in momentum and accordingly in energy and mass.

At FLNR the stability and structure of very neutron-rich nuclei of the lightest elements ($^{7\text{-}10}$He, $^{9\text{-}11}$Li, $^{12\text{-}14}$Be) were investigated using the MSP–144 magnetic analyzer, the structure of heavy isotopes of these nuclei was investigated using the missing mass method, and their binding energy and stability were determined. The data on the energy levels in the nuclei ^{7,8,9}He, ^{10,11}Li, and ^{13}Be were obtained for the first time. The experiments were simultaneously conducted in Dubna and Berlin using intense beams of ^{11}B, ^{13}C, and ^{14}C heavy ions and the MSP–144 and Q3D precision magnetic spectrometers [1, 2]. In one of the U400 channels a beam energy monochromatization system was installed. In combination with the MSP–144 magnetic analyzer it allowed producing beams with a high energy resolution (no worse than 250 keV). When it became possible to obtain relatively intense beams of radioactive nuclei at the U400M cyclotron of FLNR JINR, a system was developed for separating nuclear reaction products resulting from fragmentation of the 50-MeV/nucleon primary beam on the production target (QD spectrometer). This magnetic system allowed secondary beams of ^{6}He and ^{9}Li nuclei with an intensity of 10^4 s^{-1} to be produced and then used its in the first experiments for measuring angular distributions of the ^{6}He and ^{9}Li ions elastically scattered from various target nuclei. Data on interaction radii of these nuclei were obtained for the first time [3, 4]. Later experiments on measurement of excitation functions for complete fusion reactions with subsequent evaporation of neutrons and fission in a wide energy range (10–70 MeV) were carried out with ^{6}He beams of the DRIBs acceleration complex using the MSP–144 spectrometer. These experiments were the first to yield results indicating to a possibility of deep sub-barrier fusion of ^{6}He (neutron halo nucleus) with other heavy nuclei [5, 6]. These results stimulated further experimental and theoretical investigations of sub-barrier reactions with loosely bound nuclei in many research centers and are still a topic for discussion at practically all international conferences on physics of exotic nuclei. A series of experiments was carried out at the QD spectrometer to study energy dependences of total reaction cross sections for interaction of ^{4}He, ^{6}He, ^{6,7}Li, and ^{9}Li nuclei with Si nuclei [7]. A unique procedure based on the use of an active Si target, which is a multilayer semiconductor telescope of thin Si detectors, allowed one-by-one detection and identification of nuclear reaction products and determination of their total interaction cross section as a function of energy. It was an important step toward understanding the mechanism for interaction of loosely bound nuclei in a wide energy range.

Startup of the new DRIBs complex for acceleration of radioactive beams opened up new possibilities for those investigations [8]. Reactions with ^{6}H nuclei near the Coulomb barrier energy were further investigated at a new experimental level (relatively high intensity of radioactive beams, good momentum resolution of the MSP–144. The intensity of the ^{6}He beam on the physical target was 3–$5{\cdot}10^7$ s^{-1}, which allowed the previous experimental sensitivity to be improved by several orders of magnitude. In addition, monochromatization of the secondary beam energy by the magnetic spectrometer made it possible to measure excitation functions for fusion and neutron transfer reactions with ^{6}He nuclei with an accuracy no worse than 300 keV. The supporting evidence was obtained for the earlier observed enhancement of the cross section for complete fusion of ^{6}He nuclei in the sub-barrier energy region [9]. Measurements of the cross section for fusion with ^{4}He nuclei with the formation of the same compound nucleus as in the case of fusion with the ^{6}He nucleus revealed a great difference between the interaction mechanisms of these two neighboring nuclei. Interesting results were obtained for transfer reactions involving loosely bound nuclei (single- and two-neutron transfers from the ^{6}He nucleus [10] and deuteron transfer from the ^{6}Li nucleus [11]). In both cases the maxima of the transfer reaction cross sections corresponded to the energy equal to the Coulomb interaction barrier and were as high as one barn. These results are of great interest to nuclear physics and astrophysics. It is worth mentioning that all those experiments were conducted in close collaboration with physicists from the JINR Member States: NPI ASCR (Rez, Czech Republic), IFIN–HH (Bucharest, Romania), NINP PAS (Krakow, Poland), and others.

2. Proposed MAVR technical characteristics

The investigations described above require high Z and A selectivity of reaction products, a large solid angle of reaction product detection, a high energy resolution, and a high purification factor in separation from the primary beam. In addition, it is necessary to segregate events from the background, which is 10^{10} to 10^{12} times higher, with an energy resolution of up to 10^{-3}, i.e., it is needed to combine the high energy resolution with a high degree of purification from the primary beam. The momentum resolving power $\Delta p/p \sim 10^{-4}$ obtained by magnetic analysis allows one to have the energy resolution about 100 to 200 keV for the bombarding particles like ^{12}C, ^{16}O, ^{20}Ne at energies up to 30 MeV/A. In view of this, the experimental setup should meet the following requirements:

1. The solid angle of the setup should be no smaller than 30 msr to allow carrying out highly efficient correlation experiments.
2. It should be possible to detect nuclear reaction products, including neutrons, in a wide range of angles, among them the angle $\Theta_{lab} = 0^{\circ}$.
3. It should be possible to compensate for the kinematical spread of nuclear reaction products at a large solid angle.

Various detector techniques in combination with a high-resolution magnetic analyzer MAVR are supposed to be used for ensuring better efficiency of heavy ion beams and beams of radioactive nuclei from the DRIBs acceleration complex. The proposed magnetic analyzer will have a large solid angle for detection of reaction products (up to 30 msr), a high momentum resolution (10^{-4}), and a high dispersion in the focal plane (1.9 m). This analyzer will allow products of nuclear reactions at energies up to 30 MeV/nucleon to be detected with a high charge resolution (~1/60), which is especially important for separation of heavy nuclear-reaction products. It is planned to construct the analyzer in two stages.

At the first stage the analyzer will be equipped with a special beam collimation system, instrument transducers for precision measurement of the magnetic field (NMR teslameter), profilometers based on multiwire proportional chambers, and a time-of-flight measurement system. A new detector module based on drift chambers and VME electronics will be made and tested.

At the second stage the MAVR analyzer will be mounted in the new U400R hall. The layout of the spectrometer in the U400R experimental hall, now under construction, is shown in Fig. 3. The MAVR construction is supposed to involve the following activities.

2.1. Development of the magnetic optical system for the MAVR analyzer

The MSP–144 magnetic spectrometer is intended for analyzing and detecting products of nuclear reactions in beams of accelerated ions with the maximum magnetic rigidity up to 1.45 Tm. This broad-range stepped-poles magnetic spectrometer was developed for analysis of light products of nuclear reactions with protons. High ion-optical parameters of the spectrometer allow its electromagnet to be used for analyzing heavy nuclear-reaction products with masses A = 40–100. The magnet consists of two regions. The magnetic field in the first region is 1.46 to 1.5 times lower than in the second region. This magnet is characterized by a wide energy range of detected reaction products ($E_{min}/E_{max} = 5.2$), high dispersion-to-enlargement ratio $D_d/M_x \sim 6.3$ m, and relatively large solid angle Ω up to 5 msr. It has a high resolving power $R_d = 2000$ and a high dispersion $D_d = 1.9$ m. This spectrometer is second to those currently available in other research centers only in the solid angle.

2.2. Increasing the solid angle of the magnetic analyzer

The solid angle of the reaction product capture can be increased to 30 msr by installing two quadrupole lenses and moving the target from its current position as far as S = 1.7 m. For this purpose it is necessary to manufacture two quadrupole lenses with magnetic field gradients −1018 Gs/cm and +895 Gs/cm. A doublet of quadrupoles in front of the analyzer will decrease the distance S between the entrance to the magnet and the new position of the target as compared with a single quadrupole [12] and increase the horizontal acceptance of the analyzer by a factor of 3. The vertical size for reaction product capture in front of the entrance to the analyzer will decrease to the size of its vertical aperture, and the vertical acceptance of the separator y_0 will increase by a factor 6. The dispersion in the focal plane of the analyzer will practically stay unchanged whereas the resolving power will decrease by about a factor of 3.

The calculations [12] show that the most effective way to increase the solid angle of the existing spectrometer is to use a doublet of quadrupoles. Given the calculated geometry of the quadrupoles and the field gradients $G_1 = -1018$ Gs/cm and $G_2 = +895$ Gs/cm, the product of the angular acceptances of the separator (solid angle) will be about 30 msr. This in turn will require manufacturing a new receiving chamber for the focal detector in view of a change in the position of the focal plane.

Table 1. Characteristics of magnetic analyzers

	D (cm/%)	Δp/p (%)	Br (Tm)	dΩ (msr)	Energy resolution dE/E	TOF resolution dT
MAVR	1.9	10	1.5	30	$5*10^{-4}$	
VAMOS	2.5	10	2.3	80	$5*10^{-4}$	
SPEG	7	7	3.2	4.9	$5*10^{-4}$	
PRISMA	4.0	10	~2.0	80	$1*10^{-3}$	
Q3D(HMI)	9.5	10	1.7	10	$2*10^{-4}$	<1 ns

Table 1 presents the expected characteristics of the MAVR analyzer in comparison with the spectrometers in other research centers.

2.3. Detecting system

The MAVR magnetic analyzer is supposed to accommodate a specially developed MULTI detecting system capable of detecting and identifying nuclear reaction products by charge Q, atomic number Z, and mass A with a high absolute accuracy.

Figure 1 shows a model of the MAVR magnetic analyzer with a new reaction chamber and a doublet of lenses. The MAVR can rotate through an angle up to 112° relative to the beam axis.

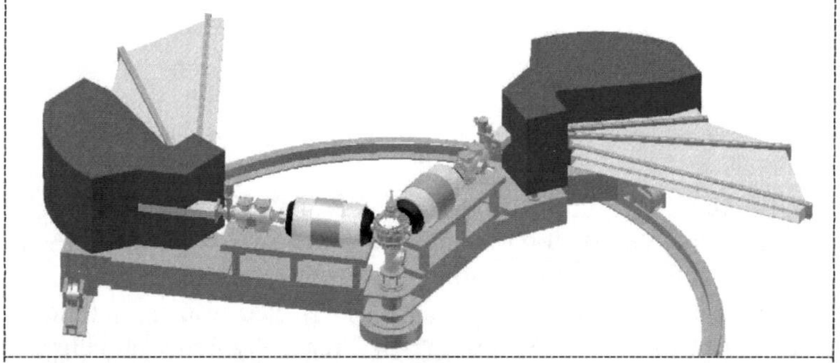

Figure 1. General view of the MAVR high-resolution magnetic analyzer.

Nuclear reaction products will be identified by A, Z, and Q through measurement of the energy loss (ΔE), time of flight (T), and total energy (E) of particles, i.e., the dependences ΔE-E and ΔE-T (or E-T), which requires reproduction of the particle motion trajectory in the analyzer. This will be done using the drift chamber developed together with the VAMOS group (GANIL).

The drift chamber to be used for measuring the particle energy loss (ΔE) and total energy (E) is schematically shown in Fig. 2. The time of flight will be measured using the parallel- plate-avalanche detector (PPAC).

Correlated reaction products arriving at the focal detector can be detected by position-sensitive semiconductor telescopes and also by a multiwire proportional chamber [13]. Since the range of problems to be solved by the proposed setup is rather wide, the focal detector will be made as a set of individual compatible modules.

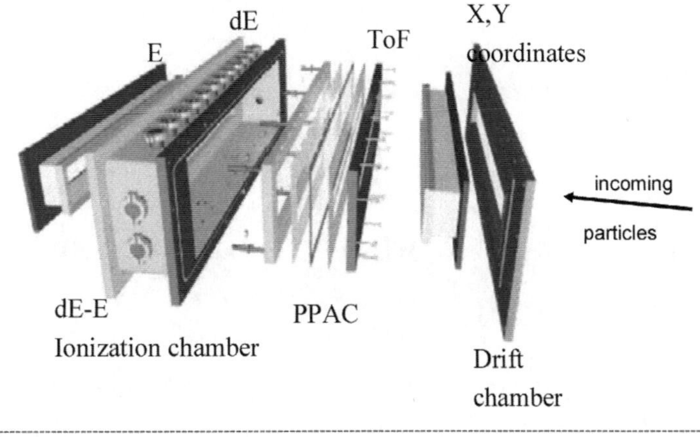

Figure 2. Schematic view of the detecting system.

3. Expected period of construction and estimated cost of work

It is planned to construct the MAVR setup in 4 years. Preparatory work and manufacture of individual units will be carried out independently of the construction of the extension to the U400R accelerator hall.

The estimated cost of the work to design and manufacture MAVR setup units and to mount and test the MAVR setup is about $750,000.

4. Collaborating organizations

INRNE, BRV, BAS (Sofia, Bulgaria), NPI, AS, CR (Rez, Czech Republic)
INP, PAS (Krakow, Poland)
GANIL (Caen, France), IPN (Orsay, France)
NRC Kurchatov Institute (Moscow, Russia)
AANL (Yerevan Phys. Inst., Yerevan, Armenia)
JYFL (Cyclotron Laboratory, University of Yuvaskyla, Finland)
IP VAST (Institute of Physics, Vietnam Academy of Science and Technology , Hanoi)

5. References

[1] R. Kalpakchieva *et al.*, Fiz. Elem. Chastits At. Yadra, 2002, 33, No. 6, 1247–1307.
[2] Yu.E. Penionzhkevich, R. Kalpakchieva, S.M. Lukyanov, Nucl.Phys. 2003, A722, 170c-175c.
[3] N.K.Skobelev *et al.*, Z.Phys. 1992, V. A341, P. 315.
[4] S.A.Fayans *et al.*, Phys.Lett. 1995, V. 357B, P. 509.
[5] 3. Yu.E.Penionzhkevich *et al.*, Phys.Atomic Nuclei, 2002, V. 65, P. 1563.
[6] Yu.E.Penionzhkevich *et al.*, Phys.Part. and Nucl.Lett., 2006, V. 3, P. 362..
[7] A.Budzanowski *et al.*, Bull. Rus. Acad.Sci., Phys. Ser., 2005, Vol.69, N.11, pp.1605-1609.
[8] Oganessian Yu.Ts., Gulbekian G.G., Proc. of the Int. Conf. “Nuclear Shells - 50 Years”. Dubna, 1999 Singapore: World Sci., 2000. P. 61.
[9] S.M. Lukyanov *et al.*, Physics Letters B, 2009, V. 670, P. 321-324.
[10] Yu.E. Penionzhkevich *et al.*, Eur. Phys. J A, 2007, V.31, P.185-194
[11] Yu.E. Penionzhkevich *et al.*, International Journal of Modern Physics E, 2008, Vol. 17, No.10, P. 2349-2353.
[12] V. I. Kazacha, On a Possibility of Increasing the Solid Angle of the MSP–144 Magnetic Separator, JINR FLNR Report, 2010, TOT-SFU-298 [in Russian].
[13] R.A. Astabatyan et al., Scientific report FLNR 2001-2002, P. 212, Ed. Popeko A.G., JINR, Dubna, 2003. Scientific report FLNR 2003-2004, Ed. Popeko A.G., JINR, Dubna, 2005.

Landau quantization and neutron emissions by nuclei in the crust of a magnetar

N Chamel[1], Y D Mutafchieva[2], Zh K Stoyanov[2], L M Mihailov[3] and R L Pavlov[2]

[1] Institute of Astronomy and Astrophysics, Université Libre de Bruxelles, CP 226, Boulevard du Triomphe, B-1050 Brussels, Belgium
[2] Institute for Nuclear Research and Nuclear Energy, Bulgarian Academy of Sciences, 72 Tsarigradsko Chaussee, 1784 Sofia, Bulgaria
[3] Institute of Solid State Physics, Bulgarian Academy of Sciences, 72 Tsarigradsko Chaussee, 1784 Sofia, Bulgaria

E-mail: nchamel@ulb.ac.be

Abstract. Magnetars are neutron stars endowed with surface magnetic fields of the order of $10^{14} - 10^{15}$ G, and with presumably much stronger fields in their interior. As a result of Landau quantization of electron motion, the neutron-drip transition in the crust of a magnetar is shifted to either higher or lower densities depending on the magnetic field strength. The impact of nuclear uncertainties is explored considering the recent series of Brussels-Montreal microscopic nuclear mass models. All these models are based on the Hartree-Fock-Bogoliubov method with generalized Skyrme functionals. They differ in their predictions for the symmetry energy coefficient at saturation, and for the stiffness of the neutron-matter equation of state. For comparison, we have also considered the very accurate but more phenomenological model of Duflo and Zuker. Although the equilibrium composition of the crust of a magnetar and the onset of neutron emission are found to be model dependent, the quantum oscillations of the threshold density are essentially universal.

1. Introduction

At the end point of stellar evolution, neutron stars are not only the most compact stars in the universe, but also the strongest magnets [1]. In particular, magnetic fields of the order of $10^{14} - 10^{15}$ G have been measured at the surface of soft gamma-ray repeaters and anomalous x-ray pulsars [2], thus dubbed *magnetars*. On the other hand, numerical simulations have shown that the internal magnetic field could be even stronger, up to about 10^{18} G [3]. The outermost layer of a neutron star is thought to consist of a solid crust, whose atoms are fully ionized by the gravitational pressure (for a review, see Ref. [4]). With increasing depth, nuclei become progressively more neutron rich by capturing electrons until at some point, neutrons start to drip out of nuclei. The presence of a neutron liquid in the crust of a magnetar is expected to leave its imprint on various observed astrophysical phenomena like sudden spin-ups [5, 6] and spin-downs [7, 8, 9, 10] (generally referred to as "glitches" and "anti-glitches" respectively), quasiperiodic oscillations detected in the giant flares from soft gamma-ray repeaters [11, 12, 13] and cooling [14].

We have recently studied the effects of a strong magnetic field on the neutron-drip transition in the crust of a magnetar [15, 16]. We have shown that the neutron-drip density and pressure increase almost linearly with the magnetic field strength in the strongly quantizing regime. In the weakly quantizing regime, the variations of the neutron-drip density with magnetic field strength exhibit typical quantum oscillations. The neutron-drip transition in a magnetar, as compared to unmagnetized neutron stars, can thus be shifted to either higher or lower densities depending on the magnetic field strength. In this paper, we explore the role of nuclear uncertainties on the neutron-drip transition considering different nuclear mass models.

2. Neutron emission in dense magnetized matter

During the gravitational collapse of the core of massive stars (whose mass lies in the range $\sim 8-10\ M_{\odot}$, where $M_{\odot}$ is the mass of the Sun), it is generally assumed that all kinds of reactions can occur so that the hot dense matter constituting the newly born neutron star remains in thermodynamic equilibrium. Eventually the neutron star interior becomes cold and fully "catalyzed" [1]. Although the temperature inside mature neutron stars with an age $\sim 10^4-10^5$ years is typically of the order of 10^7-10^8 K, matter is so highly degenerate that for most practical purposes the temperature can be set to zero. Since the pressure inside the star must vary continuously, and since the temperature is fixed, the suitable thermodynamic potential for determining the internal composition of a neutron star is the Gibbs free energy per nucleon g (this still remains the case even in the presence of a strong magnetic field, as shown in Ref. [17]). In the deep region of the outer crust on which we focus, atoms are fully ionized by the pressure, and electrons can be treated as an almost ideal Fermi gas. The main correction arises from the electrostatic lattice energy (for a discussion of various other corrections, see e.g. Ref. [17]). In each layer at pressure P, the crust is assumed to be made of only one type of ions with proton number Z and mass number A, arranged on a body-centred cubic lattice. Due to the presence of a strong magnetic field B, the electron motion perpendicular to the field is quantized into Landau levels. The magnetic field $B_\star = B/B_{\rm crit}$ can be conveniently measured in units of the critical magnetic field defined by

$$B_{\rm crit} = \left(\frac{m_e c^2}{\alpha \lambda_e^3}\right)^{1/2} \simeq 4.4\times 10^{13}\,{\rm G}\,, \tag{1}$$

where m_e is the electron mass, c the speed of light, α is the fine structure constant, and λ_e is the electron Compton wavelength. Typically, Landau quantization effects on the composition of the crust are negligible unless $B_\star \gg 1$. Further details on our crust model can be found in Refs. [15, 16].

In any region of the outer crust, the pressure is provided by the degenerate electron gas so that the electron chemical potential μ_e increases with depth until it reaches the threshold for neutron drip [16]

$$\mu_e(n_e, B_\star) + \frac{4}{3} C e^2 n_e^{1/3} Z^{2/3} = \mu_e^{\rm drip}(A,Z)\,, \tag{2}$$

where n_e is the electron number density, e is the proton electric charge, $C \approx -1.444$ is the body-centred cubic lattice structure constant, and

$$\mu_e^{\rm drip}(A,Z) \equiv \frac{-M(A,Z)c^2 + A m_n c^2}{Z} + m_e c^2\,, \tag{3}$$

with $M(A,Z)$ the nuclear mass (including the rest mass of Z protons, $A-Z$ neutrons and Z electrons [1]), and m_n the neutron mass. The equilibrium nucleus is found by minimizing the Gibbs free energy per nucleon. As discussed in Ref. [18], this nucleus is actually stable against

[1] The reason for including the electron rest mass in $M(A,Z)$ is that experimental *atomic* masses are generally tabulated rather than *nuclear* masses.

neutron emission, but unstable against electron captures accompanied by neutron emission. Equation (2) must be generally solved numerically. Nevertheless, an analytical solution can be obtained in the strongly quantizing regime for which electrons lie in the lowest Landau level and are ultrarelativistic ($\mu_e \gg m_e c^2$). In such case, the neutron-drip density and pressure are approximately given by [16]

$$n_{\rm drip} \approx \frac{A}{Z}\frac{B_\star \mu_e^{\rm drip}(A,Z)}{2\pi^2 \lambda_e^3 m_e c^2}\left[1-\frac{4}{3}C\alpha Z^{2/3}\left(\frac{B_\star}{2\pi^2}\right)^{1/3}\left(\frac{m_e c^2}{\mu_e^{\rm drip}(A,Z)}\right)^{2/3}\right], \tag{4}$$

$$P_{\rm drip} \approx \frac{B_\star \mu_e^{\rm drip}(A,Z)^2}{4\pi^2 \lambda_e^3 m_e c^2}\left[1-\frac{1}{3}C\alpha Z^{2/3}\left(\frac{4B_\star}{\pi^2}\right)^{1/3}\left(\frac{m_e c^2}{\mu_e^{\rm drip}(A,Z)}\right)^{2/3}\right], \tag{5}$$

For any given magnetic field, the composition of the outer crust and the onset of neutron drip are completely determined by nuclear masses. The nuclei expected to be found in the bottom layers of the outer crust are so neutron rich that their masses have not yet been measured. For this reason, nuclear models must be employed. We have made use of the recent series of Brussels-Montreal microscopic nuclear mass models [19, 20]. These models are based on the self-consistent Hartree-Fock-Bogoliubov (HFB) method using generalized Skyrme functionals (see e.g. Ref. [21] for a review of the latest Brussels-Montreal models). These models were primarily fitted to the 2353 measured masses of nuclei with N and $Z \geq 8$ from the 2012 Atomic Mass Evaluation [22], with a root-mean square deviation $\sim 0.5 - 0.6$ MeV. At the same time, the underlying BSk functionals were constrained to reproduce various properties of infinite homogeneous nuclear matter, most of which are summarized in Table 1. These parameters are defined by first writing the energy per nucleon of nuclear matter of density n and charge asymmetry $\eta = (n_n - n_p)/n$ (n_n and n_p are the neutron and proton densities respectively) in the form

$$e(n,\eta) \approx e(n,\eta=0) + e_{\rm sym}(n)\eta^2 \ , \tag{6}$$

in which the first term on the right-hand side is just the energy per nucleon of charge-symmetric nuclear matter; we have neglected charge-symmetry breaking terms, such as those arising from the neutron-proton mass difference. We then expand $e(n,\eta=0)$ and $e_{\rm sym}(n)$ about the equilibrium density n_0 in powers of $\epsilon = (n-n_0)/n_0$, thus

$$e(n,\eta=0) \approx a_v + \frac{1}{18}K_v \epsilon^2 , \tag{7}$$

and

$$e_{\rm sym}(n) \approx J + \frac{1}{3}L\epsilon + \frac{1}{18}K_{\rm sym}\epsilon^2 . \tag{8}$$

For pure neutron matter ($\eta = 1$) at density n_0, the pressure is approximately given by $P_n \approx L n_0/3$. Therefore, L is a measure of the stiffness of the neutron-matter equation of state. The functionals BSk22, BSk23, BSk24 and BSk25 were all fitted to the same realistic neutron-matter equation of state (as obtained from many-body calculations using realistic two- and three-nucleon forces), but were constrained to different symmetry energy coefficients, $J = 32$, 31, 30 and 29 MeV, respectively. BSk26 was constrained to the same symmetry-energy coefficient $J = 30$ MeV as BSk24, but was fitted to a softer neutron-matter equation of state. BSk27* yields the softest possible neutron-matter equation of state consistent with recent quantum Monte Carlo calculations and with the same symmetry-energy coefficient. The model HFB-27* is also the most accurate of this series. The range of values of J and L spanned by these functionals are consistent with constraints coming from the combined analysis of various experiments [23] (see also the discussion in Sec. IIIC in Ref. [19]). Moreover, the incompressibility coefficient

Table 1. Nuclear-matter properties predicted by recent Brussels-Montreal nuclear energy density functionals [19, 20].

	BSk22	BSk23	BSk24	BSk25	BSk26	BSk27*
n_0 [fm^{-3}]	0.1578	0.1578	0.1578	0.1587	0.1589	0.1586
a_v [MeV]	-16.088	-16.068	-16.048	-16.032	-16.064	-16.051
K_v	245.9	245.7	245.5	236.0	240.8	241.6
J [MeV]	32.0	31.0	30.0	29.0	30.0	30.0
L [MeV]	68.5	57.8	46.4	36.9	37.5	28.5
$K_{\rm sym}$ [MeV]	13.0	-11.3	-37.6	-28.5	-135.6	-221.4

K_v falls in the empirical range 240 ± 10 MeV [24]. Finally, the high-density equation of state of symmetric nuclear matter predicted by these functionals are compatible with the constraints inferred from the analysis of heavy-ion collision experiments [25, 26].

Taking the values of the nuclear masses from BRUSLIB [27], we have determined numerically the equilibrium composition of the outer crust of a magnetar for different magnetic field strengths. For comparison, we have also considered the more phenomenological model of Duflo and Zuker [28]. In the search of the equilibrium nucleus, we have made use of mass models only when experimental measurements from the 2012 Atomic Mass Evaluation [22] were not available. The composition at the neutron-drip transition is indicated in Table 2. For all models but HFB-22, the equilibrium nucleus is the same with or without the presence of a strong magnetic field. For model HFB-22, the equilibrium nucleus is found to be either ^{122}Kr or ^{128}Sr, depending on the magnetic field strength. The reason lies in the fact that the threshold electron chemical potentials $\mu_e^{\rm drip}$ for these two nuclides differ by less than 0.2%. As shown in Figs. 1 and 2, the oscillations of the neutron-drip density as a function of $B_\star$ are almost universal. In particular, in the presence of a nonquantizing magnetic field (i.e. more than one Landau level is populated), the shifts in the neutron-drip density are bounded, both from above and from below. The limits are approximately given by [16]:

$$\frac{n_{\rm drip}^{\rm min}}{n_{\rm drip}(B_\star=0)} \approx \frac{3}{4}\,, \qquad (9)$$

$$\frac{n_{\rm drip}^{\rm max}}{n_{\rm drip}(B_\star=0)} \approx \frac{1}{72}\left(35+13\sqrt{13}\right)\,. \qquad (10)$$

3. Conclusions

We have pursued our study of the effects of Landau quantization on the onset of neutron emission by nuclei in the crust of a magnetar [16]. In particular, we have explored the impact of nuclear uncertainties considering the recent series of Brussels-Montreal Hartree-Fock-Bogoliubov nuclear mass models, from HFB-22 to HFB-27* [19, 20]. These models yield equally good fits to essentially all experimental masses, with a root-mean-square deviation of about $0.5-0.6$ MeV. However, they lead to different predictions of nuclear-matter properties thus reflecting the current lack of knowledge of the symmetry energy, and the stiffness of the neutron-matter equation of state. For comparison, we also considered the more accurate but also more phenomenological model of Duflo and Zuker [28]. Although the equilibrium nucleus at the bottom the outer crust, and the onset of neutron emission are found to be model dependent, the oscillations of the threshold density as a function of the magnetic field strength are almost

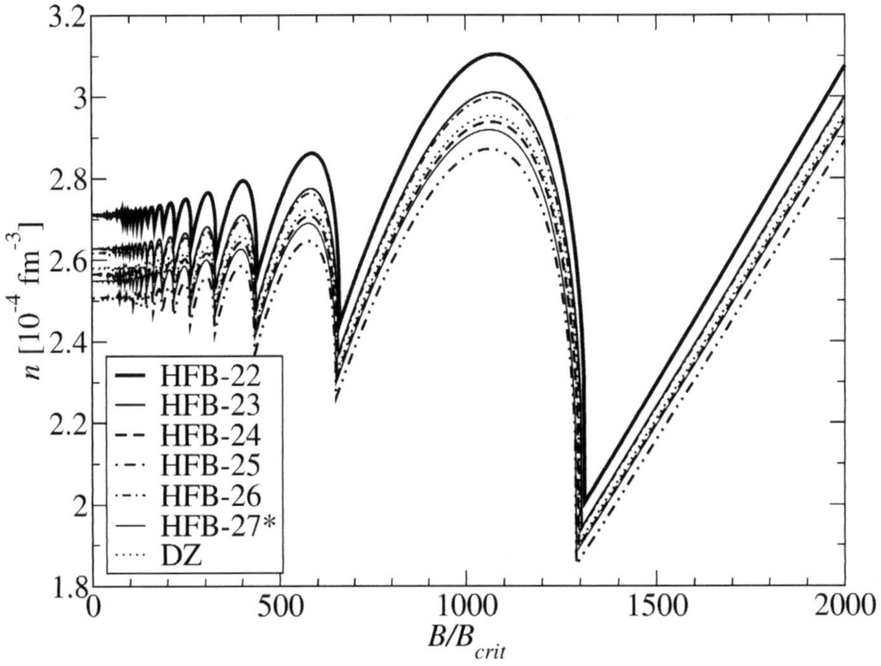

Figure 1. Threshold baryon number density for the onset of neutron drip in the crust of a neutron star as a function of magnetic field strength, as predicted by different nuclear mass models.

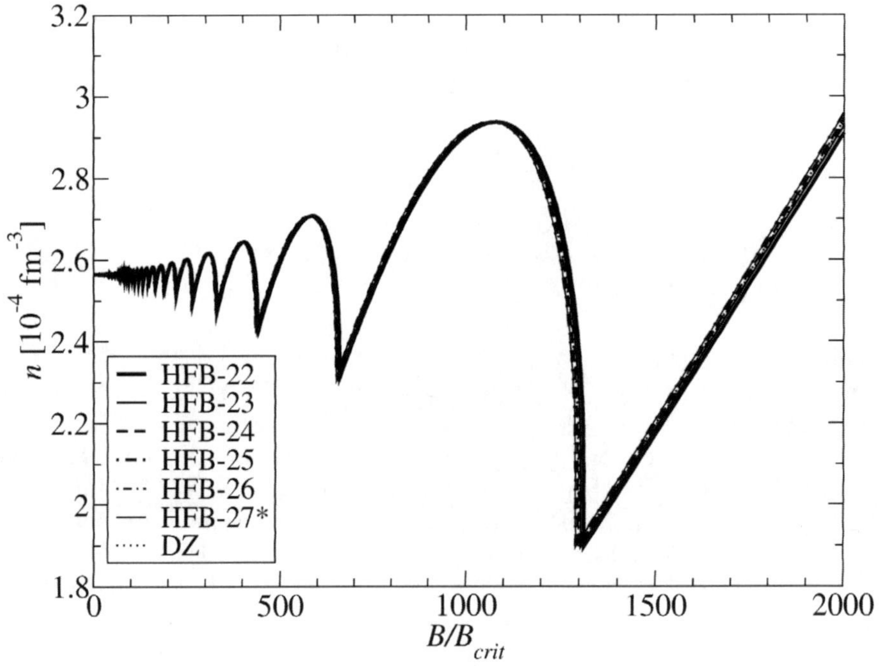

Figure 2. Same as Figure 1 after a suitable rescaling of the threshold density to that of HFB-24.

Table 2. Equilibrium nucleus and electron chemical potential at the neutron-drip transition in the crust of a magnetar, as predicted by different nuclear mass models.

	^{A}X	$\mu_e^{\rm drip}$ (MeV)
HFB-22	^{122}Kr/^{128}Sr	24.97/25.01
HFB-23	^{126}Sr	24.88
HFB-24	^{124}Sr	24.81
HFB-25	^{122}Sr	24.76
HFB-26	^{126}Sr	24.85
HFB-27*	^{124}Sr	24.76
DZ	^{118}Kr	24.87

universal. The role of the magnetic field on the neutron-drip density in magnetar crusts could thus be a priori inferred from the value of the neutron-drip density in unmagnetized neutron star crusts. On the other hand, the change of nuclear masses due to the strong magnetic field [29], which we have not taken into account, may undermine this universal behaviour.

Acknowledgments
This work was financially supported by Fonds de la Recherche Scientifique - FNRS (Belgium), Wallonie-Bruxelles-International (Belgium), the Bulgarian Academy of Sciences, the Bulgarian National Science Fund under contract No. DFNI-T02/19, and the European Cooperation in Science and Technology (COST) Action MP1304 "NewCompStar".

References

[1] Haensel P, Potekhin A Y, Yakovlev D G 2007 *Neutron Stars 1: Equation of state and structure* (Springer)
[2] Olausen S A and Kaspi V M 2014 *Astrophys. J. Suppl. Ser.* **212** 6. `http://www.physics.mcgill.ca/~pulsar/magnetar/main.html`
[3] Chatterjee D, Elghozi T, Novak J and Oertel M 2015 *Mon. Not. R. Astron. Soc.* **447** 3785
[4] Chamel N and Haensel P 2008 *Living Rev. Relativity* **11** 10. `http://www.livingreviews.org/lrr-2008-10`
[5] Dib R, Kaspi V M and Gavriil F P 2008 *Astrophys. J.* **673** 1044
[6] Gügercinoglu E and Alpar A 2014 *Astrophys. J.* **788** L11
[7] Archibald R A, Kaspi V M, Ng C -Y, Gourgouliatos K N, Tsang D, Scholz P, Beardmore A P, Gehrels N and Kennea J A 2013 *Nature* **497** 591
[8] Şaşmaz Muş S, Aydin B and Göğüş E 2014 *Mon. Not. R. Astron. Soc.* **440** 2916
[9] Duncan R C 2013 *Nature* **497** 574
[10] Kantor E M and Gusakov M E 2014 *Astrophys. J.* **797** L4
[11] Andersson N, Glampedakis K and Samuelsson L 2009 *Mon. Not. R. Astron. Soc.* **396** 894
[12] Chamel N., Page D and Reddy S 2013 *Phys. Rev. C* **87** 035803
[13] Passamonti A and Lander S K 2014 *Mon. Not. R. Astron. Soc.* **438** 156
[14] Aguilera D N, Cirigliano V, Pons J A, Reddy S, Sharma R 2009 *Phys. Rev. Lett.* **102** 091101
[15] Chamel N, Pavlov R L, Mihailov L M, Velchev Ch J, Stoyanov Zh K, Mutafchieva Y D, Ivanovich M D, Pearson J M and Goriely S 2012 *Phys. Rev. C* **86** 055804
[16] Chamel N, Stoyanov Zh K, Mihailov L M, Mutafchieva Y D, Pavlov R L, Velchev Ch J 2015 *Phys. Rev. C* **91** 065801.
[17] Chamel N and Fantina A F 2015 *Phys. Rev. D* **92** 023008
[18] Chamel N, Fantina A F, Zdunik J L and Haensel P 2015 *Phys. Rev. C* **91** 055803
[19] Goriely S, Chamel N and Pearson J M 2013 *Phys. Rev. C* **88** 024308
[20] Goriely S, Chamel N and Pearson J M 2013 *Phys. Rev. C* **88** 061302(R)
[21] Chamel N, Pearson J M, Fantina A F, Ducoin C, Goriely S and Pastore A 2015 *Acta Phys. Pol. B* **46** 349
[22] Audi G, Wang M, Wapstra A H, Kondev F G, MacCormick M, Xu X and Pfeiffer B 2012 *Chin. Phys. C* **36** 002

[23] Fantina A F, Chamel N, Pearson J M and Goriely S 2015 *AIP Conf. Proc.* **1645** 92
[24] Colò G, Giai N V, Meyer J, Bennaceur K and Bonche P 2004 *Phys. Rev. C* **70** 024307
[25] Danielewicz P, Lacey R and Lynch W G 2002 *Science* **298** 1592
[26] Lynch W G, Tsang M B, Zhang Y, Danielewicz P, Famiano M, Li Z and Steiner A W 2009 *Progress in Particle and Nuclear Physics* **62** 427
[27] `http://www.astro.ulb.ac.be/bruslib`
[28] Duflo J and Zuker A P 1995 *Phys. Rev. C* **52** 23 (R)
[29] Basilico D, Peña Arteaga D, Roca-Maza X and Coló G 2015 *Phys. Rev. C* **92** 035802

The cluster and single -particle states in ^{13}C $(\alpha,\alpha)^{13}C$ reactions

N A Mynbayev[1], A K Nurmukhanbetova[1], V Z Goldberg[2], G V Rogachev[2], M S Golovkov[3], M Koloberdin[4], I Ivanov[4], D K Nauruzbayev[1], Sh S Berdibek[1], A M Rakhymzhanov[1] and R E Tribble[2]

[1]National laboratory Astana, 53 Kabanbay batyr Ave, 010000, AstanaKazakhstan;
[2] Cyclotron Institute Texas A&M University, 3366 TAMU, College station, TX 77843-3366, USA;
[3] Joint Institute for Nuclear Research, Joliot-Curie 6, 141980 Dubna, Moscow region, Russia
[4] L.N. Gumylov Eurasian National Institute, 2 Satpayev Str, 010000, Astana, Kazakhstan.

E-mail: anurmukhanbetova@nu.edu.kz

Abstract. The excitation functions of elastic scattering of ^{13}C on alpha particle have been measured using the thick-target inverse kinematic method at the heavy ion DC-60 cyclotron. The helium gas was used as a target and also as a degrader to stop the beam. New data (including 180°degree) of the resonances close to the threshold in ^{17}O have been obtained.

1. Introduction

Many α- cluster states in light nuclei are an important issue for nuclear astrophysics and for understanding quasimolecular structure in atomic nuclei. Alpha clustering may generally play a role in helium burning in astrophysical systems. Indeed, even if astrophysical reactions involving helium do not proceed through strong α-cluster states (because of high excitation energy), these states can provide an alpha width to the states that are closer to the region of astrophysical interest [1]. α-cluster structure in nuclei are usually studied by investigation of self-conjugate *4N* nuclei [2]. Many theoretical calculations [3, 4] have suggested that clustering remains in systems composed of a collection of α particles and valence nucleons. There is very little information on α-cluster states in neutron rich nuclei [5-9], but it gives indications for the developed cluster structures with very large moments of inertia. The study of non-self-conjugate nuclei has an advantage in that one can investigate isobaric analogue states in mirror systems. Comparison of the results for mirror systems can bring new spectroscopic information and shed light on such properties as the radii of the cluster states.

The non-self-conjugate nucleus ^{17}O produced by $^{13}C(\alpha,n)^{16}O$ is important as the main source of neutrons for the s-process at low energies[10]. Two factors determine the efficiency of this reaction: the abundance of ^{13}C and the rate of the ^{13}C $(\alpha,n)^{16}O$ reaction (see [11], and references therein). Because of the very small reaction cross section below 300 keV, the rate of the $^{13}C(\alpha,n)^{16}O$ reaction at temperatures of $\sim 10^{8}$ K is uncertain [12].

There are several measurements [13] of low energy resonances in the $\alpha+^{13}C$ elastic scattering using a target with ~50% enrichment of ^{13}C. The data were obtained only at four angles, and the analysis was mainly qualitative. The difficulties of the studies of the elastic scattering of α particles on light nuclei at large angles using conventional geometry of the classical experimental approach are understandable. By using this approach, M. Heil with his colleagues obtained more complete experimental data of ^{13}C (α,n) ^{16}O reaction [14].

We measured the elastic scattering of ^{13}C on alpha particle by the thick target inverse kinematics (TTIK) method [15]. The advantage of the TTIK method is that an excitation function can be measured in a large interval using "single" beam energy. It provides also a possibility to make measurements at zero degrees (180° cms) where the resonance cross section is the largest and the potential one is minimal. It is important that the TTIK method provides for an easier access to the low energy of interaction because of the kinematic increase of the energy of light recoils.

2. Experimental procedure and results

The cyclotron DC-60 in Astana was used to study ^{13}C (α, α)^{13}C reaction with ^{13}C beam energy of 1.75MeV/A. The energy of the ^{13}C ions in the scattering chamber is 16 MeV due to the energy loss in entrance window. The setup and procedure for this experiment is described in [16]. The obtained results of excitation function of ^{13}C+ α elastic scattering at 180^0 in center of mass frame is shown in Figure 1.

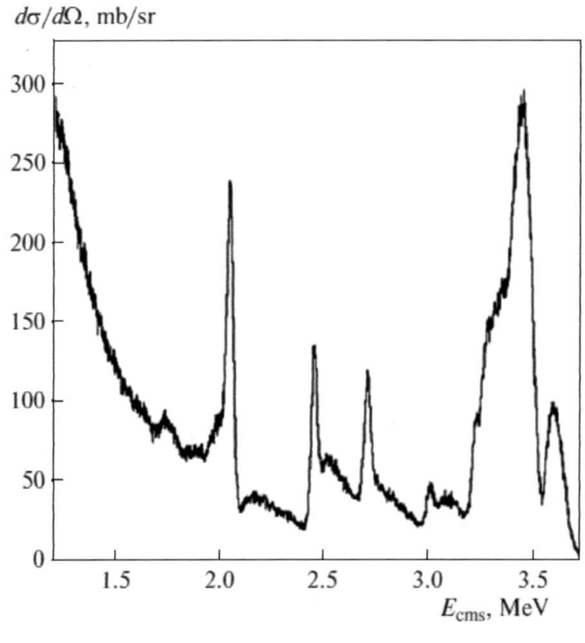

Figure 1. The excitation functions of the elastic scattering of α particles on ^{13}C.

All features of ^{13}C+ α spectrum obtained by M. Heil et. al. [14] can be seen in figure 1. α cluster states with a large spin at 2.05 and 3.47 MeV are clear in Figure 1 because of 180° center of mass system measurement. In addition, a certain new structure can be observed at low energies. The background contribution can be neglected [16].

The complete multilevel *R*-matrix analysis of the data is in progress. We determined the spin-parity, excitation energies and partial widths for 36 excited states in ^{17}O. The complete excitation function for 180° in center of mass frame for the entire energy range measured in this experiment is shown in Figure 2.

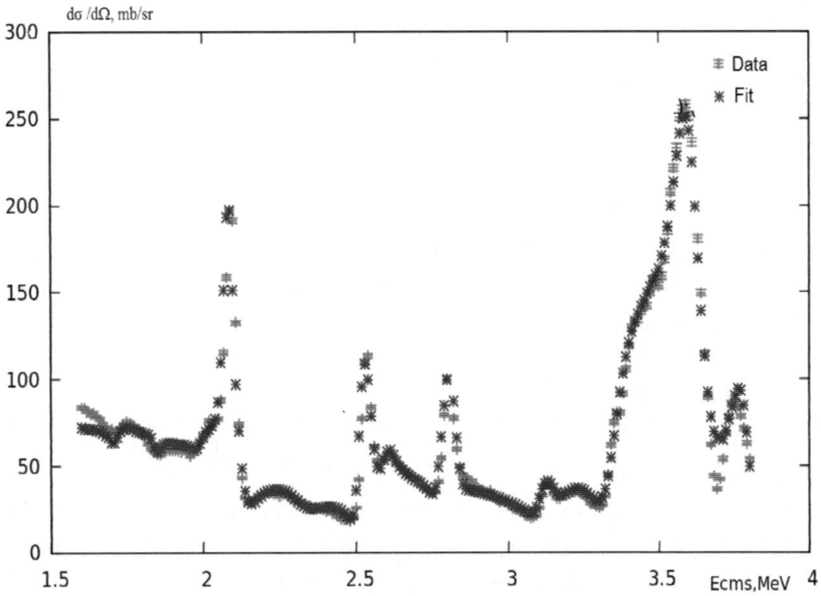

Figure 2. The excitation function for $\alpha+{}^{13}C$ elastic scattering at 180^0 in center of mass frame.

Figure 2 shows a good agreement between the fitted and experimental data. The parameters of some strong peaks, such as energy in center of mass system, its alpha width, spin, parity and the percentage of the Wigner limit for the alpha particle width of the level, are presented in Table 1. There are levels with the widths as large as 37 % for 9/2 and 47% for 7/2 and these data show that alpha clusterization is important for N not equal Z nuclei. This knowledge could be important to evaluate probabilities of astrophysical reactions induced by alpha particles at low energies. We expect that the complete R-matrix analysis of these data will give information interesting for studies in astrophysics and physics of clusters in atomic nuclei.

Table 1. The parameters of strong peaks in the fits

E_{cms}	Γ_α (keV)	J^π	Percentage of Wigner limit,%
2.081	4	$\frac{7^+}{2}$	37.31
2.532	3.2	$\frac{9^-}{2}$	61.12
2.828	4	$\frac{5^+}{2}$	4.67
3.137	18	$\frac{5^-}{2}$	3.08
3.611	32	$\frac{7^+}{2}$	9.77

3. Conclusion

The use of the TTIK method enabled us to perform an experiment to study of ${}^{13}C\ (\alpha, \alpha){}^{13}C$ reaction at 180° in the center of mass system and at low energy. The new structure below 2 MeV was obtained which was not observed before. Some parameters of strong peaks obtained by R-matrix fit are tabulated, but the complete analysis of experimental data will be given in the future.

4. Acknowledgments
The authors are grateful for financial support from 055 budget program of the Ministry of Education grants and from the US Department of Energy (grant no. DOE FG02 93ER40773)

5. References
[1] Tang X D, Rehm K E, Ahmad I, Brune C R, Champagne A, Greene J P, Hecht A A, Henderson D, Janssens R V, Jiang C L, Jisonna L, Kahl D, Moore E F, Notani M, Pardo R C, Patel N, Paul M, Savard G, Schiffer J P, Segel R E, Sinha S, Shumard B and Wuosmaa A H 2007 *Phys. Rev. Lett.* **99** 052502

[2] Freer M 2007 *Rep. Prog. Phys* **70** 2149

[3] Itagaki N and Okabe S. 2001 *Phys. Rev.* C **64** 014301

[4] Von Oertzen W 1997 *Z. Phys* A **357** 355

[5] Freer M, Casarejos E, Achouri L, Angulo C, Ashwood N I, Curtis N, Demaret P, Harlin C, Laurent B, Milin M, Orr N A, Price D, Raabe R, Soić N and Ziman V A 2006 *Phys. Rev. Lett.* **96** 042501

[6] Korsheninnikov A A., Nikolskii E Yu., Kobayashi T, Aleksandrov D V, Fujimaki M, Kumagai H,

[7] Ogloblin A A, Ozawa A, Tanihata I, Watanabe Y, Yoshida K et al 1995 Phys. Lett. B 343 53 Goldberg V. Z., Rogachev G. V., Trzaska W H., Kolata J J., Andreyev A, Angulo C, Borge M J G., Cherubini S, Chubarian G, Crowley G, Van Duppen P, Gorska M, Gulino M, Huyse M, Jesinger P, Källman K M , Lattuada M , Lönnroth T, Mutterer M, Raabe R, Romano S, Rozhkov M V, Skorodumov B B, Spitaleri C, Tengblad O, Tumino A 2004 *Phys. Rev.* C **69** 024602

[8] Buchmann L 2007 *Phys. Rev.* C **75** 012804

[9] Johnson E, Rogachev G V, Goldberg V Z, Brown S, Robson D, Crisp A M, Cottle P D, Fu C, Giles J, Green B W, Kemper K. W, Lee K, Roeder B T, Tribble R E 2009 *Eur.Phys.J.* A **42** 135

[10] Iben I 1975 *Astrophys. J.* **196** 525

[11] Goriely S and Siess L 2001 *Astron. Astrophys.* **378** L25

[12] Angulo C 1999 *Nucl. Phys.* A **656** 3

[13] Kerr G W, Morris J M and Risser J R al 1968 *Nucl. Phys.* A **110** 637

[14] Heil M, Detwiler R, Azuma R E, Couture A, Daly J, Görres J, Käppeler F, Reifarth R, Tischhauser P, Ugalde C, Wiescher 2008 *Nucl.Phys.* A **688** 499

[15] Artemov K P, Beljanin O P, Wetoshkin A L, Wolskj R, Golovkov M S, Goldberg V Z, Madeja M, Pankratov V V,Serikov I N, Timofeev V A, Shadrin VN, Szmider J A 1990 *Yad.Fiz*. **52** 634;1990 *Sov.J.Nucl.Phys*. **52**, 408

[16] Mynbayev N A, Nurmukhanbetova A K., Goldberg V Z, Golovkov M S, Rogachev G V, Dzybin V N, Tribble R E 2014 *JETP* Vol. **119** 4 663–667

Nuclei in Astrophysics

Yu. E. Penionzhkevich
Joint Institute for Nuclear Research, Dubna, Moscow region, 141980 Russia
Federal Research Nuclear University "MEPhI", Moscow, Russia

E-mail: pyuer@mail.ru

Abstract. This work is an attempt to present some problems on the evolution of the Universe: the nucleosynthesis and cosmochronology from the standpoint of physics of particles and nuclei, in particular with the use of the latest results, obtained by means of radioactive nuclear beams. The comparison is made between the processes taking place in the Universe and the mechanisms of formation and decay of nuclei, as well as of their interaction at different energies. Examples are given to show the capabilities of nuclearphysics methods for studying cosmic objects and properties of the Universe. The results of investigations in nuclear reactions, induced by radioactive nuclear beams, make it possible to analyze the nucleosynthesis scenario in the region of light elements in a new manner.

1. BASIC IDEAS ABOUT THEEVOLUTION OF THE UNIVERSE

The studies of microcosm laws where nuclear physics is involved have recently assisted in considerably extending our ideas about phenomena occurring in the macroworld ourUniverse and contributed enormously to developing the astrophysical and cosmological theories. This is concerned, first of all, with the abundance of elements and scenarios of their formation, as well as with properties of different stars and cosmic objects. Without aspiring in this review to the full description of all cosmology problems, let us dwell only upon those of them that have much in common with nuclearmatter properties manifesting themselves in nuclear interactions.

2. Model of the Expanding Universe.

According to estimates, formation of the Universe was around 10 billion years ago. It is believed that at this epoch the nowobservable Universe arose as a result of a monstrous explosion. The Big Bang, as the explosion is called, initiated the formation not only of the Universe, but also of all physical notions we know, including the notions of space and time. Some authors [1, 2] consider the evolution of the Universe in the form of four consecutive epochs, as a result of replacements of which, according to the newest models, the Universe came to its present state: $\rho \approx 10^{-30}$ g/cm^3, T = 3 K.It is supposed in these models that the Universe behaves like a perfect black body, whose initial temperature and density are very high; its initial density is higher than a nuclear one of 10^{15} g/cm^3 and a temperature is higher than 1 GeV (10^{13} K). The density of matter becomes larger than the radiation density (photon's energy density) and grows during the expansion. This corresponds to the "star epoch" that has been lasting until now. V.L. Ginzburg describes in his book [2] the problems of the expanding Universe as theinteraction of a great number of different particles: photons, electrons, neutrinos, muons, π - mesons, protons, neutrons, etc. Though, as he himself notes, this approach is rather subjective because we have no complete understanding of particle physics so far. Everything said above referred to the possible interpretation of the Universe based on strong interactions of elementary particles.

Meanwhile, one of the most interesting corollaries from the modern theory of particle physics is the fact that the Universe could experience a phase transition from one state of the matter to another. This phase transition is connected with shortrange interactions of the other class: weak interactions. Weak interactions in nuclear physics are responsible for certain radioactive-decay processes (e.g., decay of a free neutron) or of any reaction that involves neutrinos.The development of this theory made it also possible to draw important conclusions on problems concerning the study of the early Universe. (In 1998, S. Perlmutter and A. Riess from the USA and B. Schmidt from Australia predicted the acceleration of the expansion of the Universe. They were awarded the Nobel Prize in Physics for this achievement in 2011). This discovery has been made on the basis of changeability of the brightness of supernovas that end their evolution with an explosion. However, the Nobel Prize winners delivered the direct measurement of the acceleration of the expansion of the Universe, which also confirmed a theory of dark energy.However, all this requires the further comprehension and acquisition of new data and new approaches to description of the evolution of the Universe.

3. Nucleosynthesis

At this temperature in the early Universe, consisting of electrons, positrons, neutrinos, anti-neutrinos, neutrons, protons, and photons, different nuclei could form starting with deuteron and ending with helium. The heavier nuclei such as nuclei of carbon, oxygen, etc., could be synthesized only in the course of thermonuclear reactions in stars. The cause of it is that there is a certain interval of instability of light nuclei located near the lithium nucleus, and this interval cannot be surmounted in the course of the primeval nucleosynthesis. Therefore, the synthesis in the early epoch stops at the stage of the helium formation.It is believed that one of the first reactions leading to the formation of heavy nuclei is the reaction $n + p \rightarrow \alpha + \gamma$. As calculations have demonstrated, this reaction goes at the temperature $T = 10^{10}$ K, which corresponds to a ratio between the numbers of neutrons and protons in the Universe of $Nn/Np = 0.2$ and to the time ~3 s. Under these conditions, deuterium forms in sufficient quantities for producing nuclei with a mass of 3 in the following reactions:$d + n \rightarrow {}^3H + \gamma$,$d + p \rightarrow {}^3He + \gamma$,or$d + d \rightarrow {}^3H + p$,$d + d \rightarrow {}^3He + n$and finally ^{4}He can be formed as a result of the reactions:${}^3H + p \rightarrow {}^4He + \gamma$,${}^3He + n \rightarrow {}^4He + \gamma$.The binding energy of products of these reactions is larger than that of deuterium (2.225 MeV); then if a photon can form a deuterium, it can conduct other reactions too. Since the stable mass 5 does not exist, ^{4}He is the last nucleus at the initial stage of nucleosynthesis. In principle, it could form the heavier nuclei ($A = 7$) as a result of the reactions:${}^4He + {}^3H \rightarrow {}^7Li + \gamma$,${}^4He + {}^3He \rightarrow {}^7Be + \gamma$In this epoch, the formation of hydrogen and helium occurs. With this, the stage of primevalnucleosynthesis ends. The heavier nuclei form now as a result of processes related to the evolution of stars.

4. The stellar nucleosynthesis

The Universe during its evolution is enriched with ever heavier chemical elements. The abundance of chemical elements in the Universe is determined using different methods: by star radiation spectrum, by means of element analysis of terrestrial and cosmic samples (meteorites, lunar samples). The obtained curve of the element abundance is shown in Fig. 1. The curve has maxima for the silicon group and for the iron group, after which the abundance curve splits into two branches, one of which includes neutronrich isotopes and is characterized by three double peaks near the magic numbers $N = 50$, 82, and 126, while the other branch includes four less abundant protonrich isotopes.One of the nucleosynthesis stages is the formation of ^{12}C. As has been shown above, carbon can be formed as a result of the two step reaction ${}^4He + {}^4He \rightarrow {}^8Be + {}^4He \rightarrow {}^{12}C + \gamma$ This reaction makes it possible to explain the existence of carbon and the other observable isotopes together with it. At each stage of the nuclear fusion initiated by explosion of the outer shell of stars, the more and heavier nuclei form: ^{4}He, ^{12}C, ^{16}O, ^{28}S, and ^{56}Fe. In so doing, the processes of formation are accompanied by the processes of decay of these nuclei. Somewhat different is a mechanism of formation of the nuclei heavier than iron. This mechanism is explained by consecutive reactions of radioactive capture of neutrons by the irongroup elements.

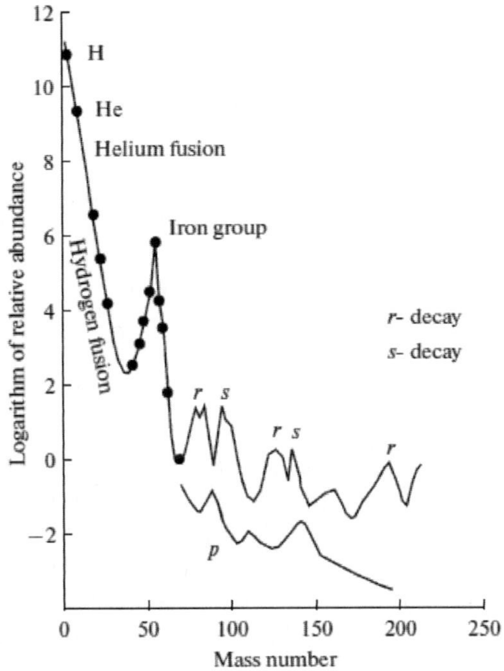

Figure 1. The curve of abundance of elements. The top curve with double (r and s) peaks corresponds to neutron- rich isotopes; the bottom curve (p) is appropriate to proton- rich isotopes.

The presence of double peaks in the curve of element abundance (see Fig. 1) evidences that there are two different processes of neutron capture, the socalled r-and s-processes. The two processes correspond to different neutron densities. In case of low densities of neutrons (s-(slow)-process) in the radiation capture (A, Z) + n (A + 1, Z), the isotope forms with a mass that is a unity larger than the targetnucleus mass. This process repeats many times and leads to the formation of neutron- poor nuclei with a mass to 200. After this, thenuclear fission occurs at a large probability, which interrupts the s- process. With high densities of neutrons (r (rapid)-process), the nucleus (A + x, Z) will absorb neutrons before it decays, and new radiation captures occur. The evidence that r- and s- processes exist is the increased concentration of isotopes with N = 50, 82, and 126. It is experimentally shown that the abundance of elements is inversely proportional to total cross sections of neutron capture. This cross section for nuclei with magic numbers is several orders of magnitude lower than the one for other neighboring nuclei. From the viewpoint of nuclear physics, this result is a manifestation of magic numbers. For astrophysics, it is the proof of s- process existence.The existence of the shell with Z = 114 and hence the increase in stability of superheavy nuclei near the double- magic nucleus $^{298}114$ (114 protons and 184 neutrons) were also predicted. Calculations performed according to the shell model showed the possibility of existence of this superheavy nucleus with a half- life to $\sim 10^8$yr [3]. Despite the fact that the accuracy of these calculations is not high and they most likely are of qualitative type, the latest experiments in synthesis of heavy isotopes of elements 112–118 have shown that the increase in stability of superheavy nuclei with respect to decay is observed [4], which is an additional confirmation of the stability of super- heavy nuclei near the shells (Fig.2). If we suppose that the longest- lived superheavy nuclei have a halflife of 10^5–10^6 yr, which do not disagree much with predictions of the theory that makes its estimations with certain accuracy, then we cannot exclude that they may be revealed in cosmic rays: witnesses of formation of elements on other, younger planets of the Universe. If we also suggest that a half- life of “long- livers” may reach tens of millions of years or more, then they could be available in the Earth, having remained intact in very small quantities from the instant of formation of elements in the Solar System to present days. Among the possible candidates, the most real ones are isotopes of element 108 (Hs), nuclei of which contain around 180 neutrons.

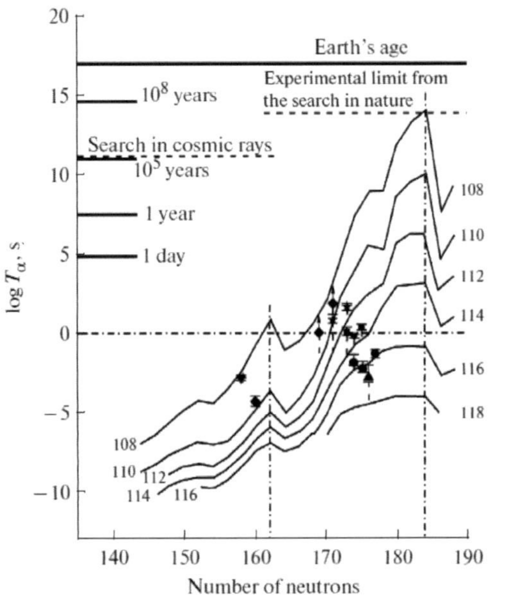

Figure 2. The α- decay half- life of superheavy elements as a function of the number of neutrons N. The solid horizontal lines present certain time marks: Earth's age. The dashed lines indicate the experimental limits of search for elements in nature and in cosmic rays.

Then the decay of a superheavy nucleus will be recorded according to a neutron flare that accompanies a spontaneous fission. Such a facility including a 4π- neutron detector has been created at the JINR Laboratory of Nuclear Reactions, and for reducing the cosmic background of neutrons, it is installed in the underground laboratory located under the Alps in the middle of the tunnel connecting France with Italy at the depth of 4000 m water equivalent . If at least one event of spontaneous fission of a superheavy nucleus is observed for a year of measurements, then it will correspond to a concentration of element 108 in the Os sample around 5.10^{-15} g/g onassumption that its half- life is 10^9 yr. Such a small value amounts only to 10^{-16} of the uranium concentration in the Earth's crust. Despite the superhigh sensitivity of the experiment, chances to detect relic superheavy nuclides are small. The absence of effect will provide only the upper limit of the long- liver's half- life at a level of $T_{1/2} \sim 3 . 10^7$ yr. Active searches in the natural objects (cosmic rays, materials, lunar samples, concentrates of heavy chemical elements of terrestrial samples) have yielded so far no positive result. At the present time, experiments are conducted at accelerators in Dubna and Darmstadt (Germany), which are directed at the artificial synthesis of superheavy elements in nuclear reactions with heavy ions, but, naturally, with shorter half- lives.

5. NUCLEAR PHYSICS EXPERIMENTS IN ASTROPHYSICS

The most fundamental problems of astrophysics, i.e., the processes of energy release in formation and explosion of stars, as well as nucleosynthesis, are most closely connected with particle physics and also with studies (using nuclear physics methods) of different nuclear characteristics and nuclear interactions at different energies including the energies near the Coulomb barrier. The primary information for solving one or another astrophysical problem is obtained from the following experimental data:

- half- lives of nuclei near the boundaries of
- nucleon stability (for r- and s-processes);
- probabilities of neutron emission after β- decay;
- characteristics of nuclear reactions leading to the synthesis of new nuclei;

- total cross sections of nuclear reactions;
- characteristics of nuclear reactions induced by exotic nuclei;
- binding energy and masses of nuclei far away from the β- stability line;
- characteristics of superneutron-rich nuclei of the lightest elements (multi-neutron systems, super- heavy isotopes of hydrogen helium (^{4,5,6,7}H),heavy isotopes of hydrogen helium (^{6,8,9,10}He), lithium (^{9,10,11,13}Li), etc.;
- nuclear temperature measurement;
- characteristics of neutrino emission from exotic nuclei (^{8}B);
- probabilities of proceeding of thermonuclear reactions with light exotic nuclei. A list of these data retrieved from nuclear physics experiments can be continued. However, it is already obvious from the listed above, how much extensive information can be obtained from nuclear experiments for solving many problems of astrophysics.

6. Exotism of nuclei

Exotic states of the nuclear matter i.e., of the nuclei in extreme states (with high spin, large deformation, high density and temperature, the neutron- or proton-rich nuclei on the boundary of nucleon stability) play an important role in studies of fundamental nuclear properties, which bring us closer to deducing the equation of state of the nuclear matter. This is undoubtedly of great significance for extrapolating microcosm characteristics to the macroworld that presents our Universe. Synthesis and study of neutron-rich isotopes have two main goals: finding the position of neutron-stability boundaries and obtaining data on properties of exotic nuclei near these boundaries. The development of accelerator technology has made it possible to obtain the accelerated beams of secondary radioactive nuclei. In this connection, new vast opportunities have opened up for studying both the structure of lightexotic nuclei themselves and the peculiarities of nuclear reactions induced by these nuclei. It is extremely important to obtain new information regarding nuclei near the nucleon-stability boundary because considerable deviations of properties of such nuclei from the widely known regularities may be expected (and are already observed). Here the nuclei in a range of small Z serve as convenient objects for investigation. However, the question of how general the corollaries made for this small number of nuclei is crucial. The experiment alone can give an answer to this question.

7. Nuclear shapes

Problems related to the fact that a deformation may lead to an increase in binding energy of nuclei have been actively discussed recently. The nuclei with the neutron number N = 20, for ground states of which a spherical shape is expected due to filling the closed shell N = 20, are of particular interest from this point of view. However the latest theoretical calculations of their binding energy predict for some of them the presence of the strong longitudinal deformation (~ 0.3) and even the existence of isomeric states. It is supposed that a corollary from this deformation is the experimentally revealed sharp growth in binding energy of two neutrons in the neutronrich nuclei ^{31}Na and ^{32}Mg, i.e., the inversion of Nilsson levels that correspond to a large deformation takes place. It is the evidence that the closed shell breaks up and N = 20is no longer the magic number. The subsequent experiments in studying the nuclei $^{33-35}$Al, ^{35}Si, and ^{36,37}P [6] have shown the possibility of finding the nuclei with the spherical and deformed shapes within the region between the magic numbers N = 20 and 28 (a domain of existence of two types of deformation). The experiments for measuring $T_{1/2}$ for the nuclei ^{27,29}F and ^{30}Ne have also shown that they are more stable than was predicted according to the shell model. The measured large value of the probability of the transfer B(E2; 0+ 2+) for the nucleus ^{32}Mg (N = 20) confirmed that a deformation can exist in magic light nuclei[7]. The detection of the latter strongly-boundneutron-rich isotopes ^{32}Ne and ^{40}Mg [8] also demonstrated the validity of the prediction that stability of neutron-rich nuclei increases as their deformation grows. The isotope ^{28}O is extremely significant in this respect. This doubly magic (N= 20, Z= 8) nucleus has not been observed so far. However the nucleus ^{29}F with the same number of neutrons but with one surplus proton (N = 20, Z = 9) proved to be nucleonstable. If pn-interaction is not a cause of this stability, then we can suppose that the effect of deformation in the ^{29}Fnucleus is more significant than in ^{31,32}Na nuclei, which causes its stability. The investigation in properties of nuclei near the magic numbers of neutrons N = 20, 28, and 50 is the most interesting problem of nuclear physics and requires the further development using diverse methods for measuring the deformation and radius of nuclei. The information on properties of these nuclei is needed for calculation of scenarios of nucleosynthesis in r-process.

8. Nuclear sizes

Determination of nuclear sizes was always a fundamental problem of nuclear physics because exact values of nuclearmatter distribution (the charge and nucleon radii) are necessary for many calculations. These distributions were mainly investigated in the experiments on electron scattering (data on charge distribution in nuclei were retrieved) and hadron scattering (nucleon distribution in a nucleus was determined). When obtaining the secondary radioactive beams became possible, the region of the nuclei extended considerably, for which sizes could be determined directly from data of experiments in measuring the cross sections of reactions induced by these nuclei. It is known that variations in the binding energy correlate with a nucleus size. This was manifested most vividly in the region of light nuclei: a set of new interesting properties were revealed which were associated with the extremely small binding energy of valence neutrons in the nuclei on the boundary of neutron stability. So, in reactions with secondary radioactive beams of He, Li, Be, and B isotopes, an extremely high value of reaction cross section was detected for certain isotopes [10]. The values of radii of nuclear matter distribution retrieved in these experiments have demonstrated their gradual growth with an increase in the number of neutrons.For the loosely bound nuclei ^{11}Li, ^{11}Be, ^{14}Be, and ^{17}B that are close to the stability boundary, these radii exceeded substantially the values determined by the standard increment in the dependence on mass ~ $A^{1/3}$ (Fig. 3) [9].
Similar results were also obtained for the region of heavier nuclei [10]. The determination of regularities in the behavior of radii as a function of the mass, isospin, and energy within a broad range makes it possible to define a structure of exotic nuclei and to predict the existence of new nuclei with neutron halo. The systematics of only 7 pairs of mirror nuclei has confirmed the existence of a neutron halo in ^{6}He and ^{8}He isotopes, predicted a halo in ^{9}Be and ^{15}C nuclei:rms - (R_n – R_p ~ 0.20–0.30 fm), and also indicates to the inversion of the s and d orbitals in the mirror pair^{17}Ne–^{17}N [15]. Thus obtained values of nucleon distribution radii allow the probabilities of their interaction with other nuclei to be determined, which is fundamental in calculations of the nucleosynthesis scenario.

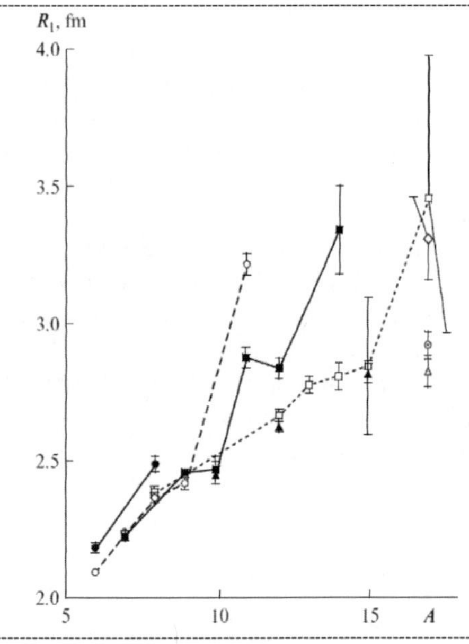

Figure 3. The radii of interaction for light elements He (●), Li (o), Be (■), B (□), C (◊), N (Δ) and F (·), retrieved from the experimental values of cross sections [9].

9. Neutron halo

Anomalous radius value of the very neutron-rich nuclei was interpreted as manifestation of the so-called neutron-halo in these nuclei, when valence neutrons in the loosely bound nuclei form a long tail in the neutron-density distribution [5]. The halo is a characteristic of the bound state lying near the threshold of particle emission. Thus, an increase in the rms radius of neutron distribution is the first signature of the halo available in exotic nuclei. Thus, the new phenomena revealed recently during the studies of lightest nuclei near the neutron stability boundary caused a need in revising some ideas on these nuclei. The data on new heavier nuclei with halo are necessary because only several nuclei with two-neutron halo (^{6}He, ^{8}He, ^{11}Li, ^{14}B, and ^{17}B) and two nuclei with single-neutron halo in all (^{11}Be and ^{19}C) have been known so far. The existence of many other halo-shaped nuclei is predicted. Vast opportunities fortheir formation and investigation are opened up when beams of radioactive nuclei are used. A problem of the order of filling the shells is important. An answer is required to the followingquestion: how do the pairing and the shells, including the deformed ones, influence the nuclear stability? More detailed data on nuclear sizes and their isospin dependence are needed. The use of secondary beams of radioactive nuclei will make it possible to find the isospin dependence of the spatial distribution of nuclear matter for many exotic nuclei. There is a question regarding existence of dineutron and tetraneutron in neutron-halo nuclei. The experimental solution to these and some other problems in the region of light neutron-rich nuclei is associated with the possibility of obtaining these nuclei in large quantities. The development of accelerator facilities producing the stable and radioactive nuclear beams has been successfully accomplishing at the recent time.

10. Nuclear reactions and nucleosynthesi

An important role in the nucleosynthesis processes is played by the nuclear reactions proceeding with the capture of protons and neutrons or heavier particles (α-particles, heavy ions) by different nuclei, including the unstable ones too. The determination of a rate of proceeding of these reactions is a very difficult experimental task. The cross section of these processes also strongly depends on a temperature of the object. In the process of the nonexplosion evolution of a star, its temperature is relatively low and the effective cross section of reactions is from a few picobarn to several nanobarn (10^{-36}–10^{-33} cm^2). With a star explosion, the temperature is very high (~10^9K) and the effective cross section is several millibarn (10^{-27} cm^2). For simulation of these processes under laboratory conditions, it is necessary to have a wide region of radioactive nuclei that play a main part in the star explosion. This opportunity has appeared recently with construction of accelerator facilities with radioactive nuclear beams. With the help of these beams, the reaction characteristics are investigated by using the targets of hydrogen and helium. Let us give several examples of these reactions. Reactions of the type of ^{8}Li(α, n)^{11}B, ^{6}He(α, n)^{9}Be, ^{1}H(6He, nγ)^{6}Li, and ^{1}H(^{9}Li, nγ)^{9}Be are fundamental for nucleosynthesis in this inhomogeneous ("big bang") process. Measurements of the reaction rates of ^{7}Be(p, γ)^{8}B, ^{7}Be(p, γ)^{9}C, ^{8}Be(p, γ)^{9}C, and ^{11}C(p, γ)^{12}N reactions are of interest for simulation of the hot proton–proton channel that may take place during the supernova explosion. Reactions with light, loosely bound nuclei, which proceed at energies close to the Coulomb barrier, are of particular interest for astrophysics. These reactions have many peculiarities that have been revealed lately by means of radioactive nuclear beams. One of these features is the enhancement in interaction cross sections in the subbarrier region of energies. This effect manifests itself most evidently for cluster nuclei (^{6,9,11}Li) [12] and also for neutron-halo nuclei (^{6,8}He) [11]. The main interaction channels for these nuclei are transfer reactions, breakup reactions, and completefusion reactions. In the case of interaction of loosely bound nuclei, the fusion process is of more intricate nature due to a large probability of breakup of these nuclei with the subsequent capture of the residue nucleus (incomplete fusion). This substantially complicates the description of interaction of these systems andleads to new unexpected effects at energies in the vicinity of the Coulomb barrier: the deepsubbarrier fusion and transfer of clusters from the loosely bound nuclei that have, as a rule, the cluster structure. So, for the neutron transfer reaction in case of interaction of ^{6}He (Fig.5b), the cross section reaches a value of several barns and has a maximum at energy near the Coulomb barrier. The large cross section of a single neutron transfer and its smooth falling to the low energy region (to 5 MeV) can be an evidence of the mechanism of interaction between a quasi-free neutron of ^{6}He nucleus and a target nucleus. These interaction peculiarities manifesting themselves by enhancement in cross section of the cluster transfer reactions and also the complete-fusion reactions in the vicinity of the Coulomb barrier are characteristic of many loosely bound cluster nuclei..Thus, from the observation of reactions with the compound-nucleus formation in the subbarrier energy region, one could draw the conclusion that a substantial (by a factor of several thousands) enhancement in cross sections of fusion reactions with the halo-shaped ^{6}He nuclei was observed in the vicinity of the Coulomb barrier. (Fig. 5) also presents the results of calculations according to the two-step model of fusion proposed in [14].

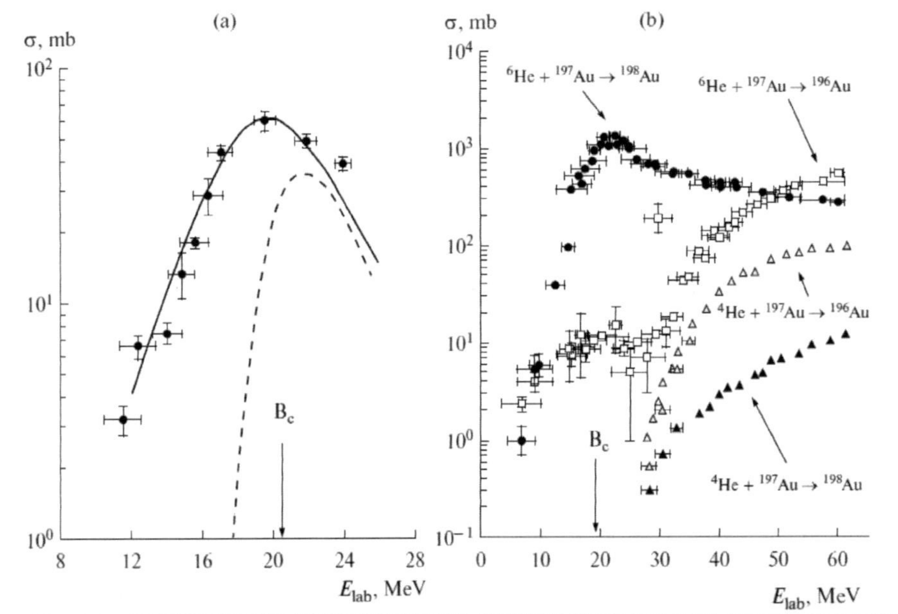

Figure 4. The energy dependences of cross sections for (a) the complet- fusion reactions of ^{6}He nuclei with ^{206}Pb with formation of the compound ^{210}Po nuclei and (b) the reactions of one- neutron transfer (●) and of one- neutron stripping (□) on the ^{197}Au nuclei with beams of ^{6}He and ^{4}He (▲ and Δ respectively).

Neutron transfer reactions for haloshaped nuclei must proceed with a large probability. Analogous conclusions on the enhancement in cross sections of fusion of ^{8}He and ^{9}Li nuclei have been recently made in [13]. The obtained results are extremely important for solving astrophysical problems, particularly for understanding of the mechanism of formation of light elements in the Universe. During the nucleosynthesis, a large cross section of interaction of cluster loosely bound nuclei (^{6}He, ^{9}Li, and ^{7}Be) can change the chains of β-decays leading to formation of different elements [23]. For example, the following channels of reactions: $^{1}H(^{6}He, n.)^{6}Li$, $^{12}C(^{6}He, 2n.)^{16}O$, $^{1}H(^{9}Li, n.)^{9}Be$, $^{3}He(^{9}Li, 2n.)^{10}B$, etc. may appear to be most probable for synthesis of light stable nuclei. This example once again confirms that the fundamental nuclear physics not only extends our knowledge of the microcosm, but also assists in development of our ideas about our ambient macroworld and makes a contribution to the adjacent fields of science and technology. The information on the structure of exotic nuclei obtained in nuclear physics experiments with the use of the stable and radioactive beams is extremely important for the solving of these or other problems of astrophysics (nucleosynthesis, cosmochronology, evolution of galaxies, formation and breakups of neutron stars and supernovas, etc.). These are only a few examples of the link between physics of atomic nuclei and physics of macroworld. Despite a small number of particles participating in the atomic nucleus formation (no more than 300), they present a unique system for simulation of the macroworld problems. Under laboratory conditions using nucleus–nucleus collisions implemented at the modern heavy ion accelerators.

11. ACKNOWLEDGMENTS

This work was supported by the Russian Foundationfor Basic Research (projects no. 07-02-00271-a, 09- 02- 91056).

12. References

[1] Valentin, Subatomic Physics: Nuclei and Particles (Hermann, North_Holland, Paris, Amsterdam, Oxford, 1981; Mir, Moscow, 1986), Vol. **1**.
[2] V. L. Ginzburg, Theoretical Physics and Astrophysics (Mir, Moscow, 1974; Pergamon, Oxford, 1979).
[3] Yu. Ts. Oganessian, Phys. At. Nucl. **69**, 932 (2006).
[4] Yu. Oganessian, Int. J. Mod. Phys. E **16**, 949 (2007).
[5] A. A. Ogloblin and Yu. E. Penionzhkevich, in *Treatiseon Heavy Ion Science*, Ed. ByD. A.Bromley (Plenum, New York, 1989), Vol. 8, p. 260.
[6] O.B.Tarasovetal.,Phys.Lett.A.**409**,64(1997).
[7] H. Sakurai et al., Phys. Rev. C **54**, R2802 (1996).
[8] T. Baumann et al., Nature (London) **449**, 1022 (2007).
[9] I. Tanihata et al., Phys. Rev. Lett. **55**, 2676 (1985).
[10] A. C. Villari et al., Phys. Lett. B **268**, 345 (1991).
[11] Yu. Penionzhkevich et al., J. Phys. G **36**, 025104(2009).
[12] W. Loveland et al., Phys. Rev. C **74**, 064609 (2006).
[13] A. Lemasson et al., Phys. Rev. Lett. **103**, 232701 (2009).
[14] V. Zagrebaev, Phys. Rev. C **67**, 061601 (2003).

Prototype with the basic architecture for the CBM-TOF inner wall tested in close to real conditions

M. Petriş, D. Bartoş, G. Caragheorgheopol, F. Constantin, M. Petrovici, L. Rădulescu, V. Simion

"Horia Hulubei" National Institute of Physics and Nuclear Engineering, Bucharest-Magurele, P.O.Box MG-6, RO-077125, Romania

I. Deppner, N. Herrmann, C. Simon

Physikalisches Institut, University of Heidelberg, Germany

J. Frühauf, M. Kiš, P-A. Loizeau

Gesellschaft für Schwerionenforschung, Darmstadt, Germany

E-mail: mpetris@nipne.ro

Abstract.
Two dimensional position sensitive timing MGMSRPC prototypes were developed for the low polar angles of the CBM - TOF wall. Four MGMSRPC counters were arranged in a staggered geometrical configuration along the z direction, with overlap along and across the strips, in order to define a basic architecture for the inner zone of the CBM-TOF wall. This configuration was tested with mixed electron-pion beam at CERN-PS and with reaction products resulted from the heavy ion induced reactions at SIS18 - GSI Darmstadt and SPS - CERN. The performance of the basic architecture in conditions close to the ones expected for their operation in the inner zone o the CBM - TOF wall at SIS100/FAIR will be presented.

1. Introduction

The Compressed Baryonic Matter (CBM) experiment is a fixed target experiment at the future experimental Facility for Antiproton and Ion Research (FAIR), with the aim to investigate the high net-baryon density matter in nucleus-nucleus collision in an energy range between 2 - 14 GeV/u at SIS100 accelerator. Due to the high interaction rates up to 10^7 interaction/s, at which the experiment is designed to run, the detectors of the innermost part of the experimental setup will be exposed to high counting rate and multiplicity environment.

The Time Of Flight (TOF) subsystem is one of the core detectors of the CBM experiment. The TOF wall in conjuction with Silicon Tracking System (STS) is foreseen to identify charged hadrons, i.e. pions, kaons and protons, in the acceptance of the system (the angular range covered by the STS detector of 2.5^0-25^0). It covers an active area of about 120 m^2, approximately rectangular in shape. A full system time resolution of at least 80 ps is needed, including all possible contributions, such as electronics jitter and the resolution of the time reference system.

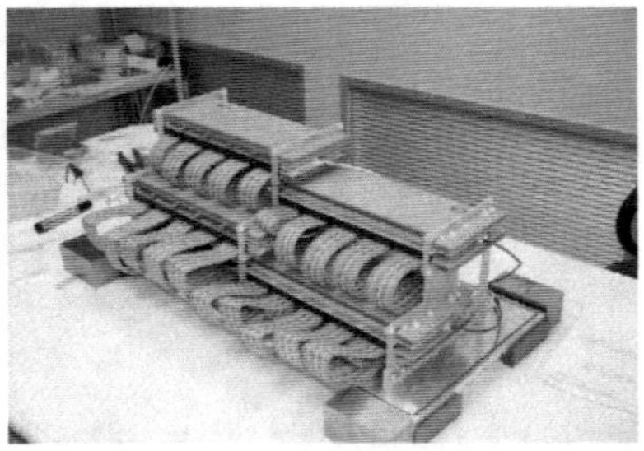

Figure 1. Sketch of the basic architecture of a module based on four staggered MGMSRPCs - left side. Photo of the four staggered MGMSRPC counters mounted on the back flange of the tight gas box - right side.

This requires a single channel time resolution better than 60 ps. At the same time the detection efficiency should be at least 95%. This performance should be maintained up to a counting rate which, very close to the beam pipe, exceeds 30 kHz/cm^2 [1]. The size of a readout channel should match an occupancy less than 5%.

Our activity has been focused on the development of a Multi-gap, Multi-strip Resistive Plate Counter (MGMSRPC) prototype for high counting rate and multiplicity environment, as it is anticipated to be in the inner zone of the CBM-TOF [2]. The double stack counter has a 10 gap structure, 5 on each stack. The gap size is 140 μm. The resistive electrodes are made from low resistivity glass, ($\sim 10^{10}$ Ωcm)[3], provided by our Chinese collaborators from CBM. The use of materials with lower resistivity than that of float glass (of $\sim 10^{12}$ - 10^{13} Ωcm, currently used in the RPC construction) is one of the way to increase the rate capabilities of RPC counters. The narrow strip pitch of 2.54 mm, combined with a strip length of 4.6 cm confers a high granularity. The in-beam test performed at the CERN PS accelerator with a mixed electron - pion beam of 6 GeV/c momentum, showed a very good performance in terms of efficiency and time resolution. The readout of each strip at both ends allows to obtain the position information along the strip with a position resolution of a few millimeters and the charge sharing based on a mean cluster size of 3 strips gives position information across the strips with a position resolution of 450 μm, better than the digital one (pitch/$\sqrt(12)$ [2].

2. A basic architecture for the inner zone of the CBM-TOF wall

Based on the positive results mentioned in the previous section, in order to decrease the number of electronic channels, a new prototype with the strip pitch equal with 7.4 mm and a strip length of 96 mm was designed and built. This new prototype has an active area of 96 x 300 mm^2, constrained by the layout of low resistivity glass plates. Two identical detectors were mounted in a tight gas box in a staggered geometry along the strips [2].

The high counting rate performed at COSY facility in Jülich, with a proton beam of 2.5 GeV/c momentum showed a very good behavior of the efficiency and time resolution as a function of counting rate [2, 4]. Up to 100 kHz/cm^2 the time resolution remains better than 70 ps and the efficiency higher than 90%. However, it should be mentioned that the detector was exposed on an area limited to the beam spot size, of about 1 cm^2.

A continuous coverage of the active area of the TOF wall requires a staggered arrangement of MGMSRPCs, with overlaps of the counter active areas on both x and y directions inside the units of the CBM-TOF wall called modules, as well between the modules themselves.

In order to configure a basic structure with such a geometry, two new identical counters were added to the previous structure. The four identical counters inside the gas tight box overlap

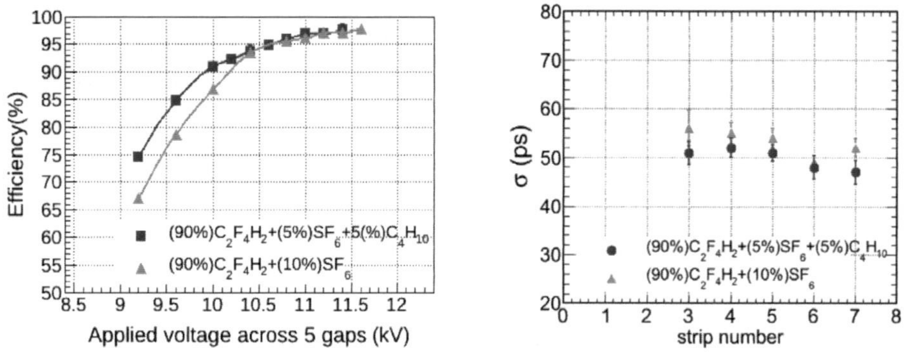

Figure 2. Working curve - left side and time resolution - right side of the prototype for a gas mixture of 85% $C_2H_2F_4$+ 5% iso-C_4H_{10}+10% SF_6 and a gas mixture without isobutane.

their active area on 16.5 mm along the strips and 17.5 mm across the strips. The arrangement of the MGMSRPCs inside the box and the cables routing is shown on the left side of Fig. 1. The right side of Fig. 1 shows a photo of the four counters mounted on the Al back plate of the box. The lateral and the front walls of the gas box are made of 10 mm thick honeycomb sheets sandwiched between two glass fiber epoxy plates of 0.4 mm which are covered on the inner side by a 0.13 mm PCB. The aluminum back plate of 12 mm thickness holds the mechanical support structure of the counters. All signals are transported through this plate via connectors on PCBs glued on the rectangular openings milled into the plate. The architecture of this prototype is designed as close as possible to a real module, the basic unit in the modular configuration of the TOF wall. A motherboard with 8 channel NINO chip [5] mounted on was used as front-end electronics (FEE) for signal processing. All FEE cards are connected on the outer side of the back plane of the housing. V1290A CAEN TDCs were used for signal digitization.

3. In-beam tests of the basic architecture prototype

3.1. In-beam test with minimum ionizing particles

The prototype performance was first tested at CERN-PS with a mixed beam of electrons and pions of 2 - 10 GeV/c momentum, in a low counting rate. The test allowed the estimation of the efficiency and the time resolution. The working curve of one of the RPC counters, shown in Fig. 2 - left side, reaches an efficiency of 97% in the plateau region. The obtained time resolution, after slewing corrections, is arround 50 ps.

The MGMSRPC performance was also investigated for two different gas mixtures flushed through the counters. A slight improvement, in both efficiency and time resolution, can be observed when 5% of isobutane is used. The effect of improving the efficiency for applied voltages lower than those corresponding to the efficiency plateau could be due to the larger average number of ionization clusters per mm produced in the gas mixture containing isobutane than in the mixture without isobutane [6]. The improvement in the time resolution is due to the quenching effect of the UV photons in the mixture containing isobutane, preventing the generation of secondary avalanches which decreases the time resolution.

3.2. In-beam test with exposure of the whole active area to reaction products

Further in-beam tests in conditions closer to the ones expected for the CBM experiment were performed with reaction products at GSI-SIS18 and CERN-SPS accelerating facilities.

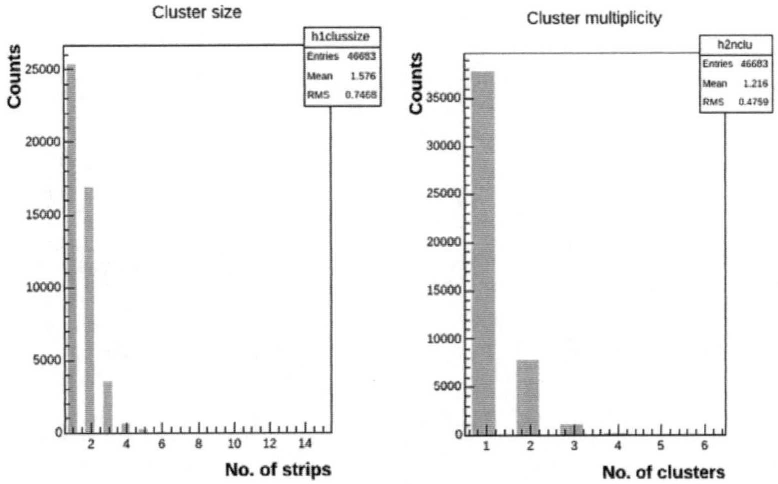

Figure 3. Cluster size - left side - and hit multiplicity - right side - distributions for one of MGMSRPC counter.

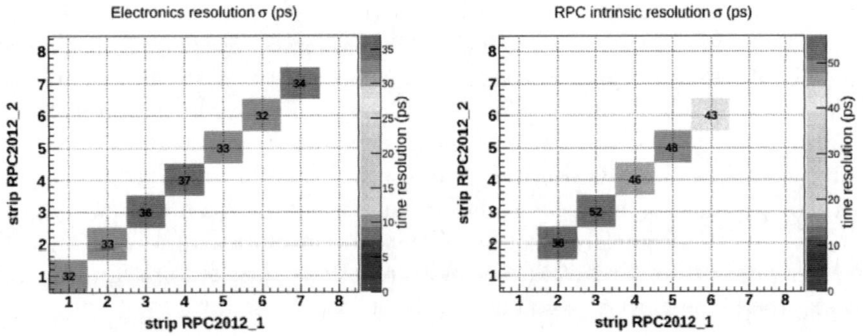

Figure 4. Time resolution of the electronic channels involved in the estimation of the resolution of the individual strips - left side. Intrinsic time resolution of the individual readout strips after quadratic subtraction of the electronic resolution - right side.

The prototype was tested using reaction products produced by an Ar beam of 18A GeV incident on a Pb target at H4 beam line of CERN SPS-facility, in a multi-hit environment. As can be seen in Fig. 4 - left side, the average cluster size is 1.6 strips per hit. The right side of Fig. 4 shows an average hit multiplicity of 1.2 hits per event.

The time of flight distribution was obtained as the coincidence between two RPCs which overlap along the strips. The intrinsic RPC time resolution after slewing corrections and quadratic subtraction of the electronic resolution is shown in Fig. 4 - right side. The value of the intrinsic time resolution for single hit events with one strip cluster size shown in Fig. 4 confirms the performance obtained in the test with minimum ionizing particles.

A high counting rate test of this prototype was performed at the SIS18 accelerator of GSI Darmstadt. The whole prototype surface was exposed to charged particles produced by Ni beam of 1.7A GeV on a 1 mm thick Pb target up to the highest intensity per spill delivered by SIS18. The detector was operated in a standard gas mixture of 85%$C_2H_2F_4$+ 5%iso-C_4H_{10}+10% SF_6

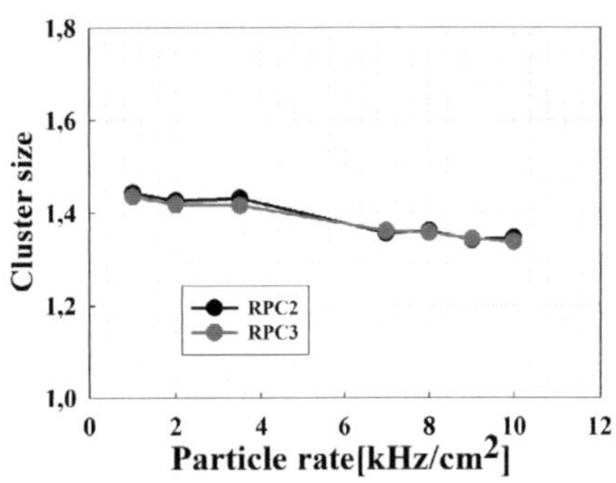

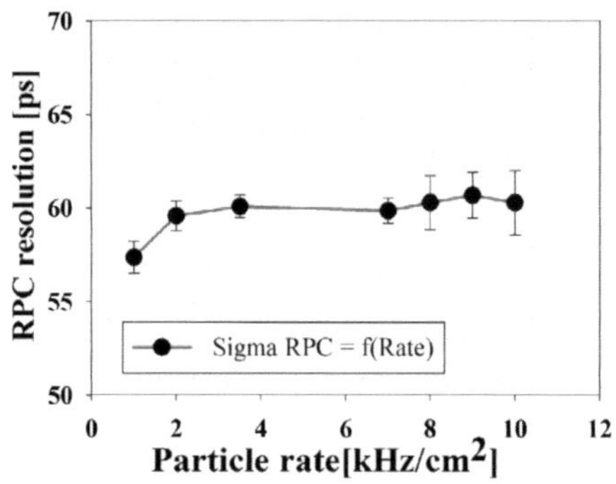

Figure 5. Cluster size - left side - and time resolution - right side - as a function of particle rate.

and at an electric field strength of 157 kV/cm. Signals of 16 strips of each counter were processed by NINO fast amplifiers, their differential outputs being converted by FPGA TDCs [7]. The cluster size as a function of the counting rate is shown in Fig. 5 - left side for two of the four counters. A decrease of about 7% is observed at 10 kHz/cm^2. The time resolution was obtained using the time difference between two MGMSRPC counters overlapped along the strips, or between two MGMSRPCs overlapped across the strips. For the first case, where the overlap is at the edge of the strip, some influence of edge effects is not excluded. After walk correction, a time resolution of ~70 ps was obtained, including electronics resolution and considering an equal contribution of the two MGMSMRPC counters. For the overlap across the strips a time resolution of ~60 ps was obtained. The time resolution as a function of counting rate is shown in Fig. 5 - right side. A slight deterioration of the time resolution of about 5% is observed up to 3 kHz/cm^2 counting rate. This value seems to remain unchanged, within the error bars, up to 10 kHz/cm^2, the highest average counting rate on the whole area of the prototype accessed in this experiment.

4. CBM-TOF inner zone design based on MGMSRPCs

The results presented in this paper recommend such an architecture as solution for the inner zone of CBM-TOF wall. A modular structure of the inner zone was designed, the modules being independent units positioned in a mechanically structure called space frame. These modules contain inside MGMSRPC counters, the front-end electronics cards being mounted on the outer side of back plane of the housing. In the current proposed design [8, 1] the CBM-TOF inner zone is covered by eight modules of three types, M1 (2 pieces), M2 (2 pieces) and M3 (4 pieces) disposed in space as it is shown in Fig. 6.

The geometric area of the inner zone is of about 14 m^2 and contains about 300 MGMSRPCs with a total of about 40000 readout channels. The design of the CBM-TOF inner zone was done such to have a full coverage with the lowest possible overlap between counters and modules, in order to minimize the number of readout channels and thus the costs and a proper matching with the design of the outer region of the TOF wall.

5. Conclusions

The prototype with the basic architecture for the inner zone of CBM-TOF wall was tested in close to real operation conditions. The MGMSRPC counters of the prototype have an efficiency of 97% and a time resolution of about 50 ps. Efficiency larger than 90% and time resolution

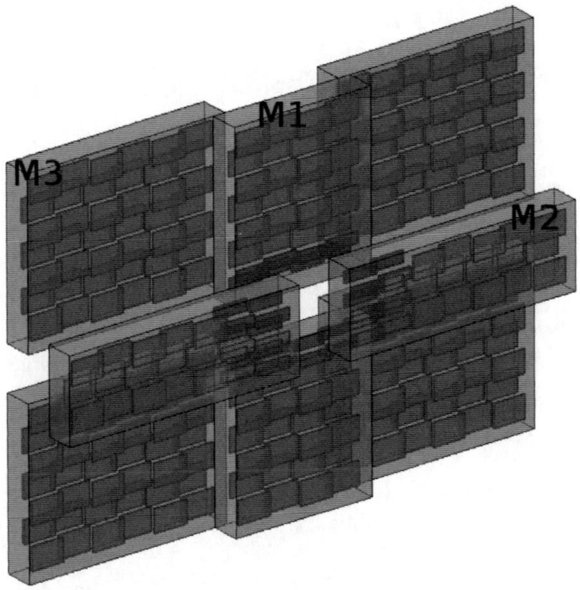

Figure 6. CBM-TOF inner zone modular configuration.

better than 70 ps up to 100 kHz/cm^2 particle flux in a focused beam and up to 10 kHz/cm^2 in exposure of the whole active area were demonstrated.

Further in-beam test with exposure on the whole active area will be dedicated to the evaluation of the prototype performance in both high counting rate and multihit environment.

Acknowledgments
This work was supported by EU-FP7/HP3-WP19 grant no.283286, project NASR/CAPACITATI/179EU, NASR/CAPACITATI/RO-FAIR/F02 and NASR/NUCLEU Project.

References

[1] CBM Collaboration, CBM-TOF Technical Design Report, October 2014.
[2] M. Petrovici et al, 2012 *Journal Of Instrumentation* **7** (JINST 7 P11003).
[3] Wang Yi, 2012 *XI Workshop on Resistive Plate Chambers and Related Detectors*, 5-10 February, Frascati.
[4] M. Petriş and M. Petrovici, Journal of Physics: Conference Series, Volume 533, 012009, 2014
[5] F. Anghinolfi et al., Nucl.Instr.and Meth. A533(2004)183
[6] W. Riegler et al. Nucl. Instr. and Meth. A, 500:144, 2003
[7] J. Frühauf et al. CBM Progress Report 2012, p.71
[8] L. Rădulescu et al., The 15th International Balkan Workshop on Applied Physics, July 2 - 4, 2015, Constanţa, Romania

Isospin-symmetry breaking and shape coexistence in A∼70 analogs

A. Petrovici
Horia Hulubei National Institute for Physics and Nuclear Engineering, R-077125 Bucharest, Romania
E-mail: spetro@nipne.ro

Abstract. Properties of proton-rich nuclei in the A∼70 region could bring insights into fundamental symmetries and interactions and may have relevance for the astrophysical scenarios on the rp-process path. Recent results concerning the interplay between isospin-symmetry-breaking and shape-coexistence effects on isospin-related phenomena in A=70 and A=74 isovector triplets obtained within the beyond-mean-field *complex* Excited Vampir variational model using an effective interaction obtained from a G-matrix based on the charge-dependent Bonn CD potential in a relatively large model space are presented.

1. Introduction

Proton-rich nuclei in the A∼70 mass region are proper candidates to get insights into fundamental symmetries and interactions and their structure and dynamics play an essential role for the astrophysical scenarios related to rapid proton capture (rp) process in the X-ray burst environment. A fundamental symmetry in nuclei, the isospin symmetry, is broken due to the Coulomb interaction between protons and in the strong interaction due to the differences in the proton-proton, neutron-neutron and neutron-proton interaction strengths because of the mass difference between the up and down quarks and electromagnetic interactions among quarks. The charge-symmetry and charge-independence breaking could be investigated studying different isospin-related phenomena like Coulomb energy differences (CED), mirror energy differences (MED), triplet energy differences (TED) as well as superallowed Fermi β-decay among the triplet T=1 nuclei. Anomalies in the Coulomb energy differences have been identified in the A∼70 mass region for nuclei supposed to manifest shape mixing at low spins [1-5]. The investigation of the structure and dynamics of nuclei near the N=Z line revealed interesting coexistence phenomena generated by the interplay between shape coexistence and mixing, competing like-nucleon and neutron-proton T=1 and T=0 pairing correlations, and isospin-symmetry-breaking interactions. In the present report first will be presented recent results on the isospin-symmetry-breaking effects on Coulomb energy differences, mirror energy differences, triplet energy differences for the A=70 and A=74 isovector triplets using the beyond-mean-field *complex* Excited Vampir variational approach based and an effective interaction obtained from a G-matrix based on the charge-dependent Bonn CD potential in a rather large model space. The structure of these nuclei has been intensively investigated both, experimentally and theoretically [5-16]. Effects of the isospin nonconserving forces on the structure of medium mass nuclei have been studied using different theoretical approaches and various effective interactions [2, 3, 4, 5, 17, 18].

Recent studies in the A∼70 mass region based on modern shell-model calculations indicated that the experimental trends in MED and TED can be reproduced adding to the Coulomb interaction some phenomenological isospin nonconserving nuclear interactions (INC), but the modern charge-dependent forces cannot account for the phenomenological strengths of the INC force [16, 18].

Superallowed Fermi β decays between 0^+ T=1 analog states providing tests of the validity of conserved-vector-current (CVC) hypothesis and the unitarity of the Cabibbo-Kobayashi-Maskawa matrix [19] have been intensively investigated in the frame of different theoretical approaches [19-24]. The present study presents recent *complex* Excited Vampir results on the superallowed Fermi β decay of the Z=N+2 members of the A=70 and A=74 isovector triplets, ^{70}Kr and ^{74}Sr nuclei, using the same theoretical procedure and ingrediences employed in the structure investigations devoted to CED, MED, and TED analysis.

Studies based on the variational approaches of the VAMPIR model family have been successfully performed for the description of a variety of coexistence phenomena in the A∼70 mass region, not only in nuclei along the valley of β-stability, but also in some exotic nuclei close to the proton drip line [24-27]. The *complex* Excited Vampir approach allows for a unified description of low- and high-spin states including in the projected mean fields neutron-proton correlations in both the T=1 and T=0 channels and general two-nucleon natural- and unnatural-parity correlations. The oblate-prolate coexistence and mixing, the variation of the deformation with mass number, increasing spin, as well as excitation energy have been nicely compared to the available experimental information. Since the Vampir approaches enable the use of rather large model spaces and of general two-body interactions, *large-scale* nuclear structure studies going far beyond the abilities of the conventional shell-model configuration-mixing approach are possible. I shall briefly describe the *complex* Excited Vampir variational procedure and define the effective Hamiltonian in the next section. In Section 3 I shall then present results on isospin-symmetry-breaking and shape-mixing effects in the A=70 and A=74 isovector triplets.

2. Theoretical framework

The *complex* Excited Vampir (EXVAM) model uses Hartree-Fock-Bogoliubov (HFB) vacua as basic building blocks, which are only restricted by time-reversal and axial symmetry. The underlying HFB transformations are essentially *complex* and do mix proton- with neutron-states as well as states of different parity and angular momentum. The broken symmetries of these vacua (nucleon numbers, parity, total angular momentum) are restored by projection techniques and the resulting symmetry-projected configurations are then used as trial wave functions in chains of successive variational calculations to determine the underlying HFB transformations. Finally the residual interaction between the orthogonal configurations is diagonalized obtaining the configuration mixing. The HFB vacua account for arbitrary two-nucleon correlations including like-nucleon as well as isovector and isoscalar neutron-proton pairing correlations.

For nuclei in the A∼70 mass region is used a ^{40}Ca core and the $1p_{1/2}$, $1p_{3/2}$, $0f_{5/2}$, $0f_{7/2}$, $1d_{5/2}$ and $0g_{9/2}$ oscillator orbits for both protons and neutrons are introduced in the valence space. We start with an isospin symmetric basis and then introduce the Coulomb shifts for the proton single-particle levels resulting from the ^{40}Ca core by performing spherically symmetric Hartree-Fock calculations using the Gogny-interaction D1S in a 21 major-shell basis [22].

The effective two-body interaction is constructed from a nuclear matter G-matrix based on the charge-dependent Bonn CD potential. In order to enhance the pairing correlations this G-matrix was modified by adding short-range (0.707 fm) Gaussians with strength of -35 MeV in the T=1 proton-proton and neutron-neutron channnel, -20 MeV in the neutron-proton T=1 channel, and -35 MeV in the neutron-proton T=0 channel. In addition, the isoscalar interaction was modified by monopole shifts for the T=0 matrix elements of the form $\langle 1p1d_{5/2}; IT=0|\hat{G}|1p1d_{5/2}; IT=0\rangle$, where 1p denotes either the $1p_{1/2}$ or the $1p_{3/2}$ orbit and $\langle 0g_{9/2}0f; IT=0|\hat{G}|0g_{9/2}0f; IT=0\rangle$, where 0f denotes either the $0f_{5/2}$ or the $0f_{7/2}$ orbitals. The Hamiltonian includes the two-body matrix elements of the Coulomb interaction between the valence protons.

3. Results and discussion

In the isovector triplets the Coulomb energy differences are defined by
$CED(J) = E_x(J, T = 1, T_z = 0) - E_x(J, T = 1, T_z = 1)$,
the mirror energy differences by
$MED(J) = E_x(J, T = 1, T_z = -1) - E_x(J, T = 1, T_z = 1)$,
and the triplet energy differences by
$TED(J) = E_x(J, T = 1, T_z = -1) + E_x(J, T = 1, T_z = 1) - 2E_x(J, T = 1, T_z = 0)$,
where E_x represents the excitation energy and $T_z = -1$ for the proton-proton pair.

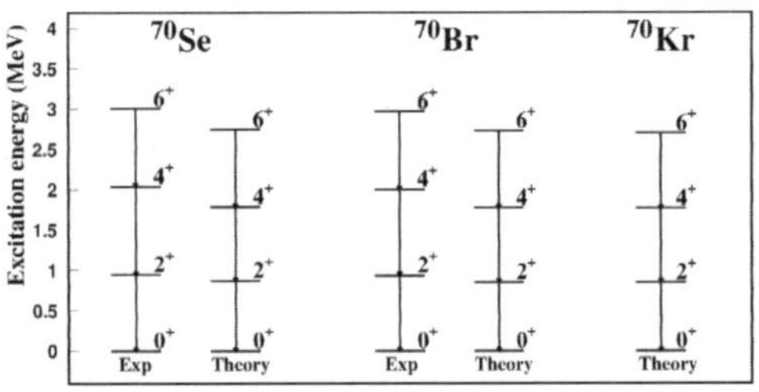

Figure 1. The *complex* Excited Vampir spectra for the analog states in ^{70}Se, ^{70}Br, and ^{70}Kr compared with the available data.

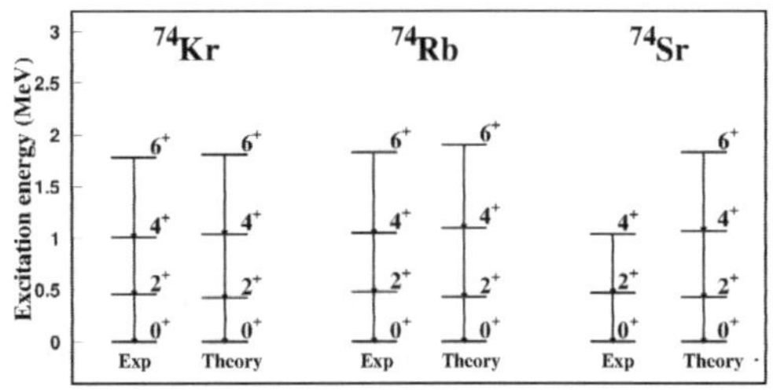

Figure 2. The *complex* Excited Vampir spectra for the analog states in ^{74}Kr, ^{74}Rb, and ^{74}Sr compared with the available data.

The results concerning the structure of the 0^+, 2^+, 4^+, and 6^+ states in ^{70}Se, ^{70}Br, and ^{70}Kr indicate a strong mixing of different prolate and oblate deformed configurations in the intrinsic system in the final wave functions. The oblate-prolate mixing is also significant but weaker in the A=74 nuclei and is changing with spin and excitation energy. In both triplets the amount of mixing is decreasing with increasing spin [5]. It is worthwhile to mention that always the yrast states are dominated by prolate deformed configurations in the intrinsic system and the first excited states by oblate components except for the lowest two 0^+ states in ^{70}Se which indicate almost egal contribution from prolate and oblate deformed EXVAM configurations.

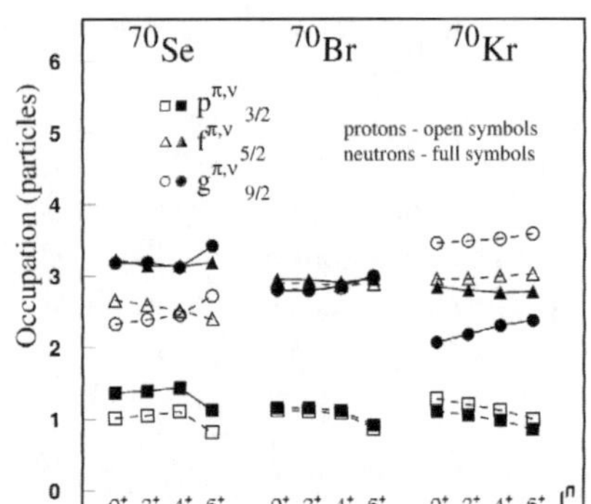

Figure 3. Occupation of valence spherical orbitals for analog states in ^{70}Se, ^{70}Br, ^{70}Kr.

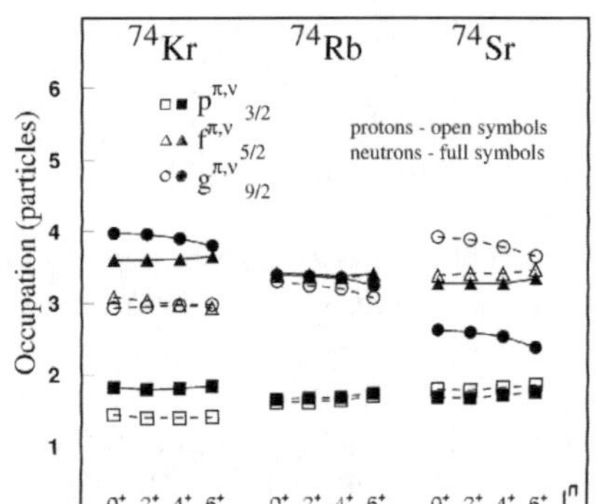

Figure 4. Occupation of valence spherical orbitals for analog states in ^{74}Kr, ^{74}Rb, ^{74}Sr.

In figures 1, 2 are presented the *complex* Excited Vampir spectra for the A=70 and A=74 analogs, respectively, compared with the available data. In figures 3, 4 are depicted the trends in the proton and neutron occupation of the valence spherical orbitals $1p_{3/2}$, $0f_{5/2}$, and $0g_{9/2}$ for the analog states in the A=70 and A=74 triplet, respectively, corroborating the evolution of the shape mixing with increasing spin.
In figure 5 are depicted the *complex* Excited Vampir results on Coulomb energy differences for the ^{70}Br - ^{70}Se and ^{74}Rb - ^{74}Kr analogs obtained using the above defined Hamiltonians based on Bonn CD potential compared with available data [6-12]. The trend manifested in the data is reproduced by the EXVAM results for the A=74 pair of nuclei as well as the anomalous behaviour revealed for the ^{70}Br - ^{70}Se case.

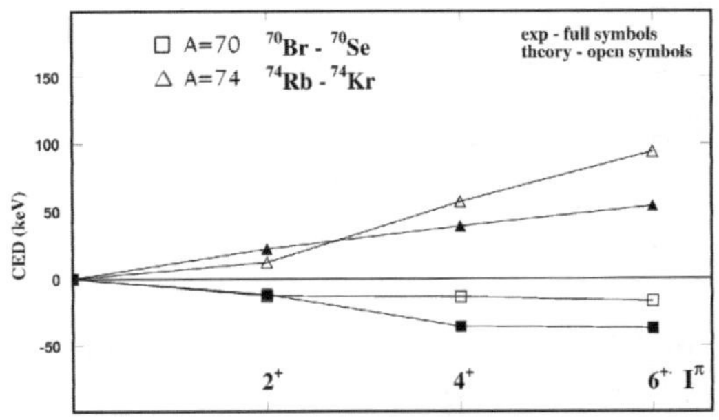

Figure 5. Comparison of the *complex* Excited Vampir results on Coulomb energy differences (CED) for A=70 and A=74 nuclei to available experimental data [6-12].

In figure 6 are presented the *complex* Excited Vampir predictions on mirror energy differences and triplet energy differences for the A=70 isovector triplet. MED manifest a negative trend, while TED indicate small positive values up to spin 4^+ and a small negative value for the spin 6^+. In figure 7 are depicted the *complex* Excited Vampir predictions on mirror energy differences and triplet energy differences for the A=74 isovector triplet. MED manifest a positive trend, while TED indicate a negative trend in agreement with the recent experimental results [16].

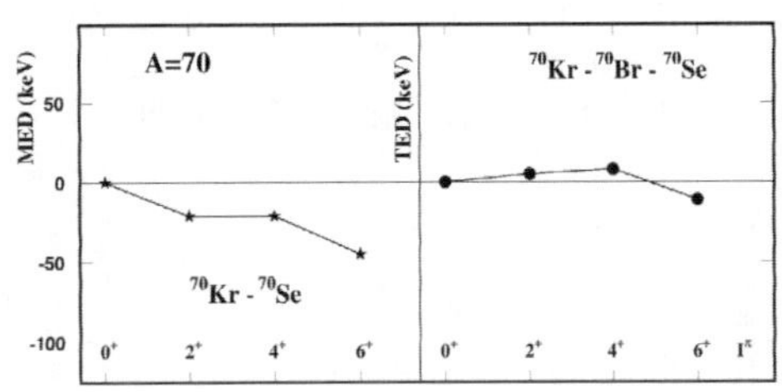

Figure 6. The *complex* Excited Vampir results for mirror (MED) and triplet (TED) energy differences in A=70 isovector triplet.

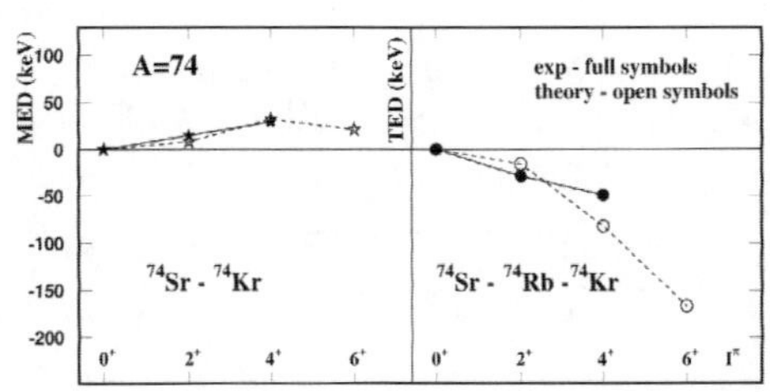

Figure 7. The *complex* Excited Vampir results for MED and TED in A=74 isovector triplet compared to data [10-12, 16].

Aiming to a unified description of the structure and dynamics of the low-spin states in the nuclei building out the two isovector triplets we extended our investigations to the isospin-mixing and shape coexistence effects on the superallowed Fermi β-decay of the Z=N+2 members of the A=70 and A=74 isovector triplets, ^{70}Kr and ^{74}Sr isotopes. We studied the Fermi β-decay for the ground state and yrast 2^+ state in ^{70}Kr to the analog states in ^{70}Br as well as for the lowest two 0^+ and 2^+ states in ^{74}Sr to the analog states in ^{74}Rb.

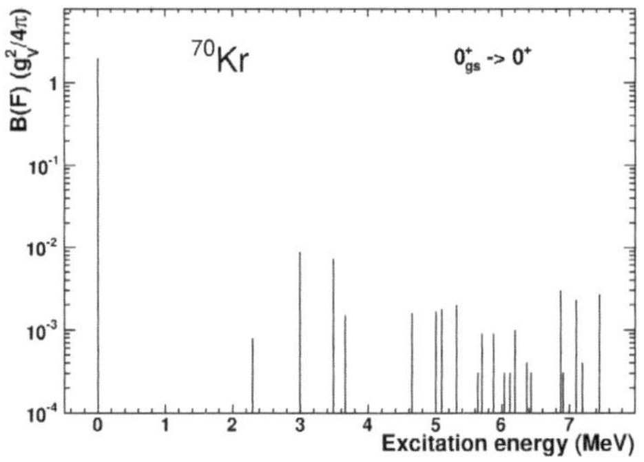

Figure 8. Fermi strength distribution for ^{70}Kr ground state decay obtained within the *complex* Excited Vampir model.

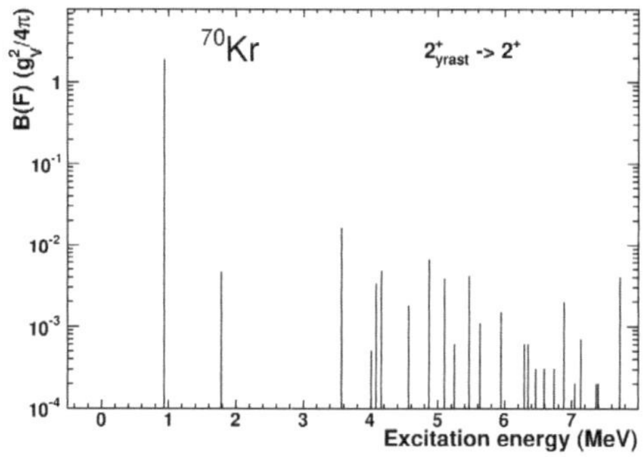

Figure 9. Fermi strength distribution for the decay of the yrast 2^+ state in ^{70}Kr obtained within the *complex* Excited Vampir model.

In figure 8 and 9 are depicted the Fermi strength distributions for the decay of the ground state and yrast 2^+ state in ^{70}Kr to the daughter states in ^{70}Br. Our results indicate for the decay of ^{70}Kr a depletion of the ground to ground transition up to 1.95% with the missing strength distributed over many excited 0^+ states. The strongest nonanalog branches are feeding the fourth and the fifth 0^+ states in ^{70}Br situated around 3 MeV excitation enegy with un upper limit of 0.4% of the total strength. For the decay of the yrast 2^+ state in ^{70}Kr the depletion of the analog branch amounts to maximum 2.98% and a significant nonanalog branch is feeding the fourth 2^+ state in ^{70}Br with an upper limit of 1.3% of the total strength.

In figure 10 and 11 are illustrated the Fermi strength distributions for the decay of the ground state and yrast 2^+ state in ^{74}Sr to the daughter states in ^{74}Rb, respectively. For the depletion of the ground to ground decay we found an upper limit of 3%, while for the yrast 2^+ decay a maximum of 3.6% of the total strength. The missing strength in the decay of the ground to ground state is distributed over many 0^+ states in ^{74}Rb with un upper limit for the branch to the second and the sixth excited 0^+ state of 0.8% from the total strength. In the decay of the yrast 2^+ state in ^{74}Sr two significant nonanalog branches feed the second ($\leq 1.3\%$) and the fourth ($\leq 0.8\%$) 2^+ states in ^{74}Rb and the rest of the strength is distributed over many 2^+ states.

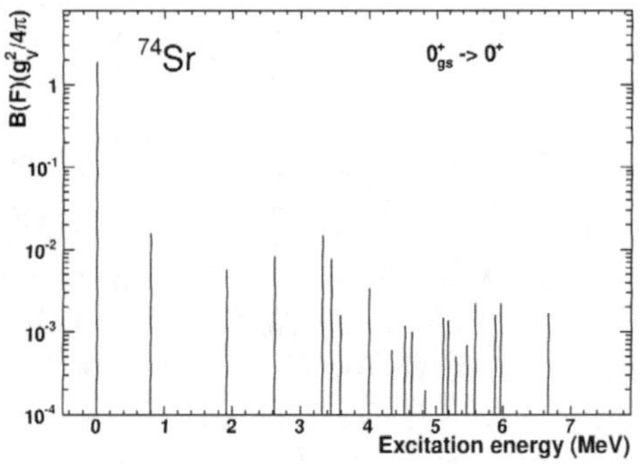

Figure 10. The same as in Fig 8, but for ^{74}Sr.

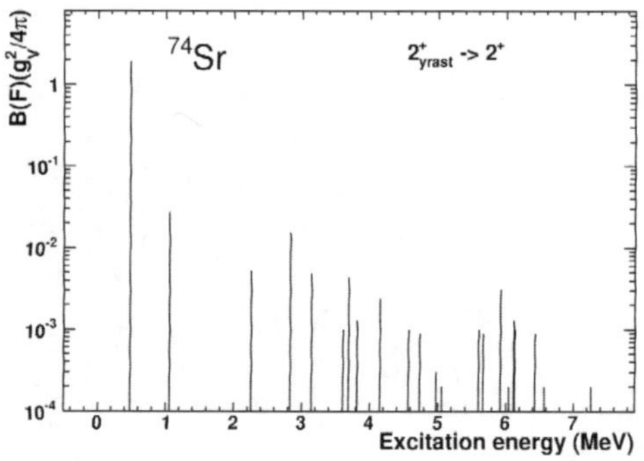

Figure 11. The same as in Fig 9, but for ^{74}Sr.

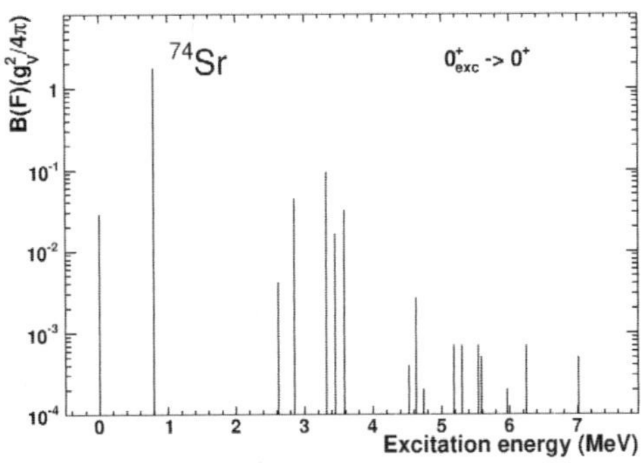

Figure 12. The same as in Fig 10, but for the first excited 0^+ state in ^{74}Sr.

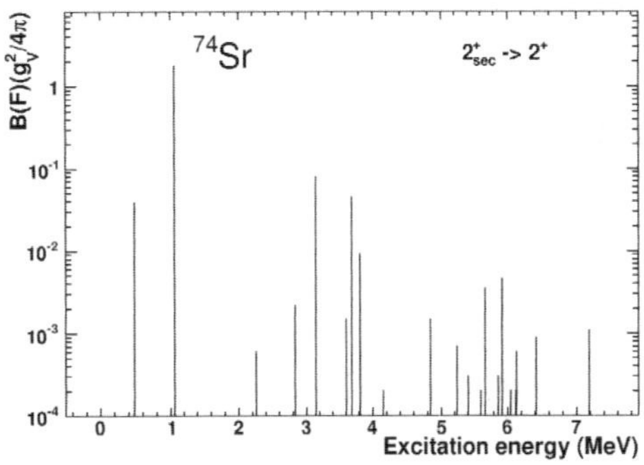

Figure 13. The same as in Fig 12, but for the second 2^+ state in ^{74}Sr.

For comparison the Fermi strength distributions for the decay of the first excited 0^+ and second 2^+ state in ^{74}Sr are illustrated in figure 12 and 13, respectively. According to *complex* Excited Vampir results the excitation energy of these states is lower than 1 MeV. Consequently, thermally populated in the X-ray burst environment their decay could have relevance for the astrophysical scenarios concerning the rp-process path.

4. Conclusions

The present report represents the first unitary beyond-mean-field treatment based on an effective two-body interaction constructed from a G-matrix starting from charge dependent Bonn CD potential of isospin-symmetry-breaking and shape coexistence effects on the structure of the A=70 and A=74 isovector triplets of nuclei as well as on the superallowed Fermi β-decay of their Z=N+2 members, ^{70}Kr and ^{74}Sr. Experimental information on the properties of the yrast and non-yrast low-lying states as well as on the predicted significant nonanalog branches are required to support our predictions.

Acknowledgments

This work was supported by the Romanian National Authority for Scientific Research, CNCS - UEFISCDI - project number PN-II-ID-PCE-2011-3-0153, CERN-ISOLDE contract 6/2012, and NUSTAR-RO contract F04/2014.

References

[1] Nara Singh B S *et al.* 2007 *Phys. Rev.* C **75** 061301
[2] Petrovici A 2010 *J. Phys. G: Nucl. Part. Phys.* **37** 064036
[3] de Angelis G *et al.* 2012 *Phys. Rev.* C **85** 034320
[4] Petrovici A 2013 *Rom. Jour. Phys.* **58** 1120
[5] Petrovici A 2015 *Phys. Rev.* C **91** 014302
[6] de Angelis G *et al.* 2001 *Eur. Phys. J.* A **12** 51
[7] Jenkins D G *et al.* 2002 *Phys. Rev.* C **65** 064307
[8] Hurst A M *et al.* 2007 *Phys. Rev. Lett.* **98** 072501
[9] Ljungvall J *et al.* 2008 *Phys. Rev. Lett.* **100** 102502
[10] Rudolph D *et al.* 1997 *Phys. Rev.* C **56** 98
[11] Chandler C *et al.* 1997 *Phys. Rev.* C **56** R2924
[12] Fischer S M *et al.* 2006 *Phys. Rev.* C **74** 054304
[13] Nichols A J *et al.* 2014 *Phys.Lett.* B **733** 52
[14] Clément E *et al.* 2007 *Phys. Rev.* C **75** 054313
[15] Iwasaki H *et al.* 2014 *Phys. Rev. Lett.* **112** 142502

[16] Henderson J *et al.* 2014 *Phys. Rev.* C **90** 051303(R)
[17] Zuker A P, Lenzi S M, Martinez-Pinedo G, and Poves A 2002 *Phys. Rev. Lett.* **89** 142502
[18] Kaneko K, Sun Y, Mizusaki T, and Tazaki S 2014 *Phys. Rev.* C **89** 031302(R)
[19] Towner I S and Hardy J C 2002 *Phys. Rev.* C **66** 035501
[20] Ormand W E and Brown B A 1995 *Phys. Rev.* C **52** 2455
[21] Sagawa H, Van Giai N, Suzuki T 1996 *Phys. Rev.* C **53** 2163
[22] Petrovici A, Schmid K W, Radu O, Faessler A 2005 *Nucl. Phys.* A **747** 44
[23] Towner I S and Hardy J C 2008 *Phys. Rev.* C **77** 025501
[24] Petrovici A, Schmid K W, Radu O, Faessler A 2008 *Phys. Rev.* C **78** 064311
[25] Petrovici A, Schmid K W, Faessler A 1999 *Nucl. Phys.* A **647** 197
[26] Petrovici A, Schmid K W, Faessler A 2000 *Nucl. Phys.* A **665** 333
[27] Petrovici A, Schmid K W, Faessler A 2003 *Nucl. Phys.* A **728** 396

First observation of the competitive double-gamma ($\gamma\gamma/\gamma$) decay process

N Pietralla[1], C Walz[1,3], H Scheit[1], T Aumann[1], R Lefol[1,2], and V Yu Ponomarev[1]

[1] Institut für Kernphysik, Technische Universität Darmstadt, D-64289 Darmstadt, Germany
[2] Department of Physics, University of Saskatchewan, Saskatoon S7N5E2, Canada

E-mail: pietralla@ikp.tu-darmstadt.de

Abstract. First observation of the competitive double-γ decay process is presented. It is a second-order electromagnetic decay mode. The 662-keV transition from the $11/2^-$ isomer of ^{137}Ba to its ground state proceeds at a fraction of 2×10^{-6} by simultaneous emission of two γ quanta instead of one. The angular correlation and energy distribution of coincident γ quanta are reproduced by a dominant $M2$-$E2$ and a minor $E3$-$M1$ contribution to the double-γ decay branch. The data are well accounted for by a calculation in the Quasiparticle Phonon Model.

1. Introduction

Progress on nuclear science has been intimately related with our experimental knowledge and theoretical understanding of radioactivity. Radioactive decays occur due to all three unified forces in nature: the strong force, the weak force, and the electromagnetic interactions. First order radioactive decay processes mediated by these interactions are known to us in terms of radioactive α, β, or γ-decay by more then a century thanks to the pioneering works of Becquerel [1], Rutherford [2], Villard [3], and others.

Besides the well-known first-order radioactive decay processes, also second order decays can play important roles in nuclear physics or particle physics, despite the fact that their matrix elements are typically several orders of magnitude weaker than first order processes due to their proportionality to a higher power in the coupling constant. One example is the double-β decay ($\beta\beta$-decay) which corresponds to a simultaneous transformation of two neutrons (two protons) into two protons (two neutrons) and corresponding leptons, i.e., two positrons (two electrons) and two electron-neutrinos (two electron-antineutrinos), for conserving charge and lepton numbers. Double-β decay processes have first been considered theoretically by Maria Göppert some 80 years ago [4] and have later on been found to be responsible for the finite lifetimes of some even-even nuclei at the edge of the valley of stability, such as e.g., ^{76}Ge, ^{100}Mo, ^{116}Cd, ^{124}Sn, ^{128,130}Te, ^{136}Xe, or ^{150}Nd, whose single β-decays are forbidden by energy conservation. The first direct detection of $\beta\beta$-decay reactions was achieved in 1987 by Elliott, Hahn and Moe [5].

While $\beta\beta$-decay processes are nuclear reactions of second order in the electroweak interaction it is surprising to find that even less data exist for nuclear decay reactions that proceed in second order in the electromagnetic interaction where two γ-quanta are simultaneously emitted

[3] Present address: Bosch AG, Reutlingen.

in a single quantum transition from one quantum state to another. In a $\gamma\gamma$-decay reaction both γ quanta with energies E_1 and E_2 share the total transition energy $E_0 = E_1 + E_2$, while the energy spectrum of the individual quanta is continuous, $0 < E_i < E_0$. These double-γ decay ($\gamma\gamma$-decay) reactions are formally analogous to $0\nu\beta\beta$-decay processes where in the latter two β-particles and in the former two γ-quanta appear in the final state and share the total transition energy. Indeed, $\gamma\gamma$-decay processes have first theoretically been postulated by Maria Göppert in her PhD thesis [6] with Max Born in Göttingen, even before discussing $\beta\beta$-decay processes.

Up to recently, $\gamma\gamma$-decay reactions in nuclei were known, only, in three particular cases, ^{16}O [7, 8] and ^{40}Ca, ^{90}Zr [9, 10], where the first excited states of these even-even nuclei have spin and parity quantum numbers 0^+ and a single-γ decay is strictly forbidden by helicity conservation. Searches for the more general situation, in which $\gamma\gamma$-decay occurs in a nuclear transition which could proceed by a single-γ decay in competition, the so-called *competitive double-γ decay* ($\gamma\gamma/\gamma$-decay), was not reported in peer-reviewed literature before. The main obstacle in previous unsuccessful attempts for the experimental discovery of the $\gamma\gamma/\gamma$-decay has always been the background stemming from the presence of the typically much more likely and hence much more abundant single-γ decays, e.g. [10, 11, 12, 13].

We will report here on the observation of the $\gamma\gamma/\gamma$-decay process. In particular, we have set out to firmly observe $\gamma\gamma/\gamma$-decay events, to study the energy distribution of the simultaneously emitted γ quanta and their angular correlation, to possibly determine their multipole order and radiation character, to measure the total double-γ-to-single-γ decay branching ratio, and thereby finally the partial $\gamma\gamma/\gamma$-decay rate. Our experimental results have initially been published elsewhere [14] and this conference contribution is predominantly based on the previously published material which at the time of the oral presentation has been under review, and on the material presented in the doctoral thesis of Dr. Christopher Walz [15]. The following section provides a brief description of the experimental set-up. Useful formulations with respect to the $\gamma\gamma/\gamma$-decay phenomenon will be compiled in section 3. An account of the data analysis and the results will be given in section 4. Then the data are confronted in section 5 with theory estimates from the Quasiparticle Phonon Model. Finally, we give a summary and outlook.

2. Experiment

We have studied the γ radiation from a standard sealed ^{137}Cs radiation source with an activity of 16.3(5) μCi. The radioactive ^{137}Cs nuclei undergo β-decay reactions with a half life of 30.07(3) years and populate to 5.3(2)% directly the stable $J^\pi = 3/2^+$ ground state of the daughter nucleus ^{137}Ba and to 94.7(2)% its $J^\pi = 11/2^-$ second excited state at an excitation energy of $E_0 = 661.659(3)$ keV [16]. This $11/2^-$ isomer has a halflife of 2.552(1) minutes. It decays to 89.86% by the emission of a 661.657(3) keV γ-ray line with multipolarity $M4$ and a reduced transition strength of $B(M4; 11/2^- \to 3/2^+) = 2.725(12)$ W.u. to the ground state and with a γ-decay branching ratio of $1.12(9)\times10^{-7}$ [17] by an $E5$ transition to the $1/2^+$ first excited state of ^{137}Ba at 283.54 keV. The remaining decay intensity of the 662-keV state is converted with a conversion coefficient of $\alpha = 0.1124$ for the 662-keV $M4$ transition.

This γ-ray source was surrounded by a ring of five large-volume LaBr$_3$:Ce scintillators at a distance of approximately 22 cm to the detector face. Every two LaBr-detectors observed the radiation source at a relative mean angle of either 72° or 144°, respectively. Compton-scattering cross-talk between the detectors was suppressed by lead walls with a typical thickness of 12 cm. A photograph of the set-up is shown in Fig. 1. The set-up was covered by scintillator bars operated in anti-coincidence mode for the data acquisition system for suppressing prompt coincidences between the LaBr detectors originating from shower events induced by cosmic rays.

The absolute photo-peak efficiency of the LaBr-detector array was measured to $\epsilon_{\rm abs} = 1.50(5)\%$ at 662 keV. An energy calibration point was continuously provided by the 662-keV γ-ray line from the ^{137}Cs source. Additional data points for the energy calibration and

Figure 1. Photograph of the experimental set-up with the anti-coincidence scintillator lid removed. Four LaBr detectors are visible and labeled 1 to 4. Heavy lead shielding reduces Compton-cross talk between the LaBr detectors.

for the definition of coincidence events were obtained from measurement runs with ^{60}Co γ-radiation standards. Single γ-ray spectra and $\gamma\gamma$-coincidence events were recorded for 1,273 hours corresponding to a total of 53 days of continuous data taking.

3. Formulation of the $\gamma\gamma/\gamma$-decay process

The total count rate $n_{\gamma\gamma/\gamma}$ of $\gamma\gamma/\gamma$-coincidence events is given by

$$n_{\gamma\gamma/\gamma} = A \cdot \frac{1}{\Gamma} \int_{4\pi} d\Omega \int_{4\pi} d\Omega' \int_0^{E_0} d\omega \frac{d^5\Gamma_{\gamma\gamma}}{d\omega d\Omega d\Omega'} \epsilon_{\text{intr}}(\omega, \Omega)\ \epsilon_{\text{intr}}(E_0 - \omega, \Omega') \tag{1}$$

where A is the activity of the source with respect to the $11/2^-$ state of ^{137}Ba at 662 keV, Γ is its total decay width, ω denotes the energy of one of the two simultaneously emitted γ quanta, $\Omega = (\theta, \phi)$ and Ω' are their angular directions, ϵ is the energy-dependent and angle-dependent absolute detection efficiency, and $\Gamma_{\gamma\gamma}$ denotes the partial decay width for double-γ decay. For an initially unoriented ensemble the quintuple-differential partial decay width in Eq. (1) can depend - besides on ω - only on the relative angle θ_{12} between the two γ quanta.

Following the work of Göppert and starting from Eqs. (A.15,A.34a) of [10] one derives [14]

$$\frac{d^5\Gamma_{\gamma\gamma}}{d\omega d\Omega d\Omega'} = \frac{\omega\omega'}{96\pi^3} \sum_{\substack{JS_1'L_1'S_1L_1 \\ S_2'L_2'S_2L_2}} P_J'(S_1'L_1'S_1L_1)P_J'(S_2'L_2'S_2L_2) \sum_l a_l^{J\xi} P_l(\cos\theta_{12}) \tag{2}$$

where ξ stands for a full set of parameters $\{S_1'L_1'S_1L_1S_2'L_2'S_2L_2\}$ specifying the parities and angular momenta of the two emitted photons with $S = 0$ for electric and $S = 1$ for magnetic transition characters. P_l are Legendre polynomials and $a_l^{J\xi}$ are further coefficients from angular momentum coupling and tabulated in the supplements of Ref. [14]. The functions

$$\begin{aligned} P_J'(S'L', SL, \omega'\omega) &= (-1)^{S+S'} 2\pi (-1)^{I_i+I_f} \omega^L \omega'^{L'} \cdot \sqrt{2L+1}\sqrt{2L'+1}\sqrt{\frac{L+1}{L}}\sqrt{\frac{L'+1}{L'}} \\ \frac{1}{(2L+1)!!(2L'+1)!!} &\cdot \sum_n \left[\left\{ \begin{array}{ccc} L & L' & J \\ I_f & I_i & I_n \end{array} \right\} \frac{\langle I_f || i^{L'-S'} M(S'L') || I_n \rangle \langle I_n || i^{L-S} M(SL) || I_i \rangle}{E_n - E_0 + \omega} \right. \\ &\left. (-1)^{L+L'+J} \cdot \left\{ \begin{array}{ccc} L' & L & J \\ I_f & I_i & I_n \end{array} \right\} \frac{\langle I_f || i^{L-S} M(SL) || I_n \rangle \langle I_n || i^{L'-S'} M(S'L') || I_i \rangle}{E_n - E_0 + \omega'} \right]. \end{aligned} \tag{3}$$

are generalised polarizabilities as given in Eq. (A.19) in Ref. [10]. They involve coherent sums of electromagnetic transition operators connecting the initial state I_i through virtual intermediate states I_n to the final state I_f by all combinations of multipoles L and L' possible by angular momentum coupling. It is convenient to define the unique-multipolar generalized polarizabilities

$$\alpha_{SLS'L'}(\omega) = \sum_n \frac{\langle I_f||i^{L'-S'}M(S'L')||I_n\rangle\langle I_n||i^{L-S}M(SL)||I_i\rangle}{E_n - E_i + \omega}. \tag{4}$$

that can be approximated for $E_n \gg E_0$ to a good accuracy by

$$\alpha_{SLS'L'} = \sum_n \frac{\langle I_f||i^{L'-S'}M(S'L')||I_n\rangle\langle I_n||i^{L-S}M(SL)||I_i\rangle}{E_n - E_0/2}. \tag{5}$$

and can be calculated in standard nuclear structure models. One obtains a full account of the partial double-γ decay width, the relative energy distribution, and the angular correlation of the $\gamma\gamma/\gamma$-coincidences as a function of the sizes of the unique-multipolar generalized polarizabilities α. In the case of a uniquely multipolar dominance, the energy distribution of the individual γ-quanta depends on the involved multipolarities L and L' according to

$$P(\omega) \propto \left[\omega^{(2L+1)}(E_0 - \omega)^{(2L'+1)} + \omega^{(2L'+1)}(E_0 - \omega)^{(2L+1)}\right]. \tag{6}$$

This energy distribution is symmetric around $\omega = E_0/2$ and vanishes for the extreme values of $\omega = 0$ or E_0. For identical multipoles $L' = L$ it maximizes at $E_0/2$. If more than one unique-multipolar generalized polarizability contribute significantly, then the energy distribution becomes a superposition including interferences according to Eqs. (2,3). From a combined analysis of the energy distribution of the simultaneously emitted γ quanta, their angular correlation, and the absolute value of the partial double-γ decay width $\Gamma_{\gamma\gamma}$ one can determine the unique-multipolar generalized polarizabilities α absolutely, apart from an irrelevant global sign.

4. Data analysis

The detected γ events that have not been vetoed by anti-coincidence with signals from the plastic detectors were analyzed with respect to their time difference. Clear signals of $\gamma\gamma$ coincidences were observed on a considerable background of random coincidences. Fig. 2 shows the time-difference spectrum for all detector combinations that share a mean relative angle of 72°. The prompt peak has a structure. It consists of an about 5 ns broad bump and a narrower peak with a width of about 2 ns. It has been verified that the broad bump is still present even when the ^{137}Cs source was removed. It most likely originates from non-vetoed showers from energetic cosmic radiation. The narrow peak is expected to contain shower background and background from Compton scattering of 662-keV single-γ decays along with the sought-after $\gamma\gamma/\gamma$ events.

Fig. 3 shows the sum-energy spectrum of all pairs of LaBr-detectors with a mean relative angle of 72°, once gated on the narrow prompt time peak as indicated in Fig. 2 and once with an appropriately scaled time gate on random background, only. A peaked excess of counts at a sum energy of about 662 keV is clearly visible.

The random coincidences can be subtracted from the prompt coincidences. The resulting background-subtracted sum-energy spectrum is shown in Fig. 4. A significant sum-energy peak is obtained above a background which otherwise smoothly decreases as a function of energy. When fitted with a Gaussian function consistent with the calibrated detector response above a smooth phenomenological background, as indicated by the solid and dotted curves in Fig. 4, we obtain a peak area of 693(95) counts with a centroid at 661.6 ± 1.6 keV. The latter value is in very good agreement with the $11/2^- \to 3/2^+_{\text{gs}}$ transition energy $E_0 = 661.66$ keV, albeit obtained here from the sum of two γ-ray energies of less than 481 keV, each. The statistical significance of this signal amounts to 7.3 standard deviations, substantially exceeding the required significance

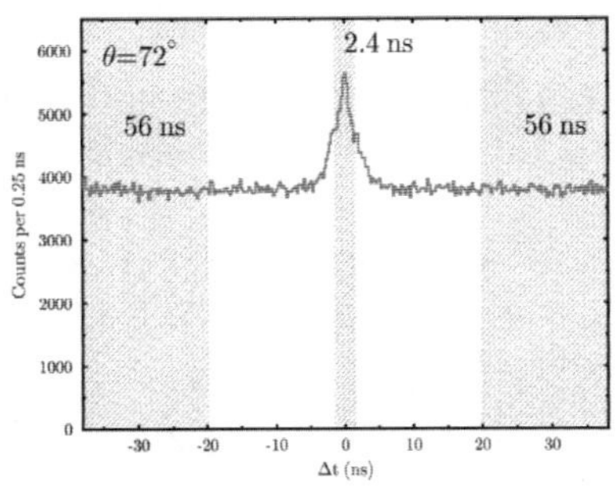

Figure 2. Distribution of time differences of events registered in either two LaBr-detectors with a relative mean observation angle of 72°. The shaded areas represent the time gates used to obtain the sum-energy spectra shown in Fig. 3.

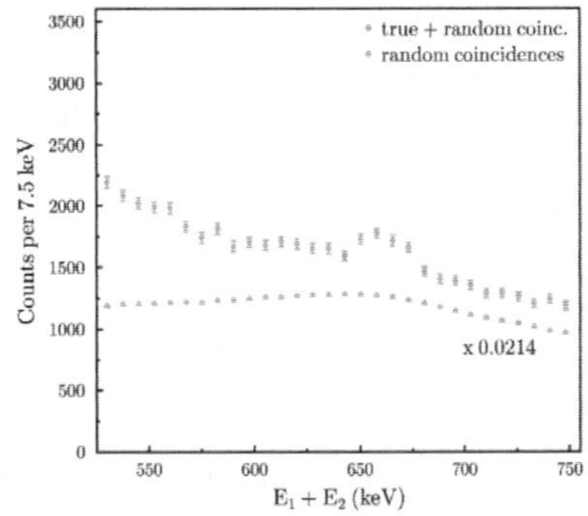

Figure 3. Histograms of events gated with the time differences indicated by the shaded areas in Fig. 2 and plotted as a function of the sum of the energies registered in both detectors. The histograms have been scaled by the relative widths of their time gates.

for a three-star discovery in particle physics. From a detailed analysis of the time difference spectrum obtained by gating on the sum-energy events rather then on the time differences, we can firmly exclude [14] that the sum-energy peak at 662 keV results from Compton scattering of a single 662-keV γ quantum from one LaBr-detector to another because of the additional time delay of $\Delta t > 0.8$ ns such a process would induce. This information fully proves that the sum-energy peak must originate from the simultaneous emission of two γ-quanta that share the total transition energy, thereby firmly establishing the discovery of the $\gamma\gamma/\gamma$-decay process.

For the group of detectors with relative observation angle of 144° a peak area of 307(78) counts has been observed at a centroid of 664.2 ± 2.8 keV. The count rate asymmetry for both detector groups points at a pronounced anisotropy of the $\gamma\gamma/\gamma$-angular correlation function.

For a quantitative analysis of the $\gamma\gamma$-angular distribution and of the distribution of energy amongst the simultaneously emitted γ quanta, we restrict ourselves to the expected two dominant multipole channels: $M2 - E2$ and $E3 - M1$. Then, Eq. (2) simplifies to

$$\frac{d^5\Gamma_{\gamma\gamma}}{d\omega d\Omega d\Omega'} = A_{qq}(\alpha^2_{M2E2}) + A_{od}(\alpha^2_{E3M1}) + A_x(\alpha_{M2E2} \cdot \alpha_{E3M1}) \tag{7}$$

which only depends on the absolute values of the two unique-multipolar generalized polarizabilities α_{M2E2} and α_{E3M1} and on their relative sign. Apart from that, the quantities A

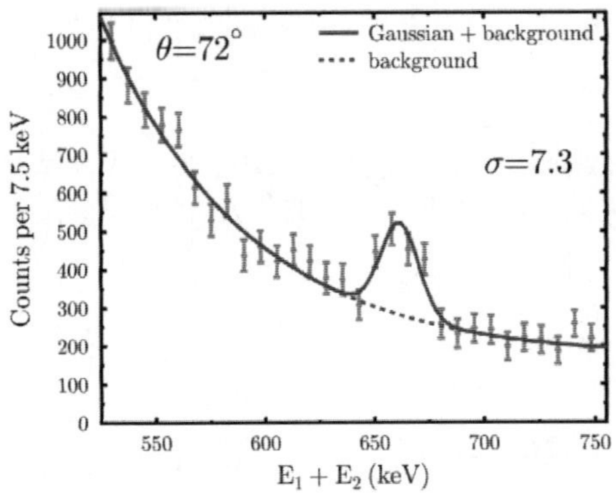

Figure 4. Sum-energy histogram for events recorded with LaBr-detectors located at a mean relative angle of 72° after subtraction of random background as indicated in Figs. 2 and 3. For further reduction of background it has been required that neither single-γ energy was smaller than 181 keV (or larger than 481 keV, correspondingly). The error bars indicate the statistical uncertainties for each channel.

for the quadrupole-quadrupole, the octupole-dipole and the interference terms are known [14] functions of the correlation angle θ_{12} according to the formulae from section 3.

From a fit of Eq. (7) to our data on the absolute angular distribution (2 absolute data points when having taken into account the known total decay rate of the $11/2^-$ isomer) and on the energy distribution (7 data points) we obtain [14] the unique-multipolar generalized polarizabilities and the total $\gamma\gamma/\gamma$-decay branching ratio as given in Table 1. Apparently, the $M2-E2$ double-γ decay mode dominates over an alternative $E3-M1$ decay which however contributes constructively to the $\gamma\gamma/\gamma$-decay of the $11/2^-$ isomer of ^{137}Ba.

5. Comparison to nuclear model calculations

We have applied the Quasiparticle Phonon Model [18] which has been developed by Professor Soloviev at Dubna. This conference commemorates the 60-years long collaboration between Bulgarian scientists and Dubna. The QPM provides a rather satisfactory description of the structure of ^{137}Ba and was used here [14] for the first time for calculating unique-multipolar generalized polarizabilities. For details of the QPM application to odd-mass nuclei we refer to [19]. The calculation is confronted with the data in Table 1.

The structures of the $3/2^+$ ground state and of the $11/2^-$ isomer of ^{137}Ba are dominated by neutron holes in the $\nu(2d_{3/2})$ and $\nu(1h_{11/2})$ orbitals outside the semi-magic core ^{138}Ba. The calculated dominance of the $M2-E2$ contribution to the $\gamma\gamma/\gamma$-branching ratio can be understood microscopically from the dominant contribution of the first excited $7/2^+$ state of ^{137}Ba at an excitation energy of 1252 keV [16]. Its structure is calculated as a superposition of a neutron hole in the $\nu(1g_{7/2})$ orbital outside the semi-magic core ^{138}Ba and the 2^+ core excitation at 1436 keV coupled to the $\nu(2d^{-1}_{3/2})$ ground state configuration of ^{137}Ba. Consequently, there exists a sizeable $M2$ matrix element between the $11/2^-$ isomer and this $7/2^+_1$ state of ^{137}Ba with a strength of the order of a single-particle transition. At the same time there exists a sizeable $E2$ matrix element from the $7/2^+_1$ state to the ground state of ^{137}Ba corresponding to the annihilation of the quadrupole core excitation. Since, in addition, the energy denominator is small the virtual contribution of the $7/2^+_1$ state dominates the α_{M2E2} coefficient and enhances it sufficiently such that it dominates the $\gamma\gamma/\gamma$-decay intensity to about 80%.

6. Summary

We have presented the first observation for the competitive double-γ decay process, which can be considered a new phenomenon of radioactive decay. This discovery was made on the 662-keV decay transition from the $11/2^-$ isomer of ^{137}Ba to its ground state. The $\gamma\gamma/\gamma$-branching ratio was found to be on the order of 10^{-6}. The observation became possible due to the employment of large-volume LaBr-detectors that feature a high detection efficiency, reasonable energy resolution together with good time resolution. Thanks to the time resolution it was possible to demonstrate that the peaks in the coincident background-subtracted sum-energy spectra cannot originate from sequential Compton scattering between the detectors which has

Table 1. Measured values of the competitive-$\gamma\gamma$ branching ratio and α_{M2E2}, α_{E3M1} coefficients and comparison to theory. The uncertainties include the statistical error from the fit (±1 standard deviation) and systematic contributions.

	exp	QPM
$\Gamma_{\gamma\gamma}/\Gamma_\gamma$ (10^{-6})	2.05(37)	2.69
α_{M2E2} (e^2fm^{4}MeV^{-1})	+33.9(2.8)	+42.60
α_{E3M1} (e^2fm^{4}MeV^{-1})	+10.1(4.2)	+9.50

been the main experimental obstacle in previous attempts for observing the $\gamma\gamma/\gamma$-branching ratio. The observed angular distribution and energy distribution of coincident γ quanta provide evidence for dominant $M2-E2$ and a minor $E3-M1$ contribution to the double-γ decay branch in competition to the $M4$ single-γ transition. The data were well accounted for by a calculation using the Quasiparticle Phonon Model.

7. Outlook

The phenomenon of double-γ decay gives experimental access to generalized nuclear polarizabilities, observables studied very little up to now, and only in the dipole sector. Our observation of $\gamma\gamma/\gamma$-decay demonstrates the feasibility to study generalized nuclear polarizabilities in a broader scope, potentially not limited to one transition of particular even-even nuclei with a first excited 0^+ state. Still, due to the long measurement time and significant difficulties in reducing background it remains to be seen how these measurements can develop.

Furthermore, theoretical work is needed for clarifying the significance of the generalized nuclear polarizabilities for the advance of physics. One obvious route of studies may address the relation of the generalized nuclear polarizabilities with the dipole polarizability of the nuclear ground state which itself has been shown to be related to the symmetry energy parameters of the nuclear equation state. It would also be very desirable to clarify if the nuclear models that are used to predict the $M^{(0\nu)}$ NMEs of $0\nu\beta\beta$-decay are capable of quantitatively describe the formally analogous generalized nuclear polarizabilities that in contrast to the $M^{(0\nu)}$ NMEs are experimentally accessible nuclear observables. The recent progress nurtures hope that a new sub-field of γ-ray spectroscopy may open up.

Acknowledgments

We thank D. J. Millener, R. J. Sutter and C. J. Lister for discussions. H.S. thanks D. Schwalm for initially raising interest in this topic. Our research was supported by the DFG having funded the LaBr detectors within the grant SFB 634 at TU Darmstadt. It was partially funded also by HIC for FAIR within the LOEWE initiative of the State of Hesse. One of us (N.P.) thanks the DAAD for the German-Bulgarian bilateral project-related personnel-exchange program with project-ID 57082997 and for support at the conference.

References

[1] Becquerel M H 1896 *Comptes Rendus de l'Académie des sciences* **122** 420
[2] Rutherford E 1899 *Philosophical Magazine, 5.Series, Vol. 47* **284** 109
[3] Villard P U 1900 *Presentation at the meeting of the French Académie des sciences* (9.4.1900)
[4] Göppert M 1935 *Phys. Rev.* **48** 512
[5] Elliott S R, Hahn A A, and Moe M K 1987 *Phys. Rev. Lett.* **59** 2020
[6] Göppert M 1930 *Über Elementarakte mit zwei Quantensprüngen* (Doctoral thesis, Universität zu Göttingen).
[7] Watson B A, Bardin T T, Becker J A, and Fisher T R 1975 *Phys. Rev. Lett.* **35** 1333
[8] Hayes A C *et al.* 1990 *Phys. Rev.* C **41** 1727
[9] Schirmer J *et al.* 1984 *Phys. Rev. Lett.* **53** 1897
[10] Kramp J *et al.* 1987 *Nucl. Phys.* A **474** 412
[11] Beusch W *et al.* 1960 *Helv Phys. Acta* **33** 363
[12] Basenko V K *et al.* 1992 *Bull. Russ. Acad.* **56** 94
[13] Lister C J *et al.* 2013 *Bull. Am. Phys. Soc.* **58**(13) DNP.CE.3
[14] Walz C, Scheit H, Pietralla N, Aumann T, Lefol R, and Ponomarev V Yu 2015 *Nature* **526** 406
[15] Walz C 2014 *The two-photon decay of the* $11/2^-$ *isomer of* ^{137}*Ba and mixed-symmetry states of* ^{92,94}*Zr and* ^{94}*Mo* (Doctoral thesis, Technische Universität Darmstadt)
[16] Browne E and Tuli J K 2007 *Nucl. Data Sheets* **108** 2173
[17] Moran K, McCutchan E A, Lister C J, Zhu S, Carpenter M P, Chowdhury P, Greene J P, Lauritsen T, Merchan E, and Shearman R 2014 *Phys. Rev.* C **90** 041303(R)
[18] Soloviev V G 1992 *Theory of Atomic Nuclei: Quasiparticle and Phonons* (Bristol, IOP)
[19] Gales S, Stoyanov Ch, and Vdovin A I 1988 *Phys. Rep.* **166** 125

Nuclear Resonance Fluorescence off ^{54}Cr: The Onset of the Pygmy Dipole Resonance

P C Ries[1], T Beck[1], J Beller[1], Krishichayan[2], U Gayer[1], J Isaak[3,4], B Löher[1,4], L Mertes[1], H Pai[1], N Pietralla[1], C Romig[1], D Savran[3,4], M Schilling[1], W Tornow[2], V Werner[1], M Zweidinger[1]

[1] Institut für Kernphysik, TU Darmstadt, Schlossgartenstraße 9, 64289 Darmstadt, Germany
[2] Triangle Universities Nuclear Laboratory, Duke Univ., P.O. Box 90308, Durham, NC 27708-0308, USA
[3] Frankfurt Institute for Advanced Studies, Ruth-Moufang-Straße 1, 60438 Frankfurt am Main, Germany
[4] GSI Helmholtzzentrum für Schwerionenforschung, Planckstraße 1, 64291 Darmstadt, Germany

E-mail: pries@ikp.tu-darmstadt.de

Abstract. Low-lying electric and magnetic dipole excitations (E1 and M1) below the neutron separation threshold, particularly the Pygmy Dipole Resonance (PDR), have drawn considerable attention in the last years. So far, mostly moderately heavy nuclei in the mass regions around $A = 90$ and $A = 140$ were examined with respect to the PDR. In the present work, the systematics of the PDR have been extended by measuring excitation strengths and parity quantum numbers of $J = 1$ states in lighter nuclei near $A = 50$ in order to gather information on the onset of the PDR. The nuclei ^{50,52,54}Cr and ^{48,50}Ti were examined via bremsstrahlung produced at the DArmstadt Superconducting electron Linear ACcelerator (S-DALINAC) with photon energies up to 9.7 MeV with the method of nuclear resonance fluorescence. Numerous excited states were observed, many of which for the first time. The parity quantum numbers of these states have been determined at the High Intensity Gamma-ray Source (HIγS) of the Triangle Universities Nuclear Laboratory in Durham, NC, USA. Informations to the methods and the experimental setups will be provided and the results on ^{54}Cr achieved will be discussed with respect to the onset of the PDR.

1. Introduction

The Pygmy Dipole Resonance (PDR) [1] is one of three components of the $E1$ transition strength below and around the neutron separation threshold besides two-phonon states [2] and the Giant Dipole Resonance (GDR). It is usually associated with an accumulation of isolated $E1$-transitions in contrast to the continuous distribution of the GDR above the particle separation threshold. Given that the GDR has been subject of intensive research since the 1940's, it is well known and is interpreted as an oscillation of the neutrons against the protons [3]. On the other hand the nature of the PDR is not yet understood. It is usually interpreted as an oscillation of an $N = Z$ core against a skin of excess neutrons. However, this notion is under current debate.

Since the PDR does not occur in the calcium isotopes [4] and is fully developed already in heavier nickel isotopes [5], the A≈50 mass region may provide crucial information on the PDR. With the investigation of this mass region, we complete a systematic trend of the PDR over several mass regions from its onset to well-developed cases. In order to provide the necessary experimental data, measurements on stable even-even chromium and titanium nuclei were conducted. Here, the $B(E1)$ values were determined via the nuclear resonance fluorescence method, which has proven to be an effective means in the past.

2. Experiment

2.1. Nuclear resonance fluorescence

In the Nuclear Resonance Fluorescence method (NRF) [6] incident photons are used to populate resonantly excited states. These states decay typically within femto- to picoseconds either direct to the ground-state, which is called elastic scattering, or through intermediate states, called inelastic scattering. The emitted photons are detected and their energy, intensity and angular distribution allow conclusions regarding cross sections, transition strength, parity and spin quantum numbers as well as branching ratios. High-resolution photon scattering experiments are feasible only below the particle threshold, where discrete states can be observed.

The experiments were performed at two facilities: at the Darmstadt High Intensity Photon Setup (DHIPS) [7] located at the S-DALINAC of the Institut für Kernphysik, TU Darmstadt, Germany and at the High Intensity γ-ray Source (HIγS) [8] at the Duke Free Electron Laser Laboratory of the Triangle Universities Nuclear Laboratory in Durham, NC, USA. The experiments will be exemplified through the nucleus ^{54}Cr, which was examined through NRF for the first time.

2.2. Darmstadt High-Intensity Photon Setup

At DHIPS the electron beam provided by the S-DALINAC is used to produce bremsstrahlung photons, impinging on a copper or gold radiator. In order to reduce the contribution of low-energy photons a beam hardener made of aluminium is placed behind the radiator. Through a massive copper block with a 12 mm bore, photons are collimated on the target located between three HPGe detectors. These are positioned at 90°, 93° and 130° polar angle with respect to the incoming photon beam. At these angles the ratio of the count rate between 90° and 130° differs the most for the spin sequences $0 \rightarrow 1 \rightarrow 0$ and $0 \rightarrow 2 \rightarrow 0$, respectively. Since only $J = 1$ and $J = 2$ states are observed, it is possible to determine the spin quantum numbers of the excited states from this counting ratio. Additional active BGO-shielding is used around the HPGe detectors to reduce Compton background in the spectra. Figure 1 shows a schematic of DHIPS. For further details on the setup, see Ref. [7]. Two energy settings for the S-DALINAC were used to produce bremsstrahlung spectra with 9.7 and 7.5 MeV endpoint energy for approximately 125 and 100 h, respectively. The target consisted of 1500 mg ^{54}Cr enriched to 95.58 %, bound in chromium monoxide and enclosed in 191.8 and 207.5 mg ^{11}B, respectively, which is used for energy and photon flux calibration.

2.3. High Intensity γ-ray Source

At HIγS photons are produced in a free electron laser utilizing electrons provided by a linac and booster synchrotron. These photons are reflected inside the laser resonator cavity and are boosted to energies suitable for NRF via Compton back scattering off the oncoming electrons. After passing a collimator, a quasi-monochromatic photon beam is provided for the experiments. For details see Ref. [8]. Since Compton scattering preserves the polarisation of the primary laser photons, the beam has nearly 100 % polarization. This allows a determination of the excited states' parity quantum numbers [9]: For positive parity the intensity of the emitted photons has a maximum in the polarization plane of the incident photons and a minimum perpendicular to it and vice versa for negative parity. Due to the quasi-monochromatic energy profile the beam has been tuned to several energy settings. Twenty settings between 5.5 and 10 MeV were

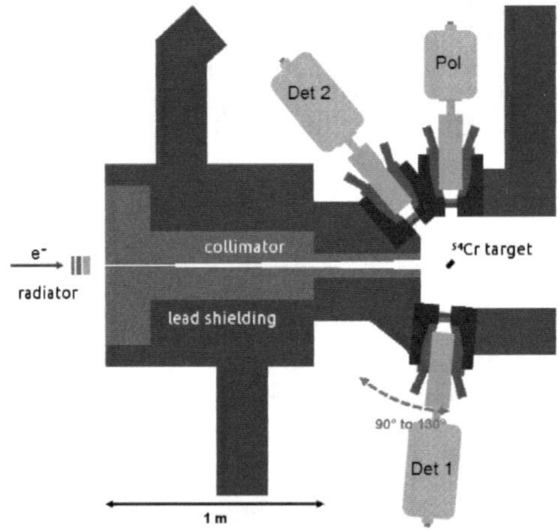

Figure 1. Schematic display of DHIPS. Lead shielding, BGO-shielding and copper collimator are shown in grey, blue and red. Figure adapted from Ref. [10]

measured, each for 4 to 5 hours. The same target of ^{54}Cr as at DHIPS was used, but without boron enclosure.

3. Analysis

In order to determine the parity quantum number of the excited states, an asymmetry ε is defined according to Ref. [9]:

$$\varepsilon = \frac{N_{\parallel} - N_{\perp}}{N_{\parallel} + N_{\perp}}, \tag{1}$$

where $N_{\parallel}$ and $N_{\perp}$ are the γ-ray intensities in and perpendicular to the polarization plane. The analysing power of the NRF reaction and hence the ideal asymmetry equal either +1 or -1. Because of the finite opening angle of the detectors this sensitivity is not fully reached. Nevertheless, a clear assignment of parity quantum numbers is possible.

The spin quantum numbers are determined through the measurements at DHIPS via the ratio w of intensities N at 90° and 130°:

$$w = \frac{N(90^\circ)}{N(130^\circ)}. \tag{2}$$

Detection efficiencies for both experiments have been obtained from a measurement of a ^{56}Co calibration source and extrapolated to 10 MeV photon energy using the detector responses simulated with the Monte-Carlo-tracking simulation tool GEANT4 [11]. The expected values for w are 2.26 for quadrupole and 0.71 for dipole transitions. The assignment of spin quantum numbers is complemented by known values taken from Ref. [12] and by the parity measurement. Since almost exclusively $E1$, $M1$ and $E2$ transitions are excited by real photons, spin $J = 1$ has been assigned to all states with negative parity.

The data taken at DHIPS are also used to determine the integrated cross section. Here the essential formula is:

$$I_{j\to i} = \frac{A_{j\to i}}{N_\gamma(E) \cdot N_T \cdot \varepsilon(E) \cdot W_{0\to j\to i}}, \tag{3}$$

where the integrated cross section $I_{j\to i}$ is calculated from the peak area $A_{j\to i}$, the photon flux $N_\gamma(E)$, the number of nuclei in the target N_T, the efficiency $\varepsilon(E)$ and the angular distribution

$W_{0\to j\to i}$. Because the photon flux is not known, the well known integrated cross sections of the excited states of the calibration standard ^{11}B are used. With that the photon flux at 2125, 4445, 5020, 7286 and 8920 keV can be measured, and after interpolating using the Schiff-formula [13] for bremsstrahlung, the cross sections for the states in ^{54}Cr can be determined. One obtains the transition width Γ from the equation:

$$\frac{\Gamma_0^2}{\Gamma} = \frac{I_{0\to j\to 0}}{\pi^2 \cdot \left(\frac{\hbar c}{E_r}\right)^2 \cdot g} \tag{4}$$

if either the ground state decay branching ratio $\frac{\Gamma_0}{\Gamma}$ has been measured or with the approximation $\frac{\Gamma_0^2}{\Gamma} \approx \Gamma$, where Γ_0 is the ground-state transition width, E_r the excitation energy and $g = \frac{2J_j+1}{2J_0+1}$ a spin dependent statistical factor. In cases where inelastic transitions have been observed, Γ_0 was corrected accordingly. From that the transition strength $B(\lambda L)$ is calculated via

$$\Gamma_0 = 8\pi \sum_{L}^{2L+1} \left(\frac{E_\gamma}{\hbar c}\right)^{2L+1} \cdot \frac{L+1}{L((2L+1)!!)^2} \cdot B(\lambda L)\downarrow . \tag{5}$$

In that way, numerous states have been observed for the first time.

4. Discussion

The measured $B(E1)$ values were compared to calculations within the framework of the quasi-particle phonon model. These calculations were capable of satisfactorily reproducing the level density and transition strengths. predicting an isoscalar character of the excited states in the energy region of interest near the neutron separation threshold. This allows a differentiation from the isovector GDR, whereby the properties usually describing the PDR are present. Similar results were already published for the closed-shell neighbouring isotope ^{52}Cr [14].

With the newly acquired data a comparison between the even-even chromium isotopes measured under the same conditions is possible. A common way to do so is the usage of the so-called Thomas-Reiche-Kuhn Energy Weighted Sum Rule (EWSR), which gives an approximation for the summed up electric dipole transition strength as function of proton and neutron number [15]. The ratio of the total observed $E1$ transition strength up to 9.7 MeV to the EWSR is almost constant for the two lighter chromium isotopes ^{50,52}Cr and increases rather drastically to ^{54}Cr as illustrated in figure 2.

This interesting finding hints on one hand at the important role of valence neutrons for the formation of the low-lying $E1$ strength and on the other hand highlights the crucial role of shell structure on the properties of the PDR questioning in its fully collective character.

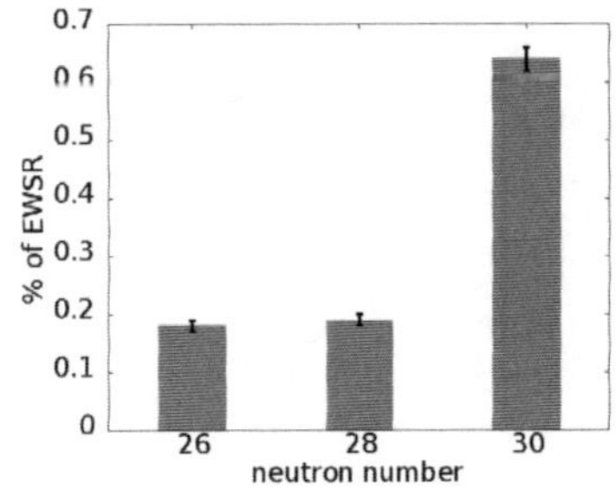

Figure 2. Exhausted percentage of the EWSR for ^{50}Cr, ^{52}Cr, and ^{54}Cr. Beyond the shell-closure at 28 neutrons, the overall strength increases noticeable [14].

Acknowledgments
The effort of the operators of the S-DALINAC in particular M. Arnold, F. Hug, and T. Krzeder and the crew at HIγS is gratefully acknowledged. This work was supported by the Deutsche Forschungsgemeinschaft through CRC 634 and by the US Department of Energy, Office of Nuclear Physics under Grant No. DE-FG02- ER41033.

References
[1] Savran D, Aumann T and Zilges A 2013 *Progress in Particle and Nuclear Physics* **70** 210 – 245 ISSN 0146-6410 URL `http://www.sciencedirect.com/science/article/pii/S0146641013000057`
[2] Andrejtscheff W, Kohstall C, von Brentano P, Fransen C, Kneissl U, Pietralla N and Pitz H 2001 *Physics Letters B* **506** 239 – 246 ISSN 0370-2693 URL `http://www.sciencedirect.com/science/article/pii/S0370269301003409`
[3] Goldhaber M and Teller E 1948 *Phys. Rev.* **74**(9) 1046–1049 URL `http://link.aps.org/doi/10.1103/PhysRev.74.1046`
[4] Hartmann T, Babilon M, Kamerdzhiev S, Litvinova E, Savran D, Volz S and Zilges A 2004 *Phys. Rev. Lett.* **93**(19) 192501 URL `http://link.aps.org/doi/10.1103/PhysRevLett.93.192501`
[5] Scheck M, Ponomarev V Y, Aumann T, Beller J, Fritzsche M, Isaak J, Kelley J H, Kwan E, Pietralla N, Raut R, Romig C, Rusev G, Savran D, Sonnabend K, Tonchev A P, Tornow W, Weller H R and Zweidinger M 2013 *Phys. Rev. C* **87**(5) 051304 URL `http://link.aps.org/doi/10.1103/PhysRevC.87.051304`
[6] Kneissl U, Pietralla N and Zilges A 2006 *Journal of Physics G: Nuclear and Particle Physics* **32** R217 URL `http://stacks.iop.org/0954-3899/32/i=8/a=R01`
[7] Sonnabend K, Savran D, Beller J, Büssing M, Constantinescu A, Elvers M, Endres J, Fritzsche M, Glorius J, Hasper J, Isaak J, Löher B, Müller S, Pietralla N, Romig C, Sauerwein A, Schnorrenberger L, Wälzlein C, Zilges A and Zweidinger M 2011 *Nuclear Instruments and Methods in Physics Research Section A: Accelerators, Spectrometers, Detectors and Associated Equipment* **640** 6 – 12 ISSN 0168-9002 URL `http://www.sciencedirect.com/science/article/pii/S0168900211005535`
[8] Weller H R, Ahmed M W, Gao H, Tornow W, Wu Y K, Gai M and Miskimen R 2009 *Progress in Particle and Nuclear Physics* **62** 257 – 303 ISSN 0146-6410 URL `http://www.sciencedirect.com/science/article/pii/S0146641008000434`
[9] Pietralla N, Berant Z, Litvinenko V N, Hartman S, Mikhailov F F, Pinayev I V, Swift G, Ahmed M W, Kelley J H, Nelson S O, Prior R, Sabourov K, Tonchev A P and Weller H R 2002 *Phys. Rev. Lett.* **88**(1) 012502 URL `http://link.aps.org/doi/10.1103/PhysRevLett.88.012502`
[10] Romig C 2015 *Investigation of Nuclear Structure with Relative Selfabsorption Measurements* Dissertation TU Darmstadt
[11] Agostinelli S and *et al* 2003 *Nuclear Instruments and Methods in Physics Research Section A: Accelerators, Spectrometers, Detectors and Associated Equipment* **506** 250 – 303 ISSN 0168-9002 URL `http://www.sciencedirect.com/science/article/pii/S0168900203013688`
[12] National Nuclear Data Center, URL: http://www.nndc.bnl.gov/, 9.1.2015
[13] Schiff L I 1951 *Phys. Rev.* **83**(2) 252–253 URL `http://link.aps.org/doi/10.1103/PhysRev.83.252`
[14] Pai H, Beller J, Benouaret N, Enders J, Hartmann T, Karg O, von Neumann-Cosel P, Pietralla N, Ponomarev V Y, Romig C, Scheck M, Schnorrenberger L, Volz S and Zweidinger M 2013 *Phys. Rev. C* **88**(5) 054316 URL `http://link.aps.org/doi/10.1103/PhysRevC.88.054316`
[15] Greiner W and Maruhn J A 1996 *Nuclear Models* (Springer, Berlin)

Towards the improvement of spin-isospin properties in nuclear energy density functionals

X. Roca-Maza[1], G. Colò[1], H. Z. Liang[2], J. Meng[3,4,5], P. Ring[3,6], H. Sagawa[2,7] and P. W. Zhao[3]

[1]Dipartimento di Fisica, Università degli Studi di Milano and INFN, sezione di Milano, via Celoria 16, I-20133 Milano, Italy
[2]RIKEN, Nishina Center, Wako, 351-0198, Japan
[3]State Key Laboratory of Nuclear Physics and Technology, School of Physics, Peking University, Beijing 100871, China
[4]School of Physics and Nuclear Energy Engineering, Beihang University, Beijing 100191, China
[5]Department of Physics, University of Stellenbosch, Stellenbosch 7602, South Africa
[6]Physik Department, Technische Universität München, D-85747 Garching, Germany
[7]Center for Mathematics and Physics, University of Aizu, Aizu-Wakamatsu, Fukushima 965-8580, Japan

E-mail: `xavier.roca.maza@mi.infn.it`

Abstract. We address the problem of improving existing nuclear Energy Density Functionals (EDFs) in the spin-isospin channel. For that, we propose two different ways. The first one is to carefully take into account in the fitting protocol some of the key ground state properties for an accurate description of the most studied spin-isospin resonances: the Gamow-Teller Resonance (GTR) [1]. The second consists in providing a strategy to build local covariant EDF keeping the main features from their non-local counterparts [2]. The RHF model based on a Lagrangian where heavy mesons carry the nuclear effective interaction have been shown to be successful in the description of spin-isospin resonances [3].

1. Introduction

Density functional theory is a successful approach extensively used in physics, chemistry and material sciences [4]. In nuclear physics, energy density functionals are usually based on effective interactions solved at the Hartree (H) or Hartree-Fock (HF) level, time-dependent (TD) versions of these schemes also exist [5, 6]. The small amplitude limit of the TDH(F) theories reduces to the well known Random Phase Approximation (RPA). Pairing correlations and deformations have been consistently taken into account within these methods. Existent functionals are of very different nature: based on finite-range or zero-range effective interactions, on relativistic or non-relativistic frameworks, assuming coupling constants that changes as a function of the baryon density in nuclei or considering point-coupling—fixed—coupling constants. In addition, very frequently in the literature, not all the operators allowed by the symmetries are included in such a model Hamiltonians or Lagrangians. All these models have been shown to be successful and reach similar accuracy in the description of bulk nuclear properties such as masses, charge radii, deformations or excitation energies in Giant Resonances. Nevertheless, there are different drawbacks to overcome. One of them, briefly reviewed in Sec.2, is the proper characterization of the spin-isospin channel in nuclear energy density functionals.

Here, we propose two different ways to, at least partially, improve current EDFs in such a channel. The first one, explained in Sec.3, takes into account some of the key ground state properties for an accurate description of one of the most studied spin-isospin resonances: the Gamow-Teller Resonance (GTR) [1]. On the other side, recent studies within the relativistic framework show the benefits of including the effects of the exchange (Fock) terms on spin-isospin resonances. The latter models usually based on the exchange of heavy mesons are, however, difficult to extend—e.g. perform full 3D calculations or include the coupling of single-particle degrees of freedom with the vibrational ones—as compared to relativistic Hartree models of the same kind. This is due to the non-locality of the Hartree-Fock potentials. Therefore, to improve current available functionals keeping the calculations as simple as possible, we proposed a way to localize relativistic Hartree-Fock models [2]. Based on the latter, we provide in Sec.4 a possible strategy to fit local and covariant EDF keeping the main features from their non-local counterparts.

2. Current energy density functional predictions on Gamow-Teller Resonances

Theoretical Gamow Teller (GT) transition strengths ($S_{\mathrm{GT}^\pm}$) are mediated by the operator $\sum_{i=1}^{A}\sigma(i)\tau_\pm(i)$, where A is the mass number and σ and τ are the spin and isospin Pauli matrices, respectively. For example, in neutron rich nuclei, the dominant transitions to S_{GT^-} will be those involving proton spin-orbit partner levels close to the Fermi surface (see left panel of Fig.1). In this respect, it is clear that an accurate description of spin-orbit splittings will be desirable for an accurate description of the GTR. Indeed, in a detailed study on the GTR and the spin-isospin Landau-Migdal parameter G_0' using several Skyrme sets [7], Bender et al. concluded that this spin-isospin coupling is not the only important quantity in determining the strength and excitation energy of the GTR in nuclei. Actually, the authors state that spin-orbit splittings together with the residual spin-isospin interaction influences the above-mentioned quantities. As an example, most of the Skyrme functionals overestimate the experimental spin-orbit splittings in heavy nuclei [8]. In addition to this work, an empirical determination of the Landau-Migdal parameters G_0 (spin-channel) and G_0' can be found in Ref. [9, 10]. The estimated values for these two parameters were based on single-particle energies obtained with a Woods-Saxon potential. Since we aim at using EDFs which usually predict different effective masses, we can just take inspiration from the empirical indications in [9, 10] in a qualitative way. Specifically, the latter means to keep the empirical hierarchy $G_0' > G_0 > 0$ found. In Ref. [11], Li-Gang Cao et al. show that the mentioned hierarchy is not a very common feature within available Skyrme functionals (cf. Fig.1 in Ref. [11]).

The studies discussed above are based on non-relativistic approaches. Only recently some efforts have been reported on the description of charge-exchange excitations in the relativistic framework (see [12] and references therein). In covariant Hartree calculations, the pion do not contribute because of parity conservation. However, for the calculation of the linear response (RPA) produced by the GT operator, the pion contribution is a crucial ingredient. Therefore, it is needed to introduce the pion term in the residual interaction—where the bare pion coupling constant is commonly adopted ($f_\pi^2/4\pi = 0.08$). Because of the derivative type of the pion-nucleon coupling, it is necessary to include the so-called Landau-Migdal (zero-range) term that accounts for the short-range correlations of the nucleon-nucleon interaction (see for example discussion around Eqs. (1) and (2) in Ref. [13]). Following this strategy, one does not find satisfactory results (cf. Fig.3 in Ref. [2]) and, commonly, the strength of the Migdal term is fitted to reproduce the excitation energy of the GTR in ^{208}Pb [12]. A possible solution to avoid the latter fitting staying at the H level is what we proposed in Ref. [2]. The idea is inspired by the good results obtained within the relativistic HF approach for spin-isospin resonances [3]. In the latter case, the terms coming from the isoscalar σ and ω mesons—that generate an isovector contribution via the Fock terms—were able to produce an accurate result for the excitation

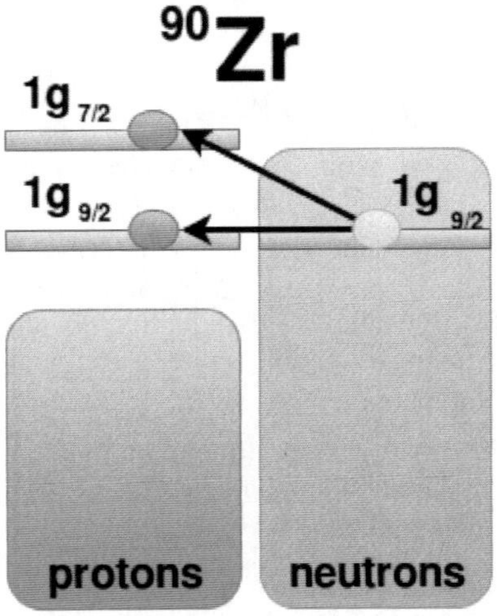

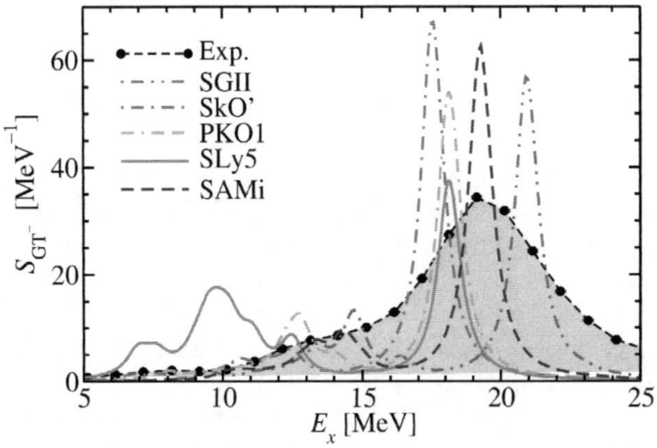

Figure 1. Left panel: schematic picture of dominant single-particle transitions excited by $\sigma\tau_-$ in ^{90}Zr. Right panel: strength of the GTR in ^{208}Pb as a function of the excitation energy. Experimental data [15] and predictions of different EDFs are shown, see the text for the details. A Lorentzian smearing parameter equal to 2 MeV is used in the calculations.

energy of the GTR in different closed-shell nuclei, without recourse of a fitting of the Migdal term [2].

Finally, it is worth mentioning that from the experimental side, the GTR exhausts 60-70% of the well known Ikeda Sum Rule (ISR) given by $\int[S_{\rm GT^-}(E) - S_{\rm GT^+}(E)]{\rm d}E = 3(N-Z)$. To explain this quenching problem, it has been proposed that the effects of the second-order (or multi-order) configuration mixing, namely $2p-2h$ correlations, have to be taken into account [14]. Since our approach is based on the $1p-1h$ RPA, we concentrate here on improving the excitation energies of the GTR in some selected doubly magic spherical nuclei. The latter quantity is expected to be accurately reproduced at the adopted approach.

3. A fitting protocol to improve the spin-isospin channel: example using a Skyrme interaction

The Skyrme EDF is one of the most successful non-relativistic models for the study of the ground state properties of nuclei along the nuclear chart [16]. Nevertheless, Skyrme functionals have shown already some drawbacks [17]. Here we would like to focus on how one may improve their spin-isospin properties. Particularly sensitive to the spin-isospin channel is the Gamow Teller Resonance (GTR) [18]. Accurate GT matrix elements determine weak-interaction transitions and are necessary for the study of double-β decay [19, 20]. A reliable description of spin-isospin resonances will impact also on our understanding of the density dependence of the nuclear symmetry energy [21]. The latter quantity plays an important role also in other areas such as heavy ion reactions or nuclear astrophysics.

A new Skyrme EDF named SAMi has been developed in Ref. [1] and briefly presented in this work. This interaction has been accurately calibrated to reproduce properties of doubly-magic nuclei and infinite nuclear matter. In this respect, the novelties introduced in the fitting protocol of SAMi are (i) the careful description of the empirical hierarchy of spin (G_0) and spin-isospin (G_0') Landau-Migdal parameters: $0 < G_0 < G_0'$ [9], a feature that most of available Skyrme forces fail to reproduce [11] and (ii) a two parameter spin-orbit potential. The latter is motivated in Ref. [7] where the importance, not only of G'_0, but also of the spin-orbit splittings were highlighted. Thus, the presented fitting protocol aims at going one step forward in setting the bases for a more precise description of spin-isospin resonances.

The presented model is based on a standard Skyrme Hartree-Fock (HF) plus charge-exchange Random Phase Approximation (RPA) described in Refs. [22, 23], to which we refer the reader for details. The Skyrme functional employed here (SAMi [1]) include the central tensor terms (J^2-terms) and, as mentioned, two spin-orbit parameters. We have chosen a set of fitted data and *pseudo*-data, inspired by the fitting protocol used to build SLy interactions [24]: the binding energies of ^{40,48}Ca, ^{90}Zr, ^{132}Sn and ^{208}Pb and the charge radii of ^{40,48}Ca, ^{90}Zr and ^{208}Pb; the spin-orbit splittings of the $1g$ and $2f$ proton levels in ^{90}Zr and ^{208}Pb, respectively; the Landau-Migdal parameters G_0 and G'_0 are fixed at the values 0.15 and 0.35, respectively, at saturation density; finally, *pseudo*-data corresponding to more fundamental microscopic calculations of the energy per particle of uniform neutron matter (e_n) at baryon density ρ between 0.07 fm^{-3} and 0.4 fm^{-3} has been also used to define the χ^2 [1].

In the right panel of Fig. 1 the strength function of the GTR (S_{GT^-}) is depicted as a function of the excitation energy (E_x) in ^{208}Pb. Experimental data are also displayed [15]. The models shown for the GTR (see [1] for the original references) are: SGII that constitutes the first attempt to provide a quantitative description of the excitation energy of the GTR; SkO' which was also build taking special care on the spin-isospin channel of the interaction and which performs better than SGII in the description of GTRs; SAMi [1], the new functional revisited here; and SLy5, a commonly used model in nuclear applications to which our fitting protocol has been inspired. In the right panel of Fig. 1 SAMi is seen to be quite accurate in the description of the E_{GTR} in ^{208}Pb. These results support the employed fitting protocol. In addition, SAMi is as accurate as previous SLy Skyrme models in the description of other ground state properties and non-charge exchange resonances [1].

4. Exploratory results on how to build a Localized Hartree-Fock model

Relativistic Hartee-Fock (RHF) models based on Lagrangians where heavy mesons mediate the nuclear effective interaction have been shown to be as accurate as those based at the Hartee level in the description of masses, radii and nuclear collective excitations. Specifically, they are successful in the description of spin-isospin resonances such as the GTR, without the need of an adjustment of the Landau-Migdal term in the residual interaction [3]. Nevertheless, these covariant models include non-local potentials, a feature that makes more difficult and computationally demanding to extend the model to account for deformation, number or angular momentum projection or cranking, within other possible extensions. In Ref. [2], some of the authors have proposed a strategy to localize a RHF model. That is, we have mapped a finite-range RHF model, based on a Lagrangian with three active interaction channels: scalar-isoscalar (σ meson), vector-isoscalar (ω meson) and vector-isovector (ρ meson); into a zero-range RH model, based on a Lagrangian where all possible interaction channels allowed by the symmetries where active. We followed the strategy: i) perform a zero-range reduction of the meson-exchange Lagrangian terms; ii) via the Fierz transformations relate the coupling constants at the HF level with those at the H level [25, 26]. The Fierz transformation employed in [2] is

$$\alpha_S^{\mathrm{H}} = +\tfrac{7}{8}\alpha_S^{\mathrm{HF}} - \tfrac{4}{8}\alpha_V^{\mathrm{HF}} - \tfrac{12}{8}\alpha_{tV}^{\mathrm{HF}} \qquad \alpha_{tS}^{\mathrm{H}} = -\frac{1}{8}\alpha_S^{\mathrm{HF}} - \frac{4}{8}\alpha_V^{\mathrm{HF}} + \frac{4}{8}\alpha_{tV}^{\mathrm{HF}}$$

$$\alpha_V^{\mathrm{H}} = -\tfrac{1}{8}\alpha_S^{\mathrm{HF}} + \tfrac{10}{8}\alpha_V^{\mathrm{HF}} + \tfrac{6}{8}\alpha_{tV}^{\mathrm{HF}} \qquad \alpha_{tV}^{\mathrm{H}} = -\frac{1}{8}\alpha_S^{\mathrm{HF}} + \frac{2}{8}\alpha_V^{\mathrm{HF}} + \frac{6}{8}\alpha_{tV}^{\mathrm{HF}}$$

$$\alpha_T^{\mathrm{H}} = -\tfrac{1}{16}\alpha_S^{\mathrm{HF}} \qquad \alpha_{tT}^{\mathrm{H}} = -\frac{1}{16}\alpha_S^{\mathrm{HF}}$$

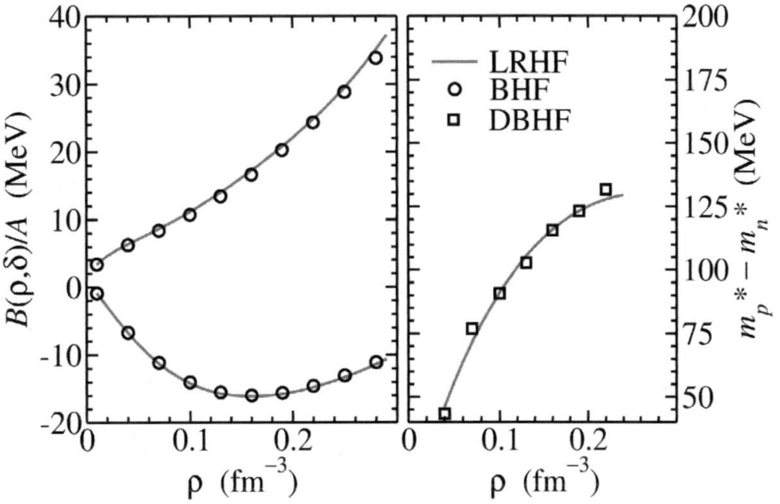

Figure 2. Left panel: binging energy in nuclear matter and pure neutron matter as a function of the density as predicted by the Brueckner-Hartree-Fock (BHF) calculations in Ref.[28] (black circles) and as fitted to the previous benchmark calculations by the Localized RHF model (LRHF). Right panel: difference between the neutron and proton Dirac effective masses as a function of the density as predicted by the Dirac-Brueckner-Hartree-Fock (DBHF) calculations in Ref.[29] (black squares) and as fitted to the previous benchmark calculations by the LRHF model.

$$\alpha^{\rm H}_{PS} = -\tfrac{1}{8}\alpha^{\rm HF}_{S} + \tfrac{4}{8}\alpha^{\rm HF}_{V} + \tfrac{12}{8}\alpha^{\rm HF}_{tV} \qquad \alpha^{\rm H}_{tPS} = -\frac{1}{8}\alpha^{\rm HF}_{S} + \frac{4}{8}\alpha^{\rm HF}_{V} - \frac{4}{8}\alpha^{\rm HF}_{tV}$$
$$\alpha^{\rm H}_{PV} = +\tfrac{1}{8}\alpha^{\rm HF}_{S} + \tfrac{2}{8}\alpha^{\rm HF}_{V} + \tfrac{6}{8}\alpha^{\rm HF}_{tV} \qquad \alpha^{\rm H}_{tPV} = +\frac{1}{8}\alpha^{\rm HF}_{S} + \frac{2}{8}\alpha^{\rm HF}_{V} - \frac{2}{8}\alpha^{\rm HF}_{tV} \qquad (1)$$

where $\alpha^{\rm H(HF)}$ is the coupling constant in the Hartree representation (Hartree-Fock after the zero-range reduction) corresponding to the interaction channels: scalar (S), vector (V), tensor (T), pseudo-scalar (PS) and pseudo-vector (PV); t indicates if the coupling constant corresponds to the isovector channel. The zero-range reduction is needed to be able to apply the Fierz transformations [25] and it is well justified due to the heavy mass of the mesons involved in the RHF model. For further details we refer the reader to Ref.[2] where this strategy has been explored, by mapping an existing model: PKO2 [27]. The result of this test was satisfactory (cf. Figs. 2–4 in Ref.[2]) and suggested us to push forward this strategy and propose a new way to build local covariant functionals based on a Localized RHF (LRHF) model. The strategy consists on a double Fierz transformation, first, one connects operators (and coupling constants) in the H representation with their HF counter-parts, second, apply again a transformation in the opposite direction. The result of this double transformation would not produce any interesting result unless one reduces the number of active channels in the initial transformation, as in Eq.(1) where the number of active channels were just three: S, V and tV. This implies that the number of fitted coupling constants might be strongly reduced and determine, via the transformation, all the other channels. In general, this reduction of the channels does not produce a *perfect* mapping. However, from an *empirical* point of view, our results in Ref.[2] suggest that such a method might be successful.

We present here a preliminary test by only considering a fit to nuclear matter properties calculated within the benchmark Brueckner-Hartree-Fock (BHF) calculations for symmetric and pure neutron matter of Ref.[28] as well as Dirac-Brueckner-Hartree-Fock (DBHF) calculations of the effective mass splitting of neutrons and protons in pure neutron matter of Ref.[29]. We use the same fitting procedure used to fit DD-MEδ [30] to nuclear matter and pure neutron matter properties and the DD-PC1 [31] ansatz for the density dependence of the coupling constants. The new *test*-functional includes all terms in the Lagarngian that contribute at the mean-field level (except for the tensor terms). As mention, we consider as free parameters the ones corresponding to the S, V and tV channels, the rest of the channels will be determined by the Fierz transformations within the same Hartree scheme. The aim of our test is twofold: i) test the feasibility of optimizing a localized RHF calculation; and ii) find a good starting point before fitting the complete localized functional also to finite nuclei. The results of this test look promising and can be seen in Fig.2 where the agreement found is remarkable. In the left panel of Fig.2, the binging energy in nuclear matter and pure neutron matter as a function of the density as predicted by BHF calculations of Ref.[28] (black circles) and as fitted to the previous benchmark calculations by the Localized RHF model (LRHF) are shown. In the right panel of the same figure, we depict the difference between the neutron and proton Dirac effective masses as a function of the density as predicted by the DBHF calculations of Ref.[29] (black squares) and as fitted to the previous benchmark calculations by the LRHF model.

5. Conclusions

We have reminded some of the problems in the spin-isospin channels of the Skyrme and RH models using as an example the GTR. We have briefly presented the benefits of the new proposed fitting protocol that cure part of the previous problems, test the new protocol and show some results when applied with a Skyrme interaction. We have briefly discussed the benefits of using a RHF model for the description of GTRs and proposed a new method to determine a localized RHF model.

Acknowledgments

[1] Roca-Maza X, Colò G and Sagawa H 2012 *Phys. Rev. C* **86**(3) 031306
[2] Liang H, Zhao P, Ring P, Roca-Maza X and Meng J 2012 *Phys. Rev. C* **86**(2) 021302
[3] Liang H, Van Giai N and Meng J 2008 *Phys. Rev. Lett.* **101**(12) 122502
[4] Gross E and Dreizler R 1995 *Density Functional Theory* (Springer)
[5] Bender M, Heenen P H and Reinhard P G 2003 *Rev. Mod. Phys.* **75**(1) 121–180
[6] Vretenar D, Afanasjev A, Lalazissis G and Ring P 2005 *Physics Reports* **409** 101 – 259
[7] Bender M, Dobaczewski J, Engel J and Nazarewicz W 2002 *Phys. Rev. C* **65**(5) 054322
[8] Bender M, Rutz K, Reinhard P G, Maruhn J A and Greiner W 1999 *Phys. Rev. C* **60**(3) 034304
[9] Wakasa T, Ichimura M and Sakai H 2005 *Phys. Rev. C* **72**(6) 067303
[10] 1999 *Physics Letters B* **455** 25 – 29
[11] Cao L G, Colò G and Sagawa H 2010 *Phys. Rev. C* **81**(4) 044302
[12] Paar N, Nikšić T, Vretenar D and Ring P 2004 *Phys. Rev. C* **69**(5) 054303
[13] Ma Z Y, Chen B Q, Van Giai N and Suzuki T 2004 *The European Physical Journal A - Hadrons and Nuclei* **20** 429–433
[14] Wakasa *et al* T 1997 *Phys. Rev. C* **55**(6) 2909–2922
[15] Wakasa *et al* T 2012 *Phys. Rev. C* **85**(6) 064606
[16] Stone J and Reinhard P G 2007 *Progress in Particle and Nuclear Physics* **58** 587 – 657
[17] Erler J, Klpfel P and Reinhard P G 2011 *Journal of Physics G: Nuclear and Particle Physics* **38** 033101
[18] Ichimura M, Sakai H and Wakasa T 2006 *Progress in Particle and Nuclear Physics* **56** 446 – 531
[19] Langanke *et al* K 2008 *Phys. Rev. Lett.* **100**(1) 011101
[20] Avignone F T, Elliott S R and Engel J 2008 *Rev. Mod. Phys.* **80**(2) 481–516
[21] Li B A, Ramos A, Verde G and Vidaña I 2014 *The European Physical Journal A* **50** 9
[22] Colò G, Van Giai N, Bortignon P F and Broglia R A 1994 *Phys. Rev. C* **50**(3) 1496–1508
[23] Colò G, Cao L, Giai N V and Capelli L 2013 *Computer Physics Communications* **184** 142 – 161
[24] Chabanat E, Bonche P, Haensel P, Meyer J and Schaeffer R 1998 *Nuclear Physics A* **635** 231 – 256
[25] Okun L B 1982 *Leptons and Quarks* (North-Holland, Amsterdam)
[26] Sulaksono A, Brvenich T, Maruhn J, Reinhard P G and Greiner W 2003 *Annals of Physics* **306** 36 – 57

[27] Long W, Sagawa H, Meng J and Giai N V 2008 *EPL (Europhysics Letters)* **82** 12001
[28] Baldo M, Maieron C, Schuck P and Vias X 2004 *Nuclear Physics A* **736** 241 – 254
[29] van Dalen E, Fuchs C, Fuchs C and Fuchs C 2007 *The European Physical Journal A* **31** 29–42
[30] Roca-Maza X, Viñas X, Centelles M, Ring P and Schuck P 2011 *Phys. Rev. C* **84**(5) 054309
[31] Nikšić T, Vretenar D and Ring P 2008 *Phys. Rev. C* **78**(3) 034318

Production of lanthanide molecular ion beams by fluorination technique

B. Roussière[1], I. Deloncle[2], N. Barré-Boscher[1], M. A. Cardona[3], M. Cheikh Mhamed[1], T. Corbin[1], E. Cottereau[1], H. Croizet[1], B. Dimitrov[4], S. Essabaa[1], G. Gavrilov[4], A. Gottardo[1], N. Goutev[4], J. Guillot[1], D. Hojman[3], C. Lau[1], S. Roccia[2], S. Tusseau-Nenez[1], D. Verney[1], M. Yavahchova[4] and the ALTO collaboration

[1] Institut de Physique Nucléaire, CNRS-IN2P3, Université Paris-Sud, Université Paris-Saclay, F-91406 Orsay Cedex, France

[2] Centre de Spectrométrie Nucléaire et de Spectrométrie de Masse, CNRS-IN2P3, Université Paris-Sud, Université Paris-Saclay, F-91405 Orsay, France

[3] Departamento de Física, Comisión Nacional de Energía Atómica, 1429 Buenos Aires, Argentina

[4] Institute for Nuclear Research and Nuclear Energy, Bulgarian Academy of Sciences, 1784 Sofia, Bulgaria

E-mail: roussier@ipno.in2p3.fr

Abstract. Systematic off-line fluorination studies on all the stable lanthanide isotopes have been performed. The results are presented as a function of various parameters such as the target temperature, the type of ion source used (hot plasma or surface ionization) and the quantity of CF_4 introduced. The first on-line measurements allowed us to determine the optimal experimental conditions for producing radioactive lanthanide isotopes.

1. Introduction

The neutron-rich lanthanide (Ln) nuclei allow to study, in exotic nuclei, dynamical phenomena typical of mid-shell. They offer the opportunity to observe octupole excitations and their coupling to quadrupole excitations. They can provide information on the very puzzling nature of the 0^+ states lying at low excitation energy [1–5]. Moreover large quadrupole deformations are expected in the ground state of the neutron-mid-shell nuclei: for example, ^{170}Dy, having the largest number of valence particles among the A < 208 nuclei, is predicted to have the most deformed ground state [6]. Finally, these nuclei present an astrophysical interest. The small peak exhibited at A = 160 in the solar r-process abundance distribution is probably due to a subtle interplay between nuclear deformation and β decay then half-lives (or deformation) determinations via β decay measurements in the neutron-rich Ln nuclei can help to constrain the r-process freeze out conditions [7]. Until now the neutron-rich Ln nuclei have been little studied since they are difficult to produce. For example, in ISOL facilities, they are produced by fission induced in thick uranium-carbide (UC_x) targets from which they are slowly released due to their low volatility. In previous works, CF_4 injection in the target leads to form Ln fluorides that are more volatile then better released than the elements [8]. However, the Ln-fluoride release efficiency depends strongly on the impurities in the target, on the amount of CF_4 introduced and

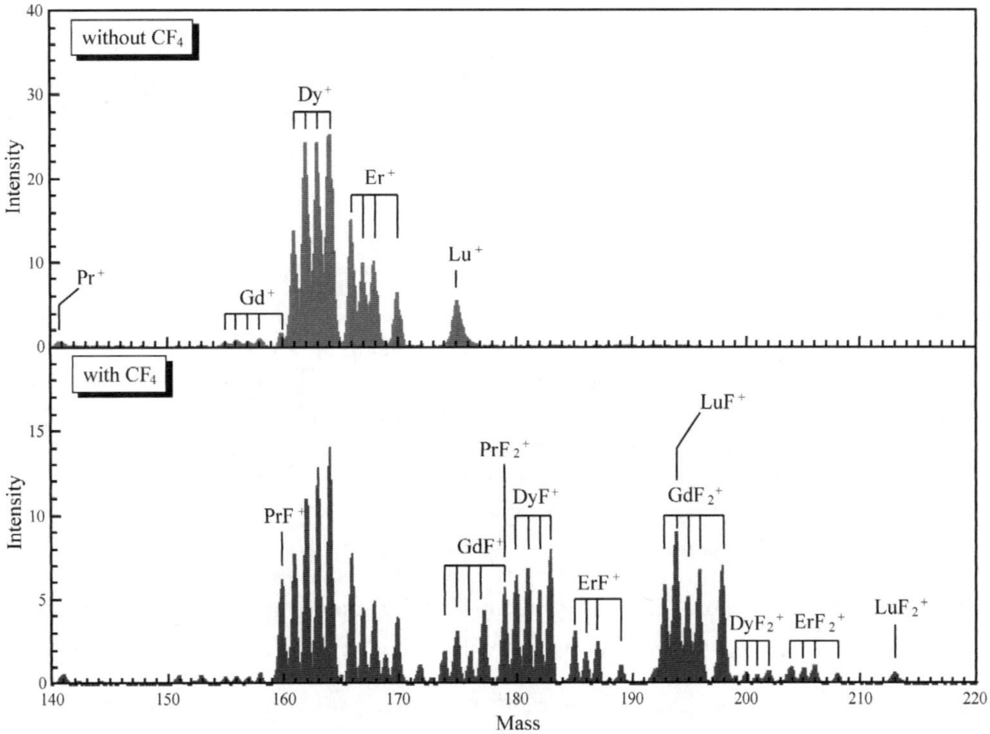

Figure 1. Mass spectra recorded in the focal plane of the off-line separator magnet.

on the temperature of the target [9,10]. In order to control and apply the fluorination technique at ALTO [11], we have performed systematic off-line fluorination studies on all the Ln stable isotopes. The experimental procedure developped, the off-line results as well as the first on-line measurements are detailed hereafter.

2. Experimental procedure

2.1. Off-line measurements

The fluorination tests on the Ln stable isotopes were performed with an off-line mass separator and either a hot plasma MK5 [12] or a surface ionization (SI) [13] ion source that are both commonly used at ALTO. Pellets containing uranium cannot be used on the off-line separator for safety reasons. The UC_x pellets used as targets at ALTO have graphite excess. Then in order to observe Ln diffusion in realistic conditions, we used pellets made of powder mixture of graphite with different sets of Ln oxides. The lanthanides were selected so that the molecular or elemental ions expected to be observed do not present any mass overlap: for example, La, Pr, Eu, Ho for the first series, La, Pr, Eu, Gd, Tb, Ho and Ce, Pr, Sm, Eu, Yb, for the second and third ones. Eu is one of the most volatile Ln elements and has only two stable isotopes, therefore it was used as a reference in all our measurements. Most of the Ln elements have been tested several times in order to control the reproducibility of the results. The powder mixture, put in a 13 mm diameter die, was pressed at 220 MPa. Each pellet weighed 275 mg and contained 50 mg of one of the Ln oxide mixes. For the tests, five pellets of a same mixture were placed inside the oven of the target and ion source system. In each case we measured the current collected in a Faraday cup located behind the separator magnet for the different masses corresponding to the Ln^+, LnF^+ and LnF_2^+ ions as a function of the target temperature, the parameters of the ion source and the quantity of CF_4 injected via a tunable calibrated leak. As a control, we registered the mass spectra obtained in the focal plane of the off-line mass-separator magnet. Figure 1 shows such spectra. For each element, one peak is observed for each stable isotope with relative intensities reflecting the isotopic ratio. Ln are observed as elemental ions without CF_4 and also as molecular ions when CF_4 is injected. The Pr and Gd intensities are very weak for the elemental ions and much stronger for the molecular ions, which proves that fluorination improves the Ln release from thick pellets.

2.2. On-line measurements

The first on-line measurement was performed with a MK5 ion source. For normalization with off-line results and in order to attest the proper running of the fluorination system, five graphite pellets doped with stable lanthanides, identical to that used in the off-line tests, were added to the UC_x pellets used as targets for the photo-fission reactions. No radioactive Ln isotopes were observed, making crucial a second experiment.

The second on-line measurement, without fluorination, was performed using a SI ion source, with a two-fold purpose. Firstly, in order to check the Ln cross sections by photo-fission, the yields of some isotopes of the most volatile Ln elements were measured. Secondly, to compare the Ln release from graphite and UC_x pellets, five UC_x pellets doped with the previous mixture of stable Ln oxides, in the same amount as in the first on-line experiment, were placed in the target oven. These UC_x Ln-doped pellets were made using a 14 mm diameter die and a pressure of 190 MPa.

The intensity of the stable ion beams was measured by a Faraday cup located behind the magnet of the on-line mass separator tuned on the Ln^+, LnF^+ and LnF_2^+ masses. Mass spectra were recorded by scanning the field of the separator magnet. The yields measurements were carried out by γ and β-γ spectrometry.

3. Experimental results

3.1. From the off-line tests

Whatever the experimental conditions, the stable Ln beam intensity exhibits high sensitivity to the target temperature and the CF_4 flow injected. Figure 2 shows the results obtained for both ion sources with target temperatures around 2000 °C. The beam intensities measured with the SI ion source are clearly higher than those measured with the MK5 ion source. Other differences are observed between the two ion sources.

With the MK5 ion source:

– for $T_{target} < 1630$ °C, only Eu is observed, as Eu^+ and as EuF^+ when CF_4 is added; the other elements begin to appear as molecular ions when $T_{target} = 1630$ °C and a CF_4 amount of 6×10^{-3} mbar·l/s is introduced in the target and ion source system;

– for $T_{target} \geq 1900$ °C and in particular at 2030 °C (see fig. 2), the more CF_4 injected, the higher the intensity of the lanthanide beams ; the CF_4 maximum quantity compatible with good running conditions of the ion source was determined to be 6×10^{-3} mbar·l/s;

– the type of ions observed preferentially (Ln^+, LnF^+ or LnF_2^+) appears to be highly correlated with the valence of the given lanthanide and thus with the type of fluoride formed (mono-, di- or tri-fluoride), for example La, Ce, Pr, Tb and Ho are preferentially observed as LnF_2^+, Eu and Sm as LnF^+, Yb as Ln^+, except Gd observed preferentially in the Gd^+ form whereas GdF_2^+ was expected. The LnF_n^+ abundance distributions generally reflect the Ln ionization energy. These variations across the lanthanide series are correlated to the Ln specific electronic configurations [14].

With the SI ion source:

– for $T_{target} \geq 1900$ °C and without CF_4, in addition to Eu, other Ln elements as Pr, Nd, Er, Lu, Dy and Tm are observed in the Ln^+ form, this is probably due to the higher ionization

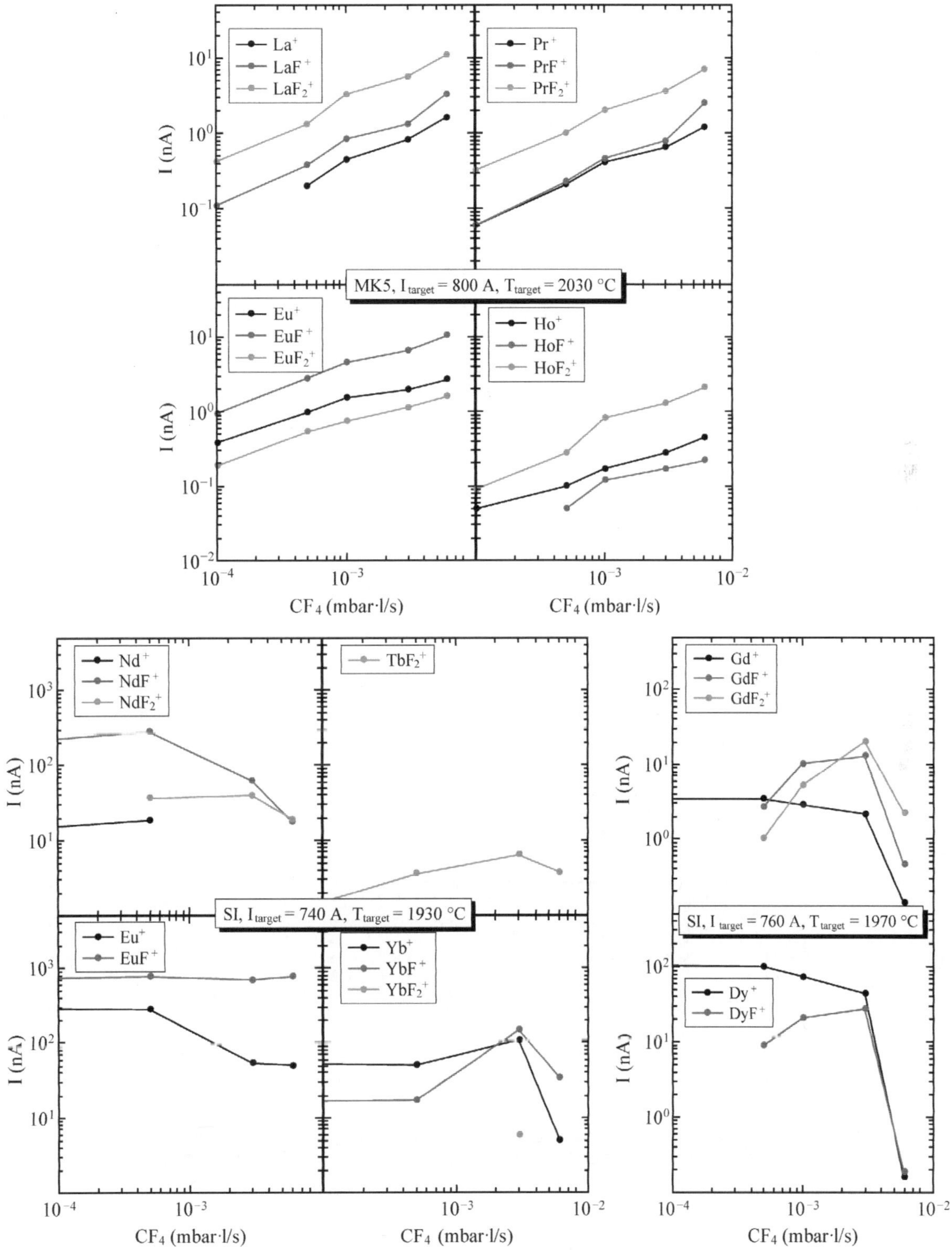

Figure 2. Ion beam intensity measured during the off-line tests as a function of the quantity of gas injected for various lanthanides in their elemental and molecular forms using MK5 or SI ion source.

Journal of Physics: Conference Series **724** (2016) 012042 doi:10.1088/1742-6596/724/1/012042

efficiency of the SI ion source (compared to the MK5 one) for Ln; with CF_4, when the gas flow is high the beam intensities collapse (see fig. 2), this feature is not well understood but has already been observed in other fluorination works [15];

– the type of ions observed preferentially (Ln^+, LnF^+ or LnF_2^+) depends strongly on the CF_4 amount introduced; for instance, with a flow lower than 3×10^{-3} mbar·l/s, Yb and Dy are observed as Ln^+, Nd as NdF^+ whereas with a CF_4 flow equal to 6×10^{-3} mbar·l/s Yb and Dy are observed as LnF^+ and Nd as NdF_2^+ (see fig. 2);

– the correlation between the valence of the given element and the type of ions preferentially observed (Ln^+, LnF^+ or LnF_2^+) is only reached under extreme running conditions ($T_{target} \sim$ 2000 °C and 6×10^{-3} mbar·l/s CF_4), as shown in fig. 2 for Yb and Gd.

One can reasonably assume that the diffusion of stable Ln isotopes in the graphite pellets and the formation of molecular compounds do not differ depending on whether the MK5 or SI ion source is used. On the contrary, the mechanisms of molecular-compound breaking and of molecular-ion formation seem to be different with the MK5 or SI ion source. The behavior of the Ln fluorinated compounds in the MK5 ion source appears to be simpler and quite consistent with what is expected from the Ln chemical properties. This is the reason why the MK5 ion source was used for the first fluorination on-line experiment.

3.2. From the on-line measurements

3.2.1. with the MK5 ion source

Figure 3 shows the stable La and Pr ion beam intensities released by the graphite pellets without and with injection of CF_4 in the target and ion source system. The LnF_2^+ beam intensities are significantly lower in the first measurement with CF_4 than in the next ones; this shows that a reaction time, equal to ∼ half an hour, is required before obtaining stable running conditions. The results observed are very similar to those obtained off-line (compare figs. 2 and 3). Indeed when CF_4 is introduced, the beam intensities measured increase strongly, by a factor of up to 100. For both elements, the LnF_2^+ molecular ions are observed preferentially, with intensity very close to that observed off-line (see figs. 2 and 3). This shows that the gas inlet system was working properly. Nevertheless the electronic components of the tunable leak were more sensitive than expected, preventing us from measuring the flow of gas introduced, this part of the system remains to be improved. However, the running conditions of the target and ion source

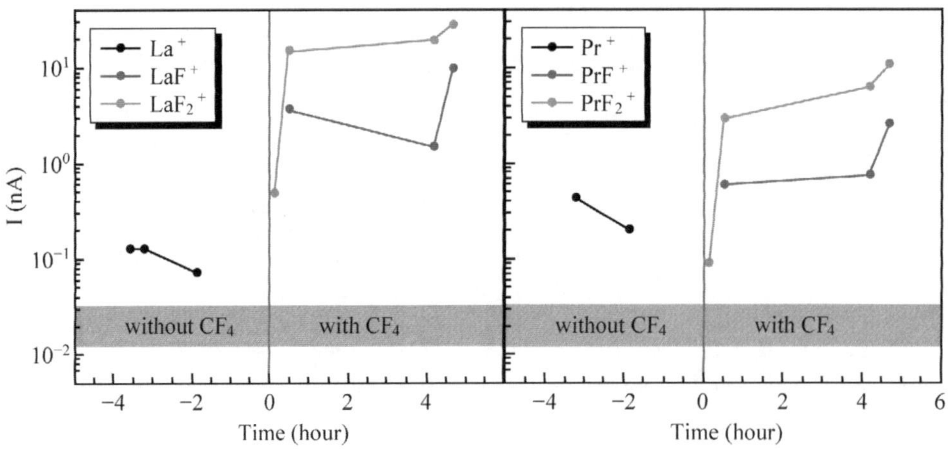

Figure 3. Ion beam intensity measured during the on-line experiment performed with a MK5 ion source for La and Pr stable isotopes without (elemental ions) and with (fluorinated ions) CF_4. Time t = 0 indicates the beginning of the CF_4 injection.

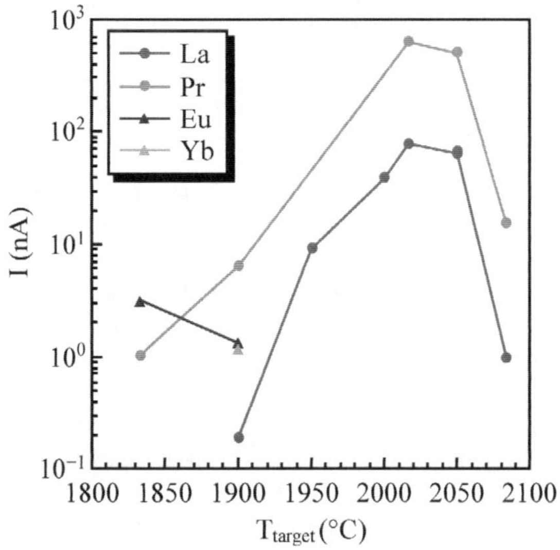

Figure 4. Ion beam intensity measured during the on-line experiment performed with the SI ion source for La, Pr, Eu and Yb stable isotopes released by UC_x pellets without CF_4.

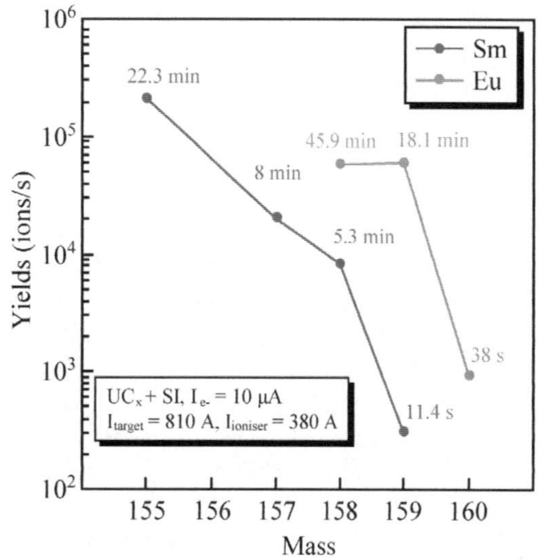

Figure 5. Yields measured for Sm and Eu isotopes released by UC_x pellets during the on-line experiment performed with the SI ion source.

system, especially the target temperature (I_{target} = 750 A, i.e. $T_{target} \sim 1950$ °C) but also the CF_4 flow, seemed to be close to that defined off-line as the optimal ones.
Concerning the radioactive species, the measurements performed to observe the Ln isotopes were all unsuccessful: not only those carried on the 158, 160 masses in order to observe some of the most volatile lanthanides, Eu and Sm, in their elemental form but also those on the 179, 182 and 191 masses in order to see ^{160}Eu, ^{144}La and ^{153}Pr isotopes in the mono and di-fluorinated molecular forms, respectively.

3.2.2. with the SI ion source
Figure 4 presents the La, Pr, Eu and Yb ion beam intensities released by UC_x pellets doped with stable Ln. It is worth noting that Eu and Yb, belonging to the most volatile lanthanides, were probably nearly totally released before the target temperature reached 1800 °C. On the contrary La and Pr were kept in the target until T_{target} = 1800 °C. Their beam intensity measured increases strongly with the target temperature. Between the two last measurements performed at T_{target} = 2050 and 2085 °C, the target was kept at high temperature (T_{target} = 2050 °C) during one week, then most of the stable lanthanides added in the UC_x pellets could have been released, which probably explains the intensity decrease from T_{target} = 2050 °C observed in fig. 4. The comparison with previous measurements indicates that the lanthanide release efficiencies are quite similar for graphite and UC_x pellets. Indeed, at a given target temperature, the intensities measured from UC_x are very similar to those obtained from graphite. Moreover the ionization efficiency of the surface ionization ion source is, depending on the element studied, greater by a factor of 50 to 100 than that obtained with the MK5 ion source.
Figure 5 shows the yields measured for the most volatile Ln isotopes released without fluorination by UC_x pellets heated at $\sim$ 2050 °C. The spectroscopic data required to extract the yields from the γ spectra, in particular the half-lives and the absolute γ intensities, were taken from refs.[16–20]. This result proves that the Ln photo-fission cross sections are sufficient to allow

the observation of Ln mass-separated ion beams provided that the target temperature is high enough. The observation of low volatile Ln radioactive isotopes should then be possible at such high temperature with fluorination.

4. Conclusions

We performed the first systematic study on fluorination of lanthanides in SI and MK5 ion sources. The off-line tests carried out using graphite pellets doped with stable lanthanides showed that adding CF_4 in the target and ion source system increases the Ln release by a factor of 10 to 100 and that most of the lanthanides are preferentially observed as di-fluorinated ions except Sm, Eu and Yb that appear as mono-fluorinated ions. These features are more pronounced with the MK5 ion source than with the SI one, which seems to indicate that the processes involved in fluorination differ between the two ion sources and are simpler within the MK5 ion source. We showed that the release properties of lanthanides from thick uranium carbide pellets are similar to that obtained from graphite pellets. With the SI ion source the Ln ionization efficiency is found to be higher by a factor of 50 to 100 compared to that obtained with the MK5 ion source. The off-line tests and the first on-line experiments allowed us to determine the experimental requirements for producing lanthanide radioactive isotopes: the target temperature is really a key parameter and has to be as high as possible ($T_{target} \geq 2050$ °C); moreover we have to use the SI ion source in order to achieve maximum efficiency. In these conditions, we succeeded in measuring, without fluorination, the yields of some isotopes of the most volatile lanthanides. Before performing a new on-line fluorination experiment, it remains to improve the gas-inlet system using a well-chosen fixed calibrated leak in order to obtain more secure running conditions.

References

[1] Bohr A and Mottelson B 1975 *Nuclear Structure* vol II (New York: W.A. Benjamin, Inc.)
[2] Casten R F and von Brentano P 1994 *Phys. Rev.* C **50** R1280, 1995 *Phys. Rev.* C **51** 3528
[3] Leviatan A and Sinai I 1999 *Phys. Rev.* C **60** 061301
[4] Zawischa D *et al.* 1978 *Nucl. Phys.* A **311** 445
[5] Sharpey-Schafer J F *et al.* 2011 *Eur. Phys. J.* A **47** 5
[6] Regan P *et al.* 2002 *Phys. Rev.* C **65** 037302
[7] Surman R *et al.* 1997 *Phys. Rev. Lett.* C **79** 1809
[8] Hoff P *et al.* 1980 *Nucl. Instrum. Methods* **172** 413
[9] Eder R *et al.* 1992 *Nucl. Instrum. Methods Phys. Res.* B **62** 535
[10] Köster U 2000 *PhD Thesis* (University of Munich)
[11] Essabaa S. *et al.* 2013 *Nucl. Instrum. Methods Phys. Res.* B **317** 218
[12] Sundell S and Ravn H 1992 *Nucl. Instrum. Methods Phys. Res.* B **70** 160
[13] Bjørnstad T *et al.* 1987 *Nucl. Instrum. Methods Phys. Res.* B **26** 174
[14] Gibson J K 1996 *J. Fluor. Chem.* **78** 65
[15] Kawase *et al.* 1992 *Nucl. Instrum. Methods Phys. Res.* B **70** 146
[16] Helmer R G 2004 *Nuclear Data Sheets* **101** 325
[17] Helmer R G 2004 *Nuclear Data Sheets* **103** 565
[18] Reich C W 2005 *Nuclear Data Sheets* **104** 1
[19] Reich C W 2005 *Nuclear Data Sheets* **105** 557
[20] Reich C W 2012 *Nuclear Data Sheets* **113** 157

Near-barrier neutron transfer in reactions $^{3,6}He + {}^{45}Sc$ and $^{3,6}He + {}^{197}Au$

V V Samarin[1], M A Naumenko[1], Yu E Penionzhkevich[1,2], N K Skobelev[1], V Kroha[3] and J Mrazek[3]

[1]Flerov Laboratory of Nuclear Reactions, Joint Institute for Nuclear Research, Dubna, Moscow region, 141980, Russia
[2]National Research Nuclear University "MEPhI", Moscow, 115409, Russia
[3]Nuclear Physics Institute, Czech Academy of Sciences, Rez, 250 68, Czech Republic

samarin@jinr.ru

Abstract. Experimental cross sections for formation of $^{196,198}Au$ isotopes in reactions $^{3,6}He + {}^{197}Au$ and cross sections for formation of $^{44,46}Sc$ isotopes in reactions $^{3,6}He + {}^{45}Sc$ have been analyzed. To calculate neutron transfer probabilities and cross sections the time-dependent Schrödinger equation for external neutrons of ^{3}He, ^{6}He, ^{45}Sc and ^{197}Au nuclei has been solved numerically. It is shown that the contribution of fusion and subsequent evaporation is significant in the case of reactions $^{3,6}He + {}^{45}Sc$, whereas in the case of reactions $^{3,6}He + {}^{197}Au$, it is negligible. Fusion-evaporation was taken into account using NRV evaporation code. Results of calculations demonstrate overall satisfactory agreement with experimental data.

1. Introduction

Low-energy reactions with He isotopes (^{3}He and halo nuclei $^{6,8}He$) attract interest due to the new possibilities of investigating the nuclear structure of light He projectiles and heavy target nuclei, *e.g.* [1]. We may explore target nuclei ^{45}Sc, ^{197}Au by weakly bound neutrons of ^{6}He and the more strongly bound neutron of ^{3}He in reactions of neutron transfer from the projectile to the target as well as study the properties of external neutrons of ^{45}Sc, ^{197}Au in the pick-up reaction $^{3}He+{}^{45}Sc$, $^{3}He+{}^{197}Au$. We may also test some theoretical models with both simple and complicated approximations by comparing them with experimental data.

2. Experiments

The experiments were performed with the extracted radioactive beams of ^{6}He obtained at the DRIBs accelerator complex, JINR, Dubna [2]. The beams of the ^{3}He nuclei were accelerated by the U-120M cyclotron of the Nuclear Physics Institute, the Academy of Sciences of the Czech Republic [3, 4]. The target assemblies consisting of gold foils of different thicknesses were installed in the reaction chamber and irradiated by the beam of particles. For the reaction with ^{6}He the gold target assemblies were installed in the focal plane of the magnetic spectrometer MSP-144, where it was possible to obtain the lower energy beam with a good resolution. The maximum energy of the accelerated ^{6}He ions was ~10 MeV/nucleon, the intensity reached $(2\text{-}5)\cdot 10^{7}$ s^{-1}. The maximum energy of ^{3}He was 24.5 MeV, the beam intensity varied in the range $(5\text{-}10)\cdot 10^{11}$ s^{-1}. After each irradiation session induced γ-activity was measured in all Au targets. The measurements were carried out using HPGe-detectors with the energy resolution of ~1.8 keV for the gamma quantum energy of 1.3 MeV. The identification of the ^{196,198}Au isotopes was performed by their gamma-radiation energies and half-lives. The measurements of the yields were performed taking into account absolute intensities of gamma-transitions and detector efficiency. The details of the experiment, data processing and results are described in [2 – 4].

3. Time-dependent model for neutron transfer

The experimental excitation functions for the formation of ^{196,198}Au isotopes in the reactions of ^{3,6}He with ^{197}Au and formation of ^{44,46}Sc isotopes in the reactions of ^{3,6}He with ^{45}Sc were analyzed and interpreted on the basis of the time-dependent Schrödinger equation [5].

3.1. Neutron states in ^{3,6}He

Information about the structure of loosely bound nuclei may be obtained by measuring their momentum distributions after breakup in a nuclear reaction. From the momentum distribution for the ^{6}He nucleus it follows that it consists of the ^{4}He core and the two-neutron cluster, see *e.g.* [1, 6]. There are several different approaches to the approximate analysis and solution of the three-body problem [7, 8]. To determine the wave function of the ground state of the ^{6}He three-body system the Schrödinger equation may also be solved by Feynman's continual integrals method [9, 10]. Within the simple approximation we use the shell model with the experimental value of the neutron separation energy 1.86 MeV and the root-mean-square charge radius 2.065 fm (the values were taken from [11]). The shell model mean field (Figure 1a) takes into account the properties of the ground state of the system of three particles ^{6}He ($\alpha + n + n$), ^{3}He ($p + p + n$) and four particles ^{4}He ($p + p + n + n$) calculated by Feynman's continual integrals method [12]. The central maximum of the mean field is the effect of the repulsive core of the nucleon-nucleon interaction. Results of calculations of the charge distributions for ^{3,4}He demonstrate overall satisfactory agreement with experimental data [13] (Figure 1b).

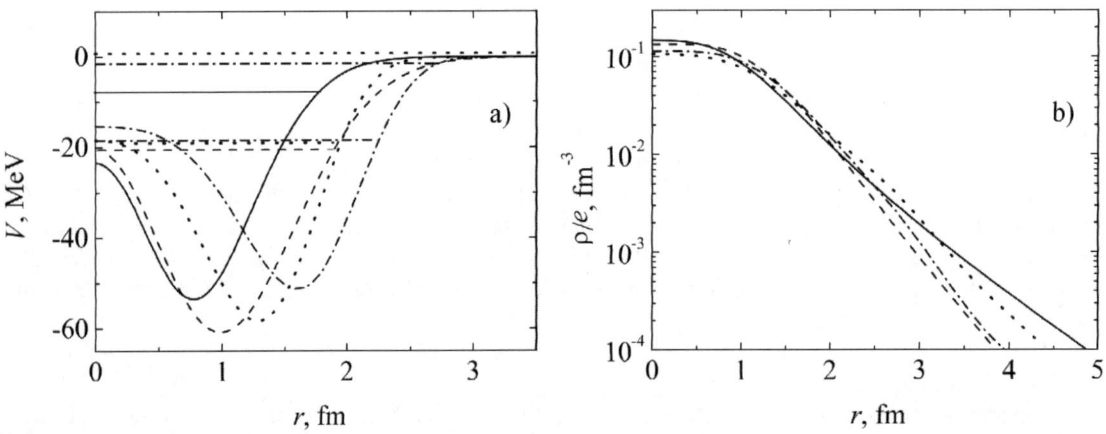

Figure 1. a) The mean field *V*(r) for neutrons and the corresponding neutron levels obtained within the shell model for ^{3}He (solid line), ^{4}He (dashed line), ^{5}He (dotted line) and ^{6}He (dash-dotted line). b) The charge densities obtained within the shell model for ^{3}He (solid line) and ^{4}He (dashed line) compared with experimental charge densities for ^{3}He (dotted line) and ^{4}He (dash-dotted line) [13].

3.2. The time-dependent Schrödinger equation (TDSE) and neutron wave functions

For theoretical description of neutron transfer during collisions of heavy atomic nuclei several semi-classical models are used [5, 14, 15]. They combine quantum description of internal one-particle and collective degrees of freedom with classical equations of motion of atomic nuclei, equation (1)

$$m_1\ddot{\boldsymbol{r}}_1 = -\nabla_{r_1} V_{12}\left(\left|\boldsymbol{r}_1 - \boldsymbol{r}_2\right|\right),\ m_2\ddot{\boldsymbol{r}}_2 = -\nabla_{r_2} V_{12}\left(\left|\boldsymbol{r}_1 - \boldsymbol{r}_2\right|\right). \quad (1)$$

Here $\boldsymbol{r}_1(t)$, $\boldsymbol{r}_2(t)$ are the centers of nuclei with the masses m_1, m_2 and $V_{12}(r)$ is the potential energy of nuclear interaction [16]. We may assume that before contact of the surfaces of spherical nuclei with the radii R_1, R_2 the potential energy of a neutron $W(\boldsymbol{r},t)$ is equal to the sum of its shell model mean fields for both nuclei.

The evolution of the components ψ_1, ψ_2 of the spinor wave function $\Psi(\boldsymbol{r},t)$ for the neutron with the mass m during the collision of nuclei is determined by the equation (2) with the operator $\hat{V}_{LS}$ of the spin-orbit interaction

$$i\hbar\frac{\partial}{\partial t}\Psi(\boldsymbol{r},t) = \left\{-\frac{\hbar^2}{2m}\Delta + W(\boldsymbol{r},t) + \hat{V}_{LS}(\boldsymbol{r},t)\right\}\Psi(\boldsymbol{r},t). \quad (2)$$

The evolution of the probability density for the external $1p_{3/2}$ shell neutrons of ^{6}He in the reaction ^{6}He + ^{197}Au was studied in Ref. [14]. In this case the neutron occupies free levels of the ^{197}Au nucleus. The neutron separation energies of ^{3}He and ^{197}Au are similar. In the case of pick-up to ^{197}Au the neutron occupies free levels only. The Pauli Exclusion Principle limits transfers to the unoccupied states. This principle was taken into account by two different approximations providing very similar results. Within the first, simple, approximation transfer to the occupied levels in the "frozen" shell structures of colliding nuclei is excluded. Within the second, more complicated, approximation the time-dependent few-body Slater determinant wave function is used [15]. In the case of pick-up to ^{3}He the neutron occupies the partially free $1s$ level of the ^{3}He nucleus and the mean field transforms from the ^{3}He shell model to the ^{4}He shell model.

The evolution of the probability density in the collisions of ^{3}He+^{45}Sc and ^{6}He+^{45}Sc is shown in Figures 2 and 3, respectively.

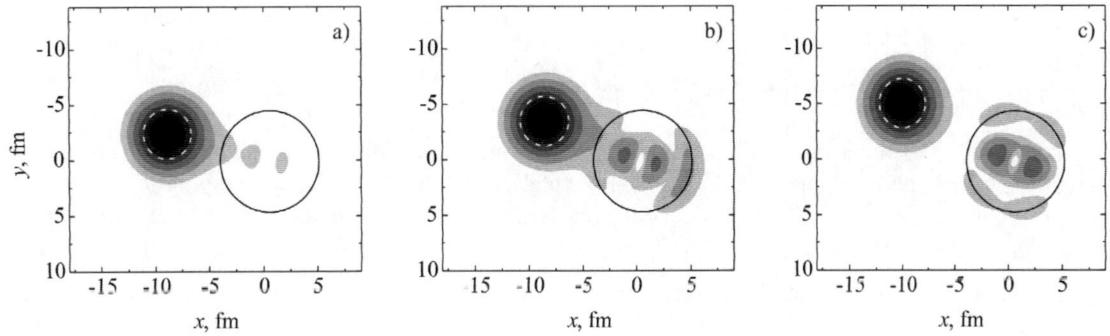

Figure 2. Example of the time evolution of the probability density for the external neutron of ^{3}He in the collision with ^{45}Sc at $E_{c.m.}$ = 6.5 MeV, impact parameter b = 1.5 fm. Radii of circumferences equal the effective radii of nuclei $\bar{R}_1 = 2.2$ fm, $\bar{R}_2 = 4.5$ fm. The course of time corresponds to the panel locations a, b, c.

After the transfer (stripping) from the ^{3,6}He nuclei (see Figure 4a) the neutron may occupy several initially vacant levels $2p_{3/2}, 2p_{1/2}, 1f_{5/2}$ and a partially vacant level $1f_{7/2}$ in the ^{45}Sc nucleus. The total probability of neutron stripping and stripping probabilities to various levels of ^{45}Sc in the reaction ^{45}Sc(^{3}He, 2p)^{46}Sc are shown in Figure 4b.

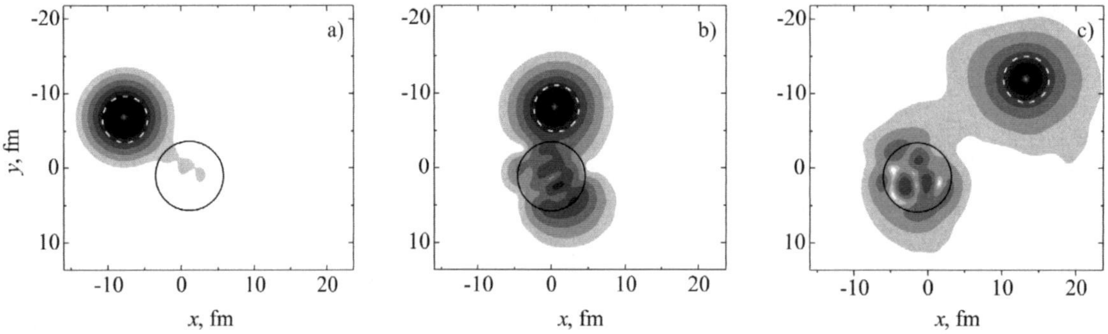

Figure 3. Example of the time evolution of the probability density for the external neutron of ^{6}He in the grazing collision with ^{45}Sc at $E_{c.m.}$ = 16 MeV, impact parameter $b=7$ fm. Radii of circumferences equal the effective radii of nuclei $\bar{R}_1 = 3.0$ fm, $\bar{R}_2 = 4.5$ fm. The course of time corresponds to the panel locations a, b, c.

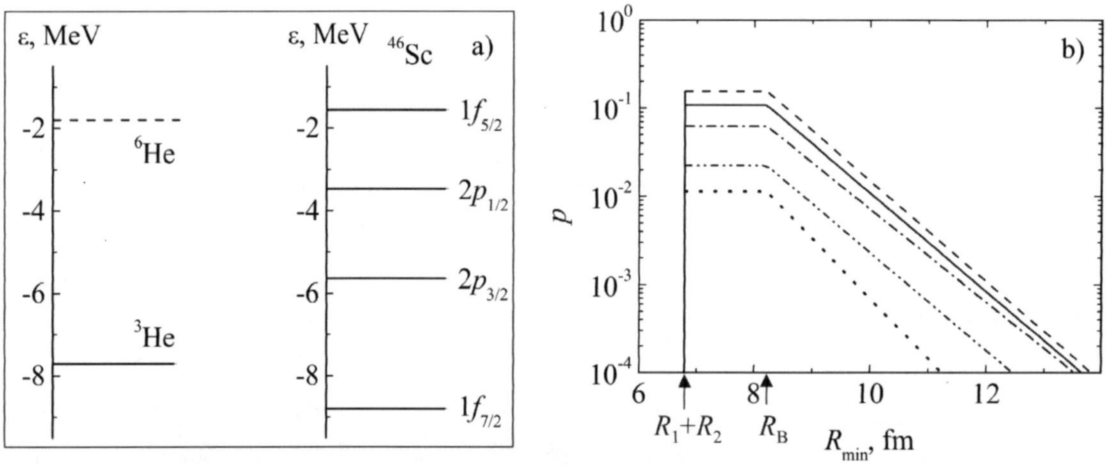

Figure 4. a) The neutron level scheme of ^{3,6}He and ^{46}Sc nuclei for the stripping reactions ^{45}Sc(^{3}He, 2p)^{46}Sc and ^{45}Sc(^{6}He, ^{5}He)^{46}Sc; b) The total probability of neutron stripping to ^{45}Sc in the reaction ^{45}Sc(^{3}He, 2p)^{46}Sc taking into account the Pauli Exclusion Principle (solid line) and without the Pauli Exclusion Principle (dashed line), the probability of stripping to the states $2p_{3/2}$ and $2p_{1/2}$ (dash-dotted line), $1f_{5/2}$ (dash-dash-dotted line) and $1f_{7/2}$ (dotted line) at $E_{c.m.}$ = 15 MeV. R_B is the barrier position.

Similarly, a weakly bound neutron of the ^{6}He nucleus may be transferred to several initially vacant top levels $2g_{9/2}$, $1i_{11/2}$, $3d_{5/2}$, $1j_{15/2}$, $4s_{1/2}$, $2g_{7/2}$, $3d_{3/2}$ in the ^{197}Au nucleus. The more strongly bound neutron of ^{3}He may be transferred to several initially vacant levels $2f_{5/2}$, $3p_{3/2}$, $3p_{1/2}$ near the Fermi level in the ^{197}Au nucleus.

3.3. Calculations of transfer cross sections

The total cross section of neutron transfer in the time-dependent model is calculated by equation (3)

$$\sigma = 2\pi \int_{b_0}^{\infty} p(b) b db . \tag{3}$$

Here b is the impact parameter, b_0 is the minimum collision impact parameter which is equal to zero for the energy less than the Coulomb barrier. The function $p(b)$ is the probability of neutron transfer during the collisions of nuclei without contact between their surfaces. Without taking into account the Pauli Exclusion Principle $p(b)$ was determined by integrating the probability density over the volume after the collision (Figures 2 and 3). With the Pauli Exclusion Principle only stripping to initially vacant levels was taken into account. In the case of ^{6}He only stripping to discrete bound states of the target was taken into account.

The comparison of the experimental data and theoretical calculations for the reactions ^{45}Sc(^{3}He, 2p)^{46}Sc and ^{197}Au(^{3}He, 2p)^{198}Au is shown in Figures 5a and 5b, respectively. Symbols are the data from Refs [3, 4].

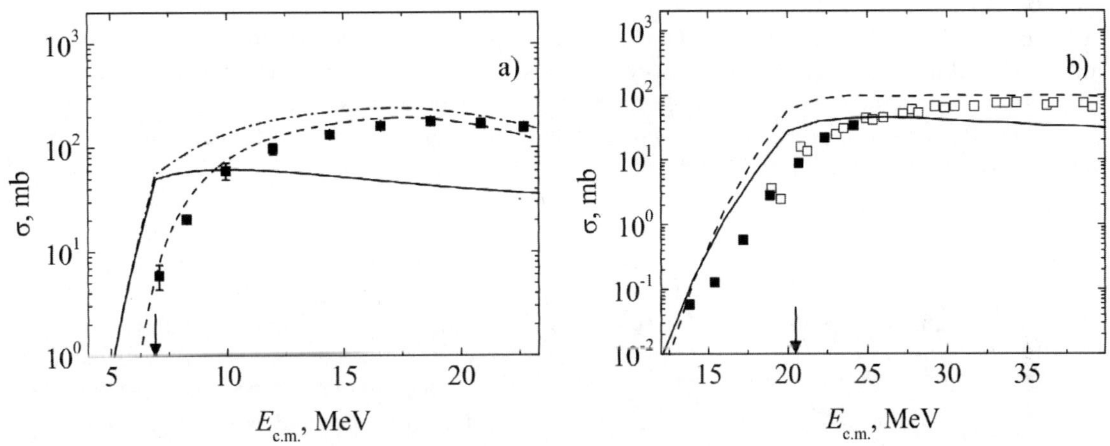

Figure 5. a) The excitation function for the reaction ^{45}Sc(^{3}He, 2p)^{46}Sc. b) The excitation function for the reaction ^{197}Au(^{3}He, 2p)^{198}Au. Symbols are the experimental data from Ref. [3, 4] (filled squares) and Ref. [17] (empty squares), dashed curves are fusion-2p-evaporation calculations using NRV evaporation code [11], solid curves are neutron transfer calculations taking into account the Pauli Exclusion Principle, dash-dotted curve is sum of fusion-2p-evaporation and neutron transfer. Here and below arrows indicate barrier heights.

The comparison of the experimental data and theoretical calculations for the reactions ^{45}Sc(^{3}He, ^{4}He)^{44}Sc and ^{197}Au(^{3}He, ^{4}He)^{196}Au is shown in Figures 6a and 6b, respectively. For neutron pick-up the energy of the neutron in ^{197}Au was chosen to coincide with the experimental neutron separation energy. As can be seen, in the reaction ^{3}He + ^{197}Au neutrons are predominantly transferred from ^{3}He to ^{197}Au. The comparison of the experimental data and theoretical calculations for the reaction ^{45}Sc(^{6}He, ^{5}He)^{46}Sc is shown in Figure 7. The comparison of the experimental data and theoretical calculations for the reaction ^{197}Au(^{6}He, ^{5}He)^{198}Au has been reported in [14].

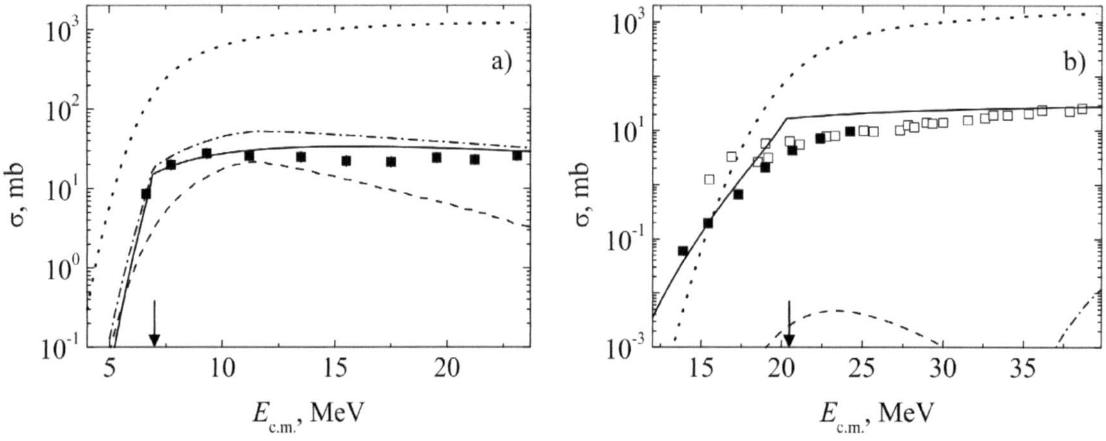

Figure 6. a) The excitation function for the reaction $^{45}Sc(^{3}He, ^{4}He)^{44}Sc$. b) The excitation function for the reaction $^{197}Au(^{3}He, ^{4}He)^{196}Au$. Symbols are the experimental data from Ref. [3, 4] (filled squares) and Ref. [17] (empty squares), dotted curves are total fusion cross sections calculated using NRV fusion code [11], dashed curves are fusion-$2p2n,\alpha$ -evaporation calculations using NRV evaporation code [11], solid curves are neutron transfer calculations taking into account the Pauli Exclusion Principle, dash-dotted curves are sums of fusion-$2p2n,\alpha$ -evaporation and neutron transfer taking into account the Pauli Exclusion Principle, dash-dot-dotted curves are neutron transfer calculations without taking into account the Pauli Exclusion Principle.

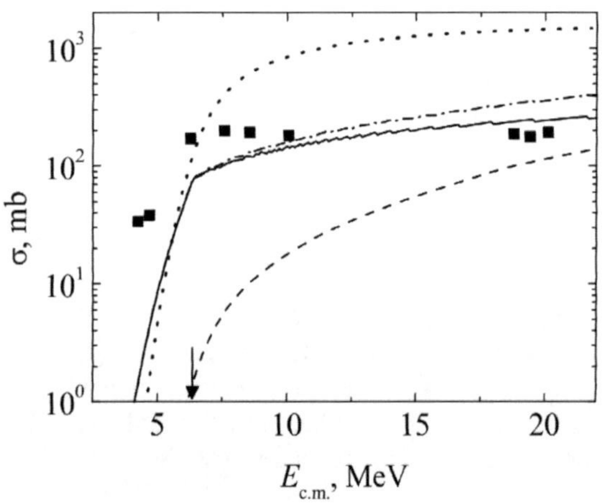

Figure 7. The excitation function for the reaction $^{45}Sc(^{6}He, ^{5}He)^{46}Sc$. Symbols are the experimental data from Ref. [18] (filled squares), dotted curve is total fusion cross section calculated using NRV fusion code [11], dashed curve is fusion-αn-evaporation calculation using NRV evaporation code [11], solid curve is neutron transfer calculation taking into account only stripping to discrete bound states, dash-dotted curve is sum of fusion-αn-evaporation and neutron transfer.

The model does not take into account the energy change associated with the Q-value of the neutron transfer. In the case of $Q>0$ it may lead to closer distances between nuclei and, thus, higher transfer probability resulting in higher transfer cross section. In the opposite case, $Q<0$, transfer probability is lower resulting in lower transfer cross section. This also implies the change of the structure of the projectile during the neutron transfer.

The effect of the Q-value and the quantum effects related to the motion of cores, *e.g.* tunneling, may be taken into account in the coupled channel approach [5, 12, 19] combined with the time-dependent Schrödinger equation method.

4. Conclusion

The results of calculations within the time-dependent Schrödinger equation method demonstrate overall satisfactory agreement with experimental data. The method may also be applied for calculation of reactions with cluster nuclei [20].

Acknowledgments
Authors thank A S Denikin and A V Karpov for fruitful discussions.

References

[1] Penionzhkevich Yu E 2011 *Phys. Atom. Nucl.* **74** 1615
[2] Penionzhkevich Yu E, Astabatyan R A, Demekhina N A *et al* 2007 *Eur. Phys. J.* A **31** 185
[3] Skobelev N K, Penionzhkevich Yu E, Voskoboynik E I *et al* 2014 *Phys. Part. and Nucl. Lett.* **11** 114
[4] Skobelev N K, Kulko A A, Penionzhkevich Yu E *et al* 2013 *Phys. Part. Nucl. Lett.* **10** 410
[5] Samarin V V 2015 *Phys. of Atom. Nucl.* **78** 128
[6] Oganessian Yu Ts, Zagrebaev V I and Vaagen J S 1999 *Phys. Rev.* C **60** 044605
[7] Kukulin V I, Krasnopolsky V M, Voronchev V T *et al* 1986 *Nucl. Phys.* A **453** 365
[8] Zhukov M V, Danilin B V, Fedorov D V *et al* 1993 *Phys. Rep.* **231** 151
[9] Feynman R P and Hibbs A R 1965 *Quantum Mechanics and Path Integrals* (New York: McGraw-Hill)
[10] Shuryak E V 1984 *Sov. Phys. Usp.* **27** 448
[11] Nuclear Reactions Knowledge Base http://nrv.jinr.ru/nrv
[12] Samarin V V 2015 *Phys. of Atom. Nucl.* **78** 861
[13] McCarthy J S, Sick I, Whitney R R 1977 *Phys. Rev.* C **15** 1396
[14] Samarin V V 2015 Time-dependent quantum models of the dynamics of neutron transfer reactions near the barrier *EPJ Web of Conf.* **86** (2015) 00040
[15] Samarin V V, Naumenko M A, Penionzhkevich Yu E *et al* 2014 *Proc. Int. Symp. on Exotic Nuclei (Kaliningrad)* (Singapore: World Scientific) p 115
[16] Winther A 1994 *Nucl. Phys.* A **572** 191
[17] Nagame Y, Sueki K, Baba S *et al* 1990 *Phys. Rev.* C **41** 889
[18] Skobelev N K, Kulko A A, Kroha V *et al* 2011 *Eur. Phys. J.* G **38** 035106
[19] Samarin V V 2015 Coupling with two-center neutron states and two-surface collective excitations at fusion reactions in the vicinity of the Coulomb barrier *EPJ Web of Conf.* **86** (2015) 00039
[20] Samarin V V 2015 Microscopic time-dependent description of alpha-cluster transfer and incomplete fusion in reactions near Coulomb barrier *Proc. Int. Conf. "Nucleus 2015" (Saint-Petersburg)* (Saint-Petersburg: BBM) p 198

Beta-decay properties of neutron-rich medium-mass nuclei

Pedro Sarriguren

Instituto de Estructura de la Materia, IEM-CSIC, Serrano 123, E-28006 Madrid, Spain

E-mail: p.sarriguren@csic.es

Abstract. β-decay properties of even-even and odd-A neutron-rich Ge, Se, Kr, Sr, Zr, Mo, Ru, and Pd isotopes involved in the astrophysical rapid neutron capture process are studied within a microscopic proton-neutron quasiparticle random-phase approximation. The underlying mean field is based on a self-consistent Skyrme Hartree-Fock + BCS calculation that includes deformation as a key ingredient. The isotopic evolution of the various nuclear equilibrium shapes and the corresponding charge radii are investigated in all the isotopic chains. The energy distributions of the Gamow-Teller strength, as well as the β-decay half-lives are discussed and compared with the available experimental information. It is shown that nuclear deformation plays a significant role in the description of the decay properties in this mass region. Reliable predictions of the strength distributions are essential to evaluate decay rates in astrophysical scenarios.

1. Introduction

Weak β-decay is a very important mechanism to understand the late stages of the stellar evolution, playing a critical role to determine both the presupernova stellar structure and the nucleosynthesis of heavier nuclei. These processes are dominated by Gamow-Teller (GT) transitions. An accurate understanding of those astrophysical processes requires input from nuclear physics. Since this information cannot be measured directly in exotic nuclei with very short half-lives, the nuclear properties must be estimated by model calculations. Obviously, nuclear physics uncertainties will finally affect the reliability of the description of those astrophysical processes. Particularly interesting from this point of view is the region of medium-mass neutron-rich nuclei that are involved in the rapid neutron capture process (r process). This process is considered as the main nucleosynthesis mechanism responsible for the production of heavy neutron-rich nuclei and for the existence of about half of the nuclei heavier than iron [1, 2].

In this work, we study the GT strength distributions and β-decay half-lives of neutron-rich nuclei calculated within a proton-neutron quasiparticle random-phase approximation (pnQRPA) based on a selfconsistent deformed Hartree-Fock (HF) mean field with Skyrme interactions including pairing correlations and residual separable forces in both particle-hole (ph) and particle-particle (pp) channels. The present nuclear model has been tested successfully reproducing very reasonably the experimental information available on both bulk and decay properties of medium-mass nuclei [3, 4, 5, 6, 7]. The study is focused on neutron-rich isotopes including $^{80-94}$Ge, $^{86-100}$Se, $^{90-104}$Kr, $^{94-108}$Sr, $^{100-116}$Zr, $^{104-120}$Mo, $^{110-124}$Ru, and $^{114-128}$Pd, although in this work we pay special attention to Kr and Sr isotopes.

2. Theoretical approach

The β-decay half-life is obtained by summing all the allowed transition strengths to states in the daughter nucleus with excitation energies lying below the corresponding Q_{β^-} energy, $Q_\beta = Q_{\beta^-} = M(A,Z) - M(A,Z+1) - m_e$, written in terms of the nuclear masses $M(A,Z)$ and the electron mass (m_e), and weighted with the phase space factors $f(Z, Q_\beta - E_{ex})$,

$$T_{1/2}^{-1} = \frac{(g_A/g_V)^2_{\text{eff}}}{D} \sum_{0<E_{ex}<Q_\beta} f\left(Z, Q_\beta - E_{ex}\right) B(GT, E_{ex}) \,, \tag{1}$$

with $D = 6200$ s and $(g_A/g_V)_{\text{eff}} = 0.77(g_A/g_V)_{\text{free}}$, where 0.77 is a standard quenching factor. The same quenching factor is included in all the figures shown later for the GT strength distributions. The bare results can be recovered by scaling the results in this paper for $B(GT)$ and $T_{1/2}$ with the square of this quenching factor.

The Fermi integral $f(Z, Q_\beta - E_{ex})$ is computed numerically for each value of the energy including screening and finite size effects [8],

$$f^{\beta^\pm}(Z, W_0) = \int_1^{W_0} pW(W_0 - W)^2 \lambda^\pm(Z, W) \mathrm{d}W \,, \tag{2}$$

with

$$\lambda^\pm(Z, W) = 2(1+\gamma)(2pR)^{-2(1-\gamma)} e^{\mp\pi y} \frac{|\Gamma(\gamma + iy)|^2}{[\Gamma(2\gamma+1)]^2} \,, \tag{3}$$

where $\gamma = \sqrt{1-(\alpha Z)^2}$; $y = \alpha Z W/p$; α is the fine structure constant and R the nuclear radius. W is the total energy of the β particle, W_0 is the total energy available in $m_e c^2$ units, and $p = \sqrt{W^2 - 1}$ is the momentum in $m_e c$ units.

The nuclear structure involved in the weak rates is described within a microscopic deformed pnQRPA based on a selfconsistent mean field obtained from Skyrme (SLy4) Hartree-Fock + BCS calculations. A residual separable interaction in both *ph* and *pp* is added to the mean field and treated within pnQRPA. More details of the calculation can be found in Refs. [9]. The GT strength for a transition from an initial state i to a final state f is given by

$$B_{if}(GT^\pm) = \langle f || \sum_j^A \sigma_j t_j^\pm || i \rangle^2 \,. \tag{4}$$

3. Results and discussion

We can see in Figures 1 and 2 the isotopic evolution of the quadrupole deformations β (upper plots) and r.m.s. charge radii r_c (lower plots) for Kr and Sr isotopes, respectively. The deformation corresponding to the ground state for each isotope is encircled. For Kr isotopes we can see first a shape transition from prolate ($\beta \approx 0.15$) to oblate ($\beta \approx -0.25$) at $A = 92-94$ and a subsequent transition at $A = 96 - 98$ from oblate to prolate ($\beta \approx 0.35$) shapes. The radii are sensitive to these transitions, although the measured radii [10] seem to favor prolate shapes in the lighter isotopes. Sr isotopes show a clear transition from oblate to strong prolate ($\beta \approx 0.4$) deformations at $A = 96 - 98$ ($N = 58 - 60$). This shape transition is well correlated with the change in the trend observed in the charge radii that shows a sizable jump between ^{96}Sr and ^{98}Sr both theoretically and experimentally [10].

In Figure 3 we can see the accumulated GT strength corresponding to the various deformed equilibrium shapes in Sr isotopes. The GT strength is plotted versus the excitation energy of the daughter nucleus below the Q_β energy. In these figures the sensitivity of the distribution to deformation can be clearly appreciated and one can understand that measurements of the GT

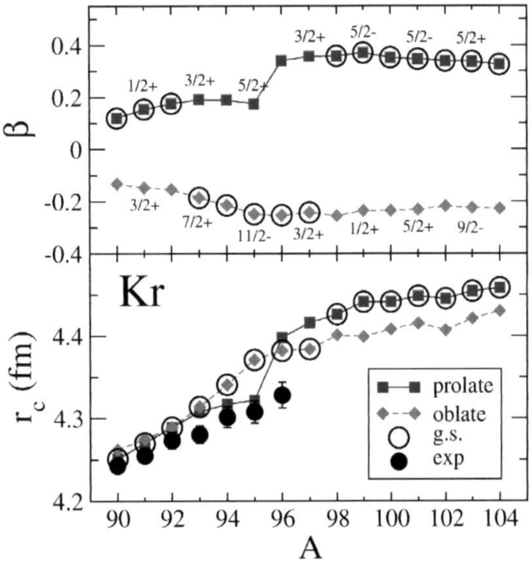

Figure 1. Isotopic evolution of the deformation β and radius r_c in Kr isotopes.

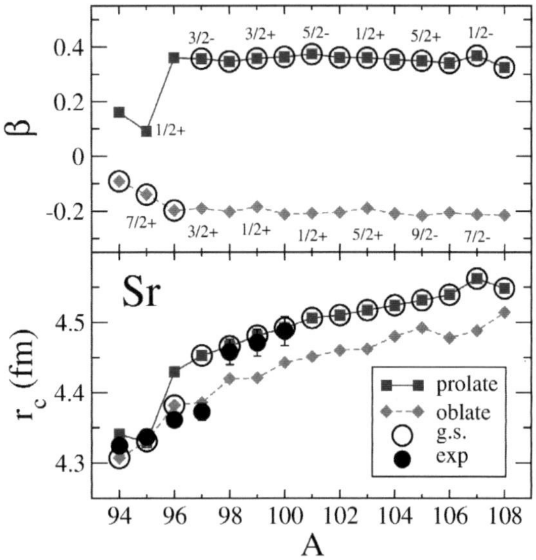

Figure 2. Same as in Fig. 1, but for Sr isotopes.

strength distribution from β-decay can be, in particular cases, an additional source of information about the nuclear deformation, as it was shown in Ref. [3].

The energy distribution of the GT strength is fundamental to constrain the underlying nuclear structure. For a theoretical model, it represents a more demanding test than just reproducing the half-life or the total GT strength that are both integral quantities obtained from these strength distributions properly weighted with phase factors (see Eq. 1) in the case of the half-lives. These quantities might be reproduced even with wrong strength distributions. The calculation of the half-lives in Eq. (1) involves knowledge of the GT strength distribution and of the β energies $(Q_\beta - E_{ex})$, which are evaluated by using Q_β values obtained from the mass differences between parent and daughter nuclei obtained from SLy4.

In Figures 4 and 5 the measured β-decay half-lives, which appear as solid dots (open dots stand for experimental values from systematics) [11, 12], are compared with the theoretical results obtained with the prolate and oblate equilibrium shapes for the Kr and Sr isotopic chains, respectively. For Kr isotopes the general tendency observed experimentally is well reproduced by the calculations, except for the lighter isotopes having larger half-lives. In Sr isotopes the trend observed experimentally is well accounted for, especially with prolate shapes. Half-lives for neutron-rich Kr, Sr, Zr, and Mo isotopes obtained from self-consistent deformed pnQRPA calculations with the Gogny D1M interaction and experimental values of Q_β [13] agree with the results in this work within the uncertainties of the calculations. The agreement is also very reasonable between our calculated half-lives and those obtained from pnQRPA calculations using deformed Woods-Saxon potentials to generate the mean field and complemented with realistic CD-Bonn residual forces [14, 15]. The agreement is also good with the results in Ref. [16] using the Skyrme force SLy4 with fully consistent residual interactions in the *ph* channel.

The impact of deformation on the decay properties is better appreciated by performing a systematic comparison of the half-lives. Figures 6 and 7 show the ratios of the calculated to experimental half-lives for two sets of data corresponding to a spherical calculation (open red dots) and to a deformed calculation at the self-consistent deformation that gives the minimum of the energy (solid black dots). In Figure 6 the ratios are plotted as a function of the quadrupole

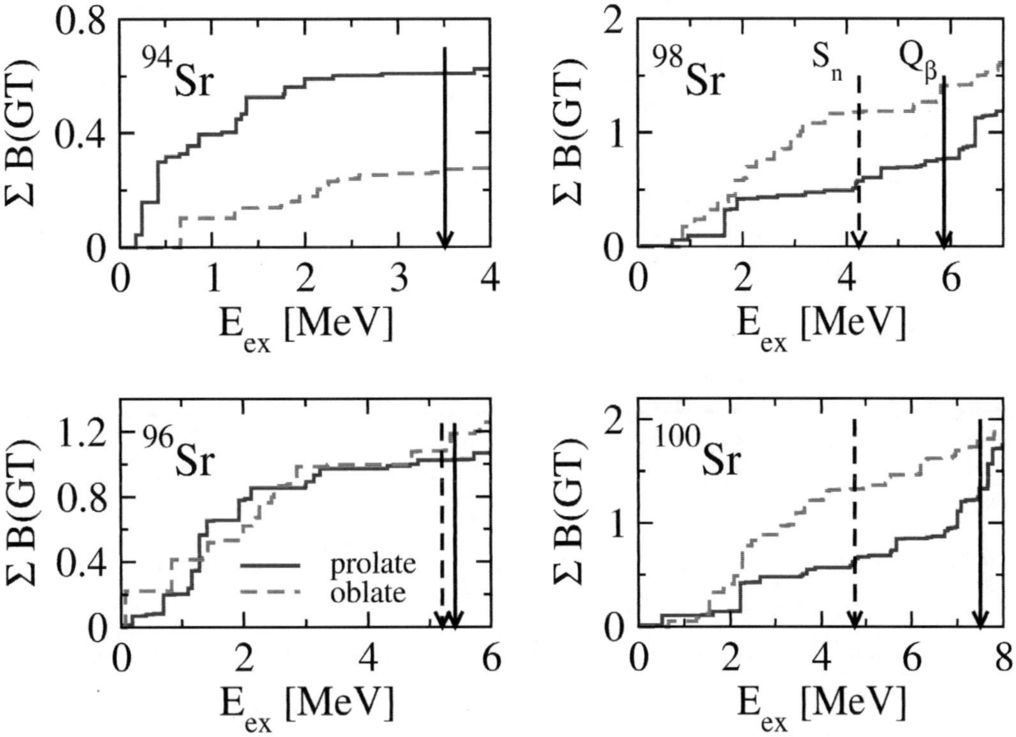

Figure 3. pnQRPA-SLy4 accumulated GT strengths as a function of the excitation energy in the daughter nucleus in some Sr isotopes calculated for the various equilibrium shapes. Q_β and S_n energies are shown by solid and dashed vertical arrows, respectively.

deformation at the minimum of the energy, whereas in Figure 7 the ratios are plotted as a function of the experimental half-lives.

One can see that deformation improves in general the description of the half-lives. Practically all the solid dots are contained within the horizontal lines defining the region of one order of magnitude agreement. On the other hand, the results from the spherical calculation are more spread out with larger discrepancy with experiment.

In order to have a quantitative estimation of the quality of the various calculations, following the analysis made in Ref. [17], the logarithms of the ratios of the calculated and experimental half-lives are introduced through the quantities

$$r = \log_{10}\left[\frac{T_{1/2}(\text{calc})}{T_{1/2}(\text{exp})}\right] . \tag{5}$$

The average position of the points, M_r, and the standard deviation, σ_r, are defined as

$$M_r = \frac{1}{n}\sum_{i=1}^{n} r_i ; \quad \sigma_r = \left[\frac{1}{n}\sum_{i=1}^{n}(r_i - M_r)^2\right]^{1/2} , \tag{6}$$

and their corresponding factors $M_r^{10} = 10^{M_r}$ and $\sigma_r^{10} = 10^{\sigma_r}$. The analysis of the results involving $n = 81$ nuclei leads to the values $M_r^{10} = 1.105$ and $\sigma_r^{10} = 10.21$ in the spherical case and $M_r^{10} = 0.937$ and $\sigma_r^{10} = 3.09$ in the deformed one. The latter values are clearly closer to unity, showing the improvement achieved with the deformed formalism. Table 1 shows the

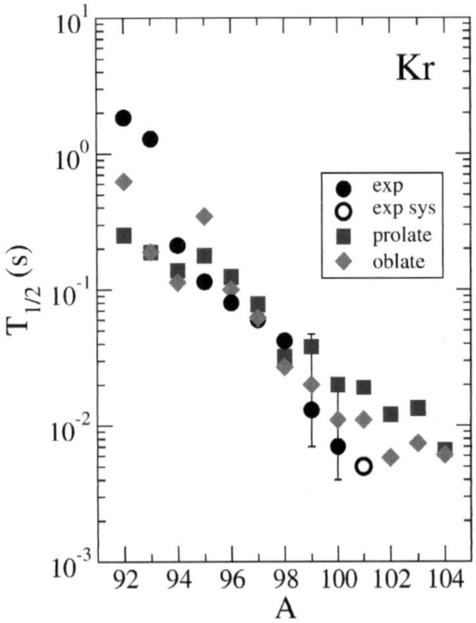

Figure 4. Experimental and calculated half-lives for Kr isotopes.

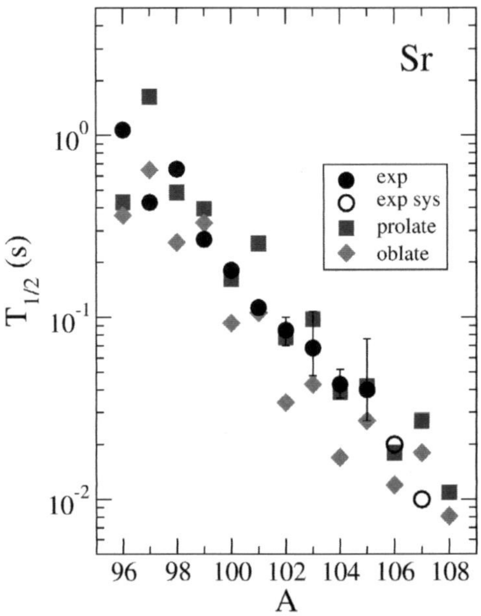

Figure 5. Same as in Fig. 4, but for Sr isotopes.

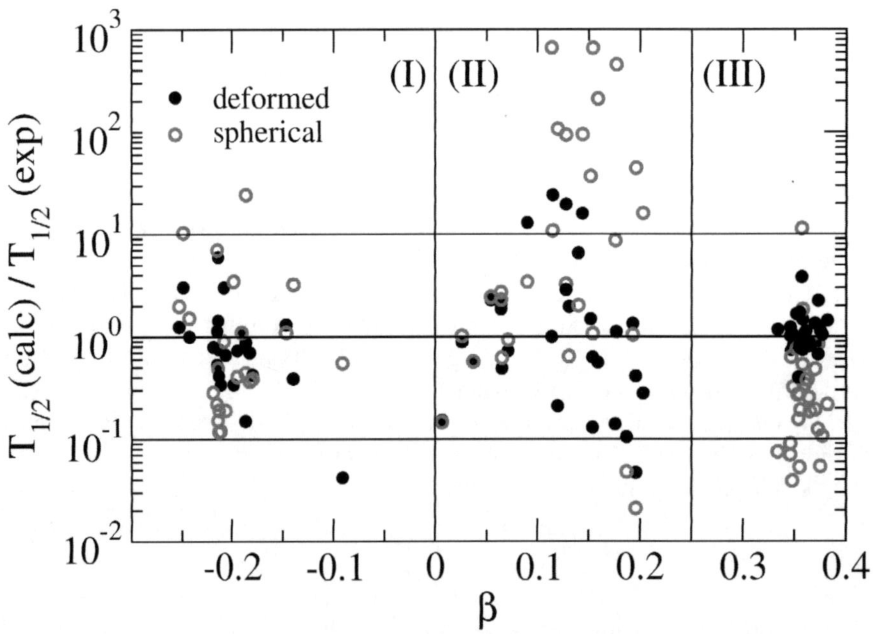

Figure 6.
Ratio of calculated to experimental β-decay half-lives plotted as a function of the quadrupole deformation at equilibrium. The results correspond to spherical (open dots) and deformed (solid dots) calculations.

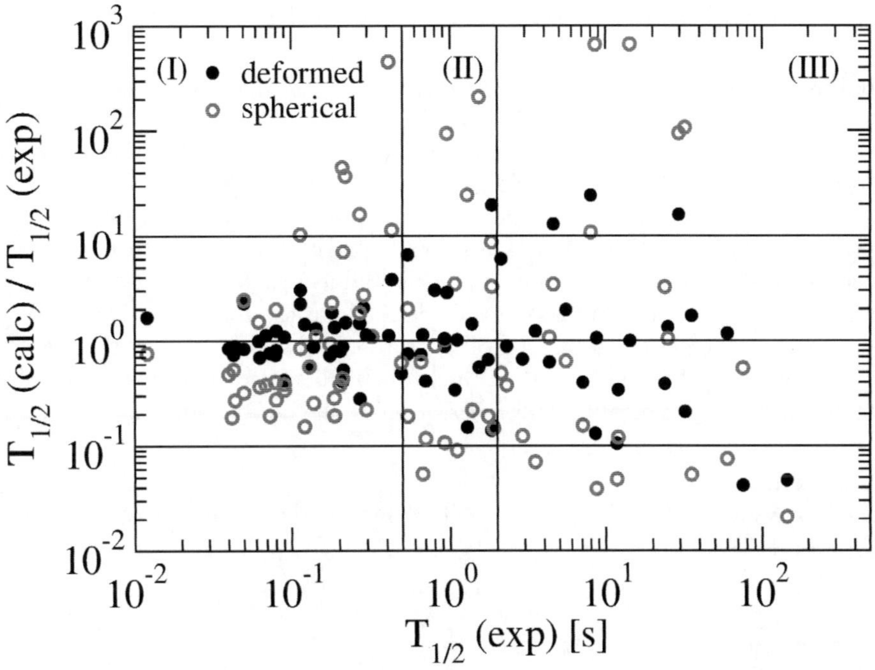

Figure 7.
Same as in Fig. 6, but plotted as a function of the experimental half-lives.

Table 1. Factors of the average position M_r^{10} and standard deviation σ_r^{10} for the ratios r (5) in different regions of half-lives and deformations.

		M_r^{10}	σ_r^{10}	points
Global	sph	1.105	10.213	81
	def	0.937	3.088	
$0.01 < T_{1/2} < 0.5$	sph	1.172	5.691	42
	def	1.053	1.758	
$0.5 < T_{1/2} < 2$	sph	1.078	10.952	18
	def	0.903	3.509	
$2 < T_{1/2} < 500$	sph	1.001	21.473	21
	def	0.843	5.376	
$-0.3 < \beta < 0$	sph	0.898	4.442	25
	def	0.744	2.647	
$0 < \beta < 0.25$	sph	5.662	15.710	28
	def	1.014	5.000	
$0.25 < \beta < 0.40$	sph	0.260	3.285	28
	def	1.062	1.518	

results of this global analysis, as well as the results corresponding to different regions in both half-lives and quadrupole deformations. The improvement of the deformed formalism is evident in the regions of well deformed nuclei and short half-lives, where the large Q_β-values make the half-lives more significant as they are sensitive to a larger region of the GT strength distribution.

4. Conclusions

We have studied the isotopic evolution of bulk and decay properties of neutron-rich medium-mass nuclei. The nuclear structure involved in the calculation of the energy distribution of the Gamow-Teller strength is described within a selfconsistent deformed HF+BCS+QRPA formalism with density-dependent effective Skyrme interactions and spin-isospin residual interactions. We find that the present pnQRPA calculation is able to reproduce the main features of the decay properties based on the experimental information available on β-decay half-lives. Predictions for the Gamow-Teller strength distributions have been made in the relevant range of energies below the Q_β-windows. A systematic comparison of the ratios of the calculated and experimental half-lives has been done using both spherical and deformed calculations, showing that the inclusion of deformation improves significantly the description of the decay properties.

Acknowledgments

This work was supported in part by MINECO (Spain) under Research Grants Nos. FIS2011–23565 and FIS2014–51971–P.

References

[1] Burbidge EM *et al.* 1959 *Rev. Mod. Phys.* **29** 547
[2] Cowan JJ *et al.* 1991 *Phys. Rep.* **208** 267
[3] Nácher E *et al.* 2004 *Phys. Rev. Lett.* **92** 232501; Poirier E *et al.* 2004 *Phys. Rev.* C **69** 034307; Pérez-Cerdán AB *et al.* 2013 *Phys. Rev.* C **88** 014324
[4] Sarriguren P *et al.* 2003 *Nucl. Phys.* **A716** 230; Sarriguren P 2013 *Phys. Rev.* C **87** 045801
[5] Sarriguren P *et al.* 2005 *Eur. Phys. J.* A **24** 193; Sarriguren P 2009 *Phys. Rev.* C **79** 044315; Sarriguren P 2009 *Phys. Lett.* B **680** 438; Sarriguren P 2011 *Phys. Rev.* C **83** 025801
[6] Sarriguren P and Pereira J 2010 *Phys. Rev.* C **81** 064314; Sarriguren P, Algora A, and Pereira J 2014 *Phys. Rev.* C **89** 034311; Sarriguren P 2015 *Phys. Rev.* C **91** 044304
[7] Sarriguren P *et al.* 2005 *Phys. Rev.* C **72** 054317; Moreno O *et al.* 2006 *Phys. Rev.* C **73** 054302; Boillos JM *et al.* 2015 *Phys. Rev.* C **91** 034311
[8] Gove NB and Martin MJ 1971 *At. Data Nucl. Data Tables* **10** 205
[9] Sarriguren P *et al.* 1998 *Nucl. Phys.* **A635** 55; Sarriguren P *et al.* 1999 *Nucl. Phys.* **A658** 13; Sarriguren P *et al.* 2001 *Nucl. Phys.* **A691** 631; Sarriguren P *et al.* 2001 *Phys. Rev.* C **64** 064306
[10] Angeli I 2004 *At. Data Nucl. Data Tables* **87** 185
[11] Audi G *et al.* 2012 *Chinese Physics* C **36** 1157
[12] Nishimura S *et al.* 2011 *Phys. Rev. Lett.* **106** 052502
[13] Martini M *et al.* 2014 *Phys. Rev.* C **89** 044306
[14] Fang FL *et al.* 2013 *Phys. Rev.* C **88** 024314
[15] Ni D and Ren Z 2014 *Phys. Rev.* C **89** 064320; 2012 *J. Phys. G: Nucl. Part. Phys.* **39** 125105
[16] Yoshida K 2013 *Prog. Theor. Exp. Phys.* 113D02
[17] Moeller P, Pfeiffer B, and Kratz KL 2003 *Phys. Rev.* C **67** 055802

Gamow-Teller decay studies with 2p-2h configurations

A. P. Severyukhin[1], V. V. Voronov[1], I. N. Borzov[2,1], N. N. Arsenyev[1] and Nguyen Van Giai[3]

[1]Bogoliubov Laboratory of Theoretical Physics, Joint Institute for Nuclear Research, 141980 Dubna, Moscow region, Russia
[2]National Research Centre "Kurchatov Institute", 123182 Moscow, Russia
[3]Institut de Physique Nucléaire, CNRS-IN2P3, Université Paris-Sud, F-91406 Orsay Cedex, France

Abstract. Starting from a Skyrme interaction with tensor terms, the β-decay rates have been studied within a microscopic model including the $2p-2h$ configuration effects. As an application we present the evolution of the neutron-rich Ni isotopes near ^{78}Ni that are important for stellar nucleosynthesis.

The study of spin-isospin excitations in neutron-rich nuclei is presently an important problem not only from the nuclear structure point of view but also because of the special role they play in many astrophysical processes. Many fundamental issues depend on our quantitative understanding of the β-decay of atomic nuclei. It is desirable to have theoretical models which can describe the data wherever they can be measured and which can predict the properties related to spin-isospin excitations in systems too short-lived to allow for experimental studies. One of the successful tools for studying charge-exchange nuclear modes is the quasiparticle random phase approximation (QRPA) with the self-consistent mean-field derived from a Skyrme-type energy-density functional (EDF), see e.g., [1, 2, 3, 4]. These QRPA calculations enable one to describe the properties of the ground states and excited charge-exchange states using the same EDF.

A comparison of such calculations with recent experimental data [5] demonstrates that the QRPA approach cannot reproduce correctly the strength distributions of the spin-isospin resonances. It is necessary to take into account the coupling with more complex configurations that results in shifting some strength upward to higher excitation energy [6, 7, 8]. Using the Skyrme EDF and the RPA, such attempts in the past [9, 10] have allowed one to understand the damping of charge-exchange resonances and their particle decay. Recently, the damping of the Gamow-Teller (GT) mode was investigated using the Skyrme-RPA plus particle-vibration coupling (PVC) [11]. However, the size of the configuration space increases very rapidly in the regions of interest of the nuclear chart, and one has to resort to severe space truncations.

It would be useful to study the effects of the $2p-2h$ configurations on the β^--decay rates of neutron-rich nuclei. It is somewhat simpler to include the PVC in QRPA calculations if one uses separable forces [7, 12]. Our tool is the QRPA with Skyrme interactions in the finite rank separable approximation (FRSA) [13, 14, 15, 16], allowing one to perform calculations in large configuration spaces. Successful applications of the method to study the electric low-energy excitations and giant resonances within and beyond the QRPA can be found in Refs [13, 14, 15, 16, 17, 18, 19, 20]. Recently, the FRSA approach was extended to charge-exchange nuclear excitations [21] and also for accommodating the tensor correlations which mimic the Skyrme-type tensor interactions [22]. Also we have generalized the approach to the coupling between one- and two-phonon components in the wave functions [23]. In the present report we describe the FRSA model and discuss the $2p-2h$ configuration effect on the β-decay rates for neutron-rich Ni isotopes.

The FRSA model for charge-exchange excitations was already introduced in Refs. [21, 22, 23]. Taking into account the basic ideas of the quasiparticle-phonon model (QPM) [12], the Hamiltonian is then diagonalized in a space spanned by states composed of one and two QRPA phonons,

$$\Psi_\nu(JM) = \left(\sum_i R_i(J\nu)Q^+_{JMi} + \sum_{\lambda_1 i_1 \lambda_2 i_2} P^{\lambda_1 i_1}_{\lambda_2 i_2}(J\nu)\left[Q^+_{\lambda_1\mu_1 i_1}\bar{Q}^+_{\lambda_2\mu_2 i_2}\right]_{JM}\right)|0\rangle\ , \tag{1}$$

where $Q^+_{\lambda\mu i}|0\rangle$ ($\bar{Q}^+_{\lambda\mu i}|0\rangle$) is the GT (electric-type) excitation having the QRPA energy $\omega_{\lambda i}$ ($\bar{\omega}_{\lambda i}$). The normalization condition for the wave functions (1) is

$$\sum_i R_i^2(J\nu) + \sum_{\lambda_1 i_1 \lambda_2 i_2} (P^{\lambda_1 i_1}_{\lambda_2 i_2}(J\nu))^2 = 1. \tag{2}$$

The amplitudes $R_i(J\nu)$ and $P^{\lambda_1 i_1}_{\lambda_2 i_2}(J\nu)$ are determined from the variational principle which leads to a set of linear equations

$$(\omega_{\lambda i} - \Omega_\nu)R_i(J\nu) + \sum_{\lambda_1 i_1 \lambda_2 i_2} U^{\lambda_1 i_1}_{\lambda_2 i_2}(Ji)P^{\lambda_1 i_1}_{\lambda_2 i_2}(J\nu) = 0, \tag{3}$$

$$(\omega_{\lambda_1 i_1} + \bar{\omega}_{\lambda_2 i_2} - \Omega_\nu)P^{\lambda_1 i_1}_{\lambda_2 i_2}(J\nu) + \sum_i U^{\lambda_1 i_1}_{\lambda_2 i_2}(Ji)R_i(J\nu) = 0. \tag{4}$$

The rank of the set of linear equations (3) and (4) is equal to the number of one- and two-phonon configurations included in the wave function of the excited states of the daughter nucleus. Its solution requires to compute the matrix elements coupling one- and two-phonon configurations

$$U^{\lambda_1 i_1}_{\lambda_2 i_2}(Ji) = \langle 0|Q_{Ji}H\left[Q^+_{\lambda_1 i_1}\bar{Q}^+_{\lambda_2 i_2}\right]_J|0\rangle. \tag{5}$$

Equations (3) and (4) have the same form as the QPM equations [7, 12], but the single-particle spectrum and the residual interaction are derived from the same Skyrme EDF [23].

To calculate the half-lives, the same ansatz as Sec. II of Ref. [24] is used with the ratio of the weak axial-vector and vector coupling constants G_A/G_V=1.25 [25]. In the allowed GT approximation, the β^--decay rate is expressed by summing the probabilities of the energetically allowed GT transitions (in units of $G_A^2/4\pi$) weighted with the integrated Fermi function

$$T^{-1}_{1/2} = \sum_m \lambda^m_{if} = D^{-1}\left(\frac{G_A}{G_V}\right)^2 \sum_m f_0(Z, A, E_i - E_{1^+_m})B(GT)_m, \tag{6}$$

$$E_i - E_{1^+_m} \approx \Delta M_{n-H} + \mu_n - \mu_p - E_m, \tag{7}$$

where λ^m_{if} is the partial β^--decay rate, D=6147 [25] is used. $\Delta M_{n-H} = 0.782$ MeV is the mass difference between the neutron and the hydrogen atom, μ_n and μ_p are the neutron and proton chemical potentials respectively, E_i is the ground state energy of the parent nucleus, and $E_{1^+_m}$

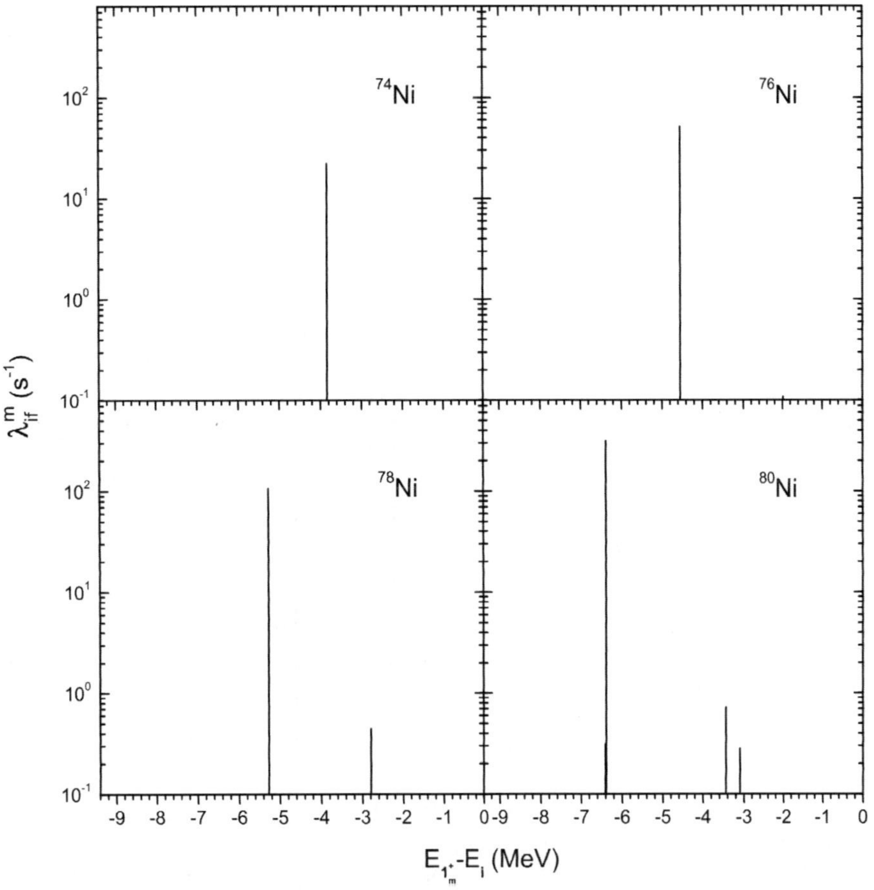

Figure 1. The β-transition rates obtained within the QRPA.

denotes a state of the daughter nucleus (Z, A). E_m are the eigenvalues of the QRPA equations or Eqs. (3) and (4) taking into account the two-phonon configurations. The wave functions allow us to determine $B(GT)_m$.

As for the parameter set in the particle-hole channel, we use the central Skyrme interaction SGII [26] and the same zero-range tensor interaction as that in Ref. [27]. Since the SGII parametrization gives reasonable values for the Landau parameters $F_0^{'} = 0.73$ and $G_0^{'} = 0.50$, one obtains a successful description of the spin-dependent properties and, in particular, experimental energies of the GT resonances of ^{90}Zr [26]. For the studied region of nuclei we use the volume pairing force fixed in Ref. [18]. The single-particle continuum is discretized by diagonalizing the HF hamiltonian on a basis of 12 harmonic oscillator shells and cutting off the single-particle spectra at the energy of 100 MeV. This is sufficient to exhaust the Ikeda sum rule $3(N-Z)$ [21, 22] as well as the sum rule for the electromagnetic excitation modes [16]. Since the tensor correlation effects within the $1p-1h$ and $2p-2h$ configuration spaces are treated, we do not need any quenching factor [6].

First, the properties of the low-lying 1^+ states in the daughter nuclei ^{74,76,78,80}Cu are studied within the one-phonon approximation. As expected, the largest contribution (>88%) in the calculated β^--decay half-life comes from the $[1_1^+]_{QRPA}$ state. To illustrate it, the β-transition rates λ_{if}^m are shown in Fig. 1. The transition energies $E_{1_m^+} - E_i$ refer to the ground state of the

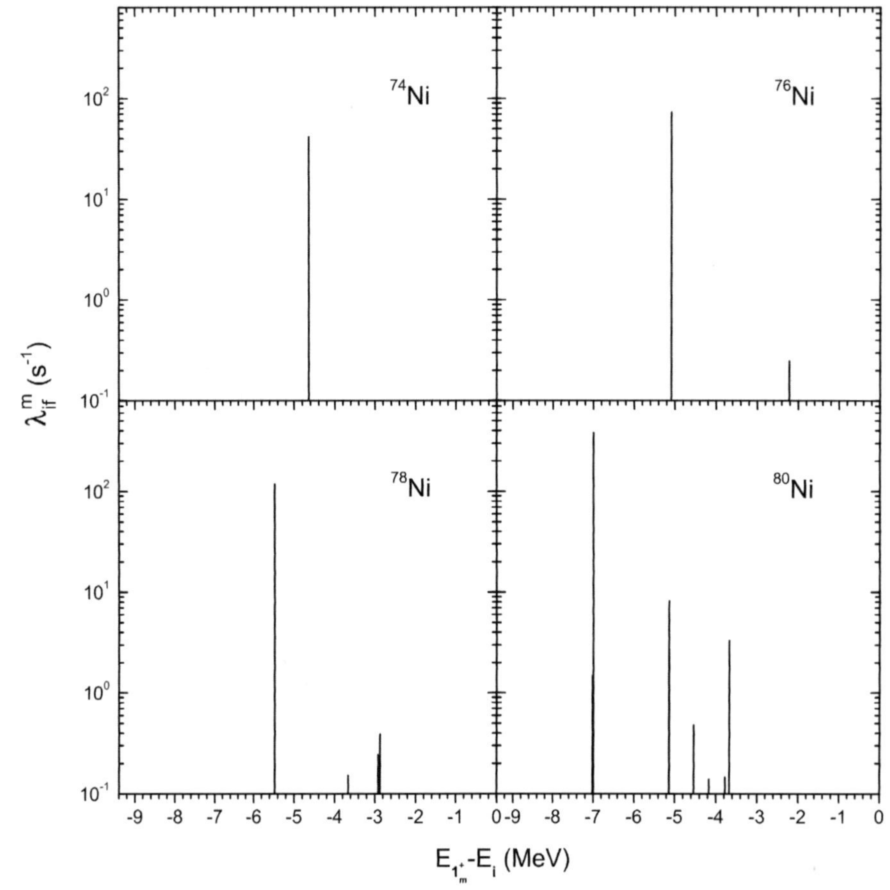

Figure 2. The β-transition rates calculated with the effects of the $2p-2h$ configurations.

parent nucleus. As shown in [23], the QRPA results indicate that the dominant configuration of the $[1^+_1]_{QRPA}$ states is $\{\pi 2p\frac{3}{2}\nu 2p\frac{1}{2}\}$ whose contribution decreases from 51% in ^{74}Ni to 39% in ^{80}Ni.

To construct the wave functions (1) of the low-lying 1^+ states we use only the $[1^+_i \otimes \lambda^+_{i'}]_{QRPA}$ terms and all electric phonons with $\lambda > 2$ vanish. All one- and two-phonon configurations with the transition energies $|E_{1^+_m} - E_i|$ up to 10 MeV are included. We have checked that the inclusion of the high-energy configurations leads to minor effects on the half-life values [23]. The $[1^+_1 \otimes 2^+_1]_{QRPA}$ and $[1^+_1 \otimes 0^+_2]_{QRPA}$ configurations are the important ingredients for the half-life description [23] since they are the main two-phonon components of the 1^+_1 wave function. Since there is a clear influence of the 2^+_1 phonon on the half-life [23, 28], we examine the properties of the 2^+_1 QRPA states of ^{74,76,78,80}Ni. The neutron amplitudes are dominant in the 2^+_1 structure and the contribution of the main neutron configuration $\{1g_{9/2}, 1g_{9/2}\}$ decreases from 79% for ^{74}Ni to 77% for ^{76}Ni when neutrons fill the subshell $1g_{9/2}$. There is a satisfactory description of 2^+_1 energies in ^{74,76}Ni [23].

The inclusion of the two-phonon configurations results in an increase of the transition energies $|E_{1^+_m} - E_i|$ and partial rates of the main GT transitions (see Fig. 2). Thus, an additional constraint on the β-strength distribution is also given by delayed neutron emission probability (P_n-value) [29]. Since the FRSA model enables one to evaluate the coupling of QRPA phonons to more complex configurations, such calculations that take into account the $2p - 2h$ fragmentation of the QRPA excitations are now in progress.

In summary, starting from the Skyrme mean-field calculations the GT strength in the Q_β-window has been studied within the FRSA model including both the tensor interaction effect and $2p - 2h$ configurations. The suggested approach enables one to perform the calculations in very large configurational spaces. We analyze the effects on the β-transition rates in the case of ^{74,76,80}Ni, in comparison to the doubly-magic nucleus ^{78}Ni. Including the 2p-2h configurations leads to the appearance of the weak fragmented satellites at low transition energy and the 1_1^+ state is moved downwards in energy to the ground state of the daughter nucleus. As a result, the β-decay half-life is decreased. Using the strong tensor correlations [27] our estimation is rather the bottom limit of these half-lives [23].

Acknowledgments
We thank Yu.E. Penionzhkevich, H. Sagawa, Ch. Stoyanov and D. Verney for useful discussions. This work is partly supported by the IN2P3-JINR agreement.

References

[1] Bender M, Dobaczewski J, Engel J and Nazarewicz W 2002 *Phys. Rev.* C **65** 054322
[2] Fracasso S and Colò G 2007 *Phys. Rev.* C **76** 044307
[3] Bai C L, Zhang H Q, Sagawa H, Zhang X Z, Colò G and Xu F R 2011 *Phys. Rev.* C **83** 054316
[4] Minato F and Bai C L 2013 *Phys. Rev. Lett.* **110** 122501
[5] Ichimura M, Sakai H and Wakasa T 2006 *Prog. Part. Nucl. Phys.* **56** 446
[6] Bertsch G F and Hamamoto I 1982 *Phys. Rev.* C **26** 1323
[7] Kuzmin V A and Soloviev V G 1984 *J. Phys.* G **10** 1507
[8] Drożdż S, Osterfeld F, Speth J and Wambach J 1987 *Phys. Lett.* B **189** 271
[9] Colò G, Nguyen Van Giai, Bortignon P F and Broglia R A 1994 *Phys. Rev.* C **50** 1496
[10] Colò G, Sagawa H, Nguyen Van Giai, Bortignon P F and Suzuki T 1998 *Phys. Rev.* C **57** 3049
[11] Niu Y F, Colò G, Brenna M, Bortignon P F and Meng J 2012 *Phys. Rev.* C **85** 034314
[12] Soloviev V G 1992 *Theory of Atomic Nuclei: Quasiparticles and Phonons* (Bristol and Philadelphia: Institute of Physics)
[13] Nguyen Van Giai, Stoyanov Ch and Voronov V V 1998 *Phys. Rev.* C **57** 1204
[14] Severyukhin A P, Stoyanov Ch, Voronov V V and Nguyen Van Giai 2002 *Phys. Rev.* C **66** 034304
[15] Severyukhin A P, Voronov V V and Nguyen Van Giai 2004 *Eur. Phys. J.* A **22** 397
[16] Severyukhin A P, Voronov V V and Nguyen Van Giai 2008 *Phys. Rev.* C **77** 024322
[17] Severyukhin A P, Voronov V V and Nguyen Van Giai 2009 *Phys. Atom. Nucl.* **72** 1733
[18] Severyukhin A P, Arsenyev N N and Pietralla N 2012 *Phys. Rev.* C **86** 024311
[19] Arsenyev N N, Severyukhin A P, Voronov V V and Nguyen Van Giai 2012 *Eur. Phys. J. Web of Conf.* **38** 17002
[20] Severyukhin A P, Arsenyev N N, Pietralla N, Werner V 2014 *Phys. Rev.* C **90** 011306(R)
[21] Severyukhin A P, Voronov V V and Nguyen Van Giai 2012 *Prog. Theor. Phys.* **128** 489
[22] Severyukhin A P and Sagawa H 2013 *Prog. Theor. Exp. Phys.* **2013** 103D03
[23] Severyukhin A P, Voronov V V, Borzov I N, Arsenyev N N and Nguyen Van Giai 2014 *Phys. Rev.* C **90** 044320
[24] Engel J, Bender M, Dobaczewski J, Nazarewicz W and Surman R 1999 *Phys. Rev.* C **60** 014302
[25] Suhonen J 2007 *From Nucleons to Nucleus* (Berlin: Springer-Verlag)
[26] Nguyen Van Giai and Sagawa H 1981 *Phys. Lett.* B **106** 379
[27] Bai C L, Zhang H Q, Zhang X Z, Xu F R, Sagawa H and Colò G 2009 *Phys. Rev.* C **79** 041301(R)
[28] Severyukhin A P, Voronov V V, Borzov I N and Nguyen Van Giai 2013 *Rom. J. Phys.* **58** 1048
[29] Borzov I N 2005 *Phys. Rev.* C **71** 065801

Microscopic analysis of ^{10,11}Be elastic scattering on protons and ^{12}C and breakup processes of ^{11}Be within the ^{10}Be+n cluster model

K Spasova[1,2], V K Lukyanov[3], D N Kadrev[1], A N Antonov[1], E V Zemlyanaya[3], K V Lukyanov[3], and M K Gaidarov[1]

[1] Institute for Nuclear Research and Nuclear Energy, Bulgarian Academy of Sciences, Sofia 1784, Bulgaria
[2] University "Ep. K. Preslavski", Shumen 9712, Bulgaria
[3] Joint Institute for Nuclear Research, Dubna 141980, Russia

E-mail: kyspasova@yahoo.co.uk

Abstract. The elastic scattering cross-sections of ^{10,11}Be on protons and ^{12}C at energy $E < 100$ MeV/nucleon using microscopically calculated optical potentials (OP) are presented. The real OP is obtained by a folding procedure with effective NN interactions, while the imaginary OP is estimated within the high energy approximation (HEA). The spin-orbit part of the OP is also included. The characteristics of the breakup processes of ^{11}Be on different nuclear targets are also considered. The cross-sections of diffractive breakup and stripping reactions of ^{11}Be on ^{9}Be, ^{93}Nb, ^{181}Ta and ^{238}U at energy $E = 63$ MeV/nucleon and the longitudinal momentum distributions of ^{10}Be fragments produced in the breakup of ^{11}Be on these nuclei are presented. The results are in a good agreement with the available experimental data, in particular the obtained widths of about 50 MeV/c are closed to the empirical ones.

In the present work we study the elastic scattering cross section of ^{10,11}Be on protons and ^{12}C at incident energies $E < 100$ MeV/nucleon using microscopically calculated OP's within the hybrid model. Following our previous works, considering processes with exotic He and Li isotopes [1]-[6], we apply this model in which ^{11}Be consists of a core of ^{10}Be and a halo formed by a motion of a neutron in its periphery. Also we estimate important characteristics of the reactions with ^{11}Be, such as the breakup cross sections and momentum distributions of fragments in breakup processes.

The real part of the OP (ReOP) is calculated by a folding of a nuclear density and the effective NN potentials and includes both direct and exchange parts. The imaginary part of the OP (ImOP) is obtained within the HEA model. There are only two or three fitting parameters N's in the hybrid model that are related to the depths of the ReOP, ImOP and the spin-orbit part of the OP and are obtained by a fitting procedure to the available experimental data. In the calculations of the microscopic OPs for the scattering of ^{10,11}Be on protons and nuclei we used realistic density distributions of ^{10}Be calculated within the quantum Monte Carlo model (QMC) [7] and of ^{10,11}Be from the generator coordinate method (GCM)[8].

The OP has the form:

$$U(r) = N_R V^F(r) + iN_I W(r) - 2\lambda_\pi^2 \left[N_R^{ls} V_R^{ls} \frac{1}{r}\frac{df_R(r)}{dr} + iN_I^{ls} W_I^{ls} \frac{1}{r}\frac{df_I(r)}{dr} \right] (\vec{l}\cdot\vec{s}), \qquad (1)$$

where $2\lambda_\pi^2 = 4$ fm^2 with the squared pion Compton wave length $\lambda_\pi^2 = 2$ fm^2. We denote the values of the ReOP and ImOP at $r = 0$ by $V_R(\equiv V^F(r = 0))$ and $W_I(\equiv W(r = 0))$. The spin-orbit part of the OP contains real and imaginary terms with the parameters V_R^{ls} and W_I^{ls} related to V_R and W_I by the $V_R^{ls} = V_R/4$ and $W_I^{ls} = W_I/4$, correspondingly. The ReOP $V^F(r)$ is a sum of isoscalar (V_{IS}^F) and isovector (V_{IV}^F) components and each of them has its direct (V_{IS}^D and V_{IV}^D) and exchanged (V_{IS}^{EX} and V_{IV}^{EX}) parts.

The ImOP is chosen either to be in the form of the microscopically calculated $W(r) = V^F(r)$ or in the form $W(r) = W^H(r)$ obtained in Ref. [1, 9] within the HEA.

The elastic scattering of light nuclei is rather sensitive to their periphery, where transfer and breakup processes also take place. In our the previous papers has been pointed out [3, 4], that the inclusion of a surface imaginary term to the OP [Eq. (1)] leads to a better agreement with the experimental data. As known, this contribution can be considered as the so-called dynamical polarization potential, which allows one to simulate the surface effects caused by the latter. In fact, the imaginary part of the ls term in our OP [see Eq. (1)] plays effectively this role. However, sometimes one needs to increase the absorption in the surface region and thus, one adds a derivative of the ImOP (surface term):

$$W^{sf}(r) = -iN_I^{sf} r \frac{dW(r)}{dr}, \tag{2}$$

where N_I^{sf} is also a fitting parameter.

All calculations of elastic scattering using the obtained OPs are performed by using the DWUCK4 code. The results for the elastic ^{10}Be+p and ^{11}Be+p scattering cross sections are given in Figs. 1 and 2, respectively, and are compared with the experimental data. In general, the account for the spin-orbit term in the volume OP gives a trend of increasing the cross sections at larger angles.

The problem of the ambiguity of the values of N's arises when the fitting procedure concerns a limited number of experimental data. We impose a physical constraint, namely choosing those ReOP and ImOP that give volume integrals with a correct dependence on the energy [12]. It is known [12] that the volume integrals (their absolute values) for the ReOP decrease with increasing the energy, while for the ImOP they increase up to a plateau and then decrease. The values of the N's parameters from the fitting procedure are given in the Table 1. It can be seen that the tendency (the decrease of J_V and the increase of J_W) is generally confirmed. A fair agreement between the calculated ^{10}Be+p angular distributions and the experimental data is obtained only when both ls- and surface contributions to the OP are included.

For a more complete analysis of the elastic scattering cross sections, we extend the incident energy region to lower energies on the example of the scattering of ^{10}Be on protons that has been recently studied by Schmitt *et al.* [13]. Moreover, this could be a test of our hybrid model at low energies. The calculated results for the differential cross sections, shown as a ratio to Rutherford scattering, are given and compared with the data [13] in Fig. 3 for energies of 7.5 MeV and 10.7 MeV. The values of the N's parameters from the fitting procedure and the corresponding total reaction cross sections and volume integrals are listed in Table 1. The results shown in Fig. 3 when including in the calculations only the ls term demonstrate a fairly good agreement with the data. The values of the parameters N_R deduced from the fitting procedure for both energies in the case of GCM density of ^{10}Be are quite large that indicates for the specific peculiarities of the elastic scattering at low energies with account for the spin-orbit term. We also calculated the ^{10}Be+p elastic scattering cross sections at the same proton energies taking into account the surface term [Eq. (2)]. In this case, only the QMC density of ^{10}Be is tested.

The elastic scattering cross sections of ^{11}Be+^{12}C (their ratios to the Rutherford one) calculated within the hybrid model are presented in the Fig. 4. The results are obtained with the GCM densities and they are in a good agreement with the available data. The values of the

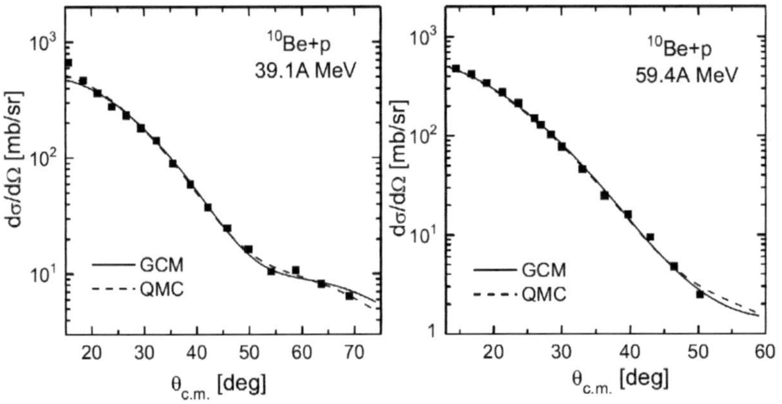

Figure 1. ^{10}Be+p elastic scattering cross sections with both ls and surface terms. Solid lines: calculations with GCM density of ^{10}Be; dashed lines: calculations with QMC density of ^{10}Be. Experimental data for 39.1 MeV/nucleon and 59.4 MeV/nucleon are from Refs. [10] and [11], respectively.

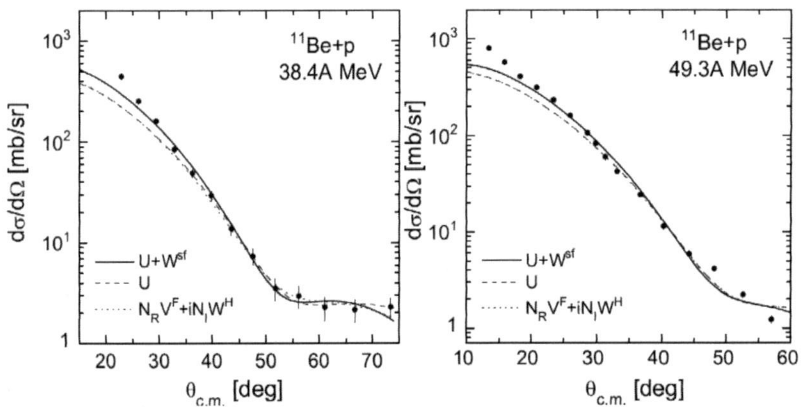

Figure 2. ^{11}Be+p elastic scattering cross sections. Calculations are performed with GCM density of ^{11}Be. Solid line: OP with both ls and surface terms [Eqs. (1) and (2)]; dashed line: OP with ls term [Eq. (1)]; dotted line: the volume part of OP from Eq. (1). Experimental data for 38.4 MeV/nucleon and 49.3 MeV/nucleon are taken from Refs. [10] and [11], respectively.

parameters N and the volume integrals J_V and J_W are given in the Table 2. It is seen from the figure that it is difficult to determine the advantage of the use for the ImOP $W = W^H$ or $W = V^F$, because the differences between the theoretical results start at angles for which the experimental data are not available.

In the paper we also consider the characteristics of breakup processes of the ^{11}Be nucleus, namely diffraction and stripping reaction cross sections and the momentum distributions of the fragments. We use a simple model in which ^{11}Be consists of a core of ^{10}Be and a halo of a single neutron. In this model the density of ^{10}Be has to be given. We use the QMC [7] and GCM [8] density distributions of ^{10}Be. The hybrid model is applied to calculate the OP of the interaction of ^{10}Be with the target, as well as OP for the n+target interaction. In the

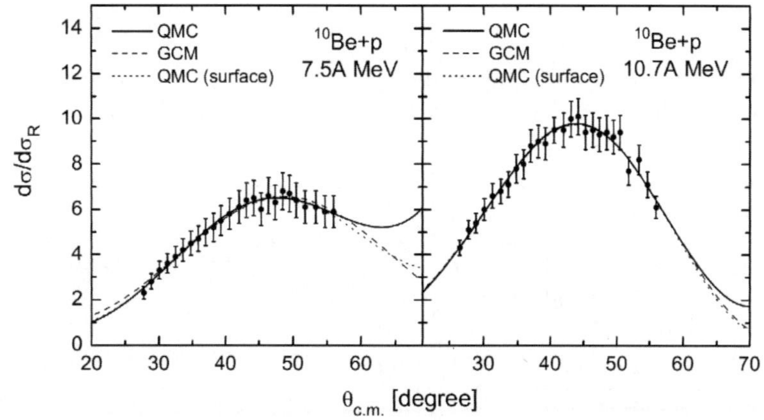

Figure 3. ^{10}Be+p elastic scattering cross sections as a ratio to Rutherford scattering at proton energies of 7.5 MeV (left panel) and 10.7 MeV (right panel). The solid and dashed lines show the results with QMC and GCM density of ^{10}Be, respectively, and with ls term in OP. The dotted lines show the QMC results obtained by accounting for both the ls- and surface terms in OP. Experimental data are taken from Ref. [13].

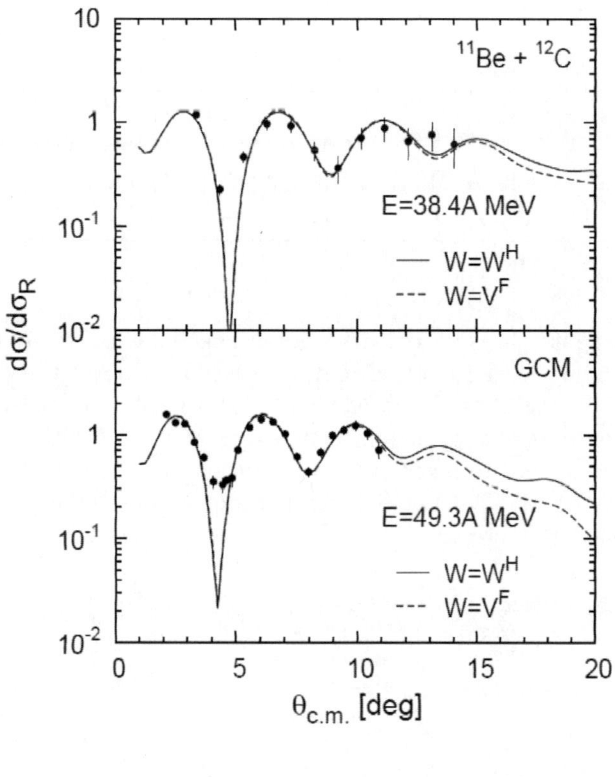

Figure 4. ^{11}Be+^{12}C elastic scattering cross sections. Solid lines: $W = W^H$; dashed lines: $W = V^F$. Experimental data are from Refs. [10] and [11], respectively.

Table 1. The renormalization parameters N_R, N_I, N_R^{ls}, N_I^{ls}, and N_I^{sf}, the total reaction cross sections σ_R (in mb), and the volume integrals J_V, $J_W^{(a)}$, and $J_W^{(b)}$ (in MeV.fm^3) as functions of the energy E (in MeV) for: (a) ^{10}Be+p with ls and $surf$ terms; (b) ^{11}Be+p with ls term; (c) ^{11}Be+p with ls and $surf$ terms.

	Model	E	N_R	N_I	N_R^{ls}	N_I^{ls}	N_I^{sf}	σ_R	J_V	$J_W^{(a)}$	$J_W^{(b)}$
(a)	GCM	39.1	0.995	0.266	0.095	0.082	0.004	298.65	394.161	117.040	122.321
	QMC		1.194	0.260	0.075	0.025	0.018	333.71	425.971	115.286	139.235
	GCM	59.4	0.970	0.000	0.365	1.000	0.373	400.26	323.404	0.000	367.802
	QMC		1.043	0.281	0.000	1.000	0.270	389.27	311.325	93.053	361.343
(a)	QMC	7.5	1.483	0.000	0.442	0.208	0.044	306.28	719.605	0.000	148.453
	QMC	10.7	1.354	0.098	0.178	1.000	0.193	636.50	632.936	100.685	695.676
(b)	GCM	38.4	0.787	0.799	0.000	0.507	0.000	458.63	324.148	355.844	355.844
		49.3	0.793	0.867	0.123	0.316	0.000	426.85	296.301	301.184	301.184
(c)	GCM	38.4	0.849	0.106	0.102	0.380	0.152	493.01	349.685	47.208	269.903
		49.3	0.801	0.000	0.213	0.394	0.200	436.46	299.280	0.000	246.162

Table 2. The renormalization parameters N_R and N_I, the total reaction cross sections σ_R (in mb), and the volume integrals J_V and J_W (in MeV.fm^3) as functions of the energy E=38.4 and 49.3 MeV/nucleon for the ^{11}Be+^{12}C elastic scattering

Nucleus	Model	E	W	N_R	N_I	σ_R	J_V	$J_W^{(a)}$
^{11}Be	GCM	38.4	W^H	0.769	0.711	127.123	216.879	287.235
			V^F	0.708	0.521	126.825	199.676	146.937
		49.3	W^H	0.820	0.883	124.406	213.754	300.193
			V^F	0.743	0.574	123.302	193.682	149.628

final step of the procedure the sum of these potentials is folded with the respective density distribution corresponding to the relative motion wave function of the clusters in ^{11}Be. The latter is obtained by solving the Schrödinger equation with the Woods-Saxon potential for a particle with a reduced mass of two clusters. The parameters of the WS potentials are obtained by a fitting procedure, namely, to reach the neutron separation energy $S_n = 504 \pm 6$ KeV.

The s-state ($l = 0, n = 1, 2$) of the relative motion of two clusters and the corresponding density distribution have the forms:

$$\phi_{00}^{(n)}(\mathbf{s}) = \phi_0^{(n)}(s)\frac{1}{\sqrt{4\pi}}; \qquad \rho_0^{(n)}(\mathbf{s}) = |\phi_{00}^{(n)}(\mathbf{s})|^2 = \frac{1}{4\pi}|\phi_0^{(n)}(s)|^2, \quad n = 1, 2. \tag{3}$$

Within the ^{10}Be+n cluster model, in order to calculate the ^{11}Be breakup in its collision with the protons and nuclear targets, one should calculate two OPs of ^{10}Be+A and n+A scattering [6]:

$$U^{(b,n)}(r) = V^{(b,n)} + iW^{(b,n)} = \int d\mathbf{s}\rho_0^{(n)}(s)\left[U_c^{(n)}(\mathbf{r} + (1/11)\mathbf{s}) + U_n^{(n)}(\mathbf{r} - (10/11)\mathbf{s})\right] \tag{4}$$

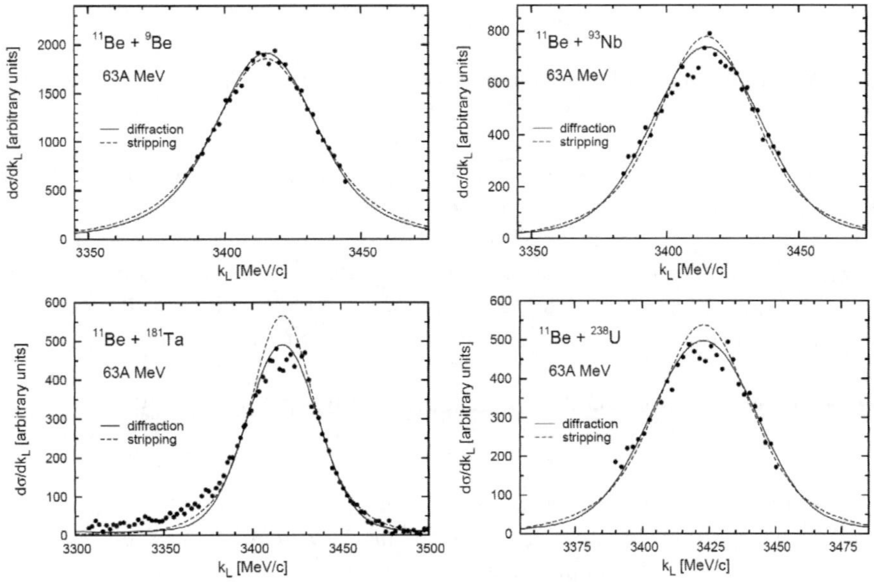

Figure 5. Cross sections of diffraction breakup and stripping reaction in ^{11}Be+ nuclei scattering at $E = 63$ MeV/nucleon. Experimental data are from Ref. [15]

In Eq. (4) $\mathbf{r} - (10/11)\mathbf{s} \equiv \mathbf{r}_n$ and $\mathbf{r} + (1/11)\mathbf{s} \equiv \mathbf{r}_c$ give the distances between the centers of each of the clusters and the target, and $\mathbf{s} = \mathbf{s}_1 + \mathbf{s}_2 = (10/11)\mathbf{s} + (1/11)\mathbf{s}$ determines the relative distance between the centers of the two clusters, where s_1 and s_2 are the distances between the centers of ^{11}Be and each of the clusters, correspondingly. The respective OPs for the ^{10}Be+A and n+A scattering are calculated within the microscopic model of OP. The differential and total cross sections (for elastic scattering, as well as for diffractive breakup and absorption) all require calculations of the probability functions $d^3P(\mathbf{b},\mathbf{k})/d\mathbf{k}$ that depend on the impact parameter $\mathbf{b}$. The general expression for the probability functions can be written as in Ref.[14]:

$$\frac{d^3P_\Omega(\mathbf{b},\mathbf{k})}{d\mathbf{k}} = \frac{1}{(2\pi)^3}\left|\int d\mathbf{s}\phi^*_{\mathbf{k}}(\mathbf{s})\Omega(\mathbf{b},\mathbf{r}_\perp)\phi^{(n)}_{00}(\mathbf{s})\right|^2, \tag{5}$$

where $\Omega(\mathbf{b},\mathbf{r}_\perp)$ depends on the two profile functions $S_i(b_i)$, $(i = c, n)$, $\phi_{\mathbf{k}}$ is the continuum wave functions, $\mathbf{k}$ is the relative momentum of both clusters in their center-of-mass frame. One can integrate over the transverse angle of the momenta φ_k to get the double-differential probability $d^2P_\Omega(\mathbf{b},\mathbf{k})/dk_L dk_\perp$.

The diffraction breakup cross section has the form

$$\left(\frac{d\sigma}{dk_L}\right)_{diff} = \int_0^\infty b_n db_n \int_0^{2\pi} d\varphi_n \int_0^\infty dk_\perp \frac{d^2P_\Omega(\mathbf{k},\mathbf{b})}{dk_L dk_\perp}. \tag{6}$$

The cross sections of the stripping reaction when the neutron leaves the elastic channel is [14]:

$$\begin{aligned}\left(\frac{d\sigma}{dk_L}\right)_{str} &= \frac{1}{2\pi^2}\int_0^\infty b_n d\,b_n d\varphi_n\left[1 - |S_n(b_n)|^2\right] \\ &\times \int \rho d\rho d\varphi_\rho |S_c(b_c)|^2 \times \left[\int_0^\infty dz\cos(k_L z)\phi_0\left(\sqrt{\rho^2+z^2}\right)\right]^2.\end{aligned} \tag{7}$$

The results for the diffraction and stripping cross sections (when a neutron leaves the elastic channel) for the reactions ^{11}Be+^{9}Be, ^{11}Be+^{93}Nb, ^{11}Be+^{181}Ta, and ^{11}Be+^{238}U are presented in the Fig. 5. We note the good agreement with the experimental data from light and heavy breakup targets. The obtained cross sections for the diffraction and stripping have a similar shape. The values of the widths are around 50 MeV in agreement with the experimental ones. Our results confirm the observations that the width almost does not depend on the mass of the target and as a result, it gives information basically about the momentum distributions of two clusters.

We can conclude that, in general, the hybrid model for microscopic calculations of the OPs gives the basic important features of the scattering cross sections and can be recommended and applied to calculate more complex processes such as breakup reactions, momentum distributions of fragments and others.

Future measurements of such reactions are highly desirable for the studies of the exotic ^{11}Be structure.

Acknowledgments

The work is partly supported by the Project from the Agreement for co-operation between the INRNE-BAS (Sofia) and JINR (Dubna). Four of the authors (D.N.K., A.N.A., M.K.G. and K.S.) are grateful for the support of the Bulgarian Science Fund under Contract No. DFNI–T02/19 and one of them (D.N.K.) under Contract No. DFNI–E02/6. The authors V.K.L., E.V.Z., and K.V.L. thank the Russian Foundation for Basic Research (Grant No. 13-01-00060) for partial support. K.S. acknowledges the support of the Project PD-08-275/2015.

References

[1] Lukyanov V K, Zemlyanaya E V, and Lukyanov K V 2006 *Phys. At. Nucl.* **69** 240
[2] Lukyanov V K, Lukyanov K V, Zemlyanaya E V, Antonov A N, Gaidarov M K 2007 *Eur. Phys. J. A* **33** 389
[3] Lukyanov V K, Zemlyanaya E V, Lukyanov K V, Kadrev D N, Antonov A N, Gaidarov M K, Massen S E 2009 *Phys. Rev. C* **80** 024609
[4] Lukyanov V K, Kadrev D N, Zemlyanaya E V, Antonov A N, Lukyanov K V, and Gaidarov M K 2010 *Phys. Rev. C* **82** 024604
[5] Lukyanov V K, Kadrev D N, Zemlyanaya E V, Antonov A N, Lukyanov K V, Gaidarov M K, and Spasova K 2013 *Phys. Rev. C* **88** 034612
[6] Lukyanov V K, Kadrev D N, Zemlyanaya E V, Spasova K, Lukyanov K V, Antonov A N, and Gaidarov M K 2015 *Phys. Rev. C* **91** 034606
[7] Pieper S C, Varga K, and Wiringa R B 2002 *Phys. Rev.* C **66** 044310
[8] Descouvemont P 1997 *Nucl. Phys.* A **615** 261
[9] Shukla P, 2003 *Phys. Rev.* C **67**, 054607
[10] Lapoux V *et al.* 2008 *Phys. Lett.* B **658** 198
[11] Cortina-Gil M D *et al.* 1997 *Phys. Lett.* B **401** 9 ; *Nucl. Phys. A* **616** 215c
[12] Romanovsky E A *et al.* 1998 *Bull. Russ. Acad. Sci.: Phys* **62** 150
[13] Schmitt K T *et al.* 2013 *Phys. Rev.* C **88** 064612
[14] Hencken K, Bertsch G, and Esbensen H 1996 *Phys. Rev.* C **54**, 3043
[15] Kelley *et al.* 1995 *Phys. Rev. Lett.* **74**, 30

Investigation of the mixed-symmetric one-quadrupole phonon $2^+_{1,ms}$ state of the heavy nucleus ^{204}Hg

R. Stegmann[1], C. Stahl[1], G. Rainovski[2], N. Pietralla[1], T. Möller[1], M. Lettmann[1], R. V. F. Janssens[3], M. Carpenter[3], S. Zhu[3]

[1] Institut für Kernphysik, TU Darmstadt, Schlossgartenstrasse 9, 64289 Darmstadt, Germany
[2] Faculty of Physics, University of Sofia St. Kliment Ohridski, 5 James Bourchier Blvd., 1164 Sofia, Bulgaria
[3] Argonne National Laboratory, 9700 S. Cass Avenue, Argonne, IL 60439, USA

E-mail: rstegmann@ikp.tu-darmstadt.de

Abstract. Low-energy quadrupole-collective excitations of isovector character, the so-called states with proton-neutron mixed-symmetric character (MSS), have been investigated in ^{204}Hg which is the sole stable nucleus in the straight vicinity of ^{208}Pb exhibiting a $2\pi - 2\nu$ structure. A ^{12}C(^{204}Hg,^{204}Hg*)^{12}C projectile Coulomb-excitation experiment at 890 MeV was performed at Argonne National Laboratory using the ATLAS accelerator. γ-rays originating from the decay of the excited states of ^{204}Hg were detected using the GAMMASPHERE spectrometer. The measured Coulomb-excitation yields provide the $B(M1)$ and the $B(E2)$ strength distributions which unambiguously reveal the 2^+_3 state at 1948 keV as the one-phonon MSS of ^{204}Hg.

1. Introduction

The appearance of collectivity from basic nuclear properties is of interest in contemporary nuclear structure research. Atomic nuclei can be considered as mesoscopic two-fluid quantum systems, which physics is determined notably by the many-body aspect, the quantum nature as well as the two-fluid character. Nuclear phenomena emerging from these properties are collectivity, shell structure and isospin. The Interacting Boson Model-2 [1, 2] formalizes this picture and predicts the appearance of two fundamental one-phonon quadrupole excitations in nuclei. Both are mixtures of the underlying proton and neutron quadrupole excitations. In weakly collective nuclei the in-phase combination of proton and neutron contributions form the so-called fully-symmetric state (FSS), whereas the anti-symmetric combination is known as mixed-symmetric state (MSS) [1, 3]. By their origin the MSSs are sensitive to the effective pn correlations in the valence shell. MSSs are uniquely identified experimentally [4, 5, 6, 7, 8] by their strong isovector $M1$ decay to the low-lying FSS. This is of importance because $M1$ transitions are forbidden between FSSs and thus are a suitable observable for unambiguous identification of MSSs. The fundamental MSS in weakly collective vibrational nuclei is the one-phonon $2^+_{1,ms}$ state [3]. The best examples of such MSSs in stable nuclei are found in the mass $A \approx 90$ region [5, 9, 10]. Recently a large number of quadrupole MSSs have been identified in the mass $A \approx 130$ region [11, 12, 13, 14, 15, 16].

Recently new experimental evidence for the formation mechanism of fully-symmetric and mixed-symmetric low-lying quadrupole-collective states was found [17]. This mechanism involves coupling of the lowest 2-quasiparticle proton and neutron excitations to the Giant Quadrupole Resonance (GQR) [17]. The $E2$ strength in the decay of FS and MS states is dominated by the GQR contributions while the $M1$ properties are determined by the valence-shell configuration. Due to normalization of the wave function this concept demands that with increasing collectivity the GQR contribution also increases, resulting in a decrease of the $M1$ strength. Thus the strongest $M1$ transitions should occur in weakly collective vibrational nuclei, which are only two proton and two neutron holes away from doubly-magic nuclei. To improve the understanding of the formation of low-lying isovector excitations cases of MSSs in the vicinity of doubly-magic nuclei have to be identified. However, the number of such nuclei which can be studied by conventional experimental methods is low. The sole stable nucleus exhibiting two proton and two neutron holes in the region around ^{208}Pb is ^{204}Hg. Preliminary results of the search for a MSS in this nucleus are presented in this contribution.

2. Experiment

An experiment was performed at Argonne National Laboratory. A pulsed (12 MHz) ^{204}Hg beam accelerated to 890 MeV was delivered by the ATLAS accelerator facility. The beam was impinging on a 1mg/cm^2 ^{nat}C target. Deexcitation γ-rays, following the projectile Coulomb-excitation reaction, were detected using the GAMMASPHERE detector array [18], consisting of 100 HPGe detectors arranged in 16 rings, at the time of the measurement. The average beam intensity was $\approx$ 1.3 pnA. GAMMASPHERE was used in singles mode, as no particle detector was present in the experiment. γ-ray counting rate during the experiment was $\approx 7 \times 10^3$ counts-per-second. A total of 4.4×10^8 events of γ-ray fold 1 or higher have been collected during 13 h of beam time. The excited ^{204}Hg nuclei were recoiling with $v/c \approx 8.4\%$ which was accounted for by Doppler correction of the obtained γ-ray spectra. The contribution of the room background

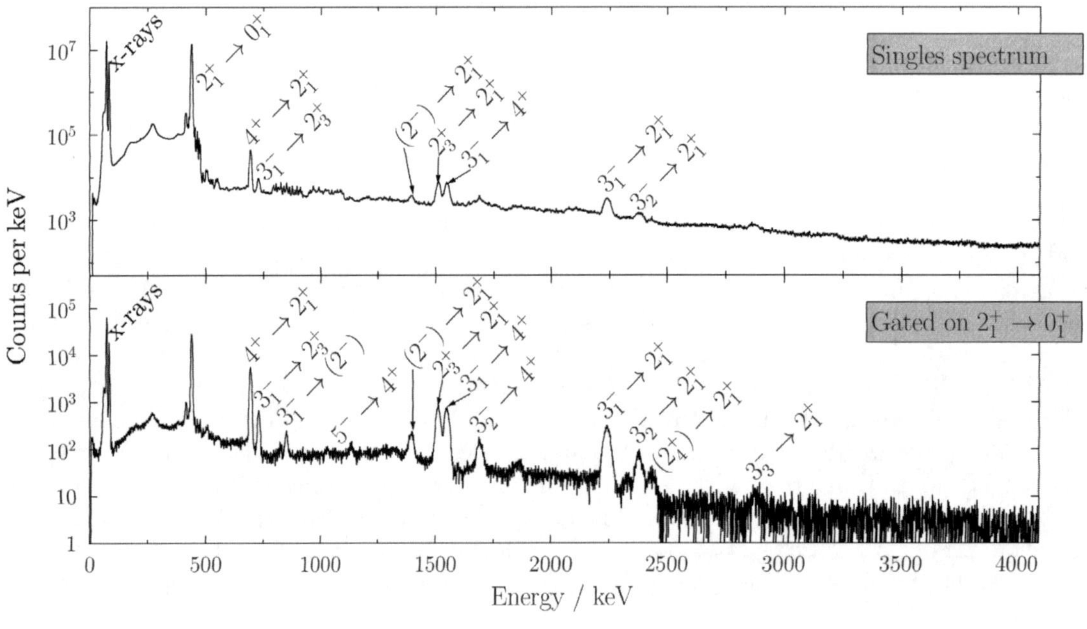

Figure 1. Doppler-corrected, background subtracted γ-ray spectra after projectile Coulomb excitation on a Carbon target. In the upper panel the singles spectrum is shown. In the lower panel a projection of the γ-γ-matrix, gated on the $2_1^+ \to 0_1^+$ transition is shown.

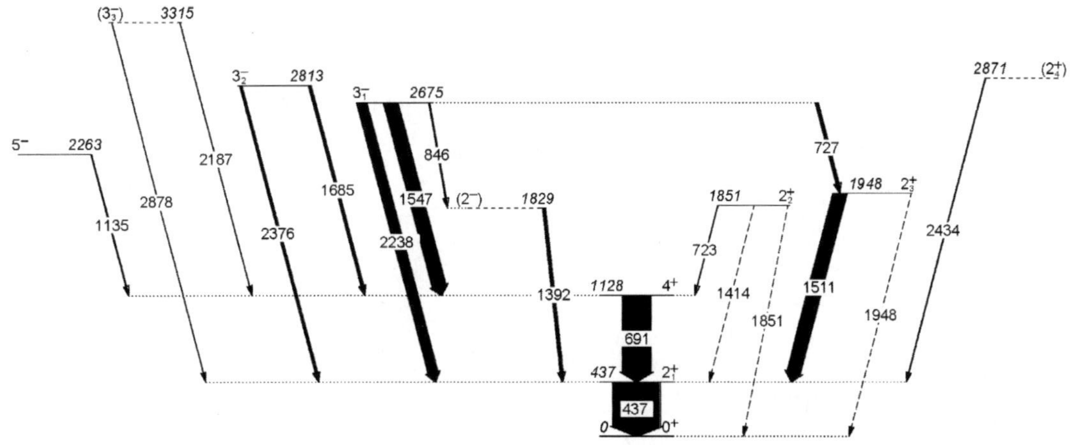

Figure 2. Deduced level scheme of ^{204}Hg. Solid lines indicate transitions observed in this experiment while dashed lines show transitions which were not observed but which intensities can be determined from beforehand known branching ratios.

was eliminated by correlating the detected γ-rays to the ATLAS radio-frequency pulses. The final spectrum is shown in the top panel of Fig. 1.

3. Data Analysis and Results

Events featuring a γ-ray fold ≥ 2 were used to construct a γ-γ-coincidence matrix. A showcase excerpt from this matrix for all γ-rays coincident to the $2_1^+ \rightarrow 0_1^+$ transition is shown in the lower panel of Fig. 1. The combined examination of the singles spectrum as well as of the coincidence matrix results in a total of 10 states decaying via emitting γ radiation observed in this experiment. Fig. 2 presents the resulting level scheme, which is in agreement with previously reported data on ^{204}Hg [19].

All prominent γ-ray transitions originate from decays of excited states of ^{204}Hg. Most of these transitions have been observed before, besides the 727-keV transition which connects the 3_1^- and the 2_3^+ state and the transition depopulating a new state at 2871 keV. Because of the dominance of $E2$ excitations in low-energy Coulomb-excitation reactions we have assumed (2^+) for the spin and the parity of this state.

Efficiency-corrected transition intensities were deduced from the singles spectrum wherever possible. The intensities of weak transitions were derived using the γ-γ-matrix. The intensities were afterwards used to determine the yield of the states excited in this experiment. Because of the quasi-4π nature of the GAMMASPHERE array these excitation yields are proportional to the Coulomb-excitation cross-section. In the analysis, excited states population yields are compared to cross-sections calculated from Coulomb-excitation theory [20]. Additional information is obtained from analyzing the angular distributions of the $2_3^+ \rightarrow 2_1^+$ and $3_1^- \rightarrow 4^+$ transitions. The latter transition exhibits the expected dipole character while the other one is characterized by a superposition of dipole and quadrupole character. For the $2_3^+ \rightarrow 2_1^+$ transition the angular distribution in combination with the orientation tensor from Coulomb-excitation theory [20] calculated using GOSIA [21] results in a preliminary multipole mixing ratio $\delta = 0.40\,(26)$ hinting at $M1$ dominance.

At this stage of the analysis only a minor subset of states consisting of the 0_1^+, 2_1^+, 2_3^+, and 3_1^- states was taken into account for the Coulomb-excitation calculations. The relative yields of the 2_3^+ and 3_1^- states have been fitted to the theory [20] using the multiple CE code CLX [22]

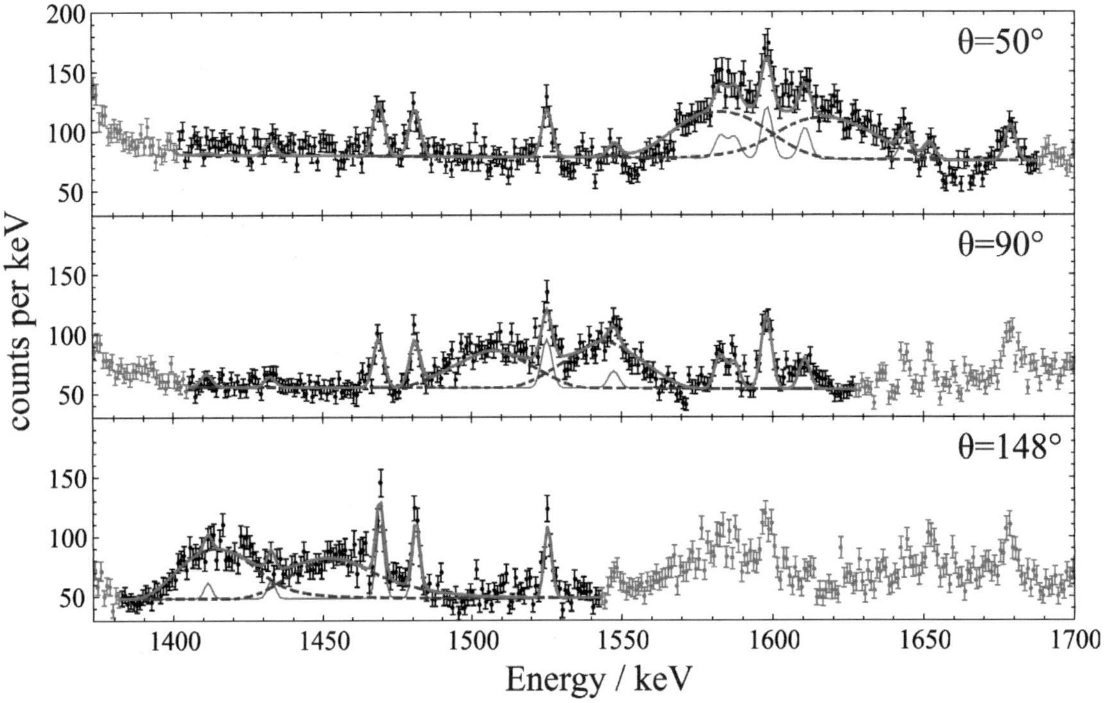

Figure 3. Example of lineshape fits in the DSAM analysis of the 1511- and 1547-keV transitions. The thick solid line shows the total fit to the spectrum, the dashed lines show the 1511-keV decay of the 2_3^+ state (blue) and the 1547-keV decay of the 3_1^- state (purple), thin solid lines show contaminants taken into account for the fit. The unshifted parts of both decays are not present under forward (top panel) and backward angles (bottom panel), indicating short lifetimes, compared to the slowing-down of the nucleus of interest in the target.

taking into account the energy loss of the beam in the target of $\approx$ 124 MeV. The scale for the calculation is set by the ground-state decay $E2$ strength $B(E2; 2_1^+ \to 0_1^+) = 11.96\,(9)$ W.u. [19]. Preliminary results from this calculation indicate an $M1$ strength for the $2_3^+ \to 2_1^+$ transition of $\sim 0.6\,\mu_n{}^2$. This large $B(M1; 2_3^+ \to 2_1^+)$ value implies that the 2_3^+ state is a single isolated one-phonon MSS.

In the same experimental campaign the ^{204}Hg beam was impinging on a thick Al target for a caDSAM measurement [23]. Data was taken for about $\approx$ 3 h. The preliminary analysis of this experiment yields a lifetime of the 2_3^+ state in the order of 100 fs in agreement with the results from the Coulomb-excitation analysis. In Fig. 3 an example of lineshape fits for the 1511- and 1547-keV transitions is presented.

4. Summary

A Coulomb-excitation experiments have been performed on the nucleus ^{204}Hg. The preliminary results from the investigations indicate the existence of a one-phonon MSS in this nucleus. This is among the first experimentally identified MSSs in the vicinity of the doubly-magic nucleus ^{208}Pb and calls for theoretical explanations about the origin of isovector states in the mass $A \approx 208$ region.

Acknowledgments
G.R. acknowledges the support from the Alexander von Humboldt foundation. This work was supported by the German-Bulgarian exchange program under grants DAAD No. PPP57082997 and BgNSF No. DNTS/01/05/2014, by the DFG under grant Pi393/2-3, and by the BMBF under grants 05P12RDCIB and 05P15RDCIA.

References
[1] Arima A, Ohtsuka T, Iachello F and Talmi I 1977 *Physics Letters B* **66** 205
[2] Iachello F and Arima A 1987 *The Interacting Boson Model* (Cambridge University Press, Cambridge)
[3] Iachello F 1984 *Phys. Rev. Lett.* **53** 1427
[4] Hamilton W D, Irbäck A and Elliott J P 1984 *Phys. Rev. Lett.* **53** 2469
[5] Pietralla N *et al.* 1999 *Phys. Rev. Lett.* **83** 1303
[6] Pietralla N *et al.* 2000 *Phys. Rev. Lett.* **84** 3775
[7] Kneissl U, Pietralla N and Zilges A 2006 *J. Phys. G: Nucl. Part. Phys.* **32** R217
[8] Pietralla N, von Brentano P and Lisetskiy A 2008 *Progress in Particle and Nuclear Physics* **60** 225
[9] Pietralla N *et al.* 2001 *Phys. Rev. C* **64** 031301
[10] Werner V *et al.* 2002 *Physics Letters B* **550** 140
[11] Rainovski G *et al.* 2006 *Phys. Rev. Lett.* **96** 122501
[12] Ahn T *et al.* 2009 *Physics Letters B* **679** 19
[13] Coquard L *et al.* 2010 *Phys. Rev. C* **82** 024317
[14] Gladnishki K A *et al.* 2010 *Phys. Rev. C* **82** 037302
[15] Danchev M *et al.* 2011 *Phys. Rev. C* **84** 061306
[16] Ahn T *et al.* 2012 *Phys. Rev. C* **86** 014303
[17] Walz C and ot 2011 *Phys. Rev. Lett.* **106** 062501
[18] Lee I Y 1990 *Nuclear Physics A* **520** 641c
[19] Chiara C and Kondev F 2010 *Nuclear Data Sheets* **111** 141
[20] Alder K and Winther A 1975 *Electromagnetic excitation: Theory of Coulomb excitation with heavy ions* (North-Holland Publishing Company, Amsterdam-Oxford)
[21] Czosnyka T, Cline D and Wu C Y 1983 *Bull. Am. Phys. Soc.* **28** 745
[22] Ower H 1980 *Coulombanregung von Hochspinzuständen in* ^{232}Th, ^{234}U *und* ^{236}U Ph.D. thesis Johann Wolfgang Goethe-Universität Frankfurt am Main
[23] Stahl C 2015 *New methods for the* γ*-ray spectroscopy with position-sensitive detector systems* Ph.D. thesis Technische Universität Darmstadt

Nuclear induces effects and mass correlations in low and multiply charged helium-like ions

Zh K Stoyanov[1], R L Pavlov[1], L M Mihailov[2], Ch J Velchev[1], Y D Mutafchieva[1], D Tonev[1] and N Chamel[3]

[1] Institute for Nuclear Research and Nuclear Energy, Bulgarian Academy of Sciences, Sofia, Bulgaria
[2] Institute of Solid State Physics, Bulgarian Academy of Sciences, Sofia, Bulgaria
[3] Institut d'Astronomie et d'Astrophysique, CP-226, Université Libre de Bruxelles, Brussels, Belgium

E-mail: zhkstoyanov@gmail.com

Abstract. The ground-state electron energies, the mass correction and mass polarization of low and multiply charged helium-like ions are analytically and numerically calculated. Approximately 3500 different kinds of ions with charge $Z = 2 \div 118$ are considered. The two-electron Schrödinger equation was solved using a discrete variational-perturbation approach developed by the authors and based on explicitly correlated wave functions. This approach takes into account the motion of the nucleus and yields accurate values for the electron characteristics. The results are presented with and without the inclusion of the mass polarization in the minimization procedure. The relative importance of mass correlations and relativistic effects in the formation of the electron energy characteristics of the helium-like ions are studied for different values of Z. The role of the inclusion of the mass polarization in the minimization procedure as an instrument to present and take into account the effects induced by the nuclear properties, structure and characteristics has been shown.

1. Introduction

Low and multiply charged helium-like ions are responsible for the properties and characteristics of low and high density plasma and for the processes occurring in it [1]. Helium-like ions exhibit specific properties due to the unscreened long-range Coulomb field of their nucleus. High-precision calculations of the electron ground-state energies of helium-like ions require to take into account the effects associated with nuclear characteristics and electron correlations. The nuclear size effects increase steadily with increasing atomic number. The nuclear motion gives an even more significant correction to the electron energy [2,3]. This leads to two specific contributions: finite-mass effect (mass correction), electron-electron (e-e) and electron-nucleus (e-n) correlations (mass polarization). While the latter increases with increasing nuclear charge number Z, the former varies very little [3].

In our previous work [4], we calculated ground-state electron energies, mass corrections and mass polarization effects of helium isoelectronic ions with nuclear charge for the main nuclides from Z=2 to Z=118.

Apart from including the mass polarization in the variational-perturbation procedure, we explored in Ref. [5] the role of the mass excess in both the mass correction and the mass polarization on the electron ground-state energies of helium-like ions with nuclear charge Z = 2 ÷ 118. Effects associated with nuclear properties and characteristics were taken into account for the first time. We have developed a perturbation method for solving the two-electron Schrödinger equation based on a variational principle using Explicitly Correlated Wave Functions (ECWF) of a generalized Hylleraas type [6]. The variational procedure leads to a generalized eigenvalue problem. This method has been applied to compute the electron ground-state energies taking into account both the mass corrections and the mass-polarization effects, as well as the inclusion of the mass polarization in the variational-perturbation procedure. This approach provides a general framework for taking into account higher-order effects like relativistic and QED corrections. This method is therefore very well-suited for high-precision calculations of plasma diagnostics.

In the current work, the approach developed by the authors is used for studying the correlation effects of helium-like electron-nuclear systems, not only along the line of stability but also for the 3500 known nuclei around it. The calculations are made with and without the inclusion of the mass polarization in the minimization procedure. Analyzing the results shows some correlations between the energy characteristics of the ground state and the properties of the nuclei of the isotopes involved in the system.

2. Method

Atoms with few electrons can be accurately described, taking into account correlations, by solving the Schrödinger equation

$$\hat{H}\Psi = E\Psi \tag{1}$$

in the ECWF approach [6].

Let us consider two electrons with position vectors r_1 and r_2 in the Cartesian coordinate system with the origin at the nuclear centre. Introducing elliptical coordinates $s = r_1 + r_2$, $t = r_1 - r_2$, $u = r_{12} = |r_1 - r_2|$, the Hylleraas-type two-electron wave functions that we consider have the form [8]:

$$|\psi\rangle = \sum_{STU} C_{STU}|STU\rangle = \sum_{STU} C_{STU}\frac{1}{\pi\sqrt{2}}e^{-s/2}s^S t^T u^U \tag{2}$$

where S, T, U are positive integers, with T even as a consequence of the symmetry requirement of the spatial wave function.

The non-relativistic Hamiltonian H is the sum of the kinetic energy T, the Coulomb attractive potential V_{en} due to the central nuclear charge, and the electron-electron repulsive potential V_{ee}. If the Hylleraas wave function (2) is subject to a coordinate scaling transformation

$$\langle\vec{r}_1\vec{r}_2|\Psi_\alpha\rangle = \sum_{STU} C_{STU}e^{-\alpha s/2}(\alpha s)^S(\alpha t)^T(\alpha u)^U \tag{3}$$

(α being a scaling factor), then the overlap, kinetic, and potential matrix elements transform as

$$\langle\Psi_\alpha|\Psi_\alpha\rangle = \frac{\langle\Psi|\Psi\rangle}{\alpha^6}, \langle\Psi_\alpha|T|\Psi_\alpha\rangle = \frac{\langle\Psi|T|\Psi\rangle}{\alpha^4}, \langle\Psi_\alpha|V|\Psi_\alpha\rangle = \frac{\langle\Psi|V|\Psi\rangle}{\alpha^5} \tag{4}$$

with $V = V_{en} + V_{ee}$. The expectation value of the energy thus scales as

$$\frac{\langle\Psi_\alpha|H|\Psi_\alpha\rangle}{\langle\Psi_\alpha|\Psi_\alpha\rangle} = \frac{\alpha^2\langle\Psi|T|\Psi\rangle + \alpha\langle\Psi|V|\Psi\rangle}{\langle\Psi|\Psi\rangle} \tag{5}$$

3. Mass corrections

The variational calculation of the ground state energy using ECWF as trial two-electron wave functions yields a lower bound to the exact energy. To obtain a better estimate, it is necessary to include mass correction and mass polarization. The first-order correction is given by [2,3]:

$$\varepsilon_1 = \frac{-\varepsilon}{1+\varepsilon} E_0 \approx -\varepsilon E_0 \tag{6}$$

where E_0 is the electron ground-state energy and $\varepsilon = m_e/M$ (m_e being the electron mass and M the nuclear mass). This correction is independent of the atomic state. The second-order perturbation correction given by

$$\varepsilon_2 = \varepsilon \int \nabla_1 \Psi^*(r_1, r_2) \nabla_2 \Psi(r_1, r_2) d^3r_1 d^3r_2 = \langle \Psi | \mathcal{E} | \Psi \rangle \tag{7}$$

depends on the atomic state as it involves the two-electron wave function. The matrix elements of ε_2 can be obtained in an analogous manner as those of the kinetic energy operator and can be found in [2,3]. After a scaling transformation, the mass polarization (7) becomes

$$\langle \Psi_\alpha | \mathcal{E} | \Psi_\alpha \rangle = \frac{\langle \Psi | \mathcal{E} | \Psi \rangle}{\alpha^4} \tag{8}$$

4. Variational method

The energy is minimized by varying the Hylleraas wave function with respect to both the expansion coefficients C_{STU} and the scaling factor α. Variation with respect to the expansion coefficients leads to the generalized eigenvalue problem

$$(\alpha^2[T] + \alpha[V])C_\alpha = E_\alpha[O]C_\alpha \tag{9}$$

where T, V, O are the kinetic, potential, and overlap (symmetric) matrices respectively and C_α is the column vector containing the expansion coefficients. The lowest eigenvalue E_α^0 obtained for a fixed value of α defines a function of one variable. Its minimum which can be found using standard techniques yields the best estimate of the ground-state energy. The variational-perturbation procedure is described in details in Refs. [4,5].

Solving the eigenvalue problem in (9) yields the values of the expansion coefficients C_{STU} and scale parameter α, corresponding to E_α^0. After subsiding these values in (8) we obtain the mass polarization ε_2 (7). In order to include ε_2 in the variational procedure, the operator of mass-polarization $\mathcal{E}$ is added to the Hamiltonian. Then the expectation value of the energy becomes

$$\frac{\langle \Psi_\alpha | H | \Psi_\alpha \rangle}{\langle \Psi_\alpha | \Psi_\alpha \rangle} = \frac{\alpha^2 \langle \Psi | T | \Psi \rangle + \alpha \langle \Psi | V | \Psi \rangle + \alpha^2 \langle \Psi | \mathcal{E} | \Psi \rangle}{\langle \Psi | \Psi \rangle} \tag{10}$$

The variational equations (9) and (10) correspondingly are replaced by

$$(\alpha^2[T + \mathcal{E}] + \alpha[V])C_\alpha = E_\alpha[O]C_\alpha \tag{11}$$

The inclusion of the mass polarization in the minimization procedure: i) improves by one order of magnitude the values for the ground-state energy of the helium-like ions; ii) gives the possibility to account for the effects induced by the properties and characteristics of the nuclei on the electron energy characteristics; iii) gives the possibility for calculating relativistic corrections according to the variational-perturbation approach.

5. Results and discussion

The discrete variational-perturbation procedure we have developed for solving the two-electron Schrödinger equation gives us the possibility to obtain highly accurate values for the ground-state energy of the electron system, the mass correction and the mass polarization. The results obtained for Z = 2 ÷ 10 are compatible with the most precise theoretical results available in the literature.

In Refs. [4,5], we have carried out a comparative analysis of the results obtained by Pekeris, Thakkar and Koga and our results (PM – Pavlov, Maruani, Mihailov *et al*) for the correlated ground state (nonrelativistic) energies E_0 in atomic units (au) for helium isoelectronic ions, omitting the mass corrections with $Z = 2 \div 10$. Using the same method determinates the same accuracy for $Z > 10$. The nuclear masses used in our computations were derived from the tables of the mass excess, using the known relation:

$$M_{\text{NUC}} = \Delta + A - Zm_e + b_e \,,$$

where Δ is the mass excess, A is the atomic number of the nuclide, Z is its charge and b_e is the module of the binding energy of the electrons.

In the current work, we have considered the most stable isotopes using the data from Ref. [8]. Whereas precise calculations for the electron ground-state energy characteristics of the helium-like ions have been already performed for $Z \leq 10$ [9 – 14], few theoretical calculations are available in the literature for $Z > 10$, for example in Refs. [9] or [15].

In figure 1 are plotted the absolute values of the ground-state energies E_0pm, E_{0BS} [3], E_{0FF} [16] E_{0Kar} [17] и E_{exp} [15]. All theoretical values are without any corrections and for E_{0FF} and E_{0Kar} the relativistic effects are not taken into account. Our results coincide with all of the presented theoretical results. As can be seen in figure 1, all theoretical results coincide for small Z and the deviations from experimental data appear at Z = 54 and increase with increasing Z due to the growing importance of relativistic effects.

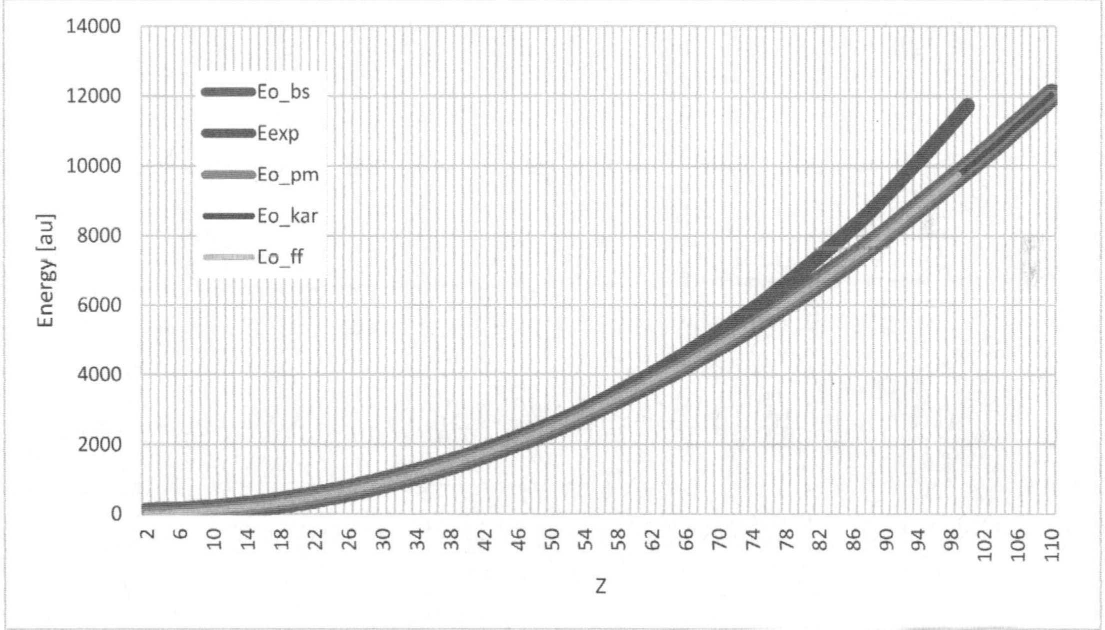

Figure 1. Comparison between ground state energies.

When one derives all theoretical results by including all possible corrections, as follows: Epm – mass corrections ε_1 and mass polarization ε_2; E_{BS} – mass corrections ε_1; E_{FF} – Hartree-Fock equations correlation factor and relativistic corrections; E_{Kar} – ε_1 and ε_2 corrections, by using two component wave function for the energy [18] the results are quasirelativistic as shown in figure 2.

It can be seen that our results Epm practically coincide with the values obtained by the Bethe-Saltpeter formula for nuclear charge $Z = 2 \div 110$. Our results coincide with those obtained using the Froese-Fisher code (relativistic case) and with the Karwowski results (again relativistic) for the nuclei with $Z \leq 54$. For $Z > 54$, our results slightly diverge from those of Froese-Fisher and of Karwowski. Moreover, the differences increase with the nuclear charge. This can be easily explained considering that the role of j-j coupling in the electron system increases with Z; for Z > 54 this coupling preponderates in regard to the L-S coupling. For Z > 76 only the j-j coupling remains. In addition, the contribution due to relativistic effects becomes progressively more important with increasing Z, while the contribution due to mass correlation effects fluctuates around some non-increasing value.

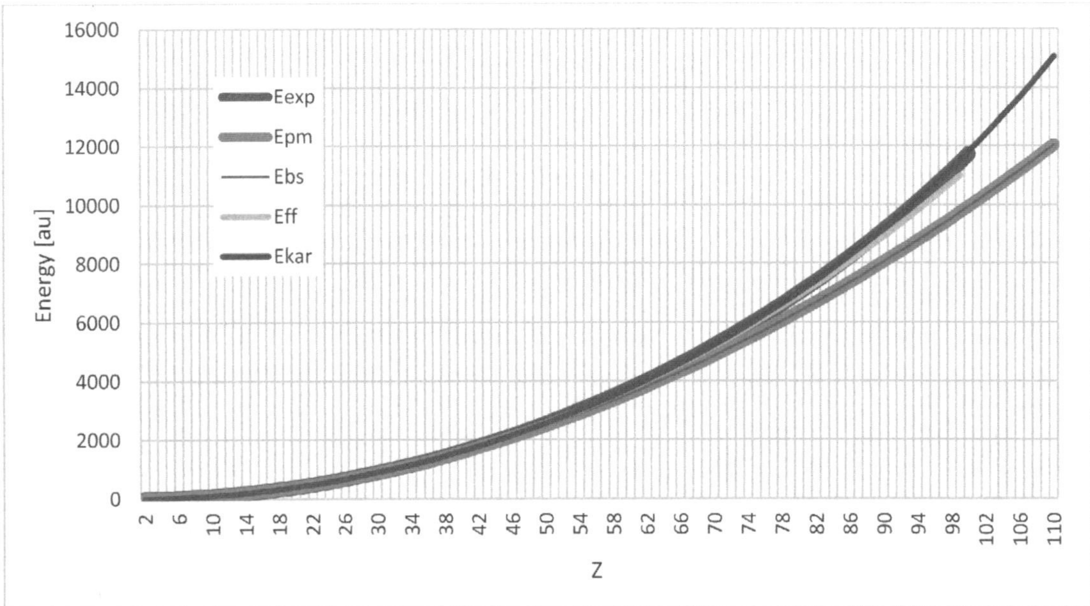

Figure 2. Comparison between ground state energies with all possible corrections.

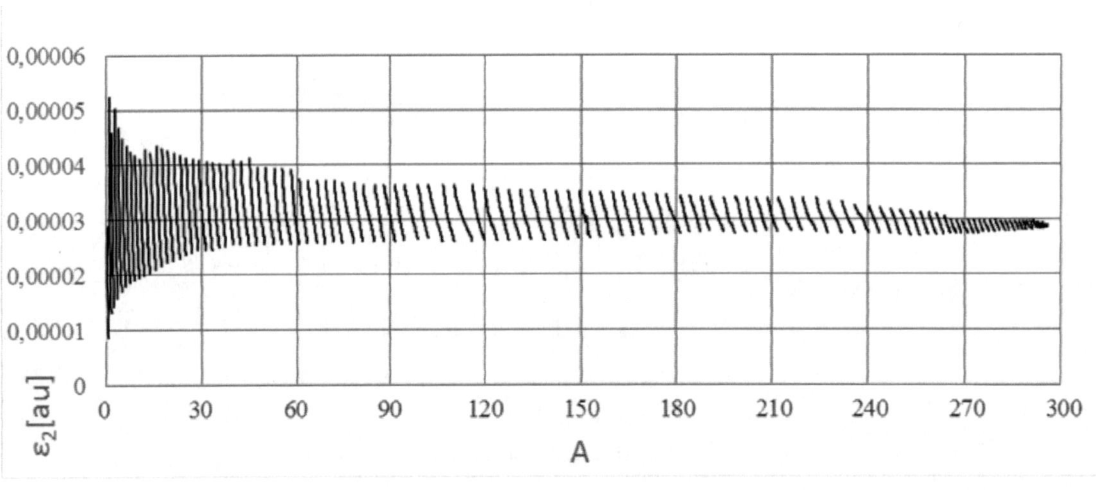

Figure 3. Polarization correction ε_2 in atomic units for different mass number A. Each vertical curve represents a single Z.

In Ref. [5] we have shown that including the mass polarization ε_2 (7) in the minimization procedure allows to establish a relation between electron-energy characteristics of helium-like electron-nuclear systems and properties of the underlying nuclei. This is the consequence of including the nuclear mass excess in the variational procedure through the operator of ε_2 in the system Hamiltonian (10).

For nuclei along the line of stability, the oscillating nature of ε_2 with Z was shown. The amplitudes of these oscillations were found to coincide or follow the nuclear magic numbers. In this way, conclusions can be drawn about the contribution of the nuclear properties and characteristics (for example shape and deformation) on the electron ground-state energy values.

Figure 3 presents the polarization correction ε_2 for each Z, depending on the mass number A. The results were obtained by studying helium-like electron-nuclear systems for about 3500 known isotopes of nuclei around the line of stability. The different length of the curves at the different Z is due to the different number of known isotopes around the line of stability. It can be seen that for each Z the values for ε_2 fluctuate around a mean value 0.00003 au depending on A. When A increases, the absolute value of ε_2 decreases with each Z.

In figure 4 are presented the mass correction ε_1 for each Z, depending on the mass number A. The different length of the curves at the different Z is again due to the different number of known isotopes around the line of stability.

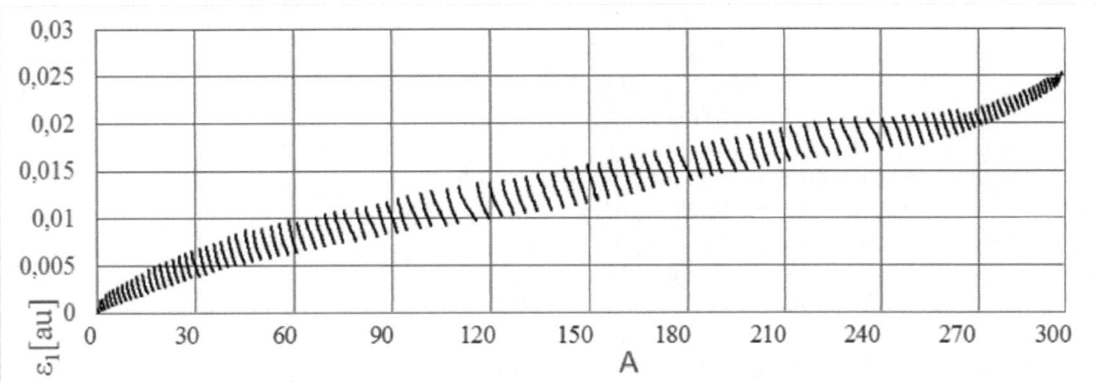

Figure 4 Mass correction ε_1 in atomic units for different mass number A. Each vertical curve represents a single Z.

Along the line of stability, the values of ε_1 increase with Z proportionally to the increase of the electron system ground-state energy (6). Moreover, for each Z the ε_1 absolute value decrease is noticeably less sensitive to the increasing of A. Therefore, it can be concluded that the effects of the nuclear properties, structure and characteristics are more explicitly included in the mass polarization ε_2 than in the mass correction ε_1. And, as with the analysis of figure 3, ε_2 demonstrates the effects of the nuclear characteristics and properties in the formation of electron energy magnitudes.

Acknowledgements
This work was financially supported by the bilateral project between Fond de la Recherche Scientifique–FNRS (Belgium), Wallonie-Bruxelles International (Belgium), and the Bulgarian Academy of Sciences; and the Bulgarian National Science Fund under contract No. DFNI-T02/19.

References

[1] Presnyakov L P, Shevelko V P and Yanev R K 1986 *Elementary processes with multiply charged ions* (Moskva: Energoatomizdat, in Russian)
[2] Bethe H A and Salpeter E E 1967 *Quantum Mechanics of One- and Two-Electron Atoms* (Springer Berlin)
[3] Gombas von P 1950 *Theorie und Loesungsmethoden des Mehrteilchenproblems der Wellenmechanik* (Basel)

[4] Pavlov R L, Maruani J, Mihailov L M, Velchev Ch J, Dimitrova-Ivanovich M, Stoianov J K and Van Neck D 2008 *Nuclear Theory* vol 27 ed S Dimitrova (Sofia: BM Trade Ltd) pp 249-260
[5] Pavlov R, Mihailov L, Velchev Ch, Dimitrova-Ivanovich M, Stoyanov Zh, Chamel N and Maruani J 2010 *J. Phys. Conf. Ser.* **253** 012075
[6] Rychlewski J (ed) 2003 *Explicitly Correlated Wave Functions in Chemistry and Physics (Progress in Theoretical Chemistry and Physics* vol 13*)* (Dordrecht: Kluwer Academic Publishers)
[7] Hylleraas E A 1928 *Z. Phys.* **48**(7–8) 469–494
Hylleraas E A 1929 *Z. Phys.* **54**(5–6) 347–366
Hylleraas E A 1930 *Z. Phys.* **60**(9–10) 624–630
Hylleraas E A 1930 *Z. Phys.* **63**(3–4) 291–292
Hylleraas E A 1930 *Z. Phys.* **65**(3–4) 209–225
[8] Tuli J 2005 *Nuclear Wallet Cards* (Upton, NY: National Nuclear Data Center) pp 1–77
[9] Pekeris C L 1958 *Phys. Rev.* **112** 1649
Pekeris C L 1959 *Phys. Rev.* **115** 1216
Pekeris C L 1962 *Phys. Rev.* **126** 1470
[10] Thakkar A J and Koga T 1994 *Phys. Rev.* A **50** 854
Koga T 1990 *J. Chem Phys.* **93**(5) 3720
Koga T 1992 *J. Chem Phys.* **96**(2) 1276
Koga T and Morishita S 1995 *Z. Phys.* D **34**(2) 71–74
Koga T 1996 *J. Chem. Phys.* **104**(16) 6308
[11] Drake G W F 1988 *Can. J. Phys.* **66** 586
[12] Schwartz H M 1956 *Phys. Rev.* **103** 110
Schwartz H M 1960 *Phys. Rev.* **120** 483
[13] Kinoshita T 1957 *Phys. Rev.* **105** 1490
Kinoshita T 1959 *Phys. Rev.* **115** 366
[14] Nakashima H and Nakatsuji H 2007 *J. Chem. Phys.* **127** 224104
[15] Kramida A, Ralchenko Yu, Reader J and NIST ASD Team 2013 *NIST Atomic Spectra Database* ver 5.1, Online: 26 August 2014 (Gaithersburg: National Institute of Standards and Technology)
[16] Froese-Fisher C 1997 *The Hartree-Fock Method for Atoms* (New York: Wiley)
[17] Karwowski J 2009 Private communication
[18] Pestka G and Karwowski J 2003 Hylleraas-CI approach to Dirac-Coulomb equation *Explicitly Correlated Wave Functions in Chemistry and Physics (Progress in Theoretical Chemistry and Physics* vol 13*)* ed J Rychlewski (Dordrecht: Kluwer Academic Publishers) pp 331-346

Cyclotron laboratory of the Institute for Nuclear Research and Nuclear Energy

D Tonev[1], N Goutev[1], L S Georgiev[1]

[1] Institute for Nuclear Research and Nuclear Energy, BAS, BG-1784, Sofia, Bulgaria

E-mail: dimitar.tonev@inrne.bas.bg

Abstract. An accelerator laboratory is presently under construction in Sofia at the Institute for Nuclear Research and Nuclear Energy, Bulgarian Academy of Sciences. The laboratory will use a TR24 type of cyclotron, which provides a possibility to accelerate a proton beam with an energy of 15 to 24 MeV and current of up to 0.4 mA. An accelerator with such parameters allows to produce a large variety of radioisotopes for development of radiopharmaceuticals. The most common radioisotopes that could be produced with such a cyclotron are PET isotopes like: ^{11}C, ^{13}N, ^{15}O, ^{18}F, ^{124}I, ^{64}Cu, ^{68}Ge/^{68}Ga, and SPECT isotopes like: ^{123}I, ^{111}In, ^{67}Ga, ^{57}Co, ^{99m}Tc. Our aim is to use the cyclotron facility for research in the fields of radiopharmacy, radiochemistry, radiobiology, nuclear physics, solid state physics, applied research, new materials and for education in all these fields including nuclear energy. The building of the laboratory will be constructed nearby the Institute for Nuclear Research and Nuclear Energy and the cyclotron together with all the equipment needed will be installed there.

1. Introduction

The Institute for Nuclear Research and Nuclear Energy (INRNE) at the Bulgarian Academy of Sciences (BAS) has been working since 2012 on the project "Cyclotron laboratory of INRNE-BAS". The project was launched in 2012 at the initiative of the Council of Ministers of the Republic of Bulgaria and the management of INRNE-BAS to solve a very old problem of the country - the lack of domestic production of short lived medical cyclotron radioisotopes. This problem has existed for decades despite repeated efforts of several generations of researchers from INRNE-BAS and the medical community to persuade the Government that such a center should be built in Bulgaria as was done a long time ago in the other Eastern European countries.

It would have been practically impossible to start-up the project without the generous contributions of our first financial donors. For the purchase of the cyclotron and construction of the laboratory 3,000,000 USD were donated from the United States Department of Energy (DOE) and 2,000,000 USD from Kozloduy Nuclear Power Plant.

In 2014 the project for a new cyclotron center was included in the updated "National Roadmap for Scientific Infrastructure" of Bulgaria. At the beginning of 2015 the project was identified as a priority of the Bulgarian-American relations in the field of science and education. For several years the management of INRNE-BAS has been actively working to insure further state support for the construction of the building of the cyclotron laboratory.

The project envisions a new cyclotron laboratory as a part of INRNE-BAS consisting of: a specialized building that meets the regulatory requirements in the field of radiation safety and GMP (good manufacturing practices) in the pharmaceutical industry; a bunker with a TR-24 cyclotron; a sector for applied research and development in radiopharmacy with enhanced educational functions; a sector for the production of 18F-FDG, called Fluorodeoxyglucose, and in future, of other radiopharmaceuticals.

The laboratory of INRNE-BAS is the novel nuclear research infrastructure in the country. Presently Bulgaria has no research nuclear facility, neither a research reactor, nor an accelerator. With the new facility we will restart our experimental research program not only in the field of nuclear physics, but also in many interdisciplinary fields related to nuclear physics. An excellent presentation of the accelerator physics is present in the work [1].

Figure 1. A picture of INRNE-BAS cyclotron TR24 during the test in Vancouver in August 2015.

In Europe there are many academic institutions where novel accelerator centers are under development. The TR24 cyclotron is an accelerator which has recently been installed at CYclotron pour la ReCherche et lEnseignement (Cyrce‘) of the Institut Pluridisciplinaire Hubert Curien (IPHC) in Strasbourg, France or will be operated soon at at the Institute of Radiopharmaceutical Cancer Research of Helmholtz-Zentrum Dresden Rossendorf (HZDR), Germany; Rzez, Czech Republic etc. The Laboratori Nazionali di Legnaro at INFN, using a 70 MeV cyclotron, plans to perform similar to our activities [2].

2. Challenges of the INRNE-BAS cyclotron project

One of the main objectives of the project is within five years to allow for regular supply of 18F-FDG to 8 regional PET/CT center at a low price, providing "full cost recovery", with a small profit margin used to sustain the laboratory. This will enable more hospitals in Bulgaria to purchase new PET/CT equipment as well as to increase the throughput of the existing scanners. This will allow the examination of 16,000 patients a year in the regional PET/CT centers. The wider access to PET/CT will improve the health and extend the life of the patients suffering from cancer. The revenue from the sale of radioisotopes and radiopharmaceuticals will be fully reinvested in the research program and in the development of the laboratory. The availability of a production sector, meeting all modern regulatory GMP requirements will ensure high quality training of nuclear pharmacists and will close the chain from research through R&D to end products with applications in medicine.

The parameters of the TR24 cyclotron (shown in Figure 1) have been selected in such a way that in the next twenty years INRNE-BAS will be able to produce a wide range of radioisotopes with applications in medicine with a relatively low initial investment and moderate maintenance costs. At these proton energies and beam current it is possible to produce commercial quantities of radioisotopes with traditional and expected future applications in medicine as: PET radioisotopes - 18F, 124I, 64Cu, 68Ge/68Ga; SPECT radioisotopes - 123I, 111In, 67Ga, 99mTc; therapeutic alpha-emitters - 225Ac/213Bi, 230U/226Th. Some part of the equipment for the production of radioisotopes will be designed and constructed in INRNE-BAS with the help of the Design and Production Workshop - Physics and will be sold on the international market subsequently.

The radioisotopes produced by the cyclotron will be used for the development of new radiolabelled molecules for diagnostics [3] and therapy purposes [4].

Figure 2. A Y-line belonging to the Sofia cyclotron.

The aim of the INRNE-BAS cyclotron laboratory is to provide a high intensity and a high quality proton beam to perform forefront research in interdisciplinary fields like medical, chemical, biological and material sciences, nuclear physics and applied research. The cyclotron can also produce very low currents (10nA) for radiobiology applications. A longer term goal of the laboratory is to become involved in preclinical studies and clinical trials through collaborations and research projects with academic partners and private companies.

The laboratory in Sofia will open a new possibility for experimental nuclear structure research in Bulgaria. We plan to investigate level-schemes of excited nuclei as well as to measure lifetimes of excited nuclear states. A proton beam will be used to bombard a target and with a multidetector system to investigate the nuclear structure. Lifetimes in the femtosecond region can be measured using Doppler-shift attenuation method (DSAM) following the inelastic proton scattering reaction. Such an approach is presented in details in the work of Hennig et al. [5]. In the setup used in Cologne University, the proton in the exit channel of the (p, p'γ) reaction is detected with charged-particle detectors placed close to the target. This approach uses the centroid-shift version of the DSAM method and profits from the coincident detection of scattered protons and de-exciting γ-rays.

Figure 3. Factory acceptance tests in Vancouver, Canada in 2015.

Utilizing this approach thirty lifetimes in ^{96}Ru are determined in the work of Hennig et al. [6].

An approach using a similar setup we plan to mount at INRNE laboratory in order to measure lifetimes for low-lying states in nuclei which could be approached by (p, p'γ) reaction.

We would like to continue our research in the fileds of investigation of phase transitions in nuclear physics [7], chirality in nuclei [8, 9] and nuclear structure of light nuclei [10].

A parallel activity of INRNE laboratory includes education and training of physicists, chemists, pharmacists and biologists for radioisotopes production, radiochemistry, quality control of radiopharmaceuticals and their uses in medical imaging, as well as training of specialist for nuclear energy. More general practical topics like radiation protection and dosimetry will also be covered by the training and education program of the laboratory. Most of the training and education activities will be carried out by involving the students in the ongoing research projects of the laboratory. Special attention will be paid to the traning of students which plan to work in the field of nuclear energy. The education of these students will guarantee the high level of nuclear safety of the Bulgarian nuclear power plant.

3. The accelerator TR24

The TR24 cyclotron is produced and commercialized by Advanced Cyclotron Systems Incorporation, located in Vancouver, Canada. The TR-24 cyclotron of INRNE-BAS, Figure 2, has the following characteristics: accelerates negatively charged hydrogen ions; variable energy of the proton beam; minimum energy of the proton beam of 15 MeV; maximum energy of the proton beam of 24 MeV; proton beam current of 400 μA, upgradeable to 1000 μA; simultaneous extraction of two proton beams; external CUSP ion source. With a very high current of up to 400 μA upgradeable to 1 mA and proton energy ranging from 15 to 24 MeV, the cyclotron can produce commercial quantities of more than 15 isotopes by irradiating simultaneously two high current targets placed at the exit ports or along high current beam lines. At the beginning one Y-like beam line, shown in Figure 2, will be mounted at one of the exit ports of our cyclotron for greater flexibility. Presently we have two target stations and three targets for the production of 18F-FDG.

4. Status of the Bulgarian National Cyclotron Center

In 2014 INRNE-BAS completed successfully the public procurement tender for the cyclotron and signed a contract for the delivery of a TR-24 cyclotron, manufactured by the Canadian company ACSI. The cyclotron passed low energy tests in the factory in August 2015. All the design parameters of the cyclotron were successfully demonstrated during the tests. An illustration for the tests is shown in Figure 3. All the components of the cyclotron were delivered and stored at INRNE-BAS on January 12-th 2016 (Figure 4).

The building for the cyclotron and the hot cells will be constructed nearby the Institute.

Figure 4. Delivery and storage of the cyclotron in INRNE-BAS.

4.1. Acknowledgments

The project has been supported by the Department of Energy of USA, the Nuclear Power Plant "Kozloduy", the Bulgarian Ministry of Education and Science. This project has been supported by the Bulgar,k ian Science Fund under Contract No. DFNI-E 01/2 and by a NUPNET-NEDENSAA project funded by the Bulgarian Ministry of Education and Science. The authors acknowledge the support and encouragement for the success of the project from colleagues and friends.

References

[1] Dinev D 2006 Particle accelerators *Marin Drinov Academic Publishing House* **ISBN - 13:978-954-322-067-0**
[2] de Angelis G, Prete G, Andrigetto A, for the SPES collaboration, *et al.* 2016 *EPJ Web of Conferences* A **107** 01001
[3] Yordanov AT, Llop J, Stimson D, Combs M, Mueller M, Soylu A, Bagci H, Shulman S 2011 *J. Nucl. Med.* **52(Suppl.1)** 1643
[4] Yordanov AT, Hens M, Pegram C, Bigner DD, and Zalutsky MR 2007 *Nucl. Med. Biol.* **34** 173
[5] Hennig A *et al.* 2015 *Nucl. Instrum. Meth. Phys. Res. A* **794** 171
[6] Hennig A *et al.* 2015 *Phys. Rev. C* **92** 064317
[7] Tonev D *et al.* 2004 *Phys. Rev. C* **69** 034334
[8] Tonev D *et al.* 2006 *Phys. Rev. Lett.* **96** 052501
[9] Tonev D *et al.* 2014 *Phys. Rev. Lett.* **112** 052501
[10] Tonev D *et al.* 2002 *Phys. Rev. C* **65** 034314

Identification of low-energy isovector octupole states in ^{144}Nd

M Thürauf[1], M Scheck[1,2,3], C Bernards[4], A Blanc[5], N Cooper[4], G Defrance[6], M Jentschel[5], J Jolie[7], O Kaleja[1], U Köster[5], T Kröll[1], P Mutti[5], G S Simpson[8], T Soldner[5], M Tezgel[1], W Urban[9], J Vanhoy[10], M Werner[1], V Werner[1,4], and K O Zell[7]

[1] Institut für Kernphysik, TU Darmstadt, Schlossgartenstr. 9, 64289 Darmstadt, Germany
[2] School of Eng. and Comp., Univ. of the West of Scotland, High Street, Paisley PA1 2BE, UK
[3] The Scottish Universities Physics Alliance, University Avenue, Glasgow G12 8QQ, UK
[4] Wright Laboratory, Yale Univ., P.O. Box 208120, New Haven, CT 06520-8120, USA
[5] Institut Laue-Langevin, 71 avenue des Martyrs, 38000 Grenoble, France
[6] Grand Accélérateur National d'Ions Lourds, Boulevard Henri Becquerel, 14000 Caen, France
[7] Institut für Kernphysik, Universität zu Köln, Zülpicher Str. 77, 50937 Köln, Germany
[8] LPSC, UJF Grenoble I, 53 avenue des Martyrs, 38026 Grenoble Cedex, France
[9] Faculty of Physcis, University of Warsaw, ul. Pasteura 5, 02-093 Warsaw, Poland
[10] Dep. of Physics, U.S. Naval Academy, 121 Blake Road, Annapolis, MD 21402, USA

E-mail: `mthuerauf@ikp.tu-darmstadt.de`

Abstract. Recently, first candidates for low-lying isovector states in the octupole sector were suggested. The unambiguous identification of those states will contribute to the decomposition of the octupole-octupole residual interaction in an isoscalar and isovector part. This will help us understand the octupole degree of freedom. In ^{144}Nd the 3^- state at 2778 keV is a good candidate for such a "mixed-symmetry" octupole state. In order to clarify the nature of this state, a ^{143}Nd(n, γ)-experiment was conducted with the EXILL-setup. Following neutron capture the 3^- states are populated and EXILL provides the opportunity to determine the multipole-mixing ratios of the $3_i^- \rightarrow 3_1^-$ transitions. For the transition from the "mixed-symmetry" octupole state to the symmetric 3_1^- state we expect a strong $M1$ component.

1. Introduction

Low-energy isovector states like the 1^+ scissors mode [1–5] in deformed nuclei or the mixed-symmetric 2^+ states [6, 7] in near-spherical nuclei are well established in those nuclei. In contrast to quadrupole mixed-symmetry states, octupole mixed symmetry states are known only sparsely in the present time, but are predicted by the sdf-IBA-2 [8]. First experimental evidence was found through an enhanced $3_i^- \rightarrow 3_1^-$-$M1$-transition in ^{94}Mo [9]. More recently, candidates for isovector octupole states were suggested from additional (n, n$'\gamma$) data on ^{92}Zr, ^{96}Mo and ^{144}Nd [10] supported also from charged particle scattering data [11]. The proof of the existence of isovector octupole states in these nuclei will help us to understand the octupole degree of freedom and the interaction of protons and neutrons in atomic nuclei. In the present work the first results for ^{144}Nd, populated via cold-neutron capture, are presented.

1.1. *Mixed-symmetry states*

The concept of mixed-symmetry states is defined in the framework of the Interacting Boson Model version 2 (IBM-2) [12]. The basic idea of this model is the coupling of the valence protons and neutrons to s-, d- and in our case to f-bosons (sdf-IBM-2 [8]). The isospin analogue in the IBM formalism is the F-spin concept: Protons and neutrons have $F = \frac{1}{2}$ with projections $F_z = +\frac{1}{2}$ for protons and $F_z = -\frac{1}{2}$ for neutrons. The wavefunction of a state with maximum F-spin $F = F_{\text{max}} = \frac{1}{2}(N_\pi + N_\nu)$ behaves symmetric under pairwise exchange of proton and neutron bosons, where N_ρ are the numbers of proton and neutron bosons ($\rho \in \{\pi, \nu\}$). This class of states is called fully-symmetric states (FSS).

States with F-spin $F < F_{\text{max}}$ are mixed-symmetry states (MSS) and a part of their wavefunction behaves anti-symmetric under proton and neutron boson exchange. Usually the mixed-symmetry states in vibrational nuclei and the scissors mode in deformed nuclei are states with $F = F_{\text{max}} - 1$. The IBM $M1$-Operator is given by [7]:

$$T(M1) = \sqrt{\frac{3}{4\pi}} \left[\frac{N_\pi\, g_\pi + N_\nu\, g_\nu}{N} L_{\text{tot}} + (g_\pi - g_\nu) \frac{N_\pi N_\nu}{N} \left(\frac{L_\pi}{N_\pi} - \frac{L_\nu}{N_\nu} \right) \right] \mu_N\,, \qquad (1)$$

where g_ρ are the g-factors and N the total number of the constituent bosons. L_ρ denotes the angular momentum operator for proton and neutron bosons. It can be shown, that $M1$ transitions from FSSs to FSSs are forbidden, but allowed for transitions from MSSs to FSSs, which can be used as an experimental signature for these class of states: For a octupole MSS we expect a strong $M1$ transition with $\langle 3_{\text{FSS}} \,|\, T(M1) \,|\, 3_{\text{MSS}} \rangle \approx 1\, \mu_{\text{N}}$ and a weak $E3$ of a few single particle units for $3^-_{\text{MSS}} \to 0^+_{\text{gs}}$.

2. Experiment

The EXOGAM Ge detector array was placed at the cold-neutron beam line PF1B at the Institut Laue-Langevin (ILL) in Grenoble, France. The cold neutrons generated in a high-flux reactor were collimated so that the beam had a diameter of 12 mm and a flux of $\Phi = 10^8 \frac{1}{\text{s}\,\text{cm}^2}$ at the target position. The target was positioned in the center of a frame, where eight EXOGAM clover HPGe-detectors were mounted at 90° with respect to the beam axis. The capsules of the EXOGAM detectors were surrounded with an active anti-Compton shield (BGO) to reduce the Compton background. A detailed description of the experimental setup can be found in Refs. [13, 14] and references therein.

In our case the target (0.8 mg) was enriched to 91 % in ^{143}Nd, which has a (n, γ) cross section of 330 b for thermal neutrons. The combination of high thermal neutron-capture cross section and the enrichment of the target made it possible to conduct the experiment in one day with sufficient statistics. For calibration a ^{152}Eu source was placed at the target position before and after the Nd-runs. The daughters of ^{152}Eu, ^{152}Gd and ^{152}Sm, have well known γ-ray intensities and energies, as well as multipole-mixing ratios.

The data acquisition system was run in triggerless timestamped mode so that every hit of a γ ray was written as an event to the data stream. This required the implementation of a new eventbuilder which can read and sort the γ events. The eventbuilder is able to identify γ rays which were in coincidence with each other and, if there were any, with the corresponding BGO events. On top of this eventbuilder an addback-algorithm for the clover-detectors was implemented to reduce the loss of full-energy events due to the Compton effect.

3. Analysis

3.1. *Angular correlations*

The eight EXOGAM-clover-HPGe detectors can be used for angular-correlation measurements. The octagonal construction of the frame gives us the opportunity to measure correlations between

Figure 1. EXOGAM setup at the cold-neutron beam line PF1B at ILL. The neutrons enter from the right side impinging on a target (here ^{143}Nd) in the center of the frame. One can see the eight EXOGAM-clover-HPGe detectors perpendicular to the beam axis which are used for the angular-correlation measurement in this work.

four angular groups: 45°, 90°, 135° and 180°. From the angular correlation measurement the transmitted angular momentum l of the γ radiation and the multipole-mixing ratio δ can be extracted. A detailed description of angular correlation techniques can be found in Ref. [15–17].

In experiments with thermal- or cold-neutrons the neutron beam doesn't characterize a z-axis of the quantum system like it would be in Coulomb excitation or (γ, γ') reactions. The angular distribution of γ rays after thermal- and cold-neutron capture is isotropic with respect to the neutron beam axis. To obtain an non-isotropic distribution of magnetic substates, γ-γ-matrices have to be sorted and a gate is set on the populating or depopulating transition. Subsequently the peak volumes of the depopulating or populating transition can be determined, which shows now a non-isotropic distribution with respect to the direction of the emitted (de)populating γ ray.

Angular correlations $\tilde{W}(\theta)$ can be parameterized as

$$\tilde{W}(\theta) = 1 + A_2 P_2 (\cos\theta) + A_4 P_4 (\cos\theta) \tag{2}$$

where $A_{\{2,4\}}$ are the correlation coefficients and $P_{\{2,4\}}(x)$ the Legendre polynomials of 2nd and 4th order. Usually the normalization of the experimental data is done with respect to the 90° angular group:

$$W(\theta) = \frac{\tilde{W}(\theta)}{\tilde{W}(90^\circ)} = \frac{1 + A_2 P_2 (\cos\theta) + A_4 P_4 (\cos\theta)}{1 - \frac{1}{2}A_2 + \frac{3}{8}A_4}. \tag{3}$$

For each γ-ray cascade the A_k can be calculated analytically depending on the spin of each state in the cascade and the δ of the first or the second transition. The experimental values of A_k have to be corrected by attenuation factors Q_k, which can be determined from a well-known cascade e.g. $0-2-0$ cascades, because here we have a strong angular correlation ($A_2^{\text{theo.}} = 0.3571$, $A_4^{\text{theo.}} = 1.1429$) and no δ is involved.

In general, the number of detector combinations is not the same for each angular group. In our case we have eight pairs in the 45°, 90° and 135° groups, but only four pairs in the 180° group. To correct for this varying number and consider eventual differences in the absolute detection efficiency a $2-0-2$ cascade was used for normalization. Owing to the reaction mechanism of

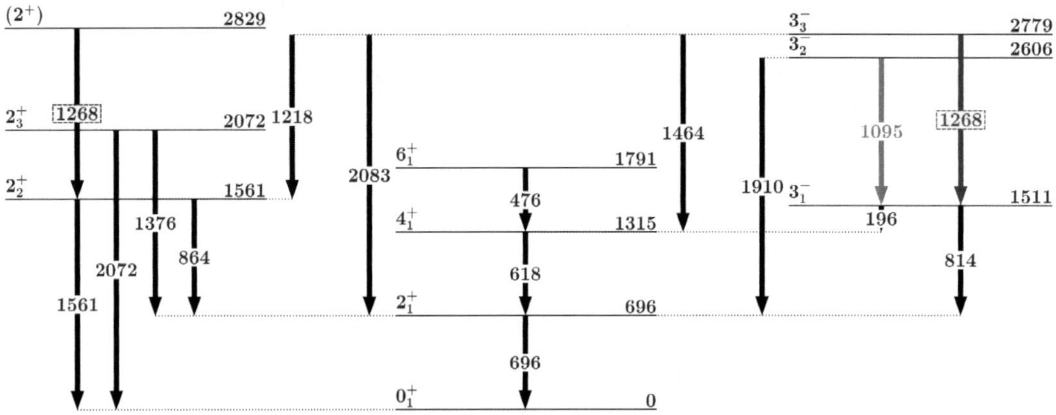

Figure 2. Partial level scheme of ^{144}Nd. The red and blue colored transitions are the transitions of interest for this work.

the ^{143}Nd(n, γ) reaction $0^{+,-}$ states are populated only weakly and cascades involving them are unsuitable. Consequently a calibration source had to be employed.

3.2. Multipole-mixing ratio

In this work the definition of the multipole-mixing ratio from Ref. [18] and the references therein is used:

$$\delta = \frac{\sqrt{3}}{10} \frac{E_\gamma}{\hbar c} \frac{\langle J_f || \hat{M}(E2) || J_i \rangle}{\langle J_f || \hat{M}(M1) || J_i \rangle}, \tag{4}$$

where $\hat{M}$ is the electro-magnetic multipole operator for $E2$ and $M1$ radiation. For the phase convention the definition of Krane and Steffen [17] is used. Determining multipole-mixing ratios via angular correlations is a necessity to identify mixed-symmetric states. For cascades with competing $M1$ and $E2$ transitions the A_k coefficients become functions of δ.

In ^{152}Sm the 3_1^+ state is populated via the electron capture of ^{152}Eu. This 3^+ state decays predominantly via $M1$ or $E2$ transitions to the 4_1^+ and the multipole-mixing ratio is well known from literature ($\delta(3_1^+ \to 4_1^+) = -6.5(3)$ [19]). Placing a gate on the 868-keV transition $4_1^+ \to 2_1^+$ and fitting the angular correlation function with δ as a free parameter leads to a χ^2-minimum at $\delta = -6.2$ which is in agreement with the literature value.

3.3. Octupole transitions in ^{144}Nd

Hicks *et al.* [20] conducted a (n, n′γ) experiment at the University of Kentucky Accelerator Laboratory in Lexington, USA, in order to clarify spectroscopic uncertainties in ^{144}Nd. The original assignment of the 3_3^- state as a candidate for a mixed-symmetric octupole state is based on this dataset [10]. The major problem with this experiment was that the 1268-keV peak is a doublet consisting of transitions from $(2_x^+) \to 2_2^+$ ($E(2_x^+) = 2829$ keV) and $3_3^- \to 3_1^-$ which cannot be separated in single γ-ray spectroscopy (see level scheme Fig. 2). To clarify the nature of the 3_3^- state the expirement at ILL was performed.

It was shown by Robinson *et al.* that the 3_2^- state at 2606 keV is not a mixed-symmetry state but member of the quadrupole-octupole-quintuplet $\left[|2_1^+\rangle \otimes |3_1^-\rangle\right]_{3^-}$ [21]. This fact was confirmed by Hicks *et al.* as well. Here a strong $B(E2)$ strength of 23(4) W.u. and a weak $B(M1)$ of $7.0(13) \times 10^{-3}\,\mu_N^2$ for the $3_2^- \to 3_1^-$ transition was measured.

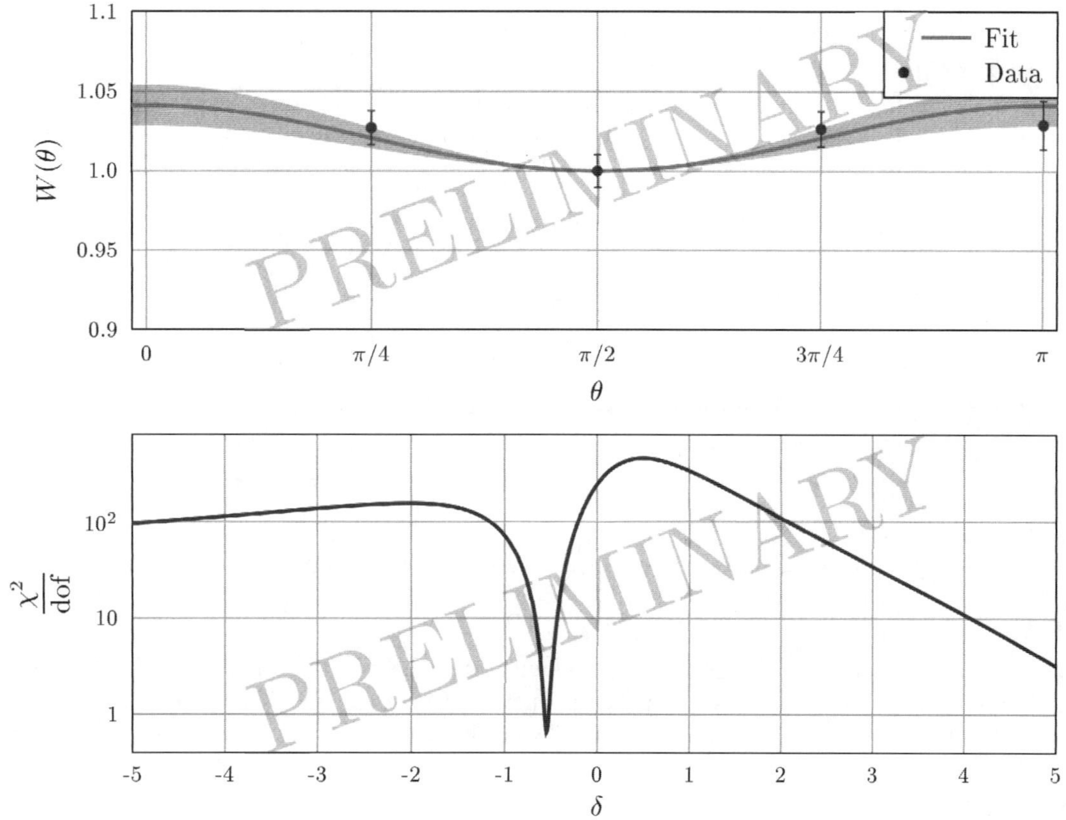

Figure 3. The upper figure shows the fitted angular correlation function to the data points of the $3^-_3 \to 3^-_1 \to 2^+_1$ cascade in ^{144}Nd. The lower picture shows the reduced χ^2 distribution of this fit with a pronounced minimum at $\delta = -0.55$.

In this work we apply angular-correlation techniques to precisely determine the multipole-mixing ratio $\delta(3^-_3 \to 3^-_1)$. The capture state $(\frac{7}{2}(f^-_{7/2}) \pm \frac{1}{2}(\mathrm{n}))$ of ^{144}Nd is a 3^- or 4^- state and the 3^-_3 is populated either directly by a $M1/E2$ transition with an unknown multipole-mixing ratio or via unknown intermediate states. Therefore, as a gate to separate the $3^-_3 \to 3^-_1$ transition from the contaminating transition the $3^-_1 \to 2^+_1$-$E1$-transition at 814 keV is favourable. This transition is of pure $E1$ character. Consequently, for each value of $\delta(3^-_3 \to 3^-_1)$ in this cascade, A_4 equals to zero, and the ellipse in the A_2-A_4-space corresponding to the different δ becomes a string. This results in two minima in the χ^2 plot with comparable χ^2 values.

In Fig. 3 the fitted angular correlation function to the data points and the χ^2 distribution is shown. The χ^2 distribution shows a minimum at $\delta = -0.55$ which is a strong indicator for the 3^-_3 state to be a mixed-symmetric octupole state. Table 1 shows an overview of all multipole-mixing ratios determined so far with the current (n, γ) dataset.

4. Outlook

In order to get the absolute $B(M1; 3^-_3 \to 3^-_1)$ strengths in ^{144}Nd the collaboration intends to measure the lifetime of the 3^-_3 state. The lifetime of states which decay via strong dipole transitions is settled in the femtosecond range. A suitable setup for such a lifetime measurement especially with non-Yrast-states is the high-resolution flat-crystal γ-ray facility, GAMS, at ILL,

Table 1. Overview of preliminary multipole-mixing ratios δ determined with the current ^{143}Nd(n, $\gamma\gamma$) dataset.

	This work	D.M. Snelling *et al.*	S.F. Hicks *et al.*
$\delta(2_2^+ \to 2_1^+)$	-0.94	-1.13^{+15}_{-2}	-0.93^{+30}_{-523}
$\delta(2_3^+ \to 2_1^+)$	0.19	0.31^{+11}_{-9}	0.6^{+4}_{-3}
$\delta(4_2^+ \to 4_1^+)$	-0.3		-0.5^{+8}_{-5}
$\delta(3_1^+ \to 2_1^+)$	0.45		0.8^{+9}_{-4}
$\delta(3_2^- \to 3_1^-)$	-2.0		-3^{+3}_{-13}
$\delta(3_3^- \to 3_1^-)$	-0.55		-0.37^{+18}_{-11}

where it will be possible to separate also the energy doublet at 1268 keV in ^{144}Nd.

Acknowledgments

This work is supported by the Deutsche Forschungsgemeinschaft through Grant No. KR 1796/2-1, BMBF Grand No. 05P12RDFN8, U.S. Department of Energy Grand No. DE-FG02-91ER-40609 and HGS-HIRe for FAIR. M. Scheck acknowledges financial support from the STFC. Financial support be the ILL to realize the fantastic opportunity of EXILL is gratefully acknowledged.

References

[1] Iudice N L and Palumbo F 1978 *Phys. Rev. Lett.* **41** 1532
[2] Richter A 1983 *Proc. of the Int. Conf. on Nuclear Physics*
[3] Iachello F and Arima A 1987 *The Interacting Boson Model* (Cambridge University Press)
[4] Caprio M A and Iachello F 2005 *Annals of Physics* **318** 454
[5] Heyde K, von Neumann-Cosel P and Richter A 2010 *Reviews of Modern Physics* **82** 2365
[6] Iachello F 1984 *Phys. Rev. Lett.* **53** 1427
[7] Pietralla N, von Brentano P and Lisetskiy A F 2008 *Progress in Particle and Nuclear Physics* **60** 225
[8] Smirnova N A, Pietralla N, Mizusaki T and van Isacker P 2000 *Nucl. Phys.* A **678** 235
[9] Fransen C *et al.* 2003 *Phys. Rev.* C **67** 024307
[10] Scheck M, Butler P A, Fransen C, Werner V and Yates S W 2010 *Phys. Rev.* **81** 064305
[11] Scheck M 2012 Experimental evidences for low-lying octupole isovector states *Journal of Physics: Conference Series* vol 366 p 012040
[12] Arima A, Ohtsuka T, Iachello F and Talmi I 1977 *Physics Letters* B **66** 205
[13] Jentschel M *et al.* *Journal of Instrumentation* to be submitted
[14] Jolie J *et al.* 2015 The (n, γ) campaigns at exill *EPJ Web of Conferences* vol 93 p 01014
[15] Frauenhelder H and Steffen R M 1965 *Alpha-, Beta-, and Gamma-Ray Spectroscopy* (North-Holland Publishing Company)
[16] Urban W *et al.* 2013 *Journal of Instrumentation* **8** P03014
[17] Krane K S, Steffen R M and Wheeler R M 1973 *Nucl. Data Tables* **11** 351
[18] Stahl C, Pietralla N, Rainovski G and Reese M 2015 *Nuclear Instruments & Methods in Physics Research* A **770** 123
[19] Lange J, Kumar K and Hamilton J H 1982 *Reviews of Modern Physics* **54** 1982
[20] Hicks S F, Davoren C M, Faulkner W M and Vanhoy J R 1998 *Phys. Rev.* C **57** 2264
[21] Robinson S J *et al.* 1999 *Physics Letters* B **465** 61

Medium-spin states of the neutron-rich ^{87,89}Br isotopes: configurations and shapes

B.M. Nyakó[1], J. Timár[1], M. Csatlós[1], Zs. Dombrádi[1], A. Krasznahorkay[1], I. Kuti[1], D. Sohler[1], T.G. Tornyi[1], M. Czerwiński[2], T. Rząca-Urban[2], W. Urban[2], P. Bączyk[2], L. Atanasova[3], D.L. Balabanski[4], K. Sieja[5], A. Blanc[6], M. Jentschel[6], U. Köster[6], P. Mutti[6], T. Soldner[6], G. de France[7], G.S. Simpson[8], C.A. Ur[9]

[1] Institute for Nuclear Research, Hungarian Academy of Sciences, H-4026 Debrecen, Hungary
[2] Faculty of Physics, University of Warsaw, ul. Pasteura 5, PL-02-093 Warsaw, Poland
[3] Department of Med. Physics and Biophysics, Medical University - Sofia, 1431, Sofia, Bulgaria
[4] ELI-NP, Horia Hulubei National Institute for R&D in Physics and Nuclear Engineering IFIN-HH, 077125 Bucharest-Magurele, Romania
[5] Université de Strasbourg, IPHC, Strasbourg, France; CNRS, UMR7178, 67037 Strasbourg, France
[6] Institut Laue-Langevin, Grenoble, France
[7] GANIL, CEA/DSM CNRS/IN2P3, Bd Henri Becquerel, BP 55027, F-14076 Caen Cedex 5, France
[8] LPSC, Université Joseph Fourier Grenoble 1, CNRS/IN2P3, Institut National Polytechnique de Grenoble, F-38026 Grenoble Cedex, France
[9] INFN, Legnaro, Italy

E-mail: timar@atomki.mta.hu

Abstract. Medium-spin excited states of the neutron-rich ^{87}Br and ^{89}Br nuclei were observed and studied for the first time. They were populated in fission of ^{235}U induced by the cold-neutron beam of the PF1B facility of the Institut Laue-Langevin, Grenoble. The measurement of γ radiation following fission has been performed using the EXILL array of Ge detectors. The observed level schemes were compared with results of large valence space shell model calculations. Both medium-spin level schemes consist of band-like structures, which can be understood as bands built on the $\pi f_{5/2}$, $\pi p_{3/2}$ and $\pi g_{9/2}$ configurations. Both nuclei have $5/2^-$ ground state spin-parity contrary to the odd-mass Br isotopes containing fewer neutrons, which have $3/2^-$ ground state spin-parity. On the basis of the properties of the $\pi g_{9/2}$ decoupled bands the deformations of the ^{87}Br and ^{89}Br fit to the systematics of nuclei in the region. ^{87}Br is close to the vibrational limit, while ^{89}Br is more rotational.

1. Introduction

Study of exotic, neutron-rich nuclei is in the forefront of the contemporary nuclear-structure research. Such investigations resulted in the observation of interesting new phenomena, like quenching of the known shell closures and formation of new ones. Thus, it is especially interesting to study neutron-rich nuclei near the shell closures. Such an interesting region is that of the nuclei near the doubly magic exotic ^{78}Ni. Besides the strength of the shell closures, the shell-model single-particle energies and residual interactions between the nucleons are also changing in these regions. Excited states of odd-mass nuclei provide important information on the single-particle energies. However, no medium-spin excited states are known for the odd-mass odd-proton nuclei with neutron number larger than 50 and proton number less than 37 near ^{78}Ni. In general, information on excited medium-spin states in this region is very scarce, due to the difficulties in finding nuclear reactions in which these states could be populated with large statistics.

Cold-neutron-induced fission of ^{235}U populates the nuclei ^{87,88,89}Br with relatively high yields, and using this reaction we can get closer to the doubly magic ^{78}Ni in studying the medium-spin states. A few low-spin excited states of ^{87}Br are known from the β decay of ^{87}Se produced by thermal neutron induced fission of ^{235}U [1]. Only tentative spins and parities could be assigned to these states on the basis of log *ft* values because the level scheme is incomplete. The ground-state spin-parity has been tentatively assigned as $5/2^-$ from β decay of this ground state to the excited levels of ^{87}Kr [2]. This assignment contradicts the previous $3/2^-$ assignment based on systematics, and raises the possibility of a deformed shape in this nucleus [3]. This provides a particular motivation to study the deformation in ^{87}Br. No excited states are known in ^{89}Br at all, and the spin-parity of the ground state is only tentatively assigned from systematics. Thus, study of medium-spin states in these nuclei looks promising in getting valuable information on this nuclear region. Therefore, we studied the excited states of the ^{87,89}Br nuclei from the cold-neutron induced fission of ^{235}U.

2. Experimental details

The experiment was performed using the PF1B cold-neutron beam of the Institut Laue-Langevin (ILL) in Grenoble [4]. The neutron beam was shaped into a 12 mm diameter pencil beam with a thermal equivalence flux of $10^8/(\mathrm{s\ cm}^2)$. Two 0.6 mg/cm^2 thick ^{235}U targets enriched to 99.7% were used. They were sandwiched between Zr and Be backings of 15 μm and 25 μm thickness, respectively, for rapid stopping of fission fragments. This enabled an almost Doppler-shift free measurement of the emitted γ rays, which were detected by the EXILL detector array [5]. It consisted of eight Compton-suppressed EXOGAM Clover detectors [6], six Compton-suppressed GASP detectors [7] and two Clover detectors of the Lohengrin spectrometer [8]. The distance between the faces of detectors and target was about 15 cm. The data were collected using a digital acquisition system with a 100 MHz clock in a triggerless mode. Altogether 15 terabytes of data were collected over the 21 day period of the experiment. During the offline analysis triggerless events, each consisting of an energy signal and the time of its registration, were arranged into coincidence events within various time windows (from 200 to 2400 ns) and sorted into 2D and 3D histograms.

3. Excited states of ^{87}Br and ^{89}Br

Prior to the present study a few excited states were known from neutron-induced fission of ^{235}U [1]. In order to assign new transitions to ^{87}Br and to build the medium-spin level scheme of this nucleus, we analysed the 3D γ-ray histograms created by applying a 200 ns coincidence time window and built in the Radware format [9]. The strategy we used for assigning new transitions to ^{87}Br, was based on the fact that the transitions from the complementary fission fragments are in prompt coincidence with each other. In the cold-neutron-induced fission of ^{235}U in most of the cases two or three neutrons and no protons are emitted from the primary fission fragments, leading to the secondary fission fragments of ^{147}La and ^{146}La, respectively, as complementary fission fragments of ^{87}Br. Fortunately, the level schemes of these nuclei are rather well known from previous fission experiments [10, 11].

As a first step, we have set double gates on several strong transition pairs of ^{147}La and ^{146}La. Fig. 1 shows example coincidence spectra produced this way. These spectra are expected to contain three types of γ rays:

(a) γ rays belonging to the respective complementary fission fragment nucleus, (b) γ rays belonging to the ^{87}Br nucleus, and (c) γ rays belonging to other nuclei and being accidentally in coincidence with the used gate pair. The first type of γ rays are denoted by stars in the figure. The second type of γ rays are expected to appear in all the gated spectra, while those of the third type are expected to appear only in one of them. In Fig. 1 four transitions are seen to appear in all the coincidence spectra, the 619-, 663-, 762- and 876-keV transitions. Thus, these transitions are expected to belong to the nucleus ^{87}Br. Based on the coincidence relationships between these transitions and also between the transitions which have been found to be in prompt coincidence with them, we could build the preliminary level scheme, plotted in Fig. 2. This level structure does not correspond to any known level structures belonging to the populated fission fragments. Furthermore, when setting double gates on the strong transitions of this structure, we systematically see the strong transitions of the ^{146}La and ^{147}La level schemes in the spectra. Based on these facts, we preliminarily assign the obtained level scheme to ^{87}Br. In this partial level scheme the levels correspond to states populated from prompt fission. Among the transitions belonging to the plotted level scheme only the 573 keV one was previously known, all the others are newly assigned to ^{87}Br. The newly observed level scheme forms three band-like structures labelled as Band 1, Band 2 and Band 3 in Fig. 2. The bandheads of Band 1 and Band 3 are very close to each other in energy. The difference between them is only 6.1 keV. The bandhead of Band 3 is the ground state of ^{87}Br. The previously known transitions, including the 573 keV one, feed the band head of Band 1, which thus can be considered as the previously assigned ground state. Two of the bands, Band 1 and Band 3, look like strongly coupled bands, while the third one, Band 2, looks like a weakly coupled or decoupled band. This is in a good agreement with the expectations in the case of a deformed shape. In that case two strongly coupled bands, based on the $\pi f_{5/2}$ and $\pi p_{3/2}$ configurations, and one weakly coupled band, based on the $\pi g_{9/2}$ configuration, are expected to appear at low excitation energy. Information to assist in spin-parity assignment of the newly assigned levels could be inferred from the measured angular correlation relations of the subsequent transitions in the γ-decay cascades. The eight EXOGAM

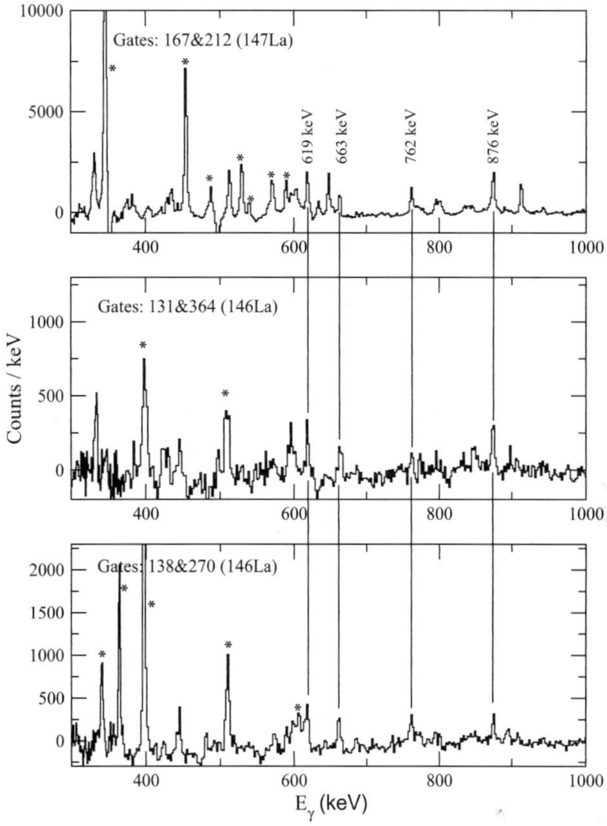

Figure 1. Coincidence spectra obtained by double gating on strong transitions of ^{146}La and ^{147}La, the complementary fission fragments of ^{87}Br. Stars denote the transitions in the respective complementary fission fragment.

clover detectors were mounted in the EXILL spectrometer in one plane perpendicular to the beam direction in an octagonal geometry.

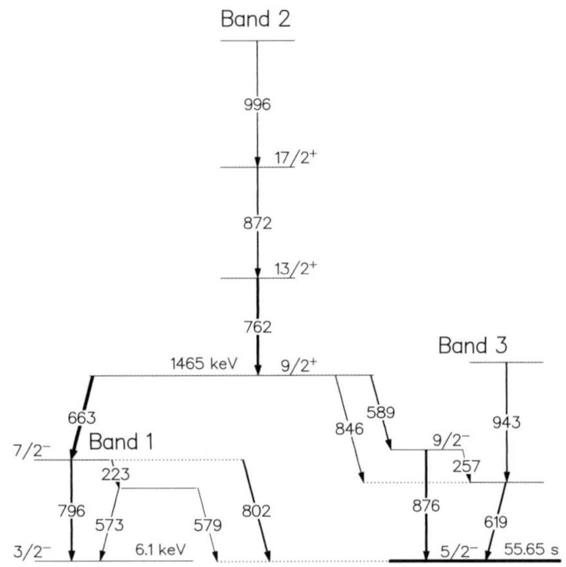

Figure 2. Preliminary partial level scheme of ^{87}Br observed in the present experiment.

The 28 detector pairs provided three different relative angles: 0°, 45° and 90° taking into account the symmetries. More details on the technique are reported in Ref. [5]. In order to derive the angular correlation relations, 3D γ-γ-angle matrices were created and analysed using the program developed in Ref. [12]. The observed angular correlations of the two lowest transitions in Band 2, i.e. the 762 keV and 872 keV transitions, are consistent with stretched quadrupole character for both of them. This indicates that Band 2 is an E2 band, probably corresponding to a weakly coupled or decoupled configuration. This assumption is confirmed by the fact that no cross-over transitions have been observed in this band. Both Band 1 and Band 3 look like ΔI=1 bands with quadrupole cross-over transitions. This assumption is also confirmed by the angular correlations observed for the 796 keV - 663 keV cascade in Band 1 and for the 876 keV - 589 keV cascade in Band 3. The observed angular correlations in the first cascade are consistent with the stretched quadrupole character of the 796 keV transition and stretched dipole character of the 663 keV transition. In Band 3, the observed angular correlations are consistent with the stretched quadrupole character for the 876 keV transition and either ΔI=0 dipole or stretched quadrupole character of the 589 keV transition. If the 589-keV transition were a stretched quadrupole then the ground-state spin would be $4\hbar$ less than that of the bandhead state of Band 2. However, in this case the 802 keV transition, feeding the ground state, would have a ΔI=3 character. This is unlikely taking into account that the 802-keV state is decaying also by quadrupole and dipole transitions. Thus, we assign a ΔI=0 dipole character to the 589-keV transition.

Similarly to ^{87}Br we could also build a preliminary level scheme of ^{89}Br, and the two level schemes were found to be very similar to each other. The levels in ^{89}Br are also grouped into three bands, which have similar characters to the ones in ^{87}Br.

4. Configurations and deformations of the observed bands

Using the derived multipole characters of the transitions together with information on the level schemes of neighbouring odd-proton nuclei and comparison with the results of shell-model calculations, we could tentatively assign spin-parity values to the observed levels. A decoupled band based on the $\pi g_{9/2}$ configuration is expected to appear at relatively low energy for both nuclei with a band-head spin-parity of $9/2^+$ from the deformed Nilsson scheme and also from the comparison with the level scheme of neighbouring odd-proton nuclei. In the case of ^{87}Br we can compare the level scheme with that of ^{89}Rb. In ^{89}Rb this bandhead is at about 1200 keV excitation energy. In ^{87}Br the bandhead energy of Band 2 is close to this. Based on these arguments we tentatively assign $\pi g_{9/2}$ configuration to Band 2 and $9/2^+$ spin-parity to its bandhead level. Assuming that spin values increase with increasing energy, and taking into account the observed multipole characters in Band 1 and Band 3, we can assign spin values of 3/2 to the bandhead of Band 1 and 5/2 to the bandhead of Band 3. Thus the ground-state spin of ^{87}Br is 5/2 in agreement with the value derived by Porquet et al. [2]. Considering the expected configurations and the determined bandhead spins, we can tentatively assign $\pi p_{3/2}$ and $\pi f_{5/2}$ configurations to Band 1 and Band 3, respectively. Therefore, the bandhead spin-parities are $3/2^-$ and $5/2^-$.

Similar tentative configurations and bandhead spin-parities could be derived for the bands in ^{89}Br. Thus, in both nuclei the ground state is the $5/2^-$ bandhead of the $\pi f_{5/2}$ band, while the first excited state is the $3/2^-$ bandhead of the $\pi p_{3/2}$ band. The two bandheads are close to each other, and the difference between their energies seems to increase with increasing neutron number.

In order to verify the suggested spin-parities and configurations, we compared them with preliminary results of the contemporary shell model calculations using a large valence space including the $1f_{5/2}$, $2p_{3/2}$, $2p_{1/2}$, $1g_{9/2}$ orbitals for protons and the $2d_{5/2}$, $3s_{1/2}$, $1g_{7/2}$, $2d_{3/2}$, $1h_{11/2}$ orbitals for neutrons, outside the ^{78}Ni core. The model and the parameters of the calculations are the same as used in Ref. [13]. The compared experimental and calculated level energies and spin-parities, as well as single-particle configurations are plotted in Fig. 3 for both nuclei. It can be seen in the figure that the tentative experimental spin-parities and configurations are rather well reproduced by the calculations. This good agreement supports the proposed assignments. The level energies are also relatively well reproduced for the low-lying states. There are, however, larger differences between the experimental and calculated level energies at the highest-spin levels of the $\pi g_{9/2}$ bands. This might be due to a larger experimental collectivity than predicted by the calculations. If the $\pi g_{9/2}$ bands are decoupled

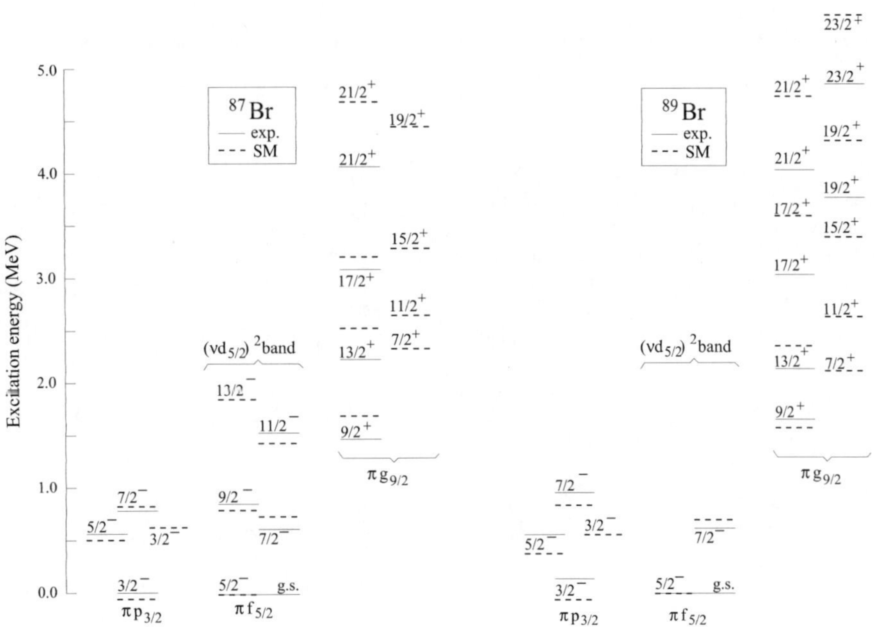

Figure 3. Comparison of the observed medium-spin level scheme of ^{87}Br and ^{89}Br with the results of large-scale shell model calculations.

bands, the energy ratios of the third and second levels relative to the bandheads play the same role as the E_{4+}/E_{2+} ratios in the even-even nuclei, i.e. they provide information about the deformation of the nucleus. These values are 2.1 and 2.5 for ^{87}Br and ^{89}Br, respectively, which are close to the E_{4+}/E_{2+} ratios of the neighbouring even-even nuclei, and indicate that ^{87}Br is close to the vibrational limit, while ^{89}Br is more rotational.

5. Conclusion

Medium-spin excited states of the neutron-rich ^{87}Br and ^{89}Br nuclei have been studied by means of in-beam γ spectroscopy of fission fragments of the ^{235}U, using the EXILL Ge detector array. The fission has been induced by the cold-neutron beam of the PF1B facility of the Institut Laue-Langevin, Grenoble. Preliminary level schemes were built for both nuclei. The observed excited states in both nuclei form three band-like structures. Angular correlation information from the experiment combined with systematics of the neighbouring odd-proton nuclei enabled tentative spin-parity assignments to the new levels and configuration assignments to the bands. They can be understood as bands built on the $\pi f_{5/2}$, $\pi p_{3/2}$ and $\pi g_{9/2}$ configurations. These assignments have been confirmed by the preliminary results of the contemporary shell model calculations using a large valence space. Both nuclei have $5/2^-$ ground state spin-parity contrary to the odd-mass Br isotopes containing fewer neutrons, which have $3/2^-$ ground state spin-parity. Properties of the $\pi g_{9/2}$ decoupled bands indicate that the deformations of the ^{87}Br and ^{89}Br fit to the systematics of nuclei in the region. The nucleus ^{87}Br is close to the vibrational limit, while ^{89}Br is more rotational.

Acknowledgements

This work was supported by the Hungarian Scientific Research Fund, OTKA (Contracts No. K100835 and No. K106035), by the Polish National Science Centre under Contract No. DEC-2013/09/B/ST2/03485 and by the Bulgarian National Science Fund under Contract number DFNI-E02/6. The authors thank the technical services of the ILL, LPSC, and GANIL for supporting the EXILL campaign. The EXOGAM Collaboration and the INFN Legnaro are acknowledged for the loan of Ge detectors.

References

[1] Zendel M, Trautmann N and Hermann G 1980 *J.Inorg.Nucl.Chem.* **42** 1387
[2] Porquet M-G et al. 2006 *Eur. Phys. J.* A **28** 153
[3] Astier A et al. 2013 *Phys. Rev.* C **88** 024321
[4] Abele H et al. 2006 Nucl. Instrum. Methods A **562** 407
[5] Blanc A et al. 2013 *EPJ Web Conf.* **62** 01001
[6] Simpson J et al. 2000 *Acta Phys. Hung. New Ser.: Heavy Ion Phys.* **11** 159
[7] Bazzacco D et al. 1999 *Proceedings of the 5th International Seminar on Nuclear Physics, edited by A. Covello (World Scientific, Singapore), pp. 417430.*
[8] Simpson G S et al. 2007 *Phys. Rev.* C **76** 041303(R)
[9] Radford D C 1995 Nucl. Instrum. Methods A **361** 297
[10] Zhu S J et al. 1999 *Phys. Rev.* C **59** 1316
[11] Hwang J K et al. 1998 *Phys. Rev.* C **58** 3252
[12] Urban W et al. 2013 *J. Instrum.* **8** P03014
[13] Czerwiński et al. 2015 *Phys. Rev.* C **92** 014328

How to Use Benchmark and Cross-section Studies to Improve Data Libraries and Models

V Wagner[1], M Suchopár[1,3], J Vrzalová[1,2], P Chudoba[1], O Svoboda[1], P Tichý[1,3], A Krása[1], M Majerle[1], A Kugler[1], J Adam[2], A Baldin[2], W Furman[2], M Kadykov[2], A Solnyshkin[2], S Tsoupko-Sitnikov[2], S Tyutyunikov[2], N Vladimirovna[2], L Závorka[2,3]
[1] Nuclear Physics Institute of the ASCR, 250 68 Řež near Prague, Czech Republic
[2] Joint Institute for Nuclear Research, Joliot-Curie 6, 141 980 Dubna, Russia
[3] Faculty of Nuclear Sciences and Physical Engineering, Czech Technical University in Prague, Břehová 7, 115 19 Prague, Czech Republic

E-mail: wagner@ujf.cas.cz

Abstract. Improvements of the Monte Carlo transport codes and cross-section libraries are very important steps towards usage of the accelerator-driven transmutation systems. We have conducted a lot of benchmark experiments with different set-ups consisting of lead, natural uranium and moderator irradiated by relativistic protons and deuterons within framework of the collaboration "Energy and Transmutation of Radioactive Waste". Unfortunately, the knowledge of the total or partial cross-sections of important reactions is insufficient. Due to this reason we have started extensive studies of different reaction cross-sections. We measure cross-sections of important neutron reactions by means of the quasi-monoenergetic neutron sources based on the cyclotrons at Nuclear Physics Institute in Řež and at The Svedberg Laboratory in Uppsala. Measurements of partial cross-sections of relativistic deuteron reactions were the second direction of our studies. The new results obtained during last years will be shown. Possible use of these data for improvement of libraries, models and benchmark studies will be discussed.

1. Introduction

Highly accurate reaction cross-section data are needed for planning of future nuclear technologies as fast breeder reactors, accelerator-driven transmutation systems or fusion reactors. Different models and codes were developed to describe and design present and future nuclear systems. Accurate measurements with simple and more complex set-ups which can simulate future nuclear technologies are needed for benchmarks and improvement of the models and codes. For this purpose the beams of neutrons and other particles in rather broad range of energies provided by different neutron sources and accelerators are used.

[1] To whom any correspondence should be addressed.

2. Relativistic deuteron reactions on copper

The experimental cross-sections of relativistic deuteron reactions on copper are almost completely missing. Situation is very different for relativistic protons. An extensive set of experimental partial cross-section data can be found in the EXFOR database for different radionuclide production in this proton case. Therefore, we have started a set of studies of relativistic deuteron reactions on copper. We used the deuteron beam produced by JINR Dubna Nuclotron accelerator. The activation method and the gamma-ray spectrometry were used for determination of partial production cross-sections. Beam integral was determined by means of ^{24}Na production in the aluminium monitor [1,2,3].

Overall 38 different radioisotopes were identified by their respective gamma lines with half-life span from less than 1 hour to hundreds of days and their independent or cumulative production cross-sections were measured using activation method. Most of our measured cross-sections are cumulative (namely ^{22}Na, ^{24}Na, ^{28}Mg, ^{38}S, ^{39}Cl, ^{43}K, ^{47}Ca, ^{43}Sc, ^{48}V, ^{48}Cr, ^{49}Cr, ^{51}Cr, ^{52}Mn, ^{56}Mn, ^{52}Fe, ^{59}Fe, ^{55}Co, ^{56}Co, ^{57}Co, ^{56}Ni, ^{57}Ni, and ^{61}Cu). We have determined independent partial cross-sections for ^{7}Be (appropriate isobars are not produced), ^{42}K (half-life of ^{42}Ar is 32.9 years), ^{44}Sc and ^{44m}Sc (half-life of ^{44}Ti is 60 years), ^{46}Sc (^{46}Ca is stable), ^{47}Sc (half-life of ^{47}Ca is 4.5 days and we used early measurements), ^{48}Sc (^{48}Ca is stable), ^{54}Mn (^{54}Cr is stable), ^{58}Co (stable isobars around) and ^{60}Co (stable isobars around). The independent cross-sections have been determined also in the case of ^{65}Ni, ^{64}Cu, ^{62}Zn, and ^{65}Zn, where other appropriate isobars are not produced.

The excitation functions were obtained. Representative set of the new experimental data for deuteron reactions are presented in Fig. 1 and 2 together with proton reaction data from the EXFOR database. The change of excitation function shape with increasing nucleon number of the produced radionuclide is nicely visible. The shape of deuteron reaction excitation functions is similar to proton ones but absolute values of the deuteron reaction cross-sections are by about 30-40 % larger than proton ones. Exception is the case of ^{7}Be production where cross-sections of deuteron reactions are about two times higher. The only one measured cross-section value of deuteron production of ^{24}Na on copper collected in the EXFOR database [4] agrees nicely with our new data (see Fig. 1a).

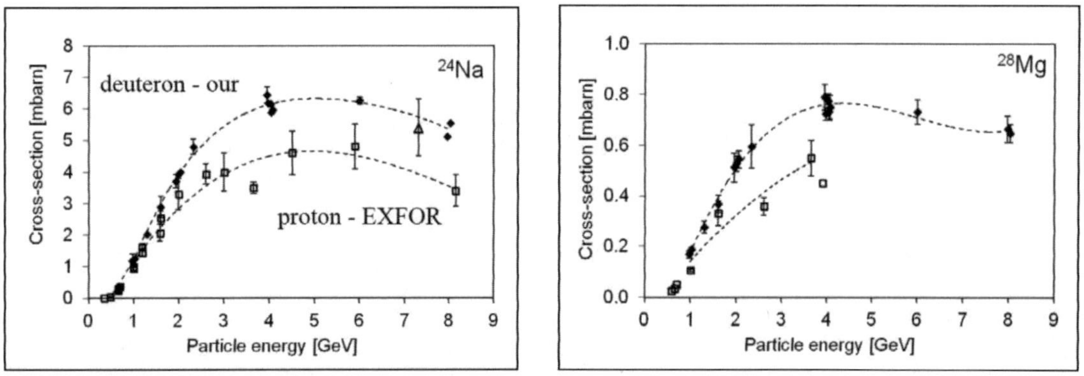

Figure 1. (a) Excitation function of ^{nat}Cu(d,x)^{24}Na reaction (b) Excitation function of ^{nat}Cu(d,x)^{28}Mg reaction. Our experimental data for deuterons are compared with EXFOR data for protons. Examples of light radionuclide production. Measured value of the ^{24}Na production cross-section existed in EXFOR database (signed by triangle) agrees with our data.

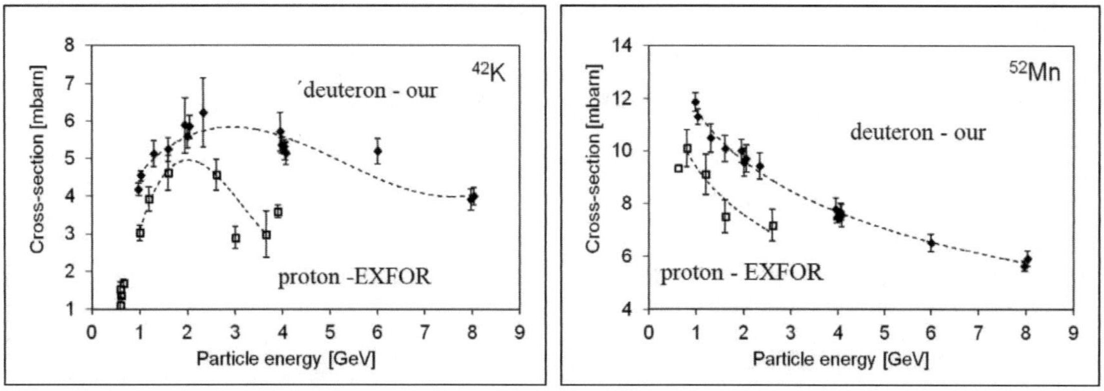

Figure 2. (a) Excitation function of $^{nat}Cu(d,x)^{42}K$ reaction, example of medium heavy radionuclide production (b) Excitation function of $^{nat}Cu(d,x)^{52}Mn$ reaction, example of target like radionuclide production. Our experimental data for deuterons are compared with EXFOR data for protons.

The obtained excitation functions were compared with simulation done by the Monte Carlo code MCNPX version 2.7.0 [5] with two intra-nuclear cascade-evaporation models INCL-ABLA and LAQGSM included in the code. The experimental data are better fitted with the LAQGSM model than the INCL-ABLA model in most cases. The simulated excitation functions are generally similar in shape but can differ significantly from the experimental data in absolute value in certain cases. Selected examples of our measured cross-sections for deuterons together with excitation functions for deuteron reactions simulated in the MCNPX code are displayed in Fig. 3 and 4. We see that the accuracy of the experimental data make possible to distinguish which model describes excitation functions better and the obtained experimental values shown possibility for improvements of models.

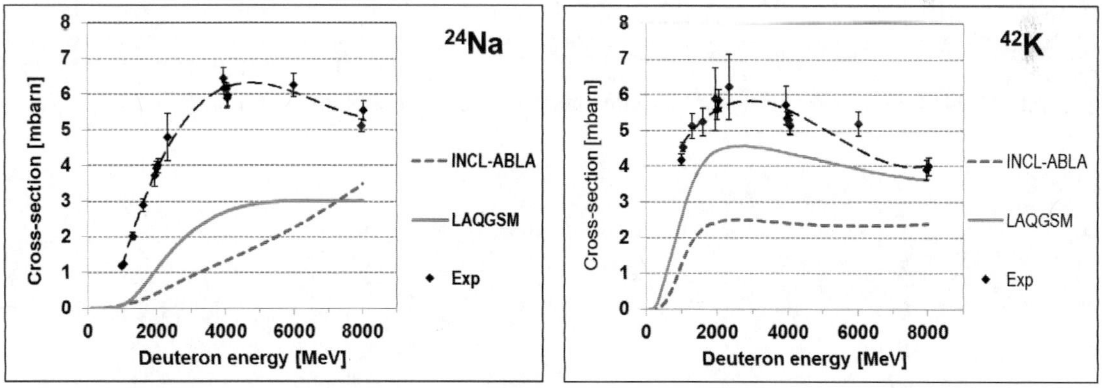

Figure 3. (a) Excitation function of $^{nat}Cu(d,x)^{24}Na$ reaction (b) Excitation of $^{nat}Cu(d,x)^{42}K$ reaction. The experimental data are compared with the model predictions (INCL-ABLA and LAQGSM are used)

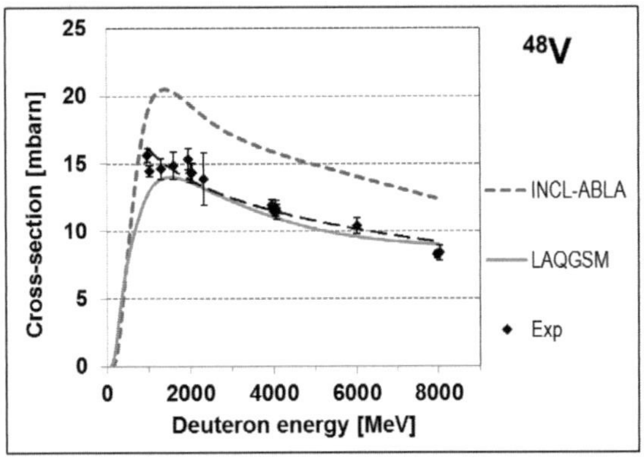

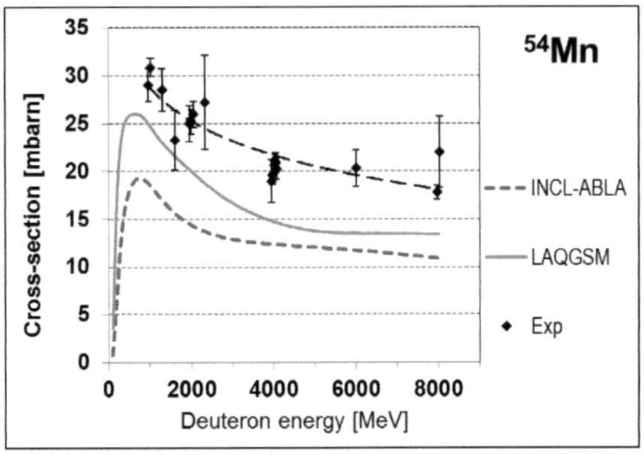

Figure 4. (a) Excitation function of $^{nat}Cu(d,x)^{48}V$ reaction (b) Excitation of $^{nat}Cu(d,x)^{54}Mn$ reaction. The experimental data are compared with the model predictions (INCL-ABLA and LAQGSM are used).

3. Excitation functions of neutron reactions

Activation detectors are a very useful tool for measurement of neutron flux. We use intensively neutron threshold detectors during our studies of accelerator driven transmutation systems in JINR Dubna [6,7]. Experimental cross-section data of used reactions are very scarce for energies above 30 MeV. The neutrons with energies above 100 MeV are produced during our experiments using the relativistic proton and deuteron beams of JINR Dubna accelerators. The threshold reactions enable to measure flux of such neutrons (see Fig 5). The decay products of radionuclides produced by reactions with very high threshold are necessary to measure in some cases. The decay chains $^{197}Au(n,9n)^{189}Au$ (0.5h) → ^{189}Pt (10.9h) → ^{189}Ir (13.2d) → ^{189}Os and $^{197}Au(n,10n)^{188}Au$ (0.5h) → ^{188}Pt (10.2d) → ^{188}Ir (41.5h) → ^{188}Os, or $^{209}Bi(n,10n)^{200}Bi$ (0.6h) → ^{200}Pb (21.5h) → ^{200}Tl (26.1h) → ^{200}Hg, $^{209}Bi(n,11n)^{199}Bi$ (0.45h) → ^{199}Pb (1.5h) → ^{199}Tl (7.4h) → ^{199}Hg and $^{209}Bi(n,12n)^{198}Bi$ (0.17h) → ^{198}Pb (2.4h) → ^{198}Tl (5.3h) → ^{198}Hg are examples of such possibilities. The production of such radionuclides by direct neutron reactions for so heavy nuclei as gold or bismuth is negligible. One example of measurement by means of bismuth activation detectors is shown on Fig. 6.

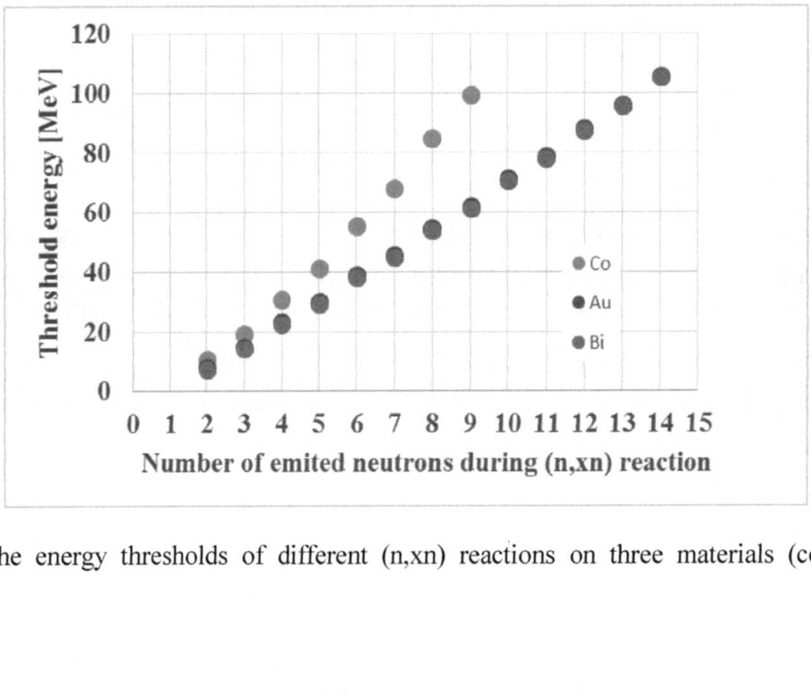

Figure 5. The energy thresholds of different (n,xn) reactions on three materials (cobalt, gold a bismuth).

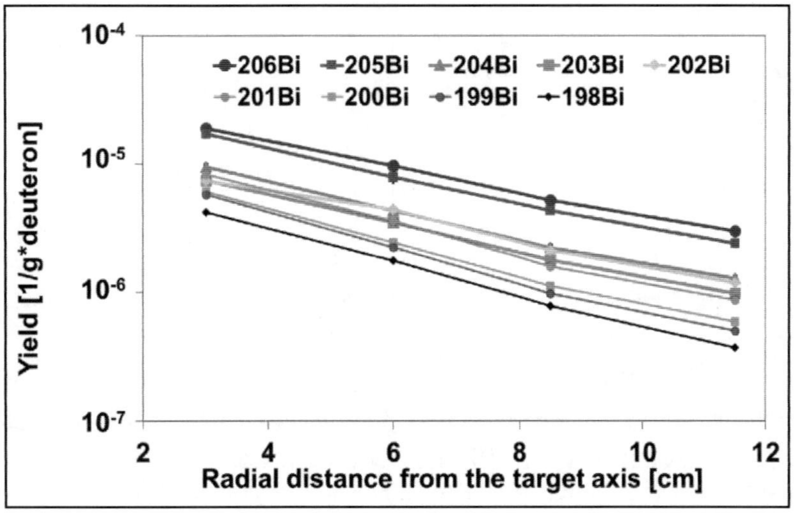

Figure 6. Example of spatial distribution of different radionuclides production on bismuth activation samples. Threshold reactions up to (n,12n) and threshold above 80 MeV were measurable during irradiation of the E+T set-up by 4 GeV deuterons. Distance of activation detectors from the front of the target was 11.4 cm

This was reason why we started extensive studies of (n,xn) reactions on materials used as activation detectors [8,9]. We used two different quasi mono-energetic neutron sources based on the reaction on ^{7}Li target. We performed experiments by means of the neutron source at Nuclear Physics Institute of ASCR in Řež and the neutron source at The Svedberg Laboratory of Uppsala University. These two neutron sources made possible to cover very broad range of neutron energies from 14 MeV up to 100 MeV. We used activation method of cross-section determination and detailed description of experiments, used techniques and discussion of all uncertainty sources are in [8,9]. The example of results for one high threshold reaction (n,7n) is on Fig 7. Our results enable to test different versions of TALYS code [10]. The Fig 8 shows that changes between the different versions of TALYS code are very small for low threshold reaction but there are big differences between the different versions for high threshold reaction.

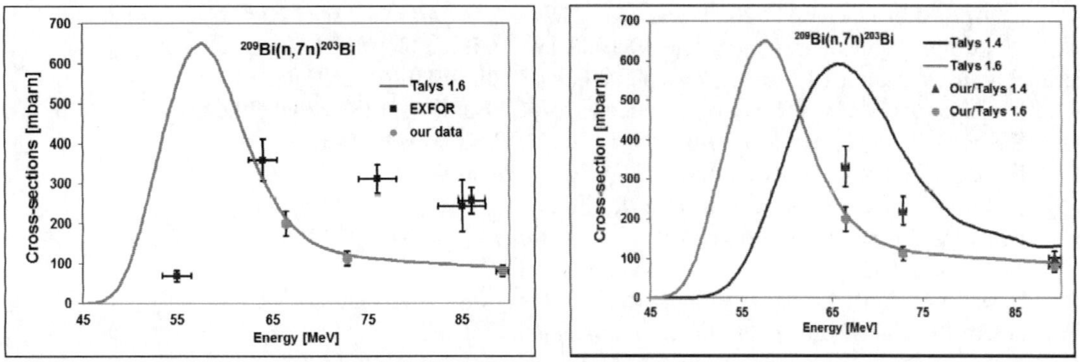

Figure 7. (a) Our data obtained during Uppsala experiment at 2010 are compared with EXFOR data and TALYS 1.6 results. (b) Comparison of predictions obtained by means of two different version of TALYS code and also results of subtractions of background estimated by these two different version of TALYS code.

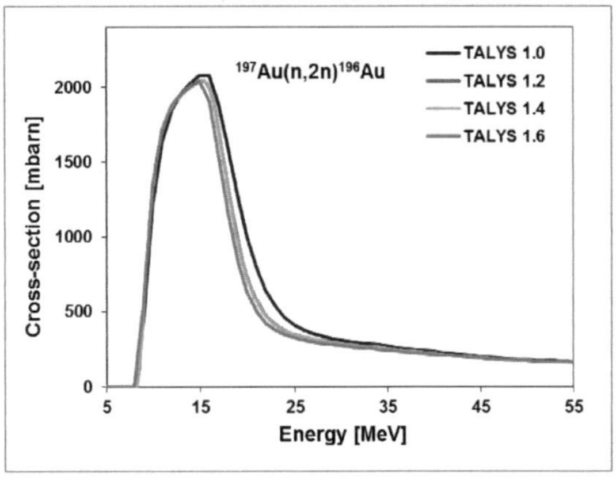

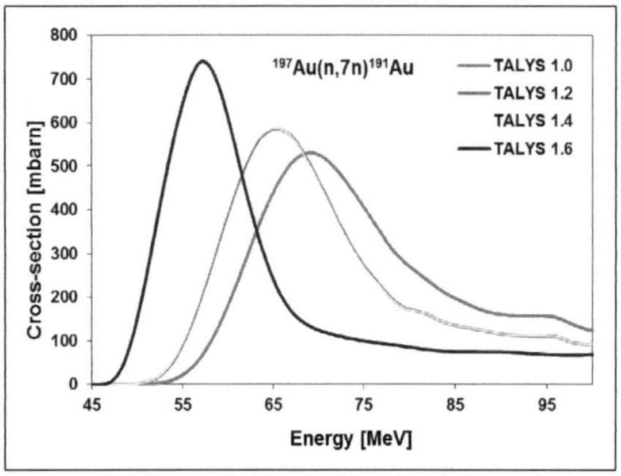

Figure 8. (a) The time evolution of TALYS code, for the low threshold reaction (n,2n) are only very small changes (b) Different situation is for the high threshold reaction (n,7n).

4. Conclusion

Accuracy of experimental cross-section data obtained in our experiments enables to test and improve models and codes (Monte Carlo and deterministic). They enable beam integral determination besides aluminium foil also by more than three dozen monitoring reactions available when copper foil is used during our accelerator-driven systems studies and also more accurate determination of high energy neutron fields inside studied set-ups by the activation detectors.

Acknowledgements

Authors are grateful to the operation crew of the JINR Nuclotron and the staff of fast neutron generators and the cyclotron at the NPI Řež for irradiation and good beam parameters. This research was financially supported by the ERINDA and CHANDA program, the Czech JINR grants and by the F4E program of the Nuclear Reaction Department of the Nuclear Physics Institute, F4E-2008-GRT-014. Some measurements were carried out at the CANAM infrastructure of the NPI ASCR Řež supported through MSMT project No. LM2011019

References

[1] Wagner V et al 2013 XX International School on Nuclear Physics, Neutron Physics and Aplications (VARNA 2013) *Journal of Physics: Conference series 533* http://iopscience.iop.org/article/10.1088/1742-6596/533/1/012052/pdf
[2] Wagner V et al 2014 *Proceedings of Science* (Baldin ISHEPP XXII **057**) http://pos.sissa.it/archive/conferences/225/057/Baldin%20ISHEPP%20XXII_057.pdf
[3] Suchopár M et al 2015 *Nucl. Instr. and Meth. in Phys. Res.* **B344** 63
[4] Experimental Nuclear Reaction Data EXFOR database https://www-nds.iaea.org/exfor/exfor.htm
[5] Pelowitz D B et al 2011 *MCNPX 2.7.0 Extensions* LA-UR-11-02295 https://mcnpx.lanl.gov/opendocs/versions/v270/v270.pdf
[6] Krása A et al 2010 *Nucl. Instr. and Meth. in Phys. Res.* **A615** (2010) 70
[7] Adam J et al 2011 *European Physical Journal* **A47** 85
[8] Vrzalová J et al 2013 *Nucl. Instr. and Meth. in Phys. Res.* **A726** 84-90
[9] Chudoba P et al *2014 Physics Procedia* **59** 114 http://www.sciencedirect.com/science/article/pii/S1875389214004933
[10] Koning A 2013 *TALYS-1.6 manual* http://www.talys.eu/fileadmin/talys/user/docs/talys1.6.pdf

Journal of Physics: Conference Series **724** (2016) 012053 doi:10.1088/1742-6596/724/1/012053

Investigation of chirality in the case of 102Rh

M. S. Yavahchova[1], D. Tonev[1], N. Goutev[1], G. de Angelis[2], P. Petkov[1], R.K. Bhowmik[3], R.P. Singh[3], S. Muralithar[3], N. Madhavan[3], R. Kumar[3], M. Kumar Raju[4], J. Kaur[5], G. Mahanto[3], A. Singh[5], N. Kaur[5], R. Garg[6], A. Sukla[7], Ts. K. Marinov[1], R. R. Mateeva[8], S. Brant[9]

[1] Bulgarian Academy of Sciences, Institute for Nuclear Research and Nuclear Energy, 72 Tzarigradsko shaussee Blvd., BG-1784 Sofia, Bulgaria
[2] INFN, Laboratori Nazionali di Legnaro, Legnaro, Italy
[3] Inter-University Accelerator Center, New Delhi, India
[4] Nuclear Physics Department, Andhra University, Visakhapatnam, India
[5] Department of Physics, Punjab University, Chandigarh, India
[6] Department of Physics and Astrophysics, Delhi University, New Delhi, India
[7] Department of Physics, Banaras Hindu University, Varanasi, India
[8] Faculty of Natural Sciences, Shumen University, Shumen, Bulgaria
[9] Department of Physics, Faculty of Science, Zagreb University, Zagreb, Croatia

E-mail: m.yavahchova@gmail.com

Abstract. Excited states in ^{102}Rh were populated in the fusion-evaporation reaction ^{94}Zr(^{11}B, 3n)^{102}Rh at a beam energy of 36 MeV using the INGA spectrometer at IUAC, New Delhi. The angular correlations and the electromagnetic character of some of the gamma-ray transitions observed were investigated in details. A new chiral candidate sister band was found. Lifetimes of exited states in ^{102}Rh were measured using the Doppler-shift attenuation technique and the derived reduced transition probabilities are compared to the predictions of the Two Quasiparticles Plus Triaxial Rotor model.

1. Introduction

Chirality is a phenomenon which is often found in nature. Examples for the existence of chirality are present in chemistry, biology, high energy physics etc. An interesting question arose last decade in nuclear physics, does chirality exists in this field? A spontaneous breaking of the chiral symmetry can take place for configurations where the angular momenta of the valence protons, valence neutrons, and the core are mutually perpendicular [1]. The projections of the angular momentum vector on the three principal axes can form either a left- or a right- handed system and therefore the system expresses chirality. Since the chiral symmetry is dichotomic, its spontaneous breaking by the axial angular momentum vector leads to a pair of degenerate I = 1 rotational bands, called chiral doublet bands. Pairs of bands, presumably due to the breaking of the chiral symmetry in triaxial nuclei, have been recently found in the mass regions $A \sim 130$ [2, 3, 4, 5, 6, 7], $A \sim 105$ [8, 9, 10, 11, 12, 13] and $A \sim 195$ [14, 15, 16], and $A \sim 80$ [17]. There is also a significant interest from theoretical point of view to investigate the chiral phenomenon [18, 19, 20, 21, 22, 23]. However, only in few cases (^{126}Cs [24], ^{128}Cs [6] and ^{198}Tl [15]) the systematic properties [25] of the chiral bands, which originate from the underlying symmetry, were confirmed including the transition from chiral vibrations to static chirality in ^{135}Nd [7]. In many cases the energy degeneracy of the chiral candidate bands was almost observed but the transition probabilities are different. This is clearly seen in the case of ^{134}Pr [25, 26, 27]. According to the work [21] and [28], the nucleus of ^{102}Rh is candidate to express multiple chirality. The main goal of the present work was to check for the existence of chirality in the mass region $A \sim 100$. In previous works [8, 11, 12], an island of chiral candidates has been proposed around ^{104}Rh. Next to the identification of the twin bands, some lifetimes have been reported for the yrast band of ^{104}Rh [12].

2. Experiment

Excited states in ^{102}Rh were populated using the reaction ^{94}Zr(^{11}B,3n)^{102}Rh at a beam energy of 36 MeV. The beam was delivered by the 15-UD Pelletron accelerator at the Inter University Accelerator Center (IUAC) in New Delhi. The target consisted of 0.9 mg/cm^2 ^{94}Zr, enriched to 96.5%, evaporated onto a 8 mg/cm^2 gold backing. The recoils were leaving the target with a mean velocity v of about 0.9% of the velocity of light, c. The de-exciting gamma-rays were registered by the Indian National Gamma Array (INGA), whose 15 Clover detectors are accommodated in a 4π geometry [29]. For the purposes of the Doppler-Shift Attenuation Method (DSAM) analysis the detectors of INGA which lie at approximately the same polar angle with respect to the beam axis were grouped in five rings. Doppler broadened line shapes were observed for transitions depopulating higher spin levels. The rings where appreciable Doppler-shifts are observed are these at angles of 32, 57, 123 and 148 degrees. Gain matching and efficiency calibration of the Ge detectors were performed using ^{152}Eu and ^{133}Ba radioactive sources before sorting the data in matrices and cubes.

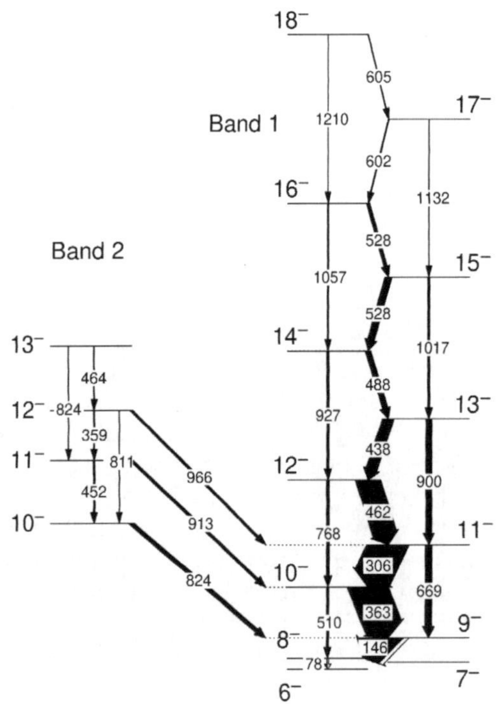

Figure 1. Partial level scheme of ^{102}Rh. Two negative-parity bands, candidates for chiral partner bands, are indicated as band 1 and band 2.

Table 1. Derived lifetimes in Band 1 and Band 2 of ^{102}Rh.

Band 1		
E_{ex1} [keV]	State [I^{π}]	τ [fs]
1576	$(11)_1^-$	458(44)
2038	$(12)_1^-$	227(58)
2477	$(13)_1^-$	320(71)
2965	$(14)_1^-$	170(25)
3494	$(15)_1^-$	117(39)
4022	$(16)_1^-$	68(14)
Band 2		
E_{ex2} [keV]	State [I^{π}]	τ [fs]
1731	$(10)_2^-$	163(52)
2183	$(11)_2^-$	156(60)
2542	$(12)_2^-$	

3. Data analysis and results

The detectors of INGA were grouped into rings positioned with respect to the beam axis. This is needed for DSAM analysis. The rings with appreciable Doppler-shifts are these at angles of 32, 57, 123 and 148 degrees. For the investigation of the level scheme and electromagnetic properties of the transitions in ^{102}Rh we have performed four different types of data analyses. On Fig. 1 is shown the level scheme of ^{102}Rh. The ordering of the transitions in the level scheme was determined according to γ - ray relative intensities, γ-γ coincidence relationships, and γ - ray energy sums. The multipolarity and the character of the transitions were deduced from the investigation of linear polarization and angular correlations measurements. For this purpose the clover detectors from the ring at 90° were used as a Compton polarimeter [30]. The lifetimes of excited states in ^{102}Rh were derived using the Doppler-shift attenuation methods. The analysis was carried out within the framework of the Differential decay method (DDCM)[31] according to the procedure outlined in [32] where details about the Monte-Carlo simulation of the slowing down process, determination of stopping powers and fitting of line shapes can be found.

For each level, lifetimes were derived independently at the four rings with appreciable Doppler shifts. The final values of the lifetimes are presented in Table 1. Thus, we have succeeded to extend the known level scheme [33] by a new $\Delta I = 1$ band with a negative parity and to determine 8 new lifetimes. The reduced transition probabilities $B(\sigma\lambda)$ deduced from lifetime data are presented in [34].

To study the bands based on the $\pi g_{9/2} \otimes \nu h_{11/2}$ configuration, we have performed two quasiparticles + triaxial rotor (TQPTR) calculations in the framework of the model presented in Ref. [36]. The Hamiltonian of this model includes the rotational energy of the core, the quasiparticle energies of the odd proton (π) and neutron (ν), and a residual proton-neutron interaction $V_{\pi\nu}$. The core is treated as a rigid body with a fixed overall quadrupole deformation ϵ and a triaxiality parameter γ. On Fig.2 is shown the comparison between the experimental and calculated B(M1) transition strenghts which leads to the conclusion that this model reproduce roughly the data in Band 1 and is consistent with the transition strenght in Band 2. The theoretical $B(E2)$ reproduce quite well the absolute values and the increasing trend of the experimental data after spin $I = 14\ \hbar$. The behaviour of the $B(E2)$ transition strengths in ^{102}Rh is very similar to what has been observed in ^{134}Pr [26]. The TQPTR calculations reveal that the optimum value of the triaxiality parameter $\gamma = 20°$ differs from the value of 30° characterizing chirality.

The present results come to supercede the preliminary level-scheme published in [35]. The absence of an appreciable staggering of the data in Band 1 indicates that the expectations for

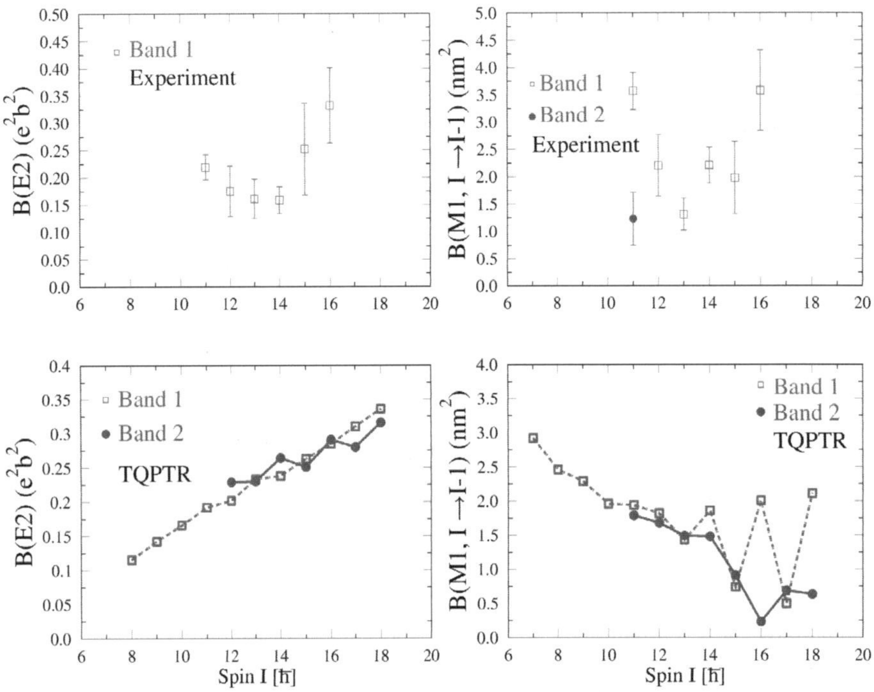

Figure 2. Experimentally derived and theoretically calculated B(E2) and B(M1) transition strengths in ^{102}Rh. In the upper panels are presented experimental B(E2) and B(M1) values for transitions in Band 1 and Band 2. In the second row, the results of TQPTR calculations are displayed.

the observation of a statc chirality in ^{102}Rh are not realized [34].

4. Conclusions

In order to investigate the level-scheme and to determine lifetimes of the excited states in the chiral candidate nucleus of ^{102}Rh we have performed an experiment at the Inter University Accelerator Center in New Delhi using the INGA spectrometer. Lifetimes in ^{102}Rh were measured by means of the Doppler-shift attenuation method. We have succeeded to extend the known level scheme [33] by a new $\Delta I = 1$ band with a negative parity and to determine 8 new lifetimes in the sub-picosecond region. For the interpretation of our results we applied two quasiparticles plus triaxial rotor model. We have succeeded to extend the level-scheme with a new $\Delta I = 1$ band with a negative parity. Our lifetime measurements and the theoretical analysis do not support static chirality in ^{102}Rh. This means that the chirality in ^{102}Rh, if it exists, has mainly a dynamical character.

5. Acknowledgments
This research has been supported by Bulgarian Science Fund under Contract No. DFNI-E 01/2 and by a NUPNET-NEDENSAA project funded by the Bulgariana Ministry of Education and Science. INGA was funded by the Department of Science and Technology, Government of India (No. IR/S2/PF-03/2003-I) and the University Grants Commission.

References
[1] Frauendorf S and Meng J 1997 *Nucl. Phys.* A **617** 131
[2] Starosta K *et al.* 2001 *Phys. Rev. Lett.* **86** 971
[3] Zhu S *et al.* 2003 *Phys. Rev. Lett.* **91** 132501
[4] Rainovski G *et al.* 2003 *Phys. Rev.* C **68** 024318
[5] Hecht A A *et al.* 2003 *Phys. Rev.* C **68** 054310
[6] Grodner E *et al.* 2006 *Phys. Rev. Lett.* **97** 172501
[7] Mukhopadhyay S *et al.* 2007 *Phys. Rev. Lett.* **99** 172501
[8] Vaman C *et al.* 2004 *Phys. Rev. Lett.* **92** 032501
[9] Joshi P *et al.* 2004 *Phys. Lett.* B **595** 135
[10] Timár J *et al.* 2004 *Phys. Lett.* B **598** 178
[11] Timár J *et al.* 2006 *Phys. Rev.* C **73** 011301
[12] Suzuki T *et al.* 2008 *Phys. Rev.* C **78** 031302
[13] Lieder E O *et al.* 2014 *Phys. Rev. Lett.* **112** 202502
[14] Lawrie E A *et al.* 2008 *Phys. Rev.* C **78** 021305
[15] Lawrie E A *et al.* 2010 *Eur. Phys. J.* A **45** 39
[16] Masiteng P L *et al.* 2014 *Eur. Phys. J.* A **50** 119
[17] Wang S Y *et al.* 2011 *Phys. Lett.* B **703** 40
[18] Dimitrov V I, Frauendorf S, and Dönau F 2000 *Phys. Rev. Lett.* **84** 5732
[19] Peng J, Meng J, and Zhang S Q 2003 *Phys. Rev.* C **68** 044324
[20] Olbratowski P *et al.* 2004 *Phys. Rev. Lett.* **93** 052501
[21] Meng J *et al.* 2006 *Phys. Rev.* C **73** 037303
[22] Ganev H G *et al.* 2009 *Phys. Rev.* C **79** 044322
[23] Ganev H G and Brant S 2010 *Phys. Rev.* C **82** 034328
[24] Grodner E *et al.* 2011 *Phys. Lett.* B **703** 46
[25] Brant S *et al.* 2008 *Phys. Rev.* C **78** 034301
[26] Tonev D *et al.* 2006 *Phys. Rev. Lett.* **96** 052501
[27] Tonev D *et al.* 2007 *Phys. Rev.* C **76** 044313
[28] Peng J *et al.* 2008 *Phys. Rev.* C **77** 024309
[29] Muralithar S *et al.* 2010 *Nucl.Instr. Meth.* A **622** 281
[30] N. Goutev *et al.* 2012 *J. Phys. Conf. Ser.* **366** 012021
[31] Böhm G *et al.* 1993 *Nucl.Instr. Meth.* A **329** 248
[32] Petkov P *et al.* 1998 *Nucl. Phys.* A **640** 293
[33] Gizon J *et al.* 1999 *Nucl. Phys.* A **658** 97
[34] Tonev D *et al.* 2014 *Phys. Rev. Lett.* **112** 052501
[35] Pancholi S C 2011 *Exotic Nuclear Excitations* Springer New York
[36] Ragnarsson I and Semmes P 1998 *Hyp. Int.* **43** 425

Using heavy-ion collisions to elucidate the asymmetric equation-of-state

Sherry Yennello & Alan McIntosh

Cyclotron Institute & Department of Chemistry, Texas A&M University, College Station, TX 77845

E-mail: yennello@comp.tamu.edu

Abstract. The nuclear equation-of-state impacts a number of nuclear properties as well as astrophysical processes. The asymmetric term of the equation-of-state, which describes the behavior away from N=Z, has significant uncertainty. Giant resonances and nuclear masses can elucidate the asymmetry energy for cold normal-density nuclei. Heavy-ion collisions can be used to probe nuclear matter at higher temperatures and densities away from saturation density. The temperatures that are attained in these nuclear collisions are predicted to depend on the isospin asymmetry. In this work we present evidence of the asymmetry dependence of the nuclear caloric curve.

1. Introduction

The equation-of-state of any substance describes the relationship between state variables such as temperature, pressure, internal energy, composition and volume. The advent of beams of rare ions in recent years have allowed nuclear scientists to explore the equation-of-state with respect to the composition (neutron-to-proton ratio) of the nuclear material. The nuclear equation-of-state (nEOS) is important for understanding astrophysical reactions and the formation of the elements. The mass-radius relationship of neutron stars and the cooling rate of neutron stars are both dependent on the nEOS[1, 2]. The nucleus has long be modeled as a drop of liquid. As such an interesting quantity would be the specific heat of nuclear material. One particular aspect of the nucleus that needs to be taken into account is that the nuclear material is a two-component liquid, thus the composition - the neutron fraction or neutron-to-proton ratio - is important to know when determining the specific heat. In recent works [3, 4, 5] a significant dependence of the temperature on the asymmetry has been demonstrated. A quantum treatment allows one to extract densities along with the temperatures, thus enabling a map of the liquid-gas co-existance curve[6]. This manuscript will describe the progress we have made on understanding the composition dependence of the relationship between the temperature and excitation energy of nuclear material - commonly referred to as the nuclear caloric curve.

2. Experiment

The experiment was performed at Texas A&M University Cyclotron Institute using beams from the K500 superconducting cyclotron. Reactions of ^{70}Zn + ^{70}Zn, ^{64}Zn + ^{64}Zn and ^{64}Ni + ^{64}Ni at 35 MeV/nucleon were studied with the fragments resulting from these collisions being detected in the NIMROD array[7]. NIMROD is a 4π particle detector housed inside a large neutron detector. This enables detection of light charged particles (LCP) and intermediate mass fragments with isotopic identification up to Z=17 as well as detection of free neutrons. With the isotopic identification of charged particles and the measurement of free neutrons the excited quasiprojectile (Q_p) is reconstructed, so that we know its mass, charge and excitation energy. Using the isotopic information from the reconstruction the composition of the reconstructed Q_p is quantified by the m_s,

$$m_s = (N - Z)/A.$$

For details of the experiment, see refs[8, 9, 10].

3. Nuclear Caloric Curve

The nuclear "caloric curve" is the relationship between the temperature of an excited nuclear system and the internal excitation energy of that system. The temperature of an excited nuclear system can be measured in various ways, using both kinetic and chemical information found in the particles emitted from a reaction. Some theoretical models have predicted that the caloric curve would be higher for more neutron-rich systems while other studies have predicted that it would be lower[11, 12, 13, 14, 15, 16]. Two previous experimental attempts have been made at detecting a composition dependence in the nuclear caloric curve [17, 18], that were unable to definitively settle this issue.

3.1. MQF Thermometer

The first thermometer we used to measure the nuclear caloric curve is the momentum quadrupole fluctuation thermometer (MQF) which is based on the kinetic information which relates fluctuations in the shape away from the ground state shape to the temperature of the excited nuclear system. Details of the method and previous uses can be found in these references [17, 19, 20] Figure 1 shows the temperature extracted versus the excitation energy of the system for five different bins in composition. As one would expect the more energy that is put into the system the higher the temperature of that system. This behavior is demonstrated for each of the composition bins. However the bins with the lowest neutron to proton composition express the highest temperatures. As more neutrons are added to the system the temperature decreases. This is systematically true across all excitation energies. One of the significant differences

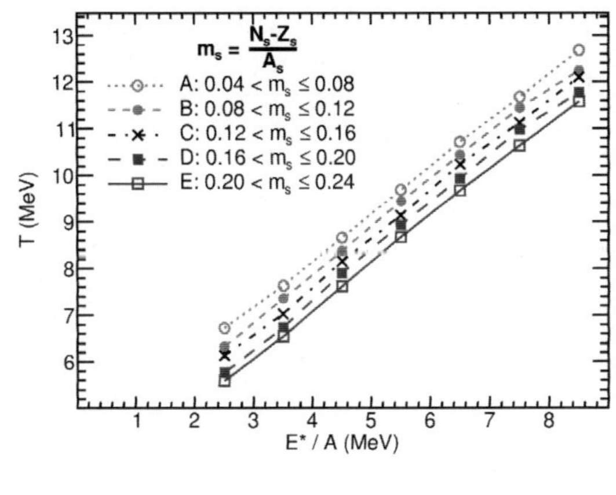

Figure 1. Temperature extracted with the MQF method as a function of excitation energy using protons for bins in m_s. ($m_s = \frac{(N-Z)}{A}$).

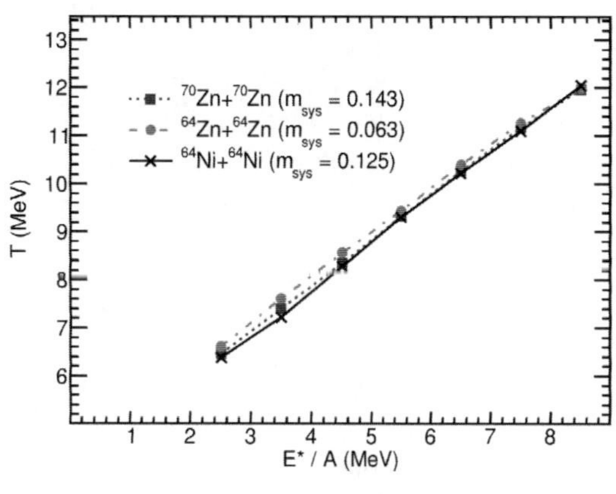

Figure 2. Temperature extracted with the MQF method as a function of excitation energy using protons from the reactions of ^{70}Zn + ^{70}Zn, ^{64}Zn + ^{64}Zn and ^{64}Ni + ^{64}Ni, ($m_s = \frac{(N-Z)}{A}$).

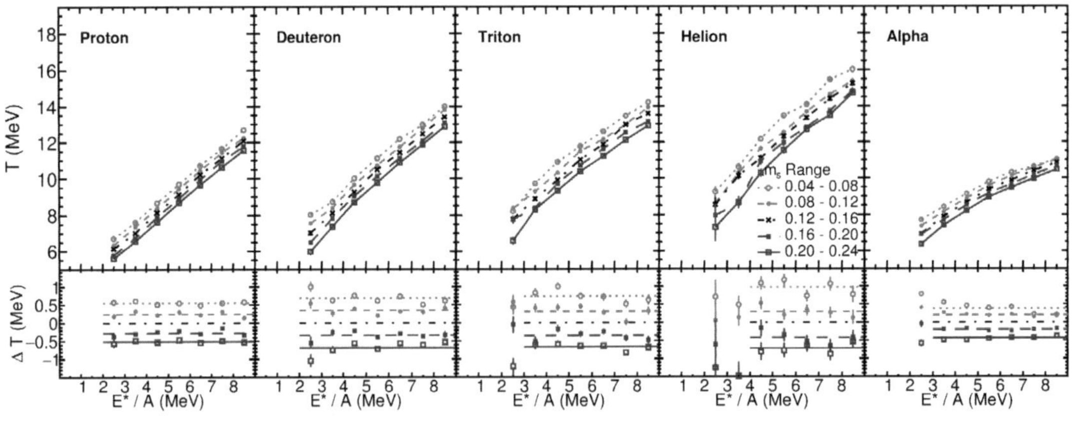

Figure 3. (Top) Temperature extracted with the Momentum Quadrupole Fluctuation method as a function of excitation energy per nucleon for light charged particles, selected on bins in asymmetry. (Bottom) Change in temperature due to a change in asymmetry as a function of excitation energy per nucleon. The middle asymmetry bin is used as the reference, ($m_s = \frac{(N-Z)}{A}$).

between the current measurement and the previous measurements is how the composition of the system was determined. In both previous measurements the composition of the system was taken as the composition of the reacting system. For the data shown in figure 1 the composition of the system was taken as the composition of the reconstructed Q_p as quantified by the m_s. Previous work has demonstrated that after the projectile interacts with the target and a hot Q_p is formed, there is a diversity in the composition of the excited Q_p [21, 22, 23] by sorting on the composition of the reconstructed Q_p the composition is more accurate and has less variation within the bin. For comparison figure 2 shows the temperature versus excitation energy for the three reacting systems with different initial compositions. As one can see the temperature rises as the excitation energy is increased just as in figure 1. However, in contrast to figure 1 the difference in the caloric curve for the three different systems in much less than when the composition was determined by the fragmenting system rather then the reacting system. The difference is still in the same direction, but the effect has been muted significantly. This analysis bridges the apparent gap between the measurement shown in figure 1 and the previous measurements where very little difference was found in the caloric curve as a function of composition.

The analysis presented in figures 1&2 was done with protons as the probe particle, but a similar analysis could be performed with other probe particles. Figure 3 shows the caloric curves for all of the light charged particles (LCP), protons, deuterons, tritons, helions, and alphas. For all of the LCP probes the temperature increases as a function of the excitation energy as one would expect. For helions the temperatures are a bit higher and for alphas a bit lower. Additionally for the alpha particles the slope of the curve is much less steep. To focus on the energy dependence of the difference the bottom panels show point by point the difference in temperature between the given m_s bin and the temperature of the central m_s bin (m_s= 0.12-0.16) The average across all excitation energies is shown by the horizontal lines. While the alpha particles exhibit a small energy dependence in the temperature difference the other probe particles are independent of excitation energy.

Figure 4 shows the temperature as a function of excitation energy for the various m_s bins using ^{7}Li and ^{9}Be as probe particles. While the behavior is similar to the LCP, the data shows more scatter due to limited statistics. These larger clusters also exhibit higher temperatures.

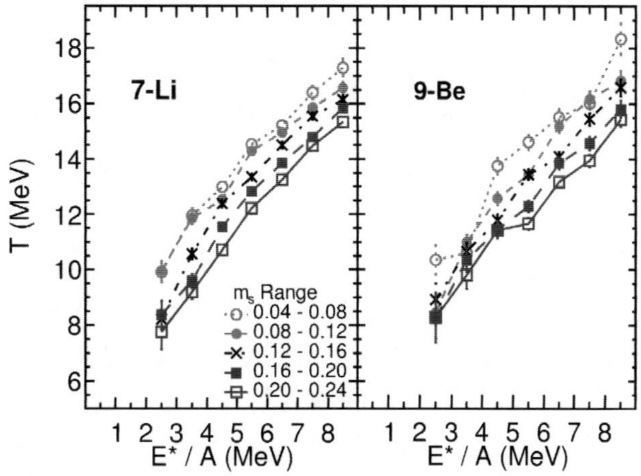

Figure 4. MQF Temperature as a function of excitation energy with ^{7}Li and ^{9}Be probes, for bins in asymmetry, ($m_s = \frac{(N-Z)}{A}$).

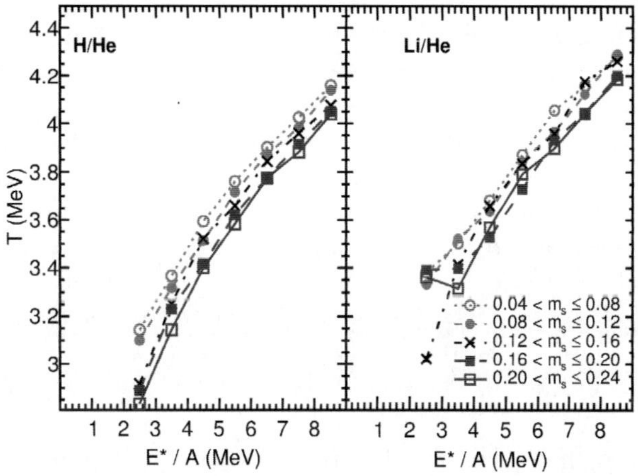

Figure 5. Temperatures calculated with the Albergo yield ratio method versus excitation energy for various bins in asymmetry, ($m_s = \frac{(N-Z)}{A}$).

This particle type dependence is similar top what was seen in previous data [17].

3.2. Chemical Thermometer

The effect of additional neutron content on the temperature expressed by an excited nuclear system as a function of excitation energy is quite clear using the MQF, which is a kinetic thermometer. Alternatively one could use the double isotope ratio or "Albergo" thermometer [24]. The Albergo thermometer compares the double ratios of pairs of isotopes that are different by one proton which are composed of isotopes that are different by 1 neutron. The temperatures extracted from the $\frac{p/d}{^3He/^4He}$ are shown on the left hand side of figure 5 while the temperatures extracted from the $\frac{^6Li/^7Li}{^3He/^4He}$ are shown on the right side. Again the temperature as a function of energy raises as one would expect and was seen with the MQF thermometer. The lowest m_s

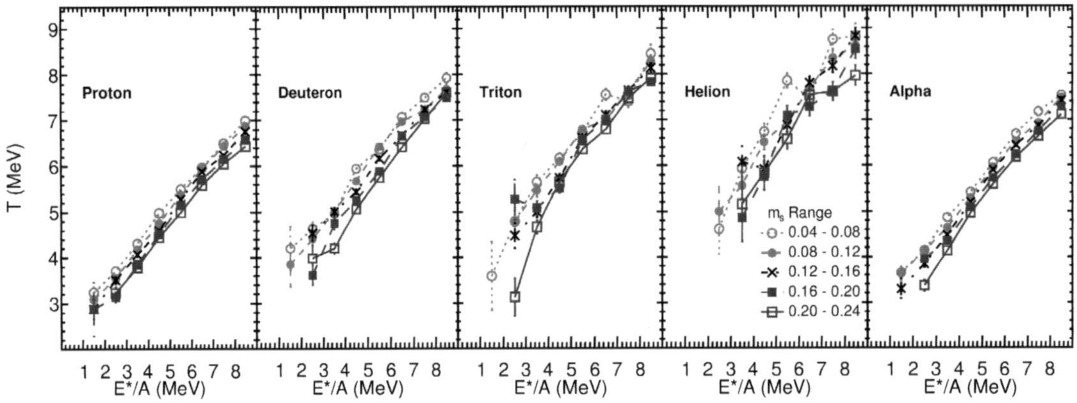

Figure 6. Slope temperatures for light charged particles versus excitation energy for bins in asymmetry, ($m_s = \frac{(N-Z)}{A}$).

bins - those with a composition closes to N=Z - express the highest temperatures relative to those with more neutron content, as a function of excitation energy.

3.3. Slope Thermometer

Another method used to determine the temperature of an excited nuclear system is to measure the kinetic energy of the particles that are emitted by that system. The kinetic energy spectra can them be fit with a Boltzman distribution and a temperature can be extracted[25]. Figure 6 shows the temperatures extracted from the kinetic energy spectra of each of the light charged particle species for each of the m_s bins. Once again the temperature rises as a function of the excitation energy for all curves. The lowest m_s bin shows the highest temperature for each excitation energy bin. This is consistent across all particle types.

4. Summary

We have measured the temperature as a function of energy - or the nuclear caloric curve - for systems of various neutron - to - proton compositions. Three different thermometer methods have been used including ones based on both kinetic and chemical information. For each of the three thermometers multiple probe particles have been used. For all systems it is found that the larger the neutron content of the system the lower the temperature is for a given excitation energy. Figure 7 summaries all the data to show that as the m_s increases the temperature decreases. The top panel of figure 7 shows the magnitude of the temperature decrease as the neutron content is increased for all the probes used with the MQF thermometer. Each of the point represents the difference in the horizontals lines in figure 3 (and the equivalent difference extracted in a similar way for the ^{7}Li and ^{9}Be data shown in figure 4). The dotted lines are linear fits to the data. The middle panel shows the decrease in temperature with increasing asymmetry for the two chemical thermometers. The bottom panel of figure 7 shows the magnitude of this asymmetry dependence as extracted from the slope temperatures. While the absolute magnitude of the effect depends on the thermometer and the particular probe particle used, there is a clear linear dependence of the decrease in in the temperature with increasing asymmetry across all probes. These experimental findings are in agreement with theoretical predictions[12, 11].

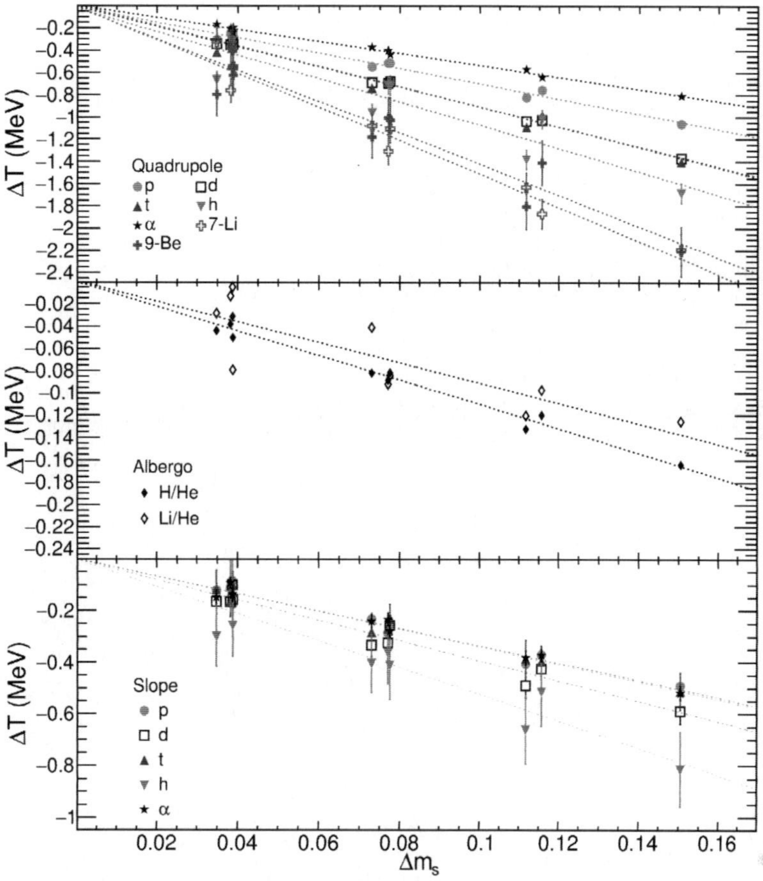

Figure 7. Change in the temperature with changing asymmetry, as determined with the MQF method (upper panel), the Albergo method (middle panel), and the slope temperature method (lower panel). The dotted lines are linear fits to the data. ($\Delta T = T_1 - T_2; \Delta m_s = \frac{(N_1-Z_1)}{A_1} - \frac{(N_2-Z_2)}{A_2}$).

Acknowledgements

This work was supported in part by the US Department of Energy under grant DE-FG02-93ER40773 and the Robert A. Welch Foundation under grant A-1266. We are indebted to the staff at the Texas A&M Cyclotron Institute for providing the high quality beams that made this work possible.

References

[1] A.W. Steiner, Phys. Rev. C **85**, 055804 (2012)
[2] F.J. Fattoyev, J. Piekarewicz, Phys. Rev. C **86**, 015802 (2012)
[3] A.B. McIntosh et al., Phys. Lett. B **719**, 337 (2013)
[4] A.B. McIntosh et al., Phys. Rev. C **87**, 034617 (2013)
[5] A.B. McIntosh et al., EPJA **50**, 35 (2014)
[6] J. Mabiala et al., Int. J. Mod. Phys. E **22**, 1350090 (2013)
[7] S. Wuenschel et al., Nucl. Inst. Meth. A **604**, 578 (2009)
[8] Z. Kohley et al., Phys. Rev. C **86**, 044605 (2012)
[9] Z. Kohley, Ph.D. thesis, Texas A&M Univ. (2010)

[10] Z. Kohley et al., Phys. Rev. C **83**, 044601 (2011)
[11] C. Hoel, L.G. Sobotka, R.J. Charity, Phys. Rev. C **75**, 017601 (2007)
[12] V.M. Kolomietz, A.I. Sanzhur, S. Shlomo, S.A. Firin, Phys. Rev. C **64**, 024315 (2001)
[13] J. Besprosvany, S. Levit, Phys. Lett. B **217**, 1 (1989)
[14] R. Ogul, A.S. Botvina, Phys. Rev. C **66**, 051601(R) (2002)
[15] J. Zhang et al., Phys. Rev. C **54**, 1137 (1996)
[16] J. Su, F.S. Zhang, Phys. Rev. C **84**, 037601 (2011)
[17] S. Wuenschel, Ph.D. thesis, Texas A&M Univ. (2009)
[18] C. Sfienti et al., Phys. Rev. Lett. **102**, 152701 (2009)
[19] S. Wuenschel et al., Nucl. Phys. A **843**, 1 (2010)
[20] H. Zheng, A. Bonasera, Phys. Rev. C **86**, 027602 (2012)
[21] D. Rowland et al., Phys. Rev. C **67**, 064602 (2003)
[22] S. Wuenschel et al., Phys. Rev. C **79**, 061602(R) (2009)
[23] S. Galanopoulos et al., Nucl. Phys. A **837**, 145 (2010)
[24] S. Albergo, S. Costa, E. Costanzo, A. Rubbino, Il Nuovo Cimento **89**, 1 (1985)
[25] D.J. Morrissey, W. Benenson, W.A. Friedman, Ann. Rev. Nucl. Part. Sci. **44**, 27 (1994)

Unstable nuclei in coherent dissociation of relativistic nuclei ^{7,9}Be, ^{10}B and ^{10,11}C

D A Artemenkov[1], V Bradnova[1], E Firu[2], N K Kornegrutsa[1], M Haiduc[2], K Z Mamatkulov[1], R R Kattabekov[1], A Neagu[2], P A Rukoyatkin[1], V V Rusakova[1], V R Sarkisyan[3], R Stanoeva[4], A A Zaitsev[1,5], P I Zarubin[1,5*] and I G Zarubina[1,5]

[1]V I Veksler and A M Baldin Laboratory of High Energy Physics, Joint Institute for Nuclear Research, Dubna, Russia
[2]Institute of Space Science, Magurele, Romania
[3]Yerevan Physics Institute, Yerevan, Republic of Armenia
[4]South-Western University, Blagoevgrad, Bulgaria
[5]P N Lebedev Physical Institute of the Russian Academy of Sciences, Moscow, Russia
*E-mail: zarubin@lhe.jinr.ru

Abstract. Contribution of the unstable nuclei ^{7}Be, ^{8}Be and ^{9}B into coherent dissociation events (“white” stars) of relativistic nuclei ^{7,9}Be, ^{10}B and ^{10,11}C is under study on the basis of a nuclear track emulsion exposed to beams of the JINR Nuclotron. Distributions over the opening angle of α-pairs indicate to a simultaneous presence of virtual $^{8}Be_{g.s.}$ and $^{8}Be_{2+}$ states in the ground states of the ^{9}Be and ^{10}C nuclei. The core ^{9}B is manifested in the ^{10}C nucleus with a probability of (30 ± 4)%, Selection of the ^{10}C “white” stars accompanied by $^{8}Be_{g.s.}$ (^{9}B) leads to the appearance in the excitation energy distribution of $2\alpha 2p$ “quartets” of the distinct peak with a maximum at 4.1 ± 0.3 MeV. $^{8}Be_{g.s.}$ decays are presented in 21% 2He + 2H and 19% in the 3He of the all ^{11}C “white” stars. $^{9}B_{g.s.}$ decays are identified in “white” stars ^{11}C → 2He + 2H constituting 14% of the ^{11}C “white” stars. The ^{9}B nucleus. is manifested in the “white” stars ^{10}B → 2He + 2H with a probability of (9 ± 1)%. For the ^{10}B case yield of $^{8}Be_{g.s.}$ nuclei with the respect to ^{9}B is about a factor of 3 higher than ^{9}B.

1. Introduction

The group of nuclei at the beginning of the table of isotopes provides a recognized laboratory of nuclear quantum mechanics allowing one to study the coexistence and evolution of cluster and shell degrees of freedom. Light nuclei can be presented as all possible superpositions of lighter nuclear cores, lightest nuclear clusters having no excited states (α-particle, triton, ^{3}He nucleus, and deuteron), and nucleons that coexist in dynamical equilibrium. Interlacing cluster and shell degrees of freedom do a “laboratory” of nuclear quantum mechanics from a group of nuclei in the beginning of the table of isotopes.

Consideration of the nucleosynthesis chains toward ^{10,11}B, ^{11,10}C and ^{12}N via the “hot breakout” ^{7}Be(^{3}He,γ)^{10}C(e^{+},ν)^{10}B assists to recognize the relationship of their structures. The ^{10}C synthesis processing due to increase of α-clustering provides an energy “window” for formation of the intermediate states with unstable nuclei ^{9}B + p, $^{8}Be_{2+}$ + 2p and ^{6}Be + α. These clusters are preserved in subsequent reactions ^{10}C(e^{+},ν)^{10}B(p,γ)^{11}C(e^{+},ν)^{11}B. The “window” of the reaction ^{7}Be(^{4}He,γ)^{11}C allows only an association of the ^{7}Be and ^{4}He clusters, also contributing to the ^{11}C and ^{10}B structure. Thus, a hidden variety of the virtual configurations in the nuclei ^{10,11}C and ^{10,11}B can be populated via electromagnetic transitions from the real ones. In turn, these nuclei provide a basis for capture reactions of protons or the He isotopes (or in neutron exchange) for synthesis of the subsequent nuclei which leads to a translation of the preceding structures.

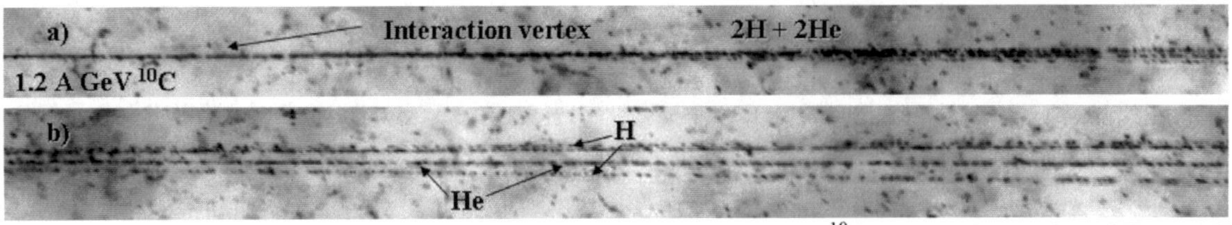

Figure 1. Macrophoto of the coherent dissociation event of 1.2 *A* GeV ^{10}C nucleus into pairs of He and H nuclei; a) primary track, approximate position of interaction vertex and appearance of fragment tracks and b) tracks of fragments are resolved; opening angles between tracks are $\Theta_{2\mathrm{He}}$ = 5.9 mrad, Θ_{HeH} = 8.6, 16.6, 3.0, 17.6 mrad, Θ_{2p} = 20.1 mrad. Both 2HeH triples in the event correspond to ^{9}B decays.

Within the BECQUEREL project [1] the cluster structure of light nuclei is studied in relativistic-fragmentation processes on the basis of the nuclear track emulsion exposed to primary and secondary beams of the JINR Nuclotron [2, 3]. Among the events of fragmentation of relativistic nuclei, those of their coherent dissociation to narrow jets of fragments are especially important for studying nuclear clustering. They do not feature tracks of either slow fragments of emulsion nuclei or charged mesons. This special feature reflects the fact that the excitation of the relativistic nucleus under investigation is minimal in the case of a glancing collision with a heavy track-emulsion nucleus. Nuclear diffraction interaction processing without nuclear density overlaps and not accompanied by an angular momentum transfer is a main underlying mechanism of excitation of coherent dissociation in nuclear track emulsions.

The experimental method in question is based on a record spatial resolution and sensitivity of nuclear track emulsion whose layers are exposed longitudinally to beams of relativistic nuclei including radioactive ones. It has already furnished unique information about cluster aspects of the structure of the whole family of light nuclei, including radioactive ones. Because of the absence of tracks of strongly ionizing particles, events of coherent dissociation were called "white" stars (example in fig. 1). The term "white" star reflects aptly a sharp "breakdown" of the ionization density at the interaction vertex upon going over from the primary-nucleus track to secondary tracks within a 6^0 cone at 1.2 *A* GeV.

This special feature generates a fundamental problem for electronic methods because more difficulties should be overcome in detecting events where the degree of dissociation is higher. On the contrary, such events in nuclear-track emulsions are observed and interpreted in the most straightforward way, and their distribution among interaction channels characterized by different compositions of charged fragments is determined exhaustively. This probabilistic distribution is a basic feature that is observed for the virtual cluster structure of the nucleus under consideration.

The probability distribution of the final configurations of fragments in white stars makes it possible to reveal their contributions to the structure of nuclei under consideration. We assumed that, in the case of dissociation, specific configurations arise at random (random-phase approximation) without sampling and that the dissociation mechanism itself does not lead to the sampling of such states via angular-momentum or isospin exchange. By and large, available results confirm the assumption that cluster features of light nuclei determine the picture of their relativistic dissociation. At the same time, events that involve the dissociation of deeply bound cluster states and which cannot arise at low collision energies are detected.

Reconstruction of the decays of relativistic ^{8}Be and ^{9}B nuclei is possible by the energy variable $Q = M^* - M$, where $M^{*2} = \sum(P_i \cdot P_k)$, M is the total mass of fragments, and $P_{i,k}$ are their 4-momenta defined under the assumption of conservation of an initial momentum per nucleon by fragments. When the identification of relativistic fragment can be reasonably supposed the quasi-invariant variable Q allows one to estimate the excitation energy of their complex ensembles uniting all angular measurements in an event. For the "white" stars of ^{9}Be and ^{10}C nuclei the assumption that He fragments correspond to ^{4}He nuclei (α), and H ones in ^{10}C – ^{1}H is justified. Then ^{8}Be and ^{9}B identification is reduced to measurements of the opening angles between the directions of fragment emission. The experimental details and development of these investigations and their illustrations are presented in the recent review article [12]. Below further progress of this research is discussed.

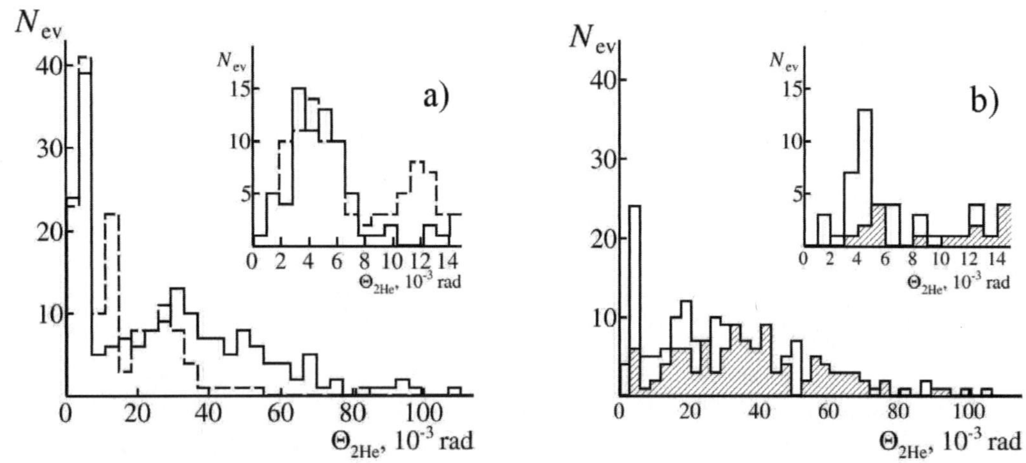

Figure 2. Distributions over the opening angle Θ_{2He} of α-particle pairs in "white" stars $^{10}C \to 2He + 2H$ (a, solid histogram) and $^{9}Be \to 2He$ (a, dashed histogram), $^{11}C \to 2He + 2H$ (b, solid histogram) and $^{11}C \to 3He$ (hatched histogram); on the insertions: the Θ_{2He} distribution in interval Θ_n.

2. The ^{8}Be and ^{9}B nuclei in dissociation of the ^{9}Be and ^{10}C nuclei

Distributions over the opening angle Θ_{2He} for pairs of He fragments of "white" stars $^{9}Be \to 2He$ and $^{10}C \to 2He + 2H$ (82% of the ^{10}C statistics) produced at energy of 1.2 *A* GeV are presented in Fig. 2. In both cases the values of Θ_{2He} of 75-80% of the pairs are distributed about equally in the intervals of $0 < \Theta_{n(arrow)} < 10.5$ mrad and $15.0 < \Theta_{w(ide)} < 45.0$ mrad. The remaining pairs are attributed to intervals $10.5 < \Theta_{m(edium)} < 15.0$ and "widest" of $15.0 < \Theta_{v(ery)w(ide)} < 45.0$ mrad. The distribution over the Q variable is directly correlated with the Θ_{2He} one. It is pointing out that "narrow" pairs of Θ_n are produced via $^{8}Be_{g.s.}$ while pairs Θ_w via $^{8}Be_{2+}$. Besides, for the ^{9}Be case there is a peak in the interval Θ_m reflecting its level $5/2^-$ (2.43 MeV). Fractions of events in the intervals Θ_n and Θ_w are equal to 0.56 ± 0.04 and 0.44 ± 0.04 for ^{9}Be, while for ^{10}C 0.49 ± 0.06 and 0.51 ± 0.06, i. e. they practically coincide. They indicate to a simultaneous presence of virtual $^{8}Be_{g.s.}$ and $^{8}Be_{2+}$ states in the ground states of the ^{9}Be and ^{10}C nuclei. Elongation above 40 mrad of the ^{10}C Θ_{2He} distribution can be due to the channel $^{4}He + ^{6}Be$.

Earlier, basing on the $Q_{2\alpha p}$ energy distribution of the triples $2\alpha + p$ from the "white" stars $^{10}C \to 2\alpha + 2p$ it is concluded that in the ^{10}C nucleus the core ^{9}B is manifested with a probability of (30 ± 4)%, and the $^{8}Be_{g.s.}$ decays are arise always through the ^{9}B decays.

Due to completeness of observation the $2\alpha 2p$ "quartets" is an interesting feature is manifested in their excitation energy distribution $Q_{2\alpha 2p}$. Fig. 3 shows distribution $Q_{2\alpha 2p}$ of all "white" stars $^{10}C \to 2He + 2H$ which appears at a first glance to be scattered. However, selection of the stars accompanied by $^{8}Be_{g.s.}$ (equally ^{9}B) leads to appearance in the $Q_{2\alpha 2p}$ distribution of the distinct peak with a maximum at 4.1 ± 0.3 MeV and RMS of 2.0 MeV. Such a value of the peak maximum is corresponding to the 4.2 MeV state established in low energy experiment and interpreted as a molecular-like state $\alpha + 2p + \alpha$ [4] while the width is determined by the accepted momentum approximation. The peak statistics present 17 ± 4 % of the total number or 65 ± 14 % of the stars containing ^{9}B decays.

Distribution over a total momentum $P_{T2\alpha 2p}$ of all $2\alpha 2p$ ensembles is described by a Rayleigh function with the parameter σ = 175 ± 10 MeV/*c* while in a case of the $^{8}Be_{g.s.}$ (^{9}B) presence it is significantly less σ = 127 ± 16 MeV/*c*. Not competing in statistics and resolution [] our observation of such a state manifesting in extra narrow $2\alpha 2p$ jets is grounded on selection of evidently glancing collisions which reduce dramatically a continuum contribution. It is worth noting the observation of a single "white" star $2\alpha 2p$ having $Q_{2\alpha 2p}$ equal to 0.77 MeV in which both $2\alpha p$ triples correspond to ^{9}B decays with $Q_{2\alpha p}$ of 0.22 and 0.67 MeV, $Q_{2\alpha}$ of 0.14 MeV and $Q_{\alpha 2p}$ of 0.64 and 0.6 MeV (fig. 1).

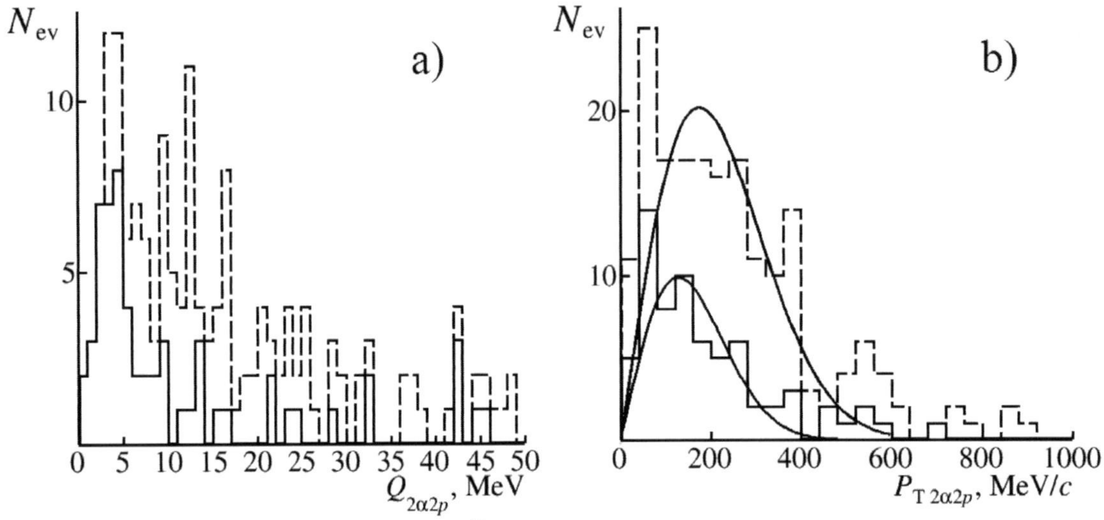

Figure 3. Distributions of all “white” stars $^{10}C \rightarrow 2He + 2H$ (dashed histogram) and the ones with the presence of $^{8}Be_{g.s.}$ (^{9}B) (solid histogram) over energy $Q_{2\alpha 2p}$ (a) and total transverse momentum $P_{T2\alpha 2p}$ (b).

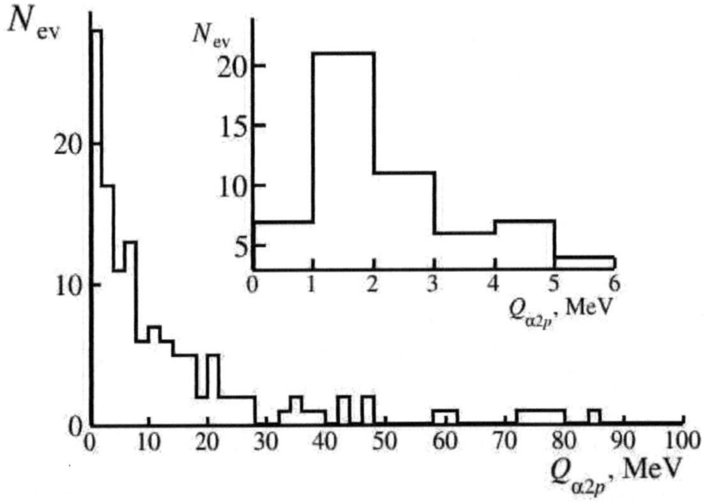

Figure 4. Distributions over energy $Q_{\alpha 2p}$ in measured “white” stars $^{7}Be \rightarrow He + 2H$; insertion; the enlarged $Q_{2\alpha p}$ distribution in the ^{6}Be region.

3. Search for ^{6}Be decays in dissociation of the ^{7}Be and ^{10}C nuclei

It is found that 289 “white” stars produced by 1.2 *A* GeV ^{7}Be nuclei [5] are distributed over the charge channels in the following way: He + 2H (54%), 2He (40%), ^{6}Li + H (1%), 4H (5%). The high statistics in the first channel allows one to estimate of the contribution of the unbound ^{6}Be nucleus in the left most peak of the distribution over $Q_{\alpha 2p}$ (fig. 4). Since angular measurements are used only to search for this narrow resonance the rough condition $Q_{\alpha 2p} < 6$ MeV is applied to estimate its contribution. 27% of 130 events in the channel He + 2H can be attributed to ^{6}Be decays in this way.

Determination of the ^{6}Be decay region in the accepted approximations gives opportunity to estimate a possible ^{6}Be contribution to the ^{10}C “white” stars. Fig. 5a shows correlation between $Q_{2\alpha}$ and $Q_{\alpha 2p}$. A peak at 3-4 MeV in the total distribution $Q_{\alpha 2p}$ is becoming profound when $^{8}Be_{g.s.}$ decays in the stars are demanded (Fig. 5b). Contrary to expectations on associated production of ^{4}He + ^{6}Be in energy continuum the dominant fraction of “^{6}Be” candidates correspond to their simultaneous decays altogether with $^{8}Be_{g.s.}$ (and, hence, ^{9}B) and $^{8}Be_{2+}$ decays. Surprisingly, but one can not separate the ^{6}Be and $^{8}Be_{g.s.}$ decays and have to assume that ^{6}Be and $^{8}Be_{g.s.}$are produced as interfering parts of $2\alpha 2p$ ensembles.

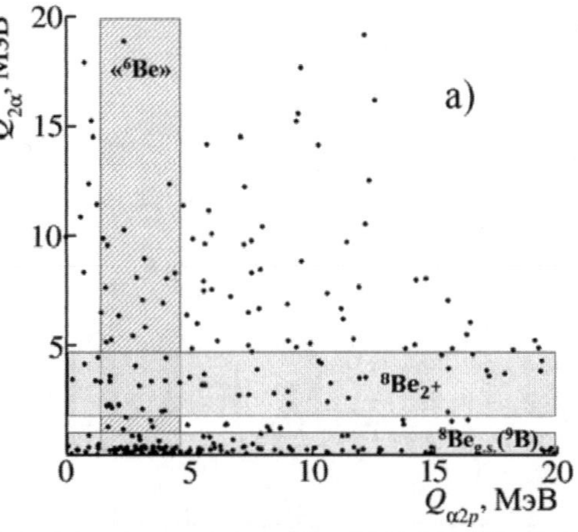

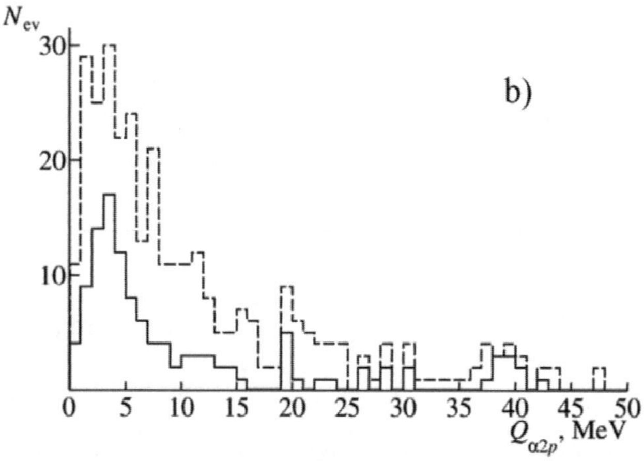

Figure 5. Distributions of "white" stars $^{10}C \rightarrow 2He + 2H$ over energy $Q_{2\alpha}$ and $Q_{\alpha 2p}$ (a, expected regions of decays ^6Be, $^8Be_{g.s.}$ (^9B) and $^8Be_{2+}$ are shown) Distributions of $\alpha 2p$ triples over energy $Q_{\alpha 2p}$ in all "white" stars $^{10}C \rightarrow 2He + 2H$ (b, dashed histogram) and with presence of $^8Be_{g.s.}$ (^9B) decays in them (b, solid histogram).

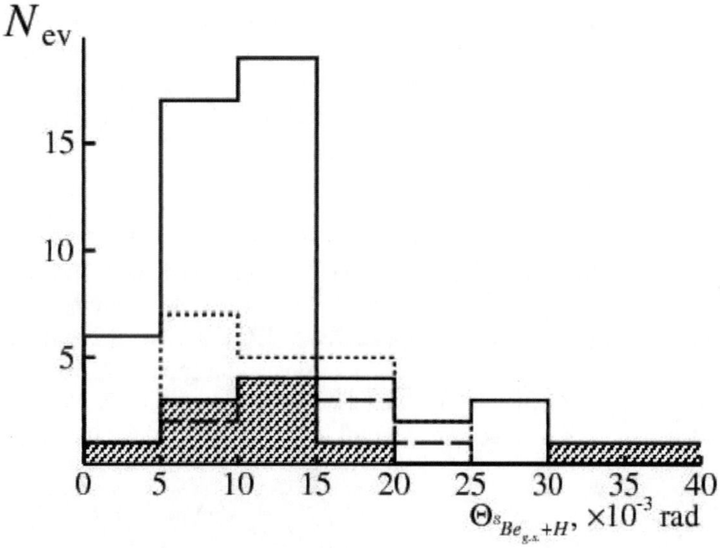

Figure 6. Distributions over the opening angle $\Theta(^8Be_{g.s.} + H)$ in "white" stars $^{10}C \rightarrow {}^8Be_{g.s.} + 2H$ (solid histogram), $^{11}C \rightarrow {}^8Be_{g.s.} + 2H$ (dashed histogram), $^{10}B \rightarrow {}^8Be_{g.s.} + H$ (hatched histogram) all found stars $^{11}C \rightarrow {}^8Be_{g.s.} + 2H$ (dotted histogram).

4. Progress of studies of dissociation of the ^{11}C and ^{10}B nuclei

It is already established that 144 "white" stars produced by ^{11}C nuclei are distributed over the charge channels in the following way: 2He +2H (50%), 3He (17%), ^{7}Be + He (13%), He + 4H (11%), B + H (5%), Li + He + H (3%), 6H (2%). The distributions of He fragments over the opening angle Θ_{2He} (Fig. 2b) show that $^8Be_{g.s.}$ decays are presented in 21% of the 2He + 2H stars and in 19% of the 3He ones. These distributions allow one to assume a strong contribution of $^8Be_{2+}$ decays but it is a subject of future consideration.

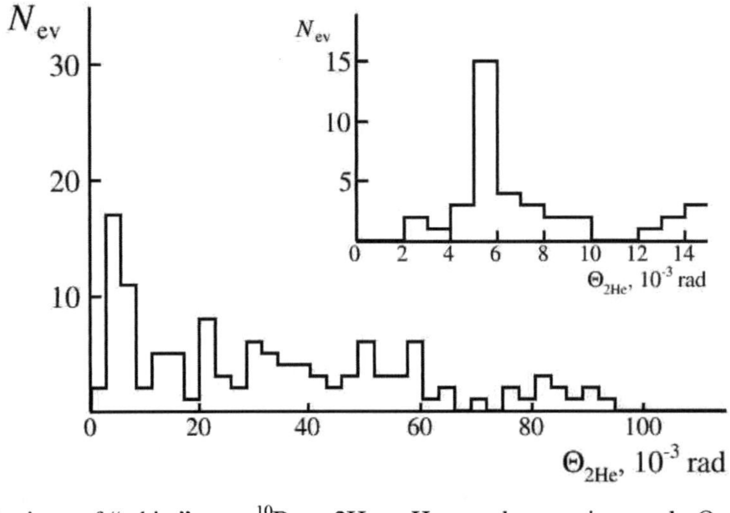

Figure 7. Distributions of "white" stars $^{10}B \rightarrow 2He + H$ over the opening angle Θ_{2He} of 2α pairs (a) and over energy $Q_{2\alpha p}$ of α2*p* triples.

The virtual $^{9}B_{g.s.}$ nucleus can exist in the ^{11}C nucleus as an independent component or as a component of a virtual core ^{10}B. Decays $^{9}B_{g.s.}$ in "white" stars $^{10}C \rightarrow 2He + 2H$ are identified in accordance with a limitation on the opening angle between directions of $^{8}Be_{g.s.}$ and each H fragments $\Theta(^{8}Be_{g.s.} + H) < 40$ mrad (Fig. 6) [3]. Application of such a condition to the "white" stars $^{11}C \rightarrow 2He + 2H$ allows one to identify 20 $^{9}B_{g.s.}$ decays constituting 30% of events in this charge channel or 12% of the ^{11}C "white" star statistics.

An analysis of the NTE exposure to 1 *A* GeV ^{10}B nuclei has pointed out that triples 2He + H (about 65%) dominate among "white" stars. However, origins of this effect have not been studied being in the "shadow" of studies with radioactive nuclei. This effect can indicate the possible presence of structures $^{9}B_{g.s.} + n$ as well as $^{9}Be + p$. In the ^{10}B nucleus a virtual ^{9}Be can exist in the superposition $^{8}Be_{g.s.} + n$ as well as $^{8}Be_{2+} + n$ one leading to 3-prong "white" stars out of $^{9}B_{g.s.}$ decays. Then the cluster configuration involving the deuteron $^{8}Be_{2+} + d$ can be a source of $^{8}Be_{2+}$ decays. Since the channel $^{10}B \rightarrow {}^{6}Li$ is observed with 10% probability some virtual ^{6}Li contribution can be expected into the 2α + *p*(*d*) channel. Thus, with attraction of existing knowledge the pattern ^{10}B dissociation via decays $^{8}Be_{g.s.}$, $^{8}Be_{2+}$ and $^{9}B_{g.s.}$ can disentangled step by step. If successful, it will lead to better understanding for the neighbouring nuclei.

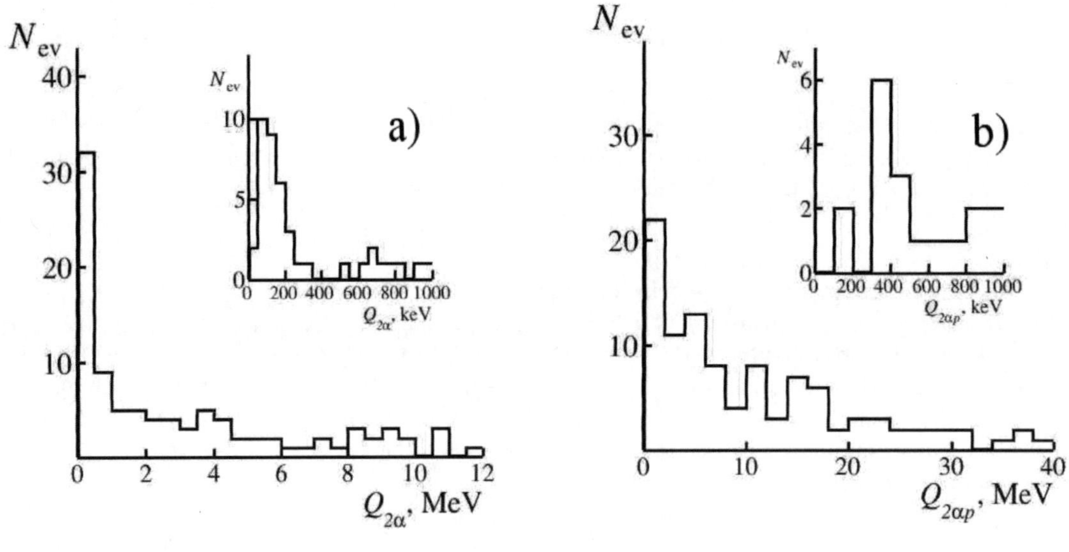

Figure 8. Distributions of "white" stars $^{10}B \rightarrow 2He + H$ over energy $Q_{2\alpha}$ of 2α pairs (a) and over energy $Q_{2\alpha p}$ of α2*p* triples.

Recently, 250 2He + H "white" stars are selected in an accelerated search. Angular measurements of the first 119 stars pointed already to 32 $^8Be_{g.s.}$ decays (fig. 7) and 11 $^9B_{g.s.}$ decays (fig. 6) accompanied by $^8Be_{g.s.}$. The distributions of the energy of α-particle pairs $Q_{2\alpha}$ and triples 2α + *p* $Q_{2\alpha p}$ from the $^{10}B \rightarrow$ 2He + H events are shown in Fig. 8a and 8b, respectively. For the moment correspondence of H to *p* and He to α is assumed. Identification of He and H isotopes by a multiple scattering method, are in progress now.

5. Conclusions

Contribution of the unstable nuclei ^{6}Be, ^{8}Be and ^{9}B into coherent dissociation events ("white" stars) of relativistic nuclei ^{7,9}Be, ^{10}B and ^{10,11}C is under study on the basis of a nuclear track emulsion exposed to beams of the JINR Nuclotron.

Distributions over the opening angle of α-pairs indicate to a simultaneous presence of virtual $^8Be_{g.s.}$ and $^8Be_{2+}$ states in the ground states of the ^{9}Be and ^{10}C nuclei. The core ^{9}B is manifested in the ^{10}C nucleus with a probability of (30 ± 4)%. $^8Be_{g.s.}$ decays in ^{10}C "white" stars always arise through the ^{9}B decays.

Selection of the ^{10}C "white" stars accompanied by $^8Be_{g.s.}$ (^{9}B) leads to appearance in the excitation energy distribution of 2α2*p* "quartets" of the distinct peak with a maximum at 4.1 ± 0.3 MeV. Distribution over the total momentum of $^8Be_{g.s.}2p$ ensembles is described by a Rayleigh function with the parameter 127 ± 16 MeV/*c*. A single 2α2*p* "white" star in which both 2α*p* triples correspond to a ^{9}B decay is observed.

On the basis of angular measurements 27% of events ^{7}Be → He + 2H can be attributed to ^{6}Be decays by the condition on the excitation energy to be less than < 6 MeV.

For ^{10}C "white" stars it have to be assumed that ^{6}Be and $^8Be_{g.s.}$ are produced as interfering parts of 2α2*p* ensembles due to impossibility of separation of the ^{6}Be and $^8Be_{g.s.}$ decays.

$^8Be_{g.s.}$ decays are presented in 21% 2He + 2H and 19% in the 3He of the all found ^{11}C "white" stars. ^{9}B decays are identified in "white" stars ^{11}C → 2He + 2H constituting 14% of the ^{11}C "white" stars. As in the ^{10}C case $^8Be_{g.s.}$ decays in ^{11}C "white" stars almost always arise through the ^{9}B decays.

The ^{9}B nucleus. is manifested in the "white" stars ^{10}B → 2He + H with a probability of (9 ± 1)%. For the ^{10}B case yield of $^8Be_{g.s.}$ nuclei is about a factor of 3 higher than ^{9}B (in distinction to the ^{10,11}C cases). Thus, there appears a ground to assume the contribution in the ^{10}B structure of virtual superposition $^8Be_{g.s.}$- $^8Be_{2+}$ like in the ^{9}Be case. This is a subject of the ongoing study

On the ground of our studies the following pattern of nuclear clustering is emerging. As the fundamental elements of its structure atomic nuclei contain possible virtual associations of the nucleons and the lightest clusters. Their superpositions form cores consisting of the unstable ^{8}Be in the ground $^8Be_{g.s.}$ and first exited $^8Be_{2+}$ states and, then, ^{9}Be and ^{9}B nuclei which, in turn, serve as composing clusters in the structure of the heavier Be, B and C isotopes. Besides, the stable ^{7}Be and, probably, unstable ^{6}Be are important in the neutron deficient nuclei. A probability balance of such superpositions of possible core states with an appropriate spin and parity determine binding and ground state parameters of the corresponding nuclei.

Acknowledgements

The authors are grateful to A. I. Malakhov (JINR), N. G. Polukhina and S. P. Kharlamov (LPI) for their support and critical discussion of the results. This work was supported by the grant from the Russian Foundation for Basic Research 16-02-00062 and grants of Plenipotentiary representatives of the governments of Bulgaria, Egypt, Romania and the Czech Republic at JINR

[1] The BECQUEREL Project WEB site: *http://becquerel.jinr.ru/*

[2] Zarubin P I 2013 "Tomography" of the cluster structure of light nuclei via relativistic dissociation *Lecture Notes in Physics* (*Clusters in Nuclei* vol 3) ed Ch Beck 875 (Springer Int. Publ.) pp 51–93 (*Preprint* arXiv: 1309.4881)

[3] Artemenkov D A *et al* 2015 Charge topology of coherent dissociation of ^{11}C and ^{12}N relativistic nuclei *Phys. At. Nucl.* **78** pp 794–799 (*Preprint* arXiv: 1411.5806)

[4] Curtis N *et al* 2008 Breakup reaction study of the Brunnian nucleus ^{10}C *Phys. Rev.* C **78** 021301(R)

[5] Kornegrutsa N K *et al* 2014 Clustering features of the ^{7}Be nucleus in relativistic fragmentation *Few Body Syst.* **55** pp 1021–1023 (*Preprint* arXiv: 1410.5162)

Journal of Physics: Conference Series **724** (2016) 012056 doi:10.1088/1742-6596/724/1/012056

SFF analysis of the small angle scattering data for investigation of a vesicle systems structure

E.V. Zemlyanaya[1], M.A. Kiselev[1], E.I. Zhabitskaya[1], A.Yu. Gruzinov[2], V.L. Aksenov[1,3], O.M. Ipatova[4], O.S. Druzhilovskaya[4]

[1]Joint Institute for Nuclear Research, 141980 Dubna, Russia
[2]National Research Center "Kurchatov Institute", Moscow, Russia
[3]Peterburg Institute of Nuclear Physics, NRC "Kurchatov Institute", Gatchina, Russia
[4]V.N. Orekhovich Research Institute for Biomedical Chemistry, Moscow, Russia

Abstract. Experimental data on the small angle synchrotron X-ray scattering (SAXS) are analyzed on a basis of Separated form factors method (SFF) for a study of the drug delivery Phospholipid Transport Nano System (PTNS). Basic parameters of polydispersed population of PTNS nanoparticles (average radius of PTNS-particles, polydispersity of radius, thickness of membrane) have been determined. The results are discussed in comparison with the SFF results for the "classical" vesicular system of dimyristoylphosphocholine (DMPC).

1. Introduction

Small angle scattering of neutrons (SANS) and of X-rays (SAXS) is well known as a powerful tool for investigation of structure of nanosystems including unilamellar vesicles (ULVs) of phospholipids in water solutions of disaccharides [1, 2, 3, 4]. Phospholipid vesicles (liposomes, nanospheres) have a a vide range of applications in modern biochemistry and pharmacology. One of important directions is a development of the drug delivery systems on the basis of phospholipds [5, 6]. In recent years, the Phospholipid Transport Nano System (PTNS) with extremely small diameter (200–300 Å) has been obtained from the soybean phosphatidylcholine in the V.N. Orekhovich Research Institute of Biomedical Chemistry (Moscow, Russia) [7]. Incorporation of drugs into PTNS is shown to sufficiently increase the therapeutic effectiveness [8]. From the point of practical applications, the information on the structure and properties of PTNS nanoparticles in different media is needed. In the recent work [9], on the basis of SAXS data collected at the Kurchatov Synchrotron Radiation Source of the National Research Center "Kurchatov Institute" (Moscow, Russia), the conclusion about vesicular structure of PTNS nanoparticles has been made and an estimation of average size of PTNS particles has been obtained.

In this contribution, we present the results of analysis of the PTNS system by means the Separated Form Factors method (SFF) . SFF is an effective approach to obtain information about a structure of vesicular systems from the small angle scattering data [4, 10, 11]. The SFF results are presented in comparison with the results of analysis on the basis of the Hollow sphere model (HS) and with the results of SFF analysis of "classical" dimyristoylphosphocholine (DMPC) vesicles in water solution of sucrose with different concentration.

2. Experiment

The measurements of the PTNS particles have been performed at room temperature at the DICSI station of the Kurchatov Synchrotron Radiation Source in Moscow, Russia. The procedure of SAXS measurement and the sample preparation is the same as in [9]. The measurements were carried out at two sample-to-detector distances of $L_{sd} = 30$ and 243.5 cm with a photon wavelength of $\lambda = 1.62\,\text{Å}$. The scattering intensity I of photons is measured in the SAXS experiment as a function of the scattering vector $q = 4\pi \sin(\theta)/\lambda$, where 2θ is the photons scattering angle. For clarity, the layout of the SAXS experiment is reproduced from [9] in Fig. 1.

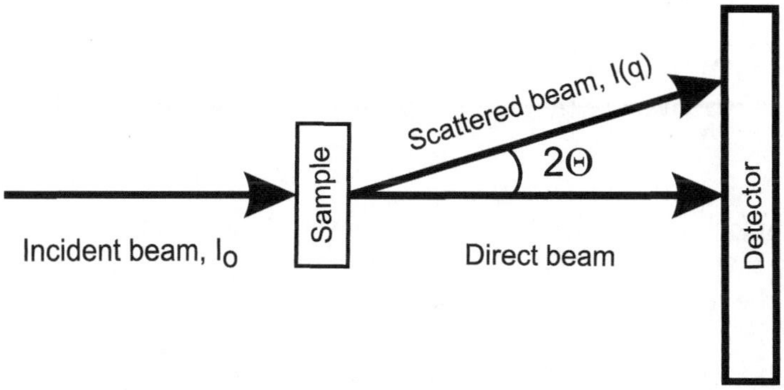

Figure 1. The layout of the small-angle scattering experiment. The incident photon beam with the intensity I_o is scattered from the sample at an angle 2θ. The intensity of the scattered beam $I(q)$ is measured by a position-sensitive detector.

The PTNS sample for the SAXS measurement is prepared via dilution of lyophilized PTNS (25% w/w) in water. The sample composition after dilution is as follows (w/w): 5% phospholipid, 20% maltose, 75% water. More details of the sample preparation are given in [9].

3. Model

In the SAXS data analysis, we employ two approaches based on the Separated Form Factor method (SFF) [11] and the Hollow Spheres model (HS) [12]. Both approaches suggest the vesicular structure of the PTNS particles as shown in Fig. 2.

Within the framework of the SFF model, in the case of the monodisperse vesicle population, the intensity $I(q)$ is given by the following expression:

$$I_{\mathrm{SFF}}(q) = n\, I_o\, F_s(q,R)\, F_b(q,d), \tag{1}$$

where I_o is the intensity of the incident beam, n — a number of vesicles in cm^3, $F_s(q,R)$ — form factor of the spherical surface with radius R, $F_b(q,d)$ — form factor of the symmetric lipid bilayer, d — thickness of the lipid bilayer of ULVs. $F_s(q,R)$ and $F_b(q,d)$ are determined as follows:

$$F_s(q,R) = \left(4\pi R^2 \frac{\sin(qR)}{qR}\right)^2, \quad F_b(q,\theta_b) = \left(\int\limits_{-d/2}^{+d/2} \rho_c(x)\cos(qx)dx\right)^2. \tag{2}$$

Here $\rho_c(x)$ is a contrast between the scattering length density of the lipid bilayer and the density of the solvent [13]. For the case of X-ray, the scattering length density is proportional to the density of electrons in the lipids or the solvent. Respectively, the contrast is proportional to

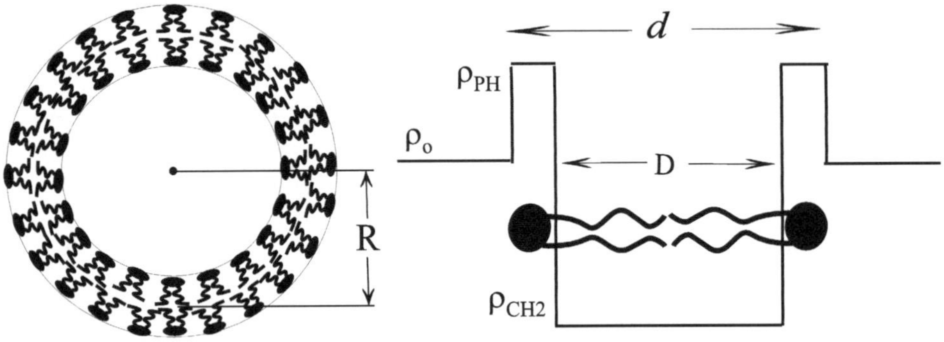

Figure 2. Left panel: structure of a phospholipid unilamellar vesicle with radius R. Right panel: the "step" distribution of the scattering length density across the bilayer.

the difference between electron density of lipid bilayer and surrounded solvent. We utilize the "step" form of the density distribution across bilayer, see Fig. 2. Here ρ_0 is the scattering length density of the solvent, ρ_{CH2} – scattering length density of the hydrocarbon chains, ρ_{PH} – scattering length density of polar head region.

In case of the HS model, the intensity $I(q)$ is determined as follows:

$$I_{HS}(q) = n\, I_o \left(\frac{4\pi}{q^3}\right)^2 \left[\sum_{i=1}^{n_s} \rho_i (A_{i+1} - A_i)\right]^2 , \tag{3}$$

where

$$A_i = \sin(qR_i) - qR_i \cos(qR_i).$$

In our calculations, we put $n_s = 3$ and determine $R_1 = R - d/2$, $R_2 = R - D/2$, $R_3 = R + D/2$, $R_4 = R + d/2$, $\rho_1 = \rho_3 = \rho_{PH} - \rho_0$, $\rho_2 = \rho_{CH2} - \rho_0$ in order to reproduce the density profile as in Fig. 2.

Polydispersity σ of the vesicular radius R relative to average radius $\tilde{R}$ is accounted utilizing the Schulz distribution with coefficient m:

$$G\left(\tilde{R}, R\right) = \frac{\tilde{R}^m}{m!} \left(\frac{m+1}{R}\right)^{m+1} \exp\left[-\frac{(m+1)\tilde{R}}{R}\right], \quad \sigma = \frac{1}{\sqrt{m+1}} \tag{4}$$

The intensity I_m for the polydispersed ULV system is obtained by means of a standard convolution procedure:

$$I_m(q) = \frac{\int_{R_{\min}}^{R_{\max}} I_{SFF,HS}(q, \tilde{R}) G\left(\tilde{R}, R\right) d\tilde{R}}{\int_{R_{\min}}^{R_{\max}} G\left(\tilde{R}, R\right) d\tilde{R}}. \tag{5}$$

The final expression of intensity $I(q)$ has the following form (details are given in [1, 2, 4, 10, 11]):

$$I(q) = I_m(q) + \frac{1}{2}\Delta^2 \frac{d^2 I_m(q)}{dq^2} + I_B, \tag{6}$$

Here I_B – parameter of incoherent background, $\Delta^2 = 3.6 \cdot 10^{-7}\,\mathrm{cm}^{-2}$ – the second momentum of a resolution spectrometer function [13].

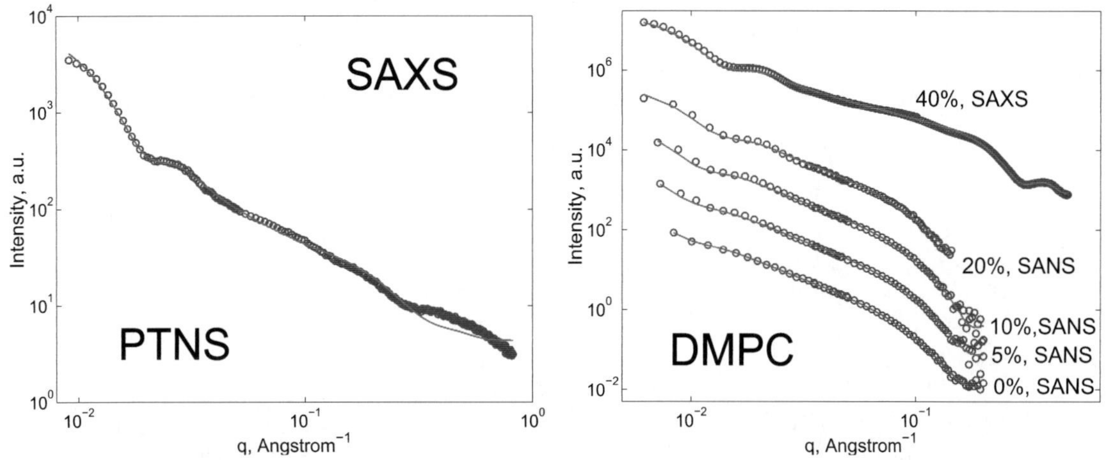

Figure 3. Left panel: the SAXS spectrum for the 25% (w/w) solution of phospholipid transport nanosystem in water. Circles – experiment, solid line – theory. Right panel: SANS and SAXS spectra for DMPC ULVs in different water concentrations of sucrose.

4. Results and Discussion

The experimental SAXS curve $I^{\text{exper}}(q)$ from 25% of PTNS solution in water is shown in Fig. 3 by circles. Plotted is the subtraction of the PTNS intensity and the pure water intensity.

Calculation on the basis SFF model with the "step" model of density distribution across bilayer (see Fig. 2) gives the following values of the average radius and bilayer thickness: $R_{\text{PTNS}} = 134.3 \pm 1.4$ Å and $d_{\text{PTNS}} = 43.5 \pm 0.5$,Å. Respective curve $I(q)$ is plotted in Fig. 3 by solid line. It is seen that the calculated intensity is in good agreement with the experimental data excluding a region of large q ($q > 0.3$ Å ^{1}).

A discrepancy between theoretical and experimental values of intensity

$$\chi^2 = \frac{1}{N-k}\sum_{j=1}^{N}\left(\frac{I(q_j) - I^{\text{exper}}(q_j)}{\delta^{\text{exper}}(q_j)}\right)^2, \tag{7}$$

(where N – number of experimental points and k – number of fitted parameters, δ – experimental errors) is equal to $\chi^2 = 2.6$. Similar calculations on the basis of the HS model give close results: $R_{\text{PTNS}} = 135.4 \pm 0.6$ Å, $d_{\text{PTNS}} = 36.9 \pm 0.2$ Å, $\chi^2 = 3.5$. Also, the calculated average radius is in agreement with the estimation 136 ± 18 Å from the dynamic light scattering experiment [7]. On the other hand, the calculated R_{PTNS} is about 15% less the estimation in [9], see Table 1.

The disagreement between experimental data and numerical results at $q > 0.3$ can be explained by fluctuations of bilayer. The fluctuations effect as well as more complex models of the density distribution has to be accounted in further analysis [10].

For comparison, we give in Table 1 parameters of the DMPC ULVs system in water solutions of sucrose with different concentrations calculated within the framework of SFF analysis of SAXS and SANS experimental data. The SANS spectra for sucrose concentration 0% - 20% in D_2O have been measured on YuMO small-angle spectrometer of IBR-2 reactor (JINR, Dubna, Russia). The SAXS measurements in the case of 40% sucrose concentration in water were performed on the A2 system of the Doris III synchrotron source (DESY, Hamburg, Germany). Details of experiment and fitting procedure are given in [10].

It is seen that the size of PTNS particles is smaller the DMPC ULVs size. Indeed, the minimal value of average radius of DMPS ULVs $R_{\text{DMPC}} = 190.6 \pm 1.6$,Åobtained in case of 40% sucrose

Table 1. Parameters of PTNS and DMPC vesicular systems. Parameters of PTNS vesicles have been obtained by means of SFF and HS analysis of SAXS specrum. Parameters of DMPC ULVs for different concentrations of sucrose in water are obtained by the SFF analysis of SANS and SAXS spectra.

System	Experiment	Model or Ref.	R, Å	d, Å	σ, %	χ^2
PTNS	SAXS	SFF	134.3±1.4	43.5±0.5	27	2.6
PTNS	SAXS	HS	135.4±1	36.9±0.2	33	3.5
PTNS	SAXS	Ref.[7]	136 ±18			
PTNS	SAXS	Ref.[9]	160±2	27.2±0.1	22	
DMPC 0%	SANS	SFF	307.5±5	49.8±1	30	1.1
DMPC 5%	SANS	SFF	270.6±5	48.1±2	31	5.9
DMPC 10%	SANS	SFF	243.4±3	47.6±1	27	6.6
DMPC 20%	SANS	SFF	213.4±2	44.5±2	26	9.1
DMPC 40%	SAXS	SFF	190.6±1.6	34.8±0.6	26	1.7
DMPC 40%	SAXS	Ref.[9]	210±2	20.5±0.1		

solution in water (SFF-analysis of SAXS data), is about 30% exceeds the numerically adjusted to SAXS data average radius of PTNS ULVs $R_{\mathrm{PTNS}} = 134.3 \pm 1.4$ Å.

The calculations show that the polydispersity σ of radius R in both DMPC and PTNS vesicle populations is about 30%, i.e. it is relatively low.

5. Summary

The SFF method is shown to be appropriate approach for analysis of SAXS and SANS data from polydispersed ULVs systems including PTNS particles. The SFF results of analysis of PTNS ULVs are shown to be in reasonable agreement with analysis on the basis of HS model. Calculated basic parameters of the polydispersed PTNS system do not contradict theoretical predictions and the SFF results of DMPC ULVs population.

The research was financed by the Russian Scientific Foundation (project No. 14-12-00516).

References

[1] Kiselev M A, Lesieur P, Kisselev A M, Lombardo D, Killany M, Lesieur S, Ollivon M 2001 *Nuclr. Inst. & Method* **A 470** 409
[2] Kiselev M, Lesieur P, Kisselev A, Lombardo D, Killany M, Lesieur S 2001 *J. Alloys and Compounds* **328** 71
[3] Kiselev M A, Wartewig S, Janich M, Lesieur P, Kisselev A M, Ollivon M, Neubert R 2003 *Chemistry and Physics of Lipids***123** (2003) 31
[4] Kiselev M A, Zbytovska J, Matveev D, Wartewig S, Gapienko I V, Perez J, Lesieur P, Hoell A, Neubert R 2005 *Colloids and Surfaces* **A 256** 1
[5] Das S, Chaudhury A 2011 *AAPS PharmSci. Tech.* **12** 62
[6] Martins S, Sarmento B, Ferreira D C, Souto E B 2007 *Int. J. Nanomedicine* **2** 595
[7] Archakov A I, Guseva M K, Uchaikin V F, Ipatova O M, Tikhonova E G, Medvedeva N V, Lisitsa A V, Prozorovskij V N, Strekalova O S, Shironin A V. Based on botanical phospholipids nanosystem for activation of biologically active compounds, and method of its manufacture (versions). Patent RU 2391966 1, Russian Federation.
[8] Ipatova O M, Torkhovskaya T I, Medvedeva N V, Prozorovsky V N, Ivanova N D, Shironin A V, Baranova V S, Archakov A I 2010 *Biomedical Chemistry* **4** 82
[9] Kiselev M A, Zemlyanaya E V, Ipatova O M, Gruzinov A Yu, Ermakova E V, Zabelin A V, Zhabitskaya E I, Druzhilovskaya O S, Aksenov V L 2015 *J of Pharmaceutical and Biomedical Analysis***114** 288
[10] Kiselev M A, Zemlyanaya E V, Zhabitskaya E I, Aksenov V L 2015 *Crystallography Reports***60** 143
[11] Kiselev M A, Zemlyanaya E V, Aswal V K, Neubert R 2006 *European Biophys. J* **35** 477
[12] Kušerka N, Kiselev M A, Balgavý P 2004 *European Biophys. J* **33** 328
[13] Feigin L A, Svergun D I 1987. Structure analysis by small-angle X-ray and neutron scattering. (Plenum Publishing Corporation, New York)

GALS – setup for production and study of multi-nucleon transfer reaction products: present status

S Zemlyanoy[1], V Zagrebaev[1], E Kozulin[1], Yu Kudryavtsev[2], V Fedosseev[3], R Bark[4] and Z Janas[5]

[1]Joint Institute for Nuclear Research, Dubna 141980, Russia
[2]Instituut voor Kern- en Stralingsfysica, Leuven, Belgium
[3]CERN, Switzerland
[4]iThemba LABS, Nat. Research Foundation, South Africa
[5]Faculty of Physics, University of Warsaw, Hoza 69, PL-00681 Warsaw, Poland

E-mail: zemlya@jinr.ru

Abstact. This is a brief report on the current status of the new GAs cell based Laser ionization Setup (GALS) at Flerov Laboratory for Nuclear Reactions (FLNR) - JINR, Dubna. GALS is planned to exploit available beams from the U-400M cyclotron in low energy multi-nucleon transfer reactions to study exotic neutron-rich nuclei located in the "north-east" region of nuclear map. Products from 4.5 to 9 MeV/nucleon heavy-ion collisions, such as ^{136}Xe on ^{208}Pb, are to be captured in a gas cell and selectively laser-ionized in a sextupole (quadrupole) ion guide extraction system.

PACS: 25.60.Je, 28.60.+s, 42.62.Fi

1. Introduction

The heavy neutron rich nuclei in the "north-east" part of the nuclide chart are from extremely importance for nuclear physics investigations, especially for the understanding of astrophysical nucleosynthesis and r-process. The closed neutron shell N = 126—the so called last "waiting point" for the r-process—is located just in this region (figure 1). Neutron shell N = 126 is a classical shell closures for nuclei along the line of stability. However, as one moves away from the line of stability, abrupt changes in nuclear structure can arise. Especially, appearance of new magic numbers and/or disappearance of traditional ones as well as the onset of deformation are expected, as e.g. already observed in the region of lighter nuclei. Study of the structural properties of nuclei along the neutron shell N = 126 could contribute to the current discussion about the quenching of shell gaps in nuclei with large neutron excess. A creation and launch of this facility will open a new field of research in low-energy heavy-ion physics, and new horizons in the study of unexplored "north-east" area of the nuclear map. It could be helpful also for finding a new way for the production of heavy and superheavy nuclei.

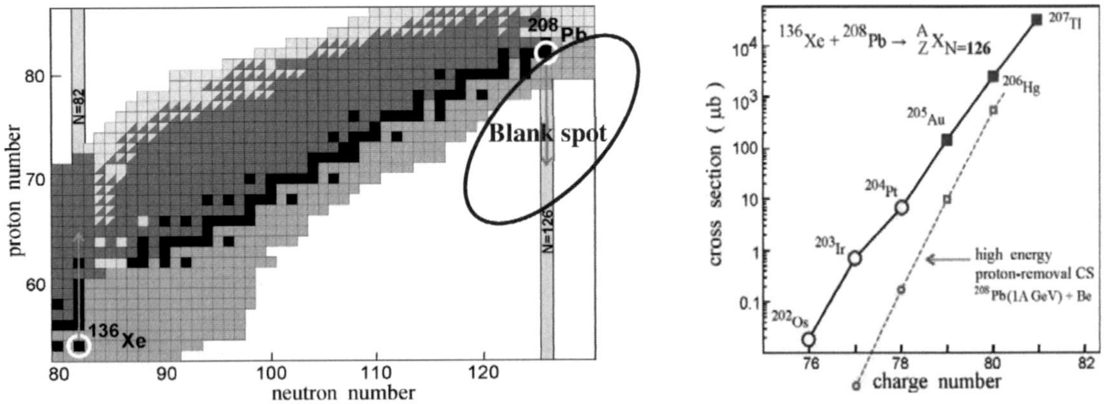

Figure 1. Left: upper "north-east" part of nuclear map. Right: cross sections for the formation of heavy nuclei in collisions of ^{198}Pt +^{238}U at center-of-mass energy of 700 MeV. Open circles indicate unknown isotopes [1]. The dashed curve illustrates the yield of the same isotopes in the proton removal reaction from closed shell nucleus ^{208}Pb.

Experiments at GALS are intended for production of heavy neutron rich nuclei in the above mentioned region by using multi-nucleon transfer reactions. As shown by Zagrebaev and Greiner [1], several tens of new nuclides in the region of N = 126 and Z ~ 75 can be produced, for example, in the near-barrier collision of ^{136}Xe with ^{208}Pb. Even higher cross sections have been predicted for the production of new neutron rich nuclei in collisions of ^{198}Pt beam with ^{238}U target. At present this seems to be the only realistically method [1-3] that could fill the "blank spot" of the nuclear map (figure 1).

2. Experimental stage

A most suitable experimental method for the purpose of this project is a combined method of Z and A/Q separation. Such method has been proposed by Van Duppen *et al* [4] in 1992 and subsequently intensively studied (first at LISOL) and developed [5-13]. Known as In-Gas Laser Ionization and Spectroscopy (IGLIS), it is based on stopping of nuclear reaction products in a gas cell and subsequent selective resonance laser ionization [5–13]. Such technique allows extracting nuclei with a given atomic number Z, while a separation of the single-ionized isotopes by their mass number can be done rather easily by a magnetic field. Half-lives of heavy neutron rich nuclei of interest (as a rule, β^-–decaying), are much longer than the extraction time of ions from a gas cell.

2.1 *Description on the experimental setup*

The schematic layout of the proposed facility can be found in [2, 3]. Here we present only a brief description of the GALS facility (figure 2). Neutron rich isotopes of heavy elements are produced in multinucleon transfer reactions with heavy ions accelerated up to 5–10 MeV/nucleon (depending on projectile-target combination). The target is a foil of about 300 μg/cm^2 thickness (or larger). It is placed at the window of the gas cell (or inside it). Nuclear reaction products recoiling out from the target as multi-charged ions are thermalized and neutralized by collisions with highly-purified argon or helium buffer gas. Then the atoms of interest (with a given Z) are ionized by means of two or three-step resonance laser irradiation and are extracted by the gas flow through the exit into the vacuum chamber as singly charged ions (Q = +1) with low energies of about 0.2 eV. Subsequently, the ions are confined in a radiofrequency

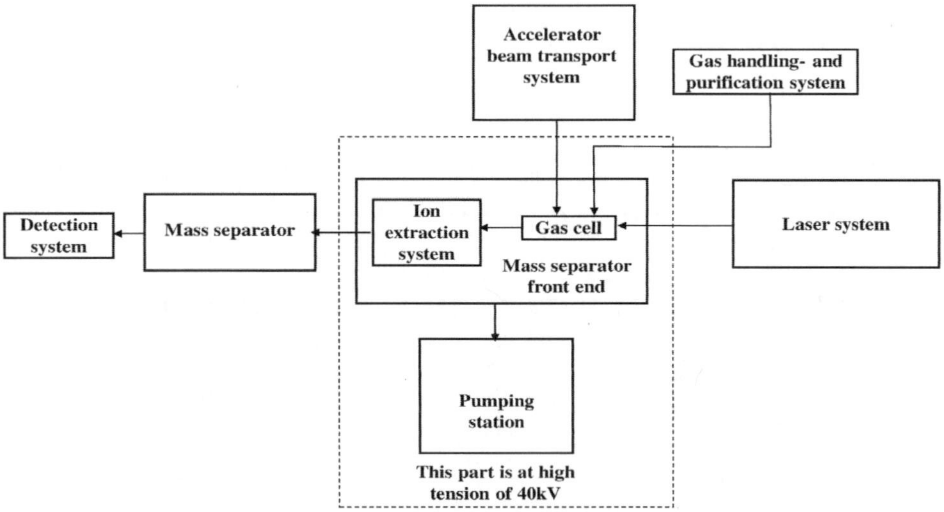

Figure 2. Block scheme of GALS-facilities.

ion guide system, which allows pumping out the residual buffer gas while transporting the ions towards the mass separator. Then the ions are accelerated up to 30–60 keV and selected by the mass–separator. In this way a low-energy beam of singly-ionized ions with a good optical quality (small emittance, energy spread less than 1 eV) is produced. This allows obtaining typical mass resolution of 1500 after the dipole magnet. After mass separation a beam of radioactive ions with a definite atomic number and a previously chosen mass value is obtained. The background due to unwanted isobar and isotope admixture is significantly suppressed that leads to an enhanced sensitivity. This gives the possibility to perform subsequently high sensitive analysis of spectroscopic and decay properties of these nuclei, as well as measurements of their spins, magnetic dipole and electric quadrupole moments and charge radii by means of laser spectroscopy. The operation principle of the different parts of the set up can be found in a number of papers (see e.g. [14] and the references therein). Below we sketch the plans of the GALS collaboration.

2.2 *Present status*

The GALS project will be realized in stage as determined by the FLNR plans and illustrated in table 1. The ion beams available at FLNR fully satisfy the requirements put from this project: the ions that can be used are quite different depending on the problem to be solved - from $^{16,18}O$ to ^{238}U, beam energies 4.5–9 MeV/nucleon (slightly above the Coulomb barrier) and beam intensity up to 10^{13} pps are accessible with a beam size at the target of 3–10 mm and beam emittance of $20\cdot\pi$·mm·mrad. Different heavy targets, including those of actinides, are expected to be used. Therefore the new facility can be developed and coupled directly to the available U-400M accelerator.

The main parts of the facility to be created and their present status are shown in table 1. Mass separator could be a standard magnet separator similar to GPS at ISOLDE II (see [2, 15]) but nevertheless substantial efforts are needed for its realization. The same refers to the gas purification system. Both, separator and gas purification systems, are in stage of construction and their realization and commissioning will be delayed to the end of 2016. The experimental halls and laser system are in the most advanced stage of realization (see figure 3).

Table 1. Status of the project as well as the planned activities in the next two years.

		2015	2016	2017
Laser system	mounting	yes		Beginning of experiments
	commissioning	yes	yes	
Front end system	mounting	yes	yes	
	commissioning		yes	
Pump station	mounting	yes	yes	
	commissioning		yes	
	preparation	yes		
Gas purification	mounting		yes	
	commissioning		yes	
	preparation	yes		
Separator, detection	mounting	yes	yes	
	commissioning		yes	

Figure 3. Laser room with installed lasers.

Table 2. Laser to be used in multistep ionization. The lasers already delivered and installed are presented in bold.

Type	Output power main&harmonic, W (2nd),$\{3^{rd}, 4^{th}\}$	Puls frequency, Hz	Puls length, ns	Wave length, nm
Dye laser	**3, (0.3)**	**10^4**	**10-30**	**213-850**
Ti:Sapphir	2, (0.2), {0.04}	10^4	30-50	680-960
Nd-YAG	**(80-100), {20-40}**	**10^4**	**10-50**	**532**
Matisse system				
Ring dye	0.8-6	cw	cw	540-900
Ti:Sapphir	0.8-6.5	cw	cw	700-1000

The choice of specific laser ionization scheme, the type and number of lasers is determined by the ionization potentials and level schemes of the elements under study. In our case a three-step scheme of ionization looks more favorable. Such a scheme allows choosing more effective optical transitions to increase the yield of resonance-ionized ions, although the use of two-step ionization scheme is as not excluded as well. Dye laser systems pumped by second and third harmonics of Nd:YAG can provide tuning in a broad spectral range: from near UV to near IR [16]. The generation of a Ti:Sapphir lasers, shifted to the red and infrared edge of the spectrum (680-960 nm), can be also used as complementary to that of the dye lasers. Thus, the installation of two laser-ionization schemes—with dye lasers and Ti:Sapphir laser—would be of a great benefit allowing to meet diverse experimental demands. Such a double RILIS scheme is already realized in CERN [17]. The parameters of the delivered (and already installed) laser as well as of the ordered ones are summarized in table 2.

At the first stage of our experiments, excitation schemes using 3-step ionization with a non-resonant transition to continuum will be tested with the available lasers: two dye lasers Credo (Sirah) and Nd:YAG. Credo laser has maximal average power of 20 W at fundamental wavelength and 2 W at the second harmonics; its line width is 1.8 GHz at pulse duration about 7 ns. It is an option that allows remote control of wavelength with stabilization to an external laser wavelength meter. The Nd:YAG laser (product of Edge Wave GmbH) generates maximal average power of 90 W and 36 W for the second and third harmonic respectively with repetition rate 10 - 15 kHz and pulse duration of 8 - 10 ns. The divergence parameter of the green beam is $M^2 = 1.4$. It needs electrical power supply of 3.6 kW including 1.6 kW for the water chiller.

A Matisse series of a ring cw-dye laser and cw-Ti:Sapphir laser pumped by cw-Nd:YAG is planned as a seeding system. Such system has an advantage in the cases when tuning of the laser wavelength is necessary, e.g. for laser spectroscopy.

3. Outlook

A new GALS facility at FLNR cyclotron U-400M is in a stage of build-up. GALS will apply the highly selective and efficient technique of stepwise resonant ionization in a gas cell (of the element of interest) with subsequently mass separation. In fact, it is a mass separator from a new type as yet not used in JINR. GALS is addressed to the production and study of new neutron rich nuclei located along the neutron shell $N = 126$. This is the so called "north-east" part of the nuclear map. Nuclei from this region are as yet "terra incognita" but from high fundamental interest mainly due to their large impact of on the astrophysical r-process. The most efficient method to produce such nuclei, as motivated in [1], are the multi-nucleon transfer reactions.

The experimental method is rather universal and allows extracting besides heavy products of multi-nucleon transfer reactions also any other nuclei with half-lives longer than a few tens of milliseconds including neutron rich fission fragments, fusion reaction products and light exotic nuclei. Such studies are already performed at other facilities, for example, in Finland [9], Belgium [10-13], and also using a different approach as that of the hot cavity ion source at CERN [16-17]. The efficiency of such facilities varies from 1%t to several tens percent depending on the half-life or the extracted ion

According to the estimation of the ion yields, at target thickness 0.3 mg/cm^2, ion beam of 0.1 pmA and setup efficiency of 10% we would be able to measure decay properties of 1 new isotope per day.

Acknowledgment. The authors would like to thank K. Marinova for helpful discussions and advices.

References

[1] Zagrebaev V and Greiner W 2008 *Phys. Rev. Lett.* **101**, 122701
[2] Zagrebaev V I, Zemlyanoy S G, Kozulin E M *et al.* 2014 *Hyperfine Interacs* **227** 181

[3] Zagrebaev V I, Zemlyanoy S G, Kozulin E M *et al* 2014 *J. Phys.Conf. Series* **533** 012025
[4] Van Duppen P , Dendooven P, Huyse M *et al* 1992 *Hypfine Interacs.* **74** 193
[5] Vermeeren L, Bijnens N, Huyse M. *et al* 1994 *Phys. Rev. Lett.* **73**, 1935
[6] Kudryavtsev Y *et al* 1996 *Nucl. Instrum. Methods Phys. Res.* B **114**, 350
[7] Kudryavstev Y *et al.* 2003 *Nucl. Instrum. Methods Phys. Res.* B **204**, 336
[8] Facina M *et al* 2004 *Nucl. Instrum. Methods* B **226**, 401
[9] Moore I D *et al* 2005 *J. Phys. G: Nucl. Part. Phys.* **31**, S1499
[10] Kudryavtsev Y *et al* 2009 *Nucl. Instrum. Methods Phys. Res.* B **267**, 2908
[11] Ferrer R *et al* 2012 *Nucl. Instrum. Methods Phys. Res.* B **291**, 29
[12] Ferrer R *et al* 2013 *Nucl. Instrum. Methods Phys. Res.* B **317**, 570
[13] Kudryavtsev Y *et al* 2013 *Nucl. Instrum. Methods Phys.* Res. B **297**, 7
[14] Campbell P, Moore I D and Pearson M R 2015 http://dx.doi.org/10.1016/j.ppnp.2015.09.003
[15] Ravn H L and Allardyce B W 1989 In: *Treatise on Heavy-Ion Science*, **8**, 363
[16] Fedosseev V N, Berg L E, Fedorov D V *et al.* 2012 *Rev. Sci. Instrum.* **83** 02A903
[17] Rothe S, Marsh B A, Mattolat C *et al* 2011 *J. Phys.: Conf. Series* **312**, 052020

Influence of different data tables on neutron induced reactions in quasi-infinite ^{238}U and ^{232}Th targets irradiated by protons with relativistic energy

P Zhivkov[1], Ch Stoyanov[1], S Tyutyunnikov[2], W Furman[2]

[1] Institute of Nuclear Research and Nuclear Energy, BAS, Bulgaria
[2] Join Institute of Nuclear Research, Dubna, Rassia

Abstract. The last decade saw the emergence of various theoretical analysis and developments of ADS (Accelerator Driving System). Different transport codes, nuclear models and nuclear cross sections have been used to predict and estimate the properties of ADS. The energy of the proton beam is supposed to range between 1 and 1.5 GeV, but some analyses suggest higher energy – up to 10 GeV. The recent papers examine the influence of the nuclear models on neutron induced reactions (n,f), (n,g), (n,xn), (n,el.) and (n,inel.). The experimental set-ups and the presumable ADS constructions consist of thousands of segments and details for example project Myrrha, Belgum [1]. The calculation of the above reactions depends on the neutron spectrum in each segment. There is a considerable difference in the size of these segments in ADS, which makes the estimation of the influence of the nuclear models and the cross sections on the integral number of neutron induced reactions more difficult. This article considers the influence of different cross section data tables on neutron induced reactions in ^{238}U or ^{232}Th targets. One nuclear model describing the high energy part of the nuclear interaction and various cross section data tagble (ENDF, ENDL, TENDL2014 and etc.) are used. All particles generated in the nuclear interaction process deposit their energy in the target volume. MCNP 6.1 transport code was used.

1. Introduction

Introduction: In the last two decades the study of the possibilities for nuclear waste transmutation, as well as the use of ^{238}U and ^{232}Th for energy production in ADS, have received considerable attention. The ADS consists of three components: 1) target comprised of heavy metals such as lead, bismuth, tungsten, fissionable nuclides or a composition of the above metals; 2) reactor part comprised of high-enriched nuclear fuel (25-30% mix of ^{238}U and ^{239}Pu) plus a cooling system; 3) proton accelerator 1-10 GeV. When the high-energy protons irradiate the target, high-energy particles are generated - neutrons, pions, protons, etc. The interaction of the high-energy particles with the uranium and thorium nuclei is described by nuclear models such as INCL4, QGSM, ISABEL, CEM, etc. All possible nuclear reaction channels are included in the models. These models are part of various transport codes such as MCNP6.1, FLUKA, GEANT4, MARS15, etc., which also include nuclear cross sections. Some of the transport codes have own neutron cross sections (neutron energy groups) like FLUKA. MCNP6.1 provides the opportunity to select cross sections for each nuclide. The standard tables of neutron induced cross reactions are defined for neutron energy up to 20 MeV or less. In MCNPX transport code it is possible to select cross sections for ^{238}U and ^{232}Th ranging between 20 and 200 MeV. The nuclear models in this transport code start when there are no cross sections. When we change only the nuclear cross sections in the input file, we can estimate their influence on the total number of fissions, captures and (n,xn) reactions. Up to now no experiments with ADS prototypes or a big target with neutron leakage less than

20% have been carried out. In order to provide a better comparison of the neutron induced reactions, we choose an infinite uranium/thorium target-sphere with a radius of 2 m. The proton source E_p=4 GeV (4π geometry) is located in the center of the sphere. With this geometry the particle leakage from the sphere is negligible.

1. Simulations

This paper aims to demonstrate the influence of the cross section data tables on the total number of fissions, captures and (n,xn) reactions in quasi–infinite ^{238}U and ^{232}Th targets. This means that all particles which participated in the interaction deliver their energy in the target volume and the particles leakage is negligible. We use ^{238}U and ^{232}Th spheres with R=2m and point proton source, E_p=4 GeV located in the centre of the sphere. The nuclear model which we use is INCL4 and the evaporation part is ABLA which are included in MCNP6.1 transport code. The cross sections data tables used in the calculations are shown in table 1 and 2. The features of the MCNP6.1 output file are extremely important. First: The number of (n,f), (n,γ) and (n,xn) reactions are calculated with data table cross sections only. It is important to note that nuclear models calculate all type of neutron induced reactions, but these reactions are not involved in neutron particle balance. Second: charged particle-and gamma-induced fissions are not involved in the output file, but they are calculated by the program.

1.1 Calculations of the total number (n,f), (p,f), (π,f), (γ,f), (n,γ) and (n,xn) reactions

We can calculate the above reactions in three different ways. First: in the MCNP6.1 output file the total balance of all particles and the integral number of the nuclear reactions per one projectile are given, but charged particle- and gamma-induced fission are not included. Second: by the FM card which gives the number of reactions in a volume/cell per one projectile. The reactions are calculated by eq. 1:

$$C\int F_i(E)\,\sigma_i dE \qquad (1)$$

where F (number of particles/cm^2) is flux and σ (1/cm^2) is cross section, C is normalization constant. The equation (1) is defined only for cross sections data tables if they are available. If there are no cross sections data tables, MCNP uses nuclear models. All nuclear reaction channels are included in the models, but it is not possible to see them in the output file. The limitations of the neutron cross sections are published in the MCNP6.1 manual. These two methods are limited and cannot calculate reactions (p,f), (π,f), (γ,f). Third: calculation of the energy spectra for all particles of interest with the F4 card measured in particles/cm^2. The number of the reactions can be calculated by eq. 2.

$$\sum_i^n F_i(E)\sigma_i(E) \qquad \mathbf{(2)}$$

Where F_i is a particle flux [particles/cm^2] for specific energy interval and σ_i is the average cross section for the same energy bin. There are experimental measurements of (n,f) cross sections for ^{238}U and ^{232}Th for incident neutron energy up to 200 MeV presented in [2], which can be used in eq. 2. By eq. (2) the total number of all reactions can be calculated.

2. Results and conclusions

The outcomes are shown in tables 1 and 2. The first column presents the maximum incident neutron energy, i.e. the energy limit of the neutron cross sections. The second column shows the total number of neutrons generated in the target-sphere. The third column presents the number of spallation neutrons with energy greater than that in column 1. Columns 4, 5 and 6 show the number of reactions occurring in the target-sphere calculated by MCNP6.1, eq. (1). If we want to calculate the total fission reactions, the eq. (2) has to be used for all particle-induced fissions. It is important to note that the neutron multiplicity (the number of neutrons irradiated per one fission) depends on the energy of the incident particles [4,5]. If the neutron fission cross section is extended, then the neutron multiplicity must also be extended to the same energy.

Table 1. Neutron-induced reactions in a big ^{238}U sphere (R=2m), E_p=4GeV

Data table	E (MeV)	N_{total}	N_{spall}	(n,xn)	(n,γ)	(n,f)
endf66d	20	556	200	12	338	60
endl92(LLNL)	30	540	206	14	342	66
endf70j	30	537	206	14	325	55
t16-2003 (LANL)	30	508	205	14	325	55
100xs(LANL)	100	491	124	30	318	56
TENDL2012	20	490	197	14	313	53
TENDL2014	200	415	90	20	270	54

MCNP 2.7 was use only with TENDL2012/14 data tables calculations and INCL4/ABLA nuclear models. MCNP6.1 was used with all other cross sections plus INCL4/ABLA.

Table 2. Neutron-induced reactions in a big ^{232}Th sphere (R=2m), E_p=4GeV

Data table	E (MeV)	N_{total}	$N_{spall.}$	(n,xn)	(n,γ)	(n,f)
uresa	20	340	206	15	218	10
endf66c	20	340	206	14	218	10
endl92	30	323	188	16	211	11
endf70c	60	332	147	23	223	13
TENDL2012	200	355	87	39	236	18
TENDL2014	200	354	87	38	236	19

In fact, the results received by MCNP6.1 for ^{232}Th target with TENDL2012 and TENDL2014 [3] cross sections coincide. On the other hand, the same data table comparison for ^{238}U shows significant differences for the reaction (n,xn) - 30%, the (n,γ) reaction and the total neutron production have a difference of 15% and the (n,f) reactions are the same which is an unexpected result. Besides, there is a difference of 17-18% in the (n,f) ^{238}U reaction calculated with endl92 and endf70j. All other reactions and parameters for both data tables almost coincide. The analysis of 100xs (defined up to 100MeV) and endf70j (defined up to 30MeV) also shows a significant difference in the (n,xn) reactions-two times, but the reactions (n,f) coincide.

The (n,f)^{232}Th cross sections TENDL2014 and ENDF70 in the 0-60 MeV energy range coincide. In the energy range from 60 to 200 MeV the (n,f) ^{232}Th cross sections decrease monotonically from 0.75 to 0.06 barns, but the experimental cross sections [2] are a plateau of 0.8 barns and are used in eq. 2. Despite the increase of the (n,f) ^{232}Th experimental cross sections in comparison with TENDL2014 the total number of (n,f) ^{232}Th reactions is lower (15%) than the one calculated by TENDL2014 in the energy range 0-200MeV. The cross sections of (n,xn)^{232}Th for TENDL2014 and ENDF70 in the energy range 1-30 MeV they actually coincide, but the total number of the reactions (n,xn) ^{232}Th differs with 40%. Calculations shows that the neutron flux in ^{232}Th target for the energy range from 1 to 200 MeV calculated with TENDL2014 is 30% higher than with ENDF70. It is the reason why the (n,f) and (nxn) reactions for ^{232}Th are 30% more with TENDL2014 calculations than with ENDF70. For ^{238}U target and ENDl92 and ENDF70 such difference in the neutron flux was observed.

The (n,γ) ^{238}U cross sections are well-known TENDL2014 and ENDF70 coincide but the total number of captures in the ^{238}U target fluctuates with 25% for different data tables. The difference is quite significant, because (n,γ)^{238}U means ^{239}Pu accumulation in the target. In contrast to the (n,γ) ^{232}Th reaction the fluctuation is 5%, which is much more acceptable. We should bear in mind that the (n,γ) ^{232}Th reaction means ^{233}U accumulation in the target. The fluctuation of the total neutron production as a function of data tables for ^{238}U and ^{232}Th targets is 30% and 5% respectively. All these calculations of neutron-induced reactions in the massive target or ADS based on ^{238}U and ^{232}Th show that the data cross sections are important. If the neutron-induced reactions are calculated in relatively small cells or regions in the targets/ADS, the above results are not visible and the calculations are within the error range.

3. References

[1] http://myrrha.sckcen.be/
[2] O Shcherbakov, A Donets at all. 2002 Neutron-Induced Fission of ^{233}U, ^{238}U, ^{232}Th, ^{239}Pu, ^{237}Np, natPb and ^{209}Bi Relative to ^{235}U in the Energy Range 1-200 MeV *Journal of Nuclear Science and Technology* September 2 pp 230-233
[3] http://www.talys.eu/
[4] D Madland 2006 Total prompt energy release in the neutron- induced fission of ^{235}U, ^{238}U, and ^{239}Pu *Nuclear Physics A* 772 pp 113–137
[5] T Ethvignot, M Devlin, H Duarte, T Granier, R Haight, B Morillon, R Nelson 2005 Neutron Multiplicity in the Fission of ^{238} U and ^{235} U with Neutrons up to 200 MeV *Physical Review Letters* 94 052701

Study of neutron-deficient isotopes of Fl in the ^{239}Pu, ^{240}Pu + ^{48}Ca reactions

A A Voinov[1a], V K Utyonkov[1], N T Brewer[2], Yu Ts Oganessian[1], K P Rykaczewski[2], F Sh Abdullin[1], S N Dmitriev[1], R K Grzywacz[2,3], M G Itkis[1], K Miernik[2,4], A N Polyakov[1], J B Roberto[5], R N Sagaidak[1], I V Shirokovsky[1], M V Shumeiko[1], Yu S Tsyganov[1], V G Subbotin[1], A M Sukhov[1], A V Sabelnikov[1], G K Vostokin[1], J H Hamilton[6], M A Stoyer[7], S Y Strauss[8]

[1]Flerov Laboratory of Nuclear Reactions, Joint Institute for Nuclear Research, RU-141980 Dubna, Russian Federation

[2]Department of Physics and Astronomy, Oak Ridge National Laboratory, Oak Ridge, TN 37831, USA

[3]Department of Physics and Astronomy, University of Tennessee, Knoxville, Tennessee 37996, USA

[4]Faculty of Physics, University of Warsaw, Warsaw, Poland

[5]Science and Technology Partnerships Directorate, Oak Ridge National Laboratory, Oak Ridge, TN 37831, USA

[6]Department of Physics and Astronomy, Vanderbilt University, Nashville, Tennessee 37235, USA

[7]Nuclear and Chemical Sciences Division, Lawrence Livermore National Laboratory, Livermore, CA 94551, USA

[8]Department of Physics, University of Notre Dame, Notre Dame, Indiana 46566, USA

Abstract. The results of the experiments aimed at the synthesis of Fl isotopes in the ^{239}Pu + ^{48}Ca and ^{240}Pu + ^{48}Ca reactions are presented. The experiment was performed using the Dubna gas-filled recoil separator at the U400 cyclotron. In the ^{239}Pu+^{48}Ca experiment one decay of spontaneously fissioning ^{284}Fl was detected at 245-MeV beam energy. In the ^{240}Pu+^{48}Ca experiment three decay chains of ^{285}Fl were detected at 245 MeV and four decays were assigned to ^{284}Fl at the higher ^{48}Ca beam energy of 250 MeV. The α-decay energy of ^{285}Fl was measured for the first time and decay properties of its descendants ^{281}Cn, ^{277}Ds, ^{273}Hs, ^{269}Sg, and ^{265}Rf were determined more precisely. The cross section of the ^{239}Pu(^{48}Ca,3n)^{284}Fl reaction was observed to be about 20 times lower than those predicted by theoretical models and 50 times less than the value measured in the ^{244}Pu+^{48}Ca reaction. The cross sections of the ^{240}Pu(^{48}Ca,4-3n)^{284,285}Fl at both ^{48}Ca energies are similar and exceed that observed in the reaction with lighter isotope ^{239}Pu by a factor of 10. The decay properties of the synthesized nuclei and their production cross sections indicate rapid decrease of stability of superheavy nuclei with departing from the neutron number N=184 predicted to be the next magic number.

[a] voinov@jinr.ru

1. Introduction

One of the outcomes of macroscopic-microscopic nuclear theory [1] predicts the existence of the island of stability of the super-heavy nuclei around Z=114 and N=184. The significant success in the synthesis and decay properties study of the superheavy nuclei was achieved in the FLNR JINR (Dubna, Russia) during last 15 years [2]. Six new super-heavy elements with Z=113-118 and more than 50 new isotopes with Z=104-118 were observed for the first time at the Dubna Gas-Filled Recoil Separator (DGFRS) in the irradiations of ^{238}U, ^{237}Np, ^{242,244}Pu, ^{243}Am, ^{245,248}Cm, ^{249}Bk and ^{249}Cf targets with accelerated ^{48}Ca ion beam delivered by U-400 cyclotron. The first super-heavy nucleus with Z=114 was discovered in June, 1999 in the complete-fusion reaction ^{244}Pu + ^{48}Ca performed by Dubna-Livermore collaboration. The detected parent nucleus and its decay chain were assigned to ^{289}Fl formation (see [2] and Refs. therein). Later, the properties of lighter isotopes $^{286-288}$Fl were determined [3-5] both as in direct cross bombardment reactions of ^{242}Pu and ^{244}Pu targets with ^{48}Ca ion beam at different beam energies corresponding to the maximum reaction yield with evaporation of 2-5 neutrons, and as descendant daughter nuclei in the decays of synthesized $^{290-293}$Lv and granddaughter of ^{294}118 in irradiations of ^{245,248}Cm and ^{249}Cf targets [5-7]. Experiments on the synthesis of Fl and Lv isotopes were repeated in other laboratories with IVO experimental setup [8-11], SHIP [12, 13], BGS [14, 15] and TASCA [16, 17]. Thus, the production cross-sections and decay properties of the isotopes of $^{286-289}$114 and $^{291-293}$116 were recently confirmed in the independent experiments.

Few attempts have been undertaken to expand the region of super-heavy nuclei with synthesis of elements 119 and 120 by using target nuclei ranging from ^{238}U to ^{249}Cf with projectiles heavier than ^{48}Ca, from ^{50}Ti to ^{64}Ni. However, no decay chains of SHN were observed in these experiments. The upper cross section limits were set at 0.07 – 1.1 pb depending on the reaction studied [18-21]. The next step in the exploring of island of stability of the SHN could be performing the nuclear reactions leading to new isotopes at the edge of the new SHE region. For example, in irradiation of ^{239}Pu, ^{240}Pu and $^{249-251}$Cf targets with ^{48}Ca beam one could observe the formation and decay of both very light $^{283-285}$Fl isotopes and two the heaviest isotopes of 118 element: ^{295}118 and ^{296}118 [19].

In this work, we present the results of experiment on the synthesis of neutron-deficient Fl isotopes in the ^{239}Pu + ^{48}Ca and ^{240}Pu + ^{48}Ca fusion-evaporation reactions. These experiments have provided evidence for the SF decay of a new isotope, ^{284}Fl, and new properties of the ^{285}Fl decay chain.

2. Experiment

The ^{48}Ca ion beam was delivered to the target with a maximum intensity of 1.3 particle μA. The beam energy was determined with a systematic uncertainty of 1 MeV by a time-of-flight system placed in front of the DGFRS.

The target materials were provided by Oak Ridge National Laboratory (ORNL) (^{240}Pu, enrichment of 98.97%) and JINR (^{239}Pu and ^{240}Pu, enrichment of 98.2 and 92%, respectively). The impurities in the ^{239}Pu and ^{240}Pu targets mainly consisted of ^{240}Pu and ^{239}Pu, respectively. The average thicknesses of the targets for the main isotopes were 0.50±0.05 mg/cm^2 for ^{239}Pu and 0.39±0.04 mg/cm^2 for ^{240}Pu (given uncertainties correspond to standard deviations of thicknesses measured for six sectors of each target). The targets were manufactured by depositing PuO_2 oxide onto 0.71-0.72 mg/cm^2 Ti foils. Each target had an area of 5.4 cm^2 in the shape of an arc segment with an angular extension of 60 deg and an average radius of 60 mm. The segments were mounted on a disk that was rotated at 1700 rpm such that the target was perpendicular to the direction of the incoming beam. In the course of the bombardment with the ^{48}Ca beam, the target layers were systematically monitored by counting α-particles from the decay of the target isotopes.

The lab-frame beam energies in the middle of the target layers, excitation energy ranges (with use of mass tables [20,21]) and beam doses for the experiments studied are summarized in Table I. For calculation of excitation-energy ranges of the resulting compound nuclei ^{287,288}Fl, we took into account the beam energy resolution, the small variation of the beam energy during irradiation, and the energy loss in the target.

Table I. Target, reaction-specific lab-frame beam energies in the middle of the target layers, corresponding excitation energy intervals, and total beam doses for the given reactions.

Target	E_{lab} (MeV)	E^* (MeV)	Beam dose
^{240}Pu	245	36.5-41.1	4.0×10^{18}
^{240}Pu	250	40.9-45.4	4.7×10^{18}
^{239}Pu	245	35.4-40.0	1.4×10^{19}

Evaporation residues (ER) recoiling from the target were separated in flight from ^{48}Ca beam ions, scattered particles and transfer-reaction products by the DGFRS. The transmission efficiency of the separator for Z=114 nuclei was estimated to be about 35%. Recoils passed through a time-of-flight system (TOF) and were implanted in the detectors. The TOF system consists of two multi-wire proportional counters (MWPC) placed at a distance of 65 mm between them. The detectors and MWPC are placed in Pentane at a pressure of about 1.5 Torr. A 0.2-mg/cm^2 Mylar foil separates the detection system from the DGFRS volume, which is filled with hydrogen at a pressure of 1 Torr.

The array of Silicon detectors at the DGFRS final focus has been modified to increase the position resolution of recorded signals and subsequently reduce the probability of observing sequences of random events that mimic decay chains of implanted nuclei. The new detection system includes a 0.3-mm thick Double-sided Silicon Strip Detector (DSSD) manufactured by Micron Semiconductor Ltd (model BB-17). This large DSSD has 1-mm wide strips, 48 at the front side and 128 at the back side, creating over 6000 1-mm^2 pixels in one Silicon wafer. Such high pixilation helps to achieve superior position resolution for recoil-correlated decay sequences reducing potential random events. The recoil implantation counter was surrounded by six single Si-detectors, MICRON model MSX-7200, each 500 microns thick and having an active area of 65 mm by 120 mm. Two pairs of these Si-box detectors were mounted at the DSSD long side, and two detectors were used to close the Si-box geometry, one at each short side. All Si counters had a minimum amount of supporting frame material. The active detection length for the DSSD escape events was extended about 120 mm, from the DSSD surface towards the separator. The DSSD was backed by the single Si-veto detector (MICRON MSX-62), of 0.5 mm thickness and 48 mm by 128 mm active size matching the respective BB-17 area. This veto counter was mounted in the frame identical to the BB-17 support about 3 mm from the back surface of the DSSD. The signals from all detectors were processed using MESYTEC preamplifiers. This new Si-detector array was designed, assembled, commissioned off-line and provided by Oak Ridge National Laboratory.

The output signals from the MESYTEC linear-logarithmic preamplifiers serving BB-17 DSSD as well as linear preamplifiers serving MSX-7200 and MSX-62 detectors were split into two branches. One of these branches was processed with analog electronics similar to those used in previous DGFRS experiments, see [22]. The analog electronics system was used to create a dedicated low-background detection scheme for the nuclei to be investigated. This detection scheme allows the beam to be switched off after a predefined event sequence. The beam interruption occurs after the detection of a recoil signal with the expected implantation energy for Z=114 evaporation residues followed by an α-like signal in the implantation detector with an energy of 9.8-11.5 MeV, in the same front and back strips, i.e., in the same 1-mm^2 DSSD pixel.

The second branch of split preamplifier signals was processed using a digital electronics system based on XIA Pixie-16 modules provided by ORNL. Such a digital pulse processing system was developed and applied successfully in earlier experiments on short-lived charged particle emitters at the Recoil Mass Separator (RMS) of ORNL's Holifield Radioactive Ion Beam Facility [23,24].

The FWHM energy resolution of the implantation detector was 34 to 78 keV, while the summed signals recorded by the side and implantation detectors had an energy resolution of 147 to 263 keV.

Other experimental conditions, including the method of calibration of the detectors, were the same as in previous DGFRS experiments (see [2] and references therein).

3. Results

According to excitation functions measured in the reactions of ^{48}Ca with target nuclei ^{238}U-^{249}Cf (see [2] and references therein), the maximum of the cross section of the fusion-evaporation reaction ^{240}Pu+^{48}Ca is expected at an excitation energy E^* of the compound nucleus ^{288}Fl of about 40 MeV. In the first experiment, performed at ^{48}Ca energy E_{lab}=245 MeV (Table I), we observed three decay chains of ^{285}Fl, the product of the 3n-reaction channel (Fig. 1). This isotope was first registered in the ^{242}Pu(^{48}Ca,5n)^{285}Fl reaction at LBNL [25]. In their single decay chain, the α particle of the parent nucleus escaped from the front of the five-sided detector box leaving the energy of 1.64 MeV in the implantation detector. Four further α decays of nuclei from ^{281}Cn to ^{269}Sg were detected with full energy by the front detector only or simultaneously with side one.

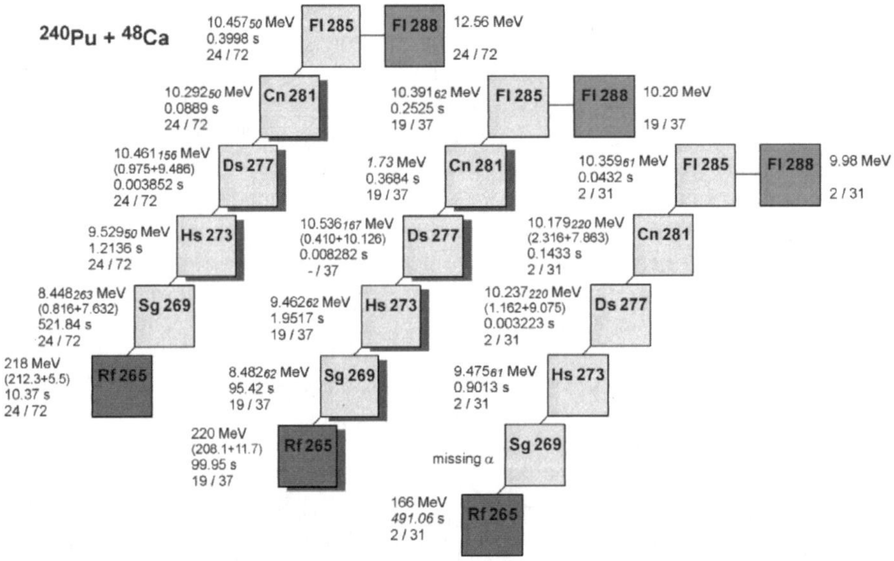

Figure 1. Decay properties of ^{285}Fl and descendant nuclei observed in the ^{240}Pu+^{48}Ca reaction at projectile energy of 245 MeV. The upper right rows for each chain show ER (in pink) energies and strip numbers (front / back). The left rows provide energies, time intervals between events and their strip numbers for α decay (in yellow) and SF (in green). Energies of summed signals are given in parentheses. Events marked with a shadow were registered during the beam-off periods. The α-particle energy errors are shown by smaller italic numbers. The time interval for an SF event following a "missing α" was measured from the preceding registered event and is shown in italics.

In all three decay chains observed in the present ^{240}Pu+^{48}Ca reaction, the α particle of ^{285}Fl was registered by the focal-plane detector with full energy which resulted in the first measurement of its energy E_α. The decay times of nuclei ^{285}Fl-^{265}Rf as well as E_α values of ^{281}Cn-^{269}Sg in three chains are in good agreement with those measured in the ^{242}Pu(^{48}Ca,5n)^{285}Fl reaction [25]. Only the energies of α particles of ^{277}Ds in the third chain and ^{273}Hs in the second and third chains are somewhat lower than values measured in [25] (by 73 and about 25 keV, respectively, taking into account energy resolutions of detectors). However, these deviations do not seem significant.

During the 527-hour ^{240}Pu + 245-MeV ^{48}Ca experiment when the beam was on the target, the total number of sequences consisting of ER-like events with E_{ER}=6-16 MeV and α-like events with E_α=10.1-10.7 MeV detected within 1.5 s in the same front and back strips of the focal-plane detector was only 57. In the first two decay chains of ^{285}Fl shown in Fig. 1, the α decay of the parent nucleus

switched the beam off. Three further α particles of ^{281}Cn-^{273}Hs in the first case and four α particles of ^{281}Cn-^{269}Sg and SF of ^{265}Rf in the second chain were registered in the absence of beam-associated background. The total duration of beam-off intervals in this run was about 4.3 h. In the first chain, the beam-off pause was not manually prolonged and decays of ^{269}Sg and ^{265}Rf were registered when the beam was switched on. For beam-off α decays with E_α=8-11 MeV, the probability of their detection as random events in any crossing of strips within period Δt=3 min was about 1.3×10^{-4} [26]. For beam-on α- and SF-like events with energies E_α=8-11 MeV and E_{SF}>130 MeV, respectively, and Δt=10 min these values were 1.2×10^{-2} and 1.3×10^{-5}, respectively. The decay of ^{281}Cn in the second chain was registered by the focal-plane detector only with energy of 1.7 MeV while the beam was stopped by ER-α_1 sequence. The probability of the random origin of an event with any energy in strips 19/37 within Δt=1 s was less than 7×10^{-3}. In the third chain, the first α particle did not stop the beam, for an unknown reason. Two subsequent α decays were detected by both the focal-plane and side detectors and could not switch the beam off. Thus, all these events as well as the fourth α particle and the SF event were observed during the beam-on time interval. However, the probability of detection of random beam-on events with E_α=8-11 MeV within Δt=5 s was 1.0×10^{-4}.
In the third chain, no α particles with E_α=8-11 MeV were found in strips 2 / 31 between decays attributed to ^{273}Hs and ^{265}Rf. This missing α event could be detected by the focal detector only with low energy-release, but the probability of detection of similar random beam-on events within the time interval of 491 s is rather large preventing the definite assignment of such an event to ^{269}Sg. Missing this α particle and incomplete detection of the energy of ^{281}Cn in the second chain are in accordance with the 85-% registration efficiency of the detectors. From the remaining 13 α particles, five decays were registered by the both implantation and side detectors which also corresponds to their detection efficiencies. In one of these events, the α-particle energy of ^{277}Ds absorbed by the focal detector was below the energy threshold, and the number of the front strip was not determined. However, the energy and corresponding strip number were registered in the back side of the detector.

In the second experiment, the energy of ^{48}Ca was increased to 250 MeV for measurement of the excitation function of the ^{240}Pu+^{48}Ca reaction and possible observation of the product of the 4n-evaporation channel, the new lightest isotope ^{284}Fl. In this experiment, decay chains of ^{285}Fl were not observed which indicates a decrease of the cross section of the 3n channel at this ^{48}Ca energy ($\sigma_{3n}\leq$1.3 pb).

However, in this run, four ER-SF chains were found which are shown in Fig. 2. In addition, 27 recoil-fission (focal-plane only events) were found with E_R=6-16 MeV, E_F>130 MeV, and Δt=100 s, with a random distribution of R-F time intervals varying from 3.3 to 96.6 s, therefore, the total number of random ER-SF chains for Δt=10 ms is 3×10^{-3}.

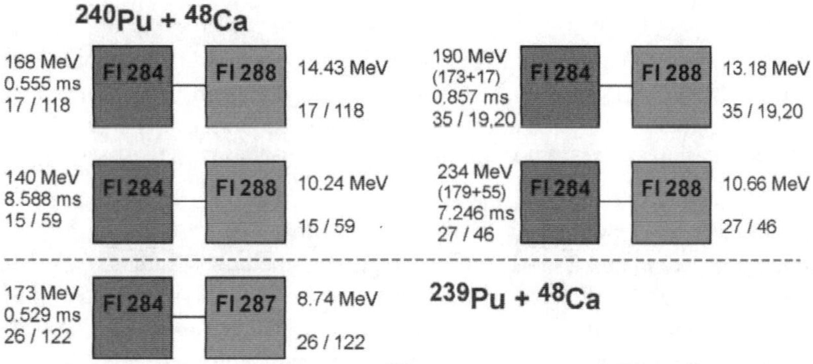

Figure 2. Decay properties of nuclei assigned to ^{284}Fl observed in the ^{240}Pu+^{48}Ca (top) and ^{239}Pu+^{48}Ca (bottom) reactions. The right rows for each chain show ER (in pink) energies and strip numbers (front / back). The left rows provide SF fragment (in green) energies, time intervals between events and their strip numbers. The decay chain shown in the right top corner was registered in two back strips.

The same nucleus ^{284}Fl could be produced in the 3*n*-evaporation channel of the reaction with the lighter Pu isotope, ^{239}Pu+^{48}Ca, and its observation could make a cross-bombardment consistency check on the synthesis of the new Fl isotope. This experiment was performed at one ^{48}Ca energy E_{lab}=245 MeV which corresponds to expected maximum of the cross section of the 3*n*-evaporation channel [27].

In the ^{239}Pu+^{48}Ca reaction short ER-SF chain with decay time of 0.53 ms was detected (see Fig. 2) as well as 30 another chains with E_R=6-16 MeV, E_F>130 MeV, and Δt=100 s (again all these fission-like events were detected by the focal-plane detector only). Like in the previous case, the number of random ER-SF chains for Δt=10 ms was estimated to be 3×10^{-3}.

The energy of the ^{48}Ca in the reaction with ^{239}Pu was the same as in the ^{240}Pu experiment where the 3*n*-reaction channel was observed. This also corresponds to the predicted cross-section maximum of the ^{239}Pu(^{48}Ca,3*n*)^{284}Fl reaction [27]. Based on these facts, one may assign this ER-SF chain to ^{284}Fl. In that case, the average half-life of this isotope calculated for five decays would be equal to $T_{1/2}=2.5^{+1.8}_{-0.8}$ ms. The production cross section was found to be $\sigma_{3n}=0.23^{+0.59}_{-0.20}$ pb, which is lower by factor of 10 than that for the reaction with ^{240}Pu. An upper cross-section limit of 0.47 pb can be set if this decay does not originate from ^{284}Fl.

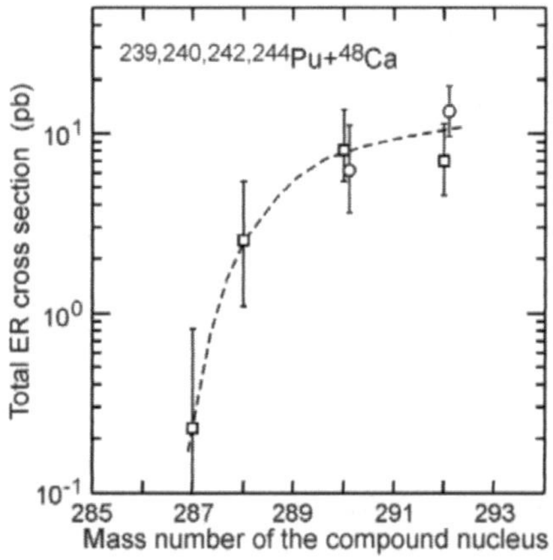

Figure 3. Maximum of the production cross sections of Fl isotopes in the ^{48}Ca-induced reactions with targets shown in figure vs. mass number of the compound nucleus. Data from DGFRS [4,5] are shown by squares as well as from BGS [25] and TASCA [16] by circles. Vertical error bars correspond to total uncertainties. The dashed line is drawn to guide the eye.

4. Conclusion

In the ^{240}Pu+^{48}Ca reaction, three decay chains of ^{285}Fl were observed. The α-particle energy of ^{285}Fl (E_α=10.41±0.05 MeV) was measured for the first time. The decay properties of isotopes ^{285}Fl, ^{281}Cn, ^{277}Ds, ^{273}Hs, ^{269}Sg and ^{265}Rf were determined more precisely. The cross section of the ^{240}Pu(^{48}Ca,3*n*)^{285}Fl reaction for 245-MeV projectiles was measured to be $2.5^{+2.9}_{-1.4}$ pb which exceeds that for ^{239}Pu by a factor of 10 but is 2-4 times lower than that for ^{244}Pu.

Two ER-SF decay chains observed in the ^{240}Pu + 250-MeV ^{48}Ca reaction with high energy-release of fission fragments (E_{SF}=190 and 234 MeV) are assigned to the new spontaneously fissioning isotope ^{284}Fl. Two more ER-SF events registered with lower energy values (E_{SF}=140 and 168 MeV) could also originate from ^{284}Fl (assignment to ^{240,242,244}Am fission isomers is not rejected also). The cross section of the ^{240}Pu(^{48}Ca,4*n*)^{284}Fl reaction ($2.6^{+3.3}_{-1.7}$ pb) is similar to that for the 3*n*-reaction channel.

In the ^{239}Pu+^{48}Ca reaction, one ER-SF chain was observed with SF decay properties comparable with those registered for ^{284}Fl in the reaction with ^{240}Pu. The production cross section for this event, presumably the product of the ^{239}Pu(^{48}Ca,3*n*)^{284}Fl reaction, of 0.2 pb is 20 times lower than the theoretically predicted value and 50 times lower than maximum fusion-evaporation cross section measured in the reaction with ^{244}Pu.

The considerable drop of the evaporation cross sections of the ^{239}Pu+^{48}Ca and ^{240}Pu+^{48}Ca reactions as well as the decline of the half-life and dominance of spontaneous fission over α decay for ^{284}Fl and increased growth of the α-decay energy of ^{285}Fl compared to the heavier Fl isotopes indicate one is approaching the neutron-deficient border of stability of SHN.

Acknowledgments
We are grateful to the JINR directorate and U400 cyclotron and ion source crews for their continuous support of the experiments. We acknowledge the support of the RFBR Grant No. 13-02-12052. Research at ORNL was supported by the U.S. DOE Office of Nuclear Physics under DOE Contract No. DE-AC05-00OR22725 with UT-Battelle, LLC. Research at LLNL was supported by LDRD Program Project No. 08-ERD-030, under DOE Contract No. DEAC52-07NA27344 with Lawrence Livermore National Security, LLC. This work was also supported by the U.S. DOE through a Grant No. DE-FG-05-88ER40407 (Vanderbilt University). These studies were performed in the framework of the Russian Federation/U.S. Joint Coordinating Committee for Research on Fundamental Properties of Matter.

References

[1] Sobiczewski A and Pomorski K 2007 *Prog. Part. Nucl. Phys.* **58**, 292.
[2] Oganessian Yu Ts and Utyonkov V K 2015 *Rep. Prog. Phys.* **78**, 036301.
[3] Oganessian Yu Ts *et al.* 2000 *Phys. Rev. C* **62**, 041604(R).
[4] Oganessian Yu Ts *et al.* 2004 *Phys. Rev. C* **70**, 064609.
[5] Oganessian Yu Ts *et al.* 2004 *Phys. Rev. C* **69**, 054607.
[6] Oganessian Yu Ts *et al.* 2000 *Phys. Rev. C* **63**, 011301(R).
[7] Oganessian Yu Ts *et al.* 2006 *Phys. Rev. C* **74**, 044602.
[8] Eichler R *et al.* 2004 *Nature* **447**, 72-5.
[9] Eichler R *et al.* 2008 A*ngew. Chem. Int. Edn.* **47**, 3262-6.
[10] Witwer D *et al.* 2010 *Nucl. Instrum. Methods Phys. Res. B* **268** 28-35
[11] Eichler R *et al.* 2010 *Radiochim. Acta* **98**, 133.
[12] Hofmann S *et al.* 2007 *Eur. Phys. J A* **32** 251-60.
[13] Hofmann S *et al.* 2012 *Eur. Phys. J A* **48** 62.
[14] Stavsetra L *et al.* 2009 *Phys. Rev. Lett.* **103**, 132502.
[15] Ellison P A *et al.* 2010 *Phys. Rev. Lett.* **105**, 182701.
[16] Gates J M *et al.* 2011 *Phys. Rev. C* **83**, 054618.
[17] Düllmann Ch E *et al.* 2010 *Phys. Rev. Lett.* **104**, 252701.
[18] Utyonkov V K *et al.* 2015 *Phys. Rev. C* **92**, 034609.
[19] Roberto J B *et al.* 2014 *Proc. Fifth Int. Conf. on Fission and Properties of Neutron-Rich Nuclei, 2012* (World Scientific, Singapore) p 287.
[20] Wang M *et al.* 2012 *Chin. Phys.* **36**, 1603.
[21] Myers W D and Swiatecki W J 1996, *Nucl. Phys. A* **601**, 141.
[22] Oganessian Yu Ts *et al.* 2013, *Phys. Rev. C* **87**, 054621.
[23] Rykaczewski K *et al.* 2002 *Nucl. Phys. A* **701**, 179.
[24] Grzywacz R *et a.l* 2003, *Nucl. Instrum. Methods Phys. Res. B* **204**, 649.
[25] Ellison P A *et al.* 2010 *Phys. Rev. Lett.* **105**, 182701.
[26] Schmidt K-H, Sahm C-C, Pielenz K, and Clerc H-G, 1984 *Z. Phys. A* **316**, 19.
[27] Zagrebaev V I, Karpov A V, and Greiner W 2012 *Phys. Rev. C* **85**, 014608.